The Laboratory Rat

Volume II
Research Applications

AMERICAN COLLEGE OF LABORATORY ANIMAL MEDICINE SERIES

Steven H. Weisbroth, Ronald E. Flatt, and Alan L. Kraus, eds.:
The Biology of the Laboratory Rabbit, 1974

Joseph E. Wagner and Patrick J. Manning, eds.:
The Biology of the Guinea Pig, 1976

Edwin J. Andrews, Billy C. Ward, and Norman H. Altman, eds.:
Spontaneous Animal Models of Human Disease, Volume I, 1979;
Volume II, 1979

Henry J. Baker, J. Russell Lindsey, and Steven H. Weisbroth, eds.:
The Laboratory Rat, Volume I: Biology and Diseases, 1979;
Volume II: Research Applications, 1980

The Laboratory Rat

Volume II
Research Applications

EDITED BY

Henry J. Baker

Department of Comparative Medicine
Schools of Medicine and Dentistry
University of Alabama in Birmingham
Birmingham, Alabama

J. Russell Lindsey

Department of Comparative Medicine
Schools of Medicine and Dentistry
University of Alabama in Birmingham
and the Veterans Administration Hospital
Birmingham, Alabama

Steven H. Weisbroth

AnMed Laboratories, Inc.
New Hyde Park, New York

ACADEMIC PRESS   1980
A SUBSIDIARY OF HARCOURT BRACE JOVANOVICH, PUBLISHERS
New York   London   Toronto   Sydney   San Francisco

ACADEMIC PRESS, INC.
111 Fifth Avenue, New York, New York 10003

United Kingdom Edition published by
ACADEMIC PRESS, INC. (LONDON) LTD.
24/28 Oval Road, London NW1 7DX

Library of Congress Cataloging in Publication Data
Main entry under title:

The Laboratory rat.

Includes bibliographical references and indexes.
CONTENTS: v. 1. Biology and diseases.--
v. 2. Research applications.
1. Rats as laboratory animals. 2. Rattus norvegious.
I. Baker, Henry J. II. Lindsey, James Russell,
Date. III. Weisbroth, Steven H. [DNLM:
1. Animals, Laboratory. 2. Rats. QY60.R6 L123]
QL737.R66613   619'.93   79-51688
ISBN 0-12-074902-5 (v. 2)

PRINTED IN THE UNITED STATES OF AMERICA

80 81 82 83   9 8 7 6 5 4 3 2 1

HENRY HERBERT DONALDSON (1857–1938)

From 1906 to 1938 this great scientist directed a multidisciplinary research program at the Wistar Institute in Philadelphia which laid the initial foundations for use of the rat in research. Thus, he is rightfully considered to be the originator of the laboratory rat. But Donaldson was more than a great scientist; he was a man of exemplary personal qualities, as revealed in the following description by Conklin:* ''Anyone who had once seen him could never forget his magnificent head, his steady sympathetic eyes, his gentle smile. With these were associated great-hearted kindness, transparent sincerity, genial humor . . . orderliness, persistence, serenity. His laboratory and library were always in perfect order, his comings and goings were as timely as the clock, he never seemed hurried and yet he worked *Ohne Hast, Ohne Rast* [no haste, no rest].'' For these reasons, the editors dedicate this text to the memory of H. H. Donaldson. (Photo by permission *J. Comp. Neurol.)*

*E. G. Conklin, ''Biographical Memoir of Henry Herbert Donaldson, 1857–1938.'' Biogr. Mem., Vol. XX. 8th Mem., pp. 229–243. Natl. Acad. Sci., Washington, D.C., 1939.

Contents

List of Contributors

Numbers in parentheses indicate the pages on which the authors' contributions begin.

Henry J. Baker (257), Department of Comparative Medicine, Schools of Medicine and Dentistry, University of Alabama in Birmingham, Birmingham, Alabama 35294

Hervé Bazin (181), Experimental Immunology Unit, Faculty of Medicine, University of Louvain, Clos Chapelle-aux-Champs 30, 1200 Brussels, Belgium

Allan R. Beaudoin (75), Department of Anatomy, The University of Michigan, Ann Arbor, Michigan 48109

Sanford P. Bishop (161), Department of Pathology, University of Alabama in Birmingham, Birmingham, Alabama 35294

G. Bruce Briggs (103), WIL Research Laboratories, Inc., 3154 Exon Avenue, Cincinnati, Ohio 45241

Joe D. Burek (149), Toxicology Research Laboratory, Health and Environmental Research, The Dow Chemical Company, Midland, Michigan 48640

Philip B. Carter (181), Trudeau Institute, Inc., Saranac Lake, New York 12983

Donald V. Cramer (213), Department of Pathology, University of Pittsburgh School of Medicine, Pittsburgh, Pennsylvania 15261

Dickson D. Despommier (225), Division of Tropical Medicine, School of Public Health, College of Physicians and Surgeons, Columbia University, New York, New York 10032

Henry L. Foster (43), The Charles River Breeding Laboratories, Inc., Wilmington, Massachusetts 01887

Thomas J. Gill, III (213), Department of Pathology, University of Pittsburgh School of Medicine, Pittsburgh, Pennsylvania 15261

Carel F. Hollander (149), Institute for Experimental Gerontology TNO, 151 Lange Kleiweg, Rijswijk, The Netherlands

Alan L. Kraus (1), Division of Laboratory Animal Medicine, University of Rochester School of Medicine and Dentistry, Rochester, New York 14642

Sam M. Kruckenberg (259), Department of Pathology, College of Veterinary Medicine, Kansas State University, Manhattan, Kansas 66502

Heinz W. Kunz (213), Department of Pathology, University of Pittsburgh School of Medicine, Pittsburgh, Pennsylvania 15261

J. Russell Lindsey (257), Department of Comparative Medicine, Schools of Medicine and Dentistry, University of Alabama in Birmingham and the Veterans Administration Hospital, Birmingham, Alabama 35294

Annie Jo Narkates (59), Institute of Dental Research, University of Alabama in Birmingham, Birmingham, Alabama 35294

Juan M. Navia (59), Institute of Dental Research and Departments of Comparative Medicine and Nutrition Sciences, University of Alabama in Birmingham, Birmingham, Alabama 35294

Frederick W. Oehme (103), Comparative Toxicology Laboratory, Kansas State University, Manhattan, Kansas 66506

John C. Peckham (119), Southern Research Institute, Birmingham, Alabama 35205

Steven H. Weisbroth (257), AnMed Laboratories, Inc., New Hyde Park, New York 11040

Christine S. F. Williams (245), Laboratory Animal Care Service, Michigan State University, East Lansing, Michigan 48824

Preface

The American College of Laboratory Animal Medicine (ACLAM), founded in 1957, is the board of the American Veterinary Medical Association for specialists in laboratory animal medicine. Along with its mission of certifying competent professionals in the field, the college maintains an aggressive continuing education program. This book is a product of that program.

In 1969 ACLAM embarked on a bold and exciting project to develop a series of comprehensive scientific texts on laboratory animals. "The Biology of the Laboratory Rabbit" was published in 1974 and "The Biology of the Guinea Pig" in 1976.

An authoritative reference work on the laboratory rat to meet the needs of modern science has been long overdue. Relatively few books dedicated exclusively to this species have ever been printed. Thirty years have now elapsed since Edmond Farris and John Griffith, assisted by twenty-nine contributors, published the second (and last) edition of "The Rat in Laboratory Investigation." Thus, this book is intended to fill a void of thirty years during which no similar text appeared despite phenomenal progress in the biomedical sciences and a parallel expansion in knowledge of the laboratory rat.

The division of this work into two volumes was dictated by the large volume of material to be included. The subjects sorted readily into two major groupings, Biology and Diseases and Research Applications, the focal points for Volumes I and II, respectively.

Volume I details the more fundamental aspects of *Rattus norvegicus* as a species, its biology and diseases. Chapter 1, for the first time, brings together in a single narrative those events and personalities involved in the development of the species as a laboratory animal. The basic biology of the rat is described in chapters on taxonomy, genetics, anatomy,

physiology, and hematology and clinical biochemistry. It will be apparent that these chapters emphasize information of the greatest importance to research applications. Chapters on nutrition, reproduction, and husbandry include aspects of basic biology but emphasize matters of practical importance in the management of rats for laboratory investigation. Considerable emphasis is given to spontaneous diseases because of their importance as complications in the use of rats as research subjects and because they may present unique models for the study of disease mechanisms. Finally, the important zoonotic hazards associated with the use of laboratory rats are described.

Volume II chronicles topics of importance to research applications of the rat. Some chapters such as Research Methodology, Gnotobiology, and Wild Rats in Research are of general interest. Others focus on the use of rats in specific areas of research, ranging from dental research to toxicology. A few research specialties in which rats are important research subjects, such as endocrinology and behavior, will be obvious by their omission. These topics were not intentionally overlooked and will, we hope, be included in another volume.

Some chapter topics were presented at symposia sponsored by ACLAM. Chapters on taxonomy, morphophysiology, hematology and clinical chemistry, nutrition, reproduction, and housing were presented at a symposium entitled "The Laboratory Rat: Biology and Use in Research" held in conjunction with the 26th Annual Scientific Session of the American Association for Laboratory Animal Science on November 19, 1975, in Boston, Massachusetts. Chapters on bacterial disease, mycoplasmoses, parasitic diseases, lesions associated with aging, and spontaneous tumors were presented at the symposium "Spontaneous Diseases of Laboratory Rats as Complications of Research" held during the 113th Annual

Meeting of the American Veterinary Medical Association on July 20, 1976, in Cincinnati, Ohio.

This book is offered to a wide range of individuals concerned with the use of rats in research. It is hoped that students in graduate and professional curricula will find the broad coverage of material useful for rapid introduction to complex subjects. Commercial and institutional organizations involved in producing rats for research use will find these volumes a rich source of practical information. Specialists in laboratory animal science will welcome its addition to the volumes on rabbits and guinea pigs. Animal care and research technicians should find topics on husbandry, reproduction, and research methodology of particular interest. Above all, investigators will find this broad-based reference work of great value.

The editors wish to express special appreciation to the contributors. They were selected from the most knowledgeable and experienced scientists in the world. As with all other texts in this series, each author has contributed all publication royalties to the American College of Laboratory Animal Medicine for the purposes of continuing education. Officers and members of the College are acknowledged for their enthusiastic support of this project. Finally, we are grateful for the patience and assistance given by those on the staff of Academic Press who have contributed so much to development of this work and to the success of the series.

Henry J. Baker
J. Russell Lindsey
Steven H. Weisbroth

Contents of Volume I
Biology and Diseases

Chapter 1

Research Methodology

Alan L. Kraus

I. INTRODUCTION

In contemporary biomedical science the rat is recognized to be one of the most popular and useful laboratory animals. Its usefulness has been enhanced by the development and refinement of literally scores of specific research techniques. This chapter describes the principal methods that, when properly used, will allow investigators to achieve a great variety of research objectives when using the rat. This body of knowledge has resulted from the efforts of many investigators who have developed innovative methods to improve the quality, reproducibility, and humaneness of research using rats. Not all techniques could possibly be mentioned or described here.

II. HANDLING AND RESTRAINT

A. General

Investigators in all disciplines are becoming increasingly aware of the specific effects of the research animal's macro- and microenvironment on body functions. Much attention is given to designing animal research facilities with the capability of maintaining rigid environmental controls. Animal care technicians have long recognized the advantages of adapting rats to the laboratory animal care routine and making them more gentle by frequent handling. Investigators also must know and use proper methods of handling and restraint to minimize discomfort and stress in order to achieve reproducible and meaningful

data from their experimental subjects. It has long been known that restraint alone can produce gastric ulcers, cardiac necrosis, or hypothermia in rats (282, 302). Even the time of day that restraint is applied has been shown to alter the "perceived stress" and the extent and severity of induced gastric ulcers. The difference in number of rat pups weaned by mothers who are handled manually versus those manipulated with forceps dramatically attests further to the kinds of differences that can result from handling techniques. An early review of the literature concerning the handling of experimental animals as a factor in animal research was written by Bernstein and Elrick (54). More recently the Committee on Physiological Effects of Environmental Factors on Animals published its findings on this subject (126).

B. Manual Restraint

Rats of most stocks and strains can be picked up, manipulated, and restrained without significant hazard. Some stocks or strains have acquired reputations for being particularly docile, while others are thought to be particularly aggressive. Sprague-Dawley and Wistar-Lewis rats are known for their docility, whereas some consider the Long-Evans or hooded rats more fractious. Most agree that Fischer 344 rats must be handled carefully. As described below, the handling and restraint of wild rats are an entirely different matter.

The use of forceps or gloves for routine handling of most laboratory rats is not advised since these techniques tend to make the animals less gentle. Proper manual restraint with

Fig. 1. A method of manual restraint of the rat.

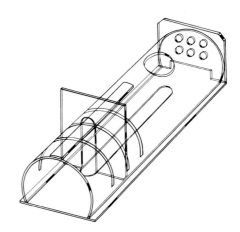

Fig. 2. A commercially available mechanical restraint device.

ungloved hands will prevent the rat from biting under most circumstances. One must confidently grasp the rat from above about the thorax with the thumb brought up behind the chin. This should effectively prevent the operator from being bitten (Fig. 1). The rat is held firmly but squeezing the neck and thorax must be avoided or the rat will react by struggling and, possibly, biting.

Young rats may be picked up by the base of the tail. However, older and heavier rats should not be lifted in this manner because the skin may actually pull off leaving subcutaneous tissues exposed. Furthermore, rats have been known to turn, climb up their tails, and bite the holder.

C. Restraint Apparatus

Many types of restraint devices, cages, and other apparatus for rats have been developed for general or specific needs. Commonly used glass and plasticware such as Erlenmeyer flasks and large (50-ml) syringe barrels have been adapted for use as rat and mouse restrainers. Rodents readily enter the orifice of such labware. Thus restrained, rectal or vaginal swabs can be taken and with the addition of a stopper notched to accommodate the tail, tail bleeding and tail vein injection can be greatly facilitated (229). For economy and convenience, improvised devices can be fabricated from readily available components (5, 20, 113, 126, 355). A commercially available plastic restraining device which is used widely to facilitate injection or withdrawal of materials from many routes is shown in Fig. 2.

A cage originally designed by Bollman has been in use for many years for relatively short-term experiments involving collection of specimens from cystostomy, thoracic duct cannulae, pancreatic juice or bile, and other procedures (61). A commercially available cage similar to the Bollman cage is depicted in

Fig. 3. While the Bollman cage was not adjustable, the modified Bollman design is adjustable to accommodate rats of various sizes. The use of Plexiglas adjustable Bollman restraint cage was reported by Thompson (609). Others have reported on the design and construction of restraining cages used for similar purposes (32, 154). Girardet (238) recently described a very simple and inexpensive restraining device made from galvanized wire mesh that functions in a fashion similar to the Bollman cage or its modifications.

Renaud (514) described a restraint technique used in studying stress-induced cardiac necrosis, gastric ulceration, and adrenal enlargement. Sholkoff et al. (561) reported on a restraint board that provides rigorous immobilization and orientation of position for studies with radiopharmaceuticals.

Many restraining boards or tables have been developed for rodent surgery, and several are marketed commercially (246, 327). One such commercially available rodent operating table is depicted in Fig. 4. Specialized devices, such as stereotaxic

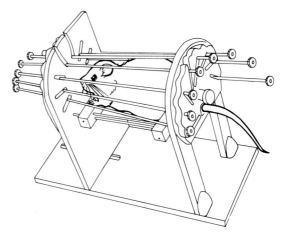

Fig. 3. A commercially available "Bollman-type" restraint device.

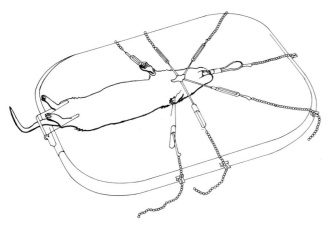

Fig. 4. A commercially available rat surgical table. The figure depicts surgical retraction of wound margins in the midventral neck region.

instruments, have long been used for procedures such as nuerosurgery and electrode implantation (302).

A rodent restraint jacket developed specifically for semen collection and artificial insemination has been described by Lawson *et al.* (373). Restraint devices to facilitate intracerebral innoculation, electrocardiographic recording (472), and oral examination (326) have also been described.

D. Handling and Restraint of Wild Rats

Wild rats are among the most aggressive and potentially dangerous of experimental animals. They are fearless, will attack viciously without provocation, and are known to be vectors of numerous well known, serious zoonotic diseases. Consequently, many of the handling and restraint methods used with their domesticated counterparts, as previously described, are totally inadequate for wild rats. Several devices designed for safe handling of wild rats in the laboratory have been described (97, 191, 342, 634).

Andrews *et al.* (18) reported on the capture and handling of wild Norway rats in a study designed to ascertain the least stressful means of capture and subsequent biochemical estimation of adrenocortical function. Sherman live traps were used for capture in the field. Animals were released into plastic pillow cases and given doses of anesthetics intraperitoneally through the plastic bags. Although it is possible to pick up young (less than 100 gm body weight) wild rats while wearing thick leather gloves with gauntlets, this is not generally recommended (512).

A simple and effective device described by Redfern and Rowe (512) consisted of a "strong, dark, closely woven cloth bag measuring about 75 by 45 cm." Wild rats are induced to enter the dark bag. Once inside, wild rats rarely attempt to bite and usually become quite. Rats can be weighed or injected

inside the bag. Another device used for many years is the Emlen restraint sleeve (183). Fall (198) recently described the use of dim red light to calm wild rats and facilitate laboratory adaptation, rearing, and handling procedures. Finally Keeler (335) reported on a simple and easily constructed floor net for recapture of escaped wild or domesticated rodents in an animal facility.

III. IDENTIFICATION METHODS

In many experimental situations it is not absolutely necessary to identify animals individually as cage cards containing relevant information may suffice. However, the common occurrence of accidents, such as the simultaneous escape of several "white rats" or inadvertant switching of cage cards from different experimental groups, underscores the need and value of individual animal identification.

A. Color Patterns

The distribution of the pigmented hair on each hooded rat (Long-Evans), is sufficiently unique that individual identification can be accomplished by sketching the color pattern or shading a preprinted outline of the dorsum of the rat. Alternately, photographs can be taken for permanent identification of individuals. Photographs taken at approximately weaning time when compared with those taken at 3 months of age reportedly presented no problems of proper identification (524).

B. Ear Notching and Punching

Several schemes for identifying individual rats by means of ear notching and/or punching have been described (95, 633). The ear is generally divided into three areas which are notched and/or punched in one or more areas to designate a given numeral. One ear is designated for the tens and the other the units, thus allowing sequential numbering from 1 to 99. Higher numbers can be achieved by utilizing additional punches or notches or combining with toe clipping, straight ear slits, and tail clipping. By using such a combined system, Walker (633) illustrated a rat whose number was 201,111! From a practical standpoint, however, such accessions are hardly ever necessary. Occasionally, fighting or partial cannabilism destroys or obliterates the carefully produced identification marks. Therefore, it is best to keep the system simple (Fig. 5). While toe clipping has been described and used throughout the world as

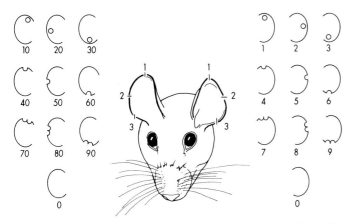

Fig. 5. A commonly used method of individual animal identification by using ear notching and punching. One ear is selected to represent tens, the other units.

an identification method in rats and mice, its use should be discouraged since it is unnecessary and causes mutiliation.

C. Ear Tags

Small numbered plastic or metal ear tags are commercially available. They are attached by means of a special pair of pliers. Like the ear notch and punch system they are subject to being lost when torn out as a result of fighting or cannibalism.

D. Tattooing

Beach (42) and Geller and Geller (231) described techniques for tattooing and India ink injection. While Beach tattooed numerals on the ears of weanling rats, Geller and Geller used the injection of small quantities of India ink into the palmar and plantar surfaces of the feet as a means of identifying newborn rats. This technique is recommended for newborn rats over the toe clipping technique.

E. Dye and Other Color Markings

Although dye marking of rodents has been commonly used in the past (166, 594), the use of indelible, colored, felt-tipped markers has generally replaced the use of dyes. By using markings and stripes on the back, legs, and/or tail, one can devise a numerical sequence for individual identification of rats. However, care must be taken to remark rats before the marks fade and become indistinguishable. For any marking or identification system to succeed, one must depend upon conscientious and responsible animal care and research technicians trained in the methodology employed.

IV. SPECIMEN COLLECTION

A. Blood

Many techniques for obtaining blood specimens from rats have been described. The technique of choice for a given investigation depends on several factors: (a) the volume required, (b) the frequency of bleeding, (c) whether or not anesthetics can be used, (d) the effect of the method chosen on blood parameters to be examined, (e) whether or not mixed or separate venous or arterial blood is required, and (f) whether or not the sample must be taken aseptically. Table I lists all of the blood collection techniques reviewed; however, the more commonly used techniques will be described.

Investigators are becoming increasingly aware of the real and/or potential role of the technique of specimen (blood) collection in causing a multitude of experimental variables. Stress associated with the method of blood collection, the choice of anesthetic, and normal circadian variations can dramatically alter various clinicopathological indexes. Blood hormone levels (e.g., corticosterone, prolactin, epinephrine, growth hormone, insulin, and plasma renin) may be affected by the stress of the method of collection or the anesthetic used (46, 77, 468). The effects of sampling technique on various serum enzymes have been reported also (218). Blood glucose levels have been shown to be related to method of blood collection and choice of anesthetic agent (43, 349). Effects of various anesthetics (ether, pentobarbital, and fentanyl-droperidol) on acid–base balance, hemoglobin concentration, packed cell volume, plasma protein, and calcium and magnesium levels recently have been compared to these same clinicopathologic indexes in blood from unanesthetized rats. In addition, hemoglobin and packed cell volume measurements have been reported to be higher in blood collected from the tail vein than in blood obtained from either the heart or aorta (663). Multiple intraperitoneal injections or trauma in the form of repeated bleeding from the retroorbital sinus have been shown to cause a significant drop in the peripheral white blood cell count in rats (240). In addition, circadian variations in hormone levels, serum enzymes, metabolic products, and cellular elements among others require that investigators recognize and compensate for these differences (46). For example, blood should not be collected for estimation of serum corticosterone in the morning one day and late afternoon the next.

While serially decapitating rats and subsequently measuring corticosterone blood levels, Ader and Friedman (8) and Carney and Walker (100) found a fivefold increase in levels of corticosterone in those rats awaiting and "sensing" impending danger that were sacrificed at the "trough" of the adrenocortical cycle. Less dramatic but significant differences were also found in rats sacrificed at the crest of their daily cycle.

The examples cited above suggest that a thorough review of

Table I

Blood Collection Methods in the Rat

Site or method	Comments	Reference
Adrenal vein	Terminal, specialized technique	538, 570
Aorta	Terminal collection, usually; or chronic catheterization	101, 407, 498, 591, 632
Brachial vessels	Terminal collection usually	666
Cardiac puncture	Routinely used, not without risk	13, 85, 194, 441, 266
Carotid artery,	Terminal collection usually; or chronic catheterization	148, 245, 321
Decapitation	Terminal; used when anesthetics are contraindicated	441
Femoral artery	Infusion route; incision required	148, 182
Femoropopliteal or lateral marginal vein	Best used for injection; cutdown may be necessary	252, 441
Hypophyseal portal	Terminal; specialized technique	48
Jugular vein	Collection; infusion route	148, 334, 487, 515, 607, 619
Lingual vein	Best used for injection	251
Metatarsal vein, dorsal	Routinely used; safe	465
Orbital plexus	Routinely used; safe	105, 278, 370, 466, 521, 595
Penile vein, dorsal	Best utilized for injection	460, 530
Renal artery	Terminal; specialized technique	43
Saphenous vein, proximal	Routine; may require incision	252
Tail amputation	Routine; safe	10, 85, 187, 315, 383, 441, 455, 465, 497, 531, 663
Tarsal vein, recurrent		
Toe clip	Generally not recommended	187, 441
Vena cava	Terminal bleeding technique	498

the literature in specific areas of interest should be performed prior to the design of the experimental plan to examine what is known about experimental variables introduced by the method of bleeding, anesthetic technique, and time of sampling.

1. Cardiac Puncture

Cardiac puncture is perhaps the most commonly used method of blood collection in rats. The anesthetized rat is placed in dorsal recumbency and the apex beat felt in the region of the fourth, fifth, or sixth ribs. Hair over the area is clipped and the skin treated with a suitable antiseptic. A 20- to 26-gauge 1/2-inch needle attached to a syringe is introduced through the thoracic wall at approximately a 45° angle to the horizontal and vertical axis of the rat's body (Fig. 6). The tip of the needle will transmit the throbbing of the heart which should be entered with further pressure on the syringe. Ideally, the wall of the left ventricle is penetrated using this method, and the strongly pulsating blood will flow into the syringe. The right ventricle may inadvertently be penetrated, but the flow and pulse will generally not be as rapid or strong. If very little blood is obtained, the needle tip may have positioned within the myocardium, the heart may have been completely penetrated, or the tip may be in the atrium. Gently retracting the needle and repositioning the tip for more accurate penetration

of the left ventricle may be required. It is generally possible to secure approximately 5 ml of blood from juvenile or adult rats in 20 sec to 1 min. Even in experienced hands, cardiac tamponade and a variety of other complications resulting in morbidity and mortality may result in some animals (85).

Additional techniques for cardiac puncture in neonatal rats (266, 451) and mice (194) are useful in certain situations and, particularly, for smaller rats. The techniques described by Grazer utilizes a 15-inch length of polyethylene tubing and a 30-gauge needle and adapter. The needle is inserted at 90° to the anesthetized neonate just left of the midline and approxi-

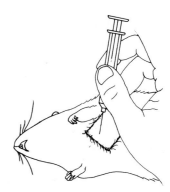

Fig. 6.   A technique of cardiac puncture in the rat.

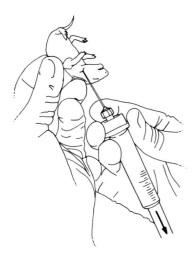

Fig. 7. Cardiac puncture in the neonatal rat using a Vacutainer system. [Redrawn from Gupta (266).]

Fig. 8. Orbital bleeding technique.

mately 1 mm above the costal margin. The newborn can withstand withdrawal of 0.075 ml of blood or injections of a like volume. In this technique a tuberculin (1.0 ml) syringe is usually used (248). The technique described by Falabella (194) for mice also utilizes polyethylene tubing. Collection is made directly into a test tube or other receptacle.

Gupta (266) utilizing a Vacutainer system and 21 gauge needle described a successful cardiac bleeding technique for neonates. The rat pup is hand held in dorsal recumbency with head downward. The needle is slowly and gradually inserted into the thorax through the thoracic inlet until a trace of blood is seen at the orifice of the needle within the collecting tube and held in that position until the desired amount is collected (Fig. 7). A volume of 0.2 ml can be collected from neonates 3 to 5 days old and 0.7 ml from pups approximately 2 weeks old.

2. Retroorbital Plexus

In 1913 Pettit proposed the use of the "sinus caverneaux" for blood collection in the mouse, rat, and guinea pig. Since that time numerous articles have appeared describing techniques using the retroorbital plexus as a reliable, convenient, and safe technique for securing blood from small laboratory rodents and rabbits (105, 252, 278, 370, 466, 521, 583, 595). The technique lends itself well to multiple or serial procedures requiring relatively small quantities of blood. While laboratory-made or adapted Pasteur pipettes have been used as collection devices, commercially available disposable heparinized polished-tipped pipettes are generally used. Standard microhematocrit tubes can be used. However, if larger quantities of blood are required, larger (e.g., 13 × 100 mm) tubes should be used. The rat is manually restrained and the loose skin of the head is tightened with the thumb and middle finger. The index finger is used to make the eye protrude

slightly by further traction on the skin adjacent to the eye. The thumb is pressed firmly, but gently, just behind the angle of the jaw causing constriction of the venous return via the internal jugular vein and subsequent engorgement of the retroorbital plexus (530). The tip of the pipette is then gently but firmly inserted at the medial canthus past the ventrolateral surface of the eye into the retroorbital plexus (Fig. 8). This procedure is facilitated by gently rotating the pipette as it is advanced. Once the capillaries of the venous plexus are ruptured, blood wells up in the periorbital space. It may be necessary to slightly withdraw the tip of the pipette to initiate good flow of the tube by capillary action. Once the bleeding is initiated, multiple tubes may be collected sequentially. Bleeding generally stops spontaneously once the normal ocular pressure is allowed to impinge on the venous plexus.

While the technique described above approaches the plexus via the medial canthus, other investigators prefer an approach via the lateral canthus, claiming less eye injury and complications from epistaxis (583). The authors state that this technique will yield 4 to 6 ml of blood from a 115- to 130-gm rat and that more than 8 ml can be withdrawn safely at one time without killing the rat.

Another report described the successful use of the retroorbital plexus for exsanguinating unanesthetized mice (105). This procedure utilizes 6 to 8 cm lengths of siliconized polyethylene tubing that was slightly beveled at the tip. One end was introduced into the venous plexus and the other placed in a collecting tube. The author stated that the procedure permitted recovery of up to 50% of the total blood volume (up to 3% of the total body weight) in the mouse. This technique also can be performed in the rat.

One may wish to become proficient using this technique with anesthetized rats; however, anesthesia is not generally required or deemed necessary.

3. Dorsal Metatarsal Vein

The dorsal metatarsal vein of the rat may be used for both

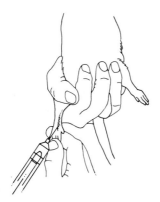

Fig. 9.   Technique of bleeding using the dorsal metatarsal vein.

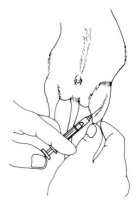

Fig. 10.   Technique of bleeding using the proximal saphenous vein.

injection and withdrawal of blood (465). Although the procedure can be performed by one person, it is easier if one person restrains the rat while the other performs the phlebotomy (465). If one wishes to withdraw a blood specimen from the right hindleg, the assistant restrains the rat in his left hand with its body held in an upright position. The palm of the right hand supports the buttocks with the tail held between the middle and ring finger. The thumb is placed on the lateral side of the leg below the knee. With simultaneous gentle pressure of both the thumb and the index finger on the medial aspect of the leg, the leg is extended and the dorsal metatarsal vein is visualized.

The operator grasps the metatarsal region of the right foot, flexes the first phalangeal joint, and adducts the leg slightly. The area should be clipped and swabbed with alcohol making the vein even more visible and distinct. A 22 to 23 gauge needle is inserted into the vein proximal to the first phalangeal joint and the specimen withdrawn (Fig. 9). The amount of blood obtainable using this technique is usually in the range of 0.1 to 0.2 ml.

4.   Proximal Saphenous Vein

The proximal saphenous vein runs superficially along the medial aspect of the thigh (Fig. 10). The unanesthetized rat is manually restrained and the area clipped and swabbed with alcohol. The assistant compresses the inguinal area causing the vein to dilate and fill with blood. The operator pierces the vessel with a 23 gauge needle or nicks it with a lancet. Blood is collected in capillary tubes as it forms droplets on the skin surface.

5.   Femoropopliteal Vein or Lateral Marginal Vein

The femoropopliteal vein courses over the caudolateral aspect of the thigh (Fig. 11). In anesthetized rats it is prepared for venipuncture by first clipping the hair and swabbing the area with alcohol. A small elliptical piece of skin is snipped out using scissors directly over the vein. Alternately a small incision may be made with a scalpel over the vessel. The vessel is

punctured using a 22 to 25 gauge needle and blood withdrawn or substances injected (441). Some authors state that percutaneous venipuncture can be performed successfully at this site (252).

6.   Tail

Various means of obtaining blood specimens from the tail of the rat have been described (10, 85, 252, 315, 388, 455, 663). The techniques chosen will depend primarily upon the quantity of the blood required and the frequency of blood sampling.

The techniques vary from amputation of the tip of the tail (85, 497, 531, 663) to incision of lateral tail vein (388, 455), or caudal artery (315), or chronic cannulation techniques of both the veins or artery.

a. Amputation or Tail Clipping.   When relatively small quantities of blood are required a small portion (less than 5

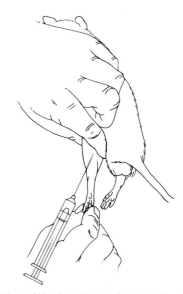

Fig. 11.   Technique of bleeding using the femoropopliteal or lateral marginal vein.

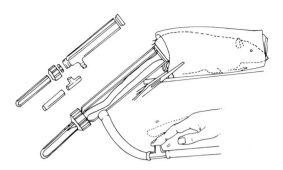

Fig. 12. A vacuum-assisted method of blood collection from the lateral tail vein. [Redrawn from Nerenberg and Zedler (455).]

mm) of the tip of the tail is clipped off. If more than a few drops is required, the tail can be massaged gently from the base toward the severed end. Prior warming of the tail or swabbing with xylol to cause vasodilation can be used to facilitate collection of even greater quantities of blood. Up to 3 ml can be obtained by this technique. To facilitate tail bleeding even further, several additional techniques involving the use of warming chambers (388, 598) and vacuum-assisted collecting devices have been developed (383, 455, 497, 531, 598, 663).

Warming of the tail to cause vasodilation can be accomplished by placing the tail in warm (48°C) water or using a conventional light bulb (252). More sophisticated techniques involve placing the rat into a warm (40°C) incubator for a few minutes, or utilizing a metal shoebox cage modified to accept a conventional electric hair dryer and a thermometer (388). A technique using a standard plastic shoebox cage and a heating pad has been described by Stuhlman et al. (598).

Vacuum devices vary from the commercially available heparinized disposable (Unopette*) macrosampler (663) with a 1-ml capacity to various improvised devices.

b. Incision of Lateral Tail Vein. Nerenberg and Zedler (455) recently described a method of incising the lateral tail vein (about 1 cm) after coating the area to be incised with petroleum jelly. The tail is then placed into a modified Liebig condensor jacket which is connected to a vacuum line (Fig. 12). Utilizing this apparatus rats have been bled sequentially without difficulty. Up to 10 ml of blood (approximately 4% of body weight) have been taken from rats without causing death.

c. Puncture of the Caudal Artery. A technique for collecting blood from the caudal artery described by Hurwitz (315) has the advantage that arterial pressure produces a good blood flow without the need for adjunct suction or vacuum devices. Ether anesthesia is recommended over pentobarbital, chloralose, or urethane because caudal blood flow is diminished with these agents. In this technique a standard

*Becton-Dickinson and Company, Rutherford, New Jersey.

heparinized microhematocrit tube (Caraway type) is inserted into the clear polyethylene hub of a 1-inch 24 gauge needle. The rat is lightly anesthetized with ether and placed in dorsal recumbency. The needle is inserted into the caudal artery which runs midventrally in the tail and a few millimeters beneath the skin. Approximately 0.35 ml of blood can be collected in one tube. This technique offers the advantage of convenience, safety, and procurement of blood with minimal hemolysis. It is reported that one person can bleed approximately 20 rats per hour using this technique (315).

d. Cannulation of the Caudal Artery or Lateral Vein. For certain studies, cannulation of the caudal artery or vein offers distinct advantages. Once implanted, cannulae can be used for repeated samples or injections over a period of time. Blood volume may be maintained by infusion of appropriate fluids or heparinized blood from a donor, and the rat can, if desired, be allowed to awaken from anesthesia while still restrained. This technique has been described as being particularly useful in studies of drug degradation over periods of 5–6 hrs. Usually, this technique is used in terminal experiments (10). For a description of long-term cannulation techniques, see Section V, C.

7. Jugular Vein

While puncture of the jugular vein may be performed for routine collection of blood, this vein usually is used as a site for chronic indwelling catheters which allow repeated or serial blood sampling and injection.

A technique for jugular venipuncture in the unanesthetized mouse has been described by Kassel and Levitan (334), and for the anesthetized rat by Renaud (515) and Philips et al. (487). Collection of blood from the jugular vein can be performed with or without skin incision. In either case, a 20 gauge 1-inch needle is used to collect 1 to 2 ml of blood per 100 gm of body weight. By alternating veins, five blood collections or injections can be performed with little difficulty. A trained technician can perform between 10 and 20 venipunctures per hour (515).

Phillips' (487) technique for jugular venipuncture is similar but is facilitated by the use of a specially constructed restraining board which can be used for rats weighing 100 to 600 gm. He suggests use of 3/4-inch 23 gauge needle and claims that a trained technician can bleed 40 rats per hour using his technique, collecting 0.25 to 0.6 ml of blood per rat.

Techniques for cannulation of the jugular vein as a means of serial sampling of blood and giving intravenous injections have been described (607, 641). Cannulation techniques are useful particularly in pharmacokinetic studies and in those where labile and otherwise variable blood levels or hormones are being investigated. While decapitation usually is used to collect blood needed to assess various labile hormones (for example, corticosterone or prolactin), only one specimen can be obtained at a fixed moment in time. The advantages of cannulation techniques for obtaining blood specimens from awake, unanesthetized animals without interfering with behavioral or physiological processes are obvious. The cannulation technique of Upton (619) is relatively simple and rapid. Silicone polymer tubing is inserted caudally into the jugular vein to the right atrium. The tubing is sutured in place and brought out following a subcutaneous route to the center of the back above the scapulae. The translucent, seamless siliconized tubing used had a 0.51 mm i.d. and 0.94 mm o.d.* For details of the surgical technique and sampling procedures the reader is referred to the original description by Upton.

8. Carotid Artery

The technique for cannulization of the rat carotid has been described and well illustrated by D'Amour and Blood (148). A semiautomatic blood sampling device, utilizing chronic carotid and jugular catheters was described by Graham (245). This relatively simple and inexpensive device permits 20-μl blood samples to be taken as rapidly as 1/min from awake rats.

A simple method for repeated blood sampling developed for rabbits and cats by Jacobs and Adriaenssens (321) utilizing a modified heparin-lock could undoubtedly be further modified for use in adult rats. Blood specimens are obtained by inserting a 20 gauge needle through the skin and into the heparin-lock via its rubber cap.

With all chronic cannulation techniques, cannulas must be flushed and filled with heparin and saline periodically (usually daily) to ensure patency.

9. Aorta and Vena Cava

Terminal bleeding by severing the abdominal aorta and/or the vena cava in the anesthetized rat is a common procedure (252, 407). The technique of permanent cannulation of the aorta and/or vena cava in rats has been described by several authors (101, 498, 632). The advantage offered by the chronically cannulated aorta and/or vena cava preparation includes the ability to withdraw samples from unanesthetized subjects and the ability to measure arterial–venous differences in blood concentrations of the substance of interest. Dramatic differences between anesthetized and unanesthetized rat plasma renin concentration was cited by Carvalho *et al.* (101) as one reason for utilizing this technique. In addition, direct measurement of arterial and venous pressure may be monitored. Blood sampling and administration of substances intraarterially and intravenously are also greatly facilitated.

*Silastic 602-135, Dow-Corning International Limited.

10. Decapitation

While seemingly primitive, barbaric, and certainly aesthetically unpleasing, decapitation performed by trained personnel using proper equipment is considered to be a humane and effective method of killing rats as well as facilitating the collection of a relatively large quantity of mixed venous and arterial blood from unanesthetized animals (517). Ader and Friedman (8) demonstrated the increasing blood corticosterone level in rats sequentially decapitated.

Blood is collected by immediately placing the rat following decapitation neck down over a suitable funnel and collecting the blood in test tubes or other receptacles.

11. Other Techniques

Techniques for catheterizing the femoral artery for chronic collection of blood as well as for angiography were described by Eklund and Olin (182).

A method of terminal bleeding of anesthetized mice by severing the brachial arteries is adaptable to rats (666).

Specialized techniques for catheterizing and collecting blood from the adrenal vein (538, 570), renal artery (43), and hypophyseal portal system also have been described (48).

B. Urine

Urine may be collected from rats by a variety of means. The design and use of so-called metabolism cages will be described separately in Section IV, D. Techniques for urine collection to be described here include (1) reflex emptying under periodic stimulation or massage, (2) bladder centesis, (3) cystostomy or urinary bladder fistula, (4) "free-catch," (5) Urethral catheterization, and (6) external drainage catheterization (Fig. 13).

1. Reflex Emptying under Periodic Stimulation or Massage

While studying the mechanism of thirst in rats, Adolph *et al.* (9) utilized a "restraint frame" or device which resembles the commercially available Plexiglas rat restrainers (Fig. 2) but which had a collection funnel located under the male rats for collection of urine. He collected urine by "quickly raising and lowering the frame and animal." Details of the many conditions found to influence drinking and collection of urine are described (9).

Micturation also may be stimulated by applying gentle but firm suprapubic pressure in rats of all ages. The urine can then be collected in capillary pipettes or caught in suitable receptacles (168, 287). By this means 150 to 200 μl of urine may be collected in capillary pipettes. Although the authors state that volume of urine cannot be estimated directly by this method,

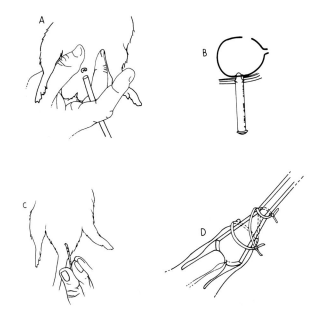

Fig. 13. Methods for obtaining urine specimens from rats. (A) Manual pressure and collection into capillary tube. (B) Urinary cystostomy technique. (C) Catheterization of female rats with a No. 4 coude catheter. Catheter itself is also depicted. (D) External drainage catheter used with male rats. [Redrawn from White (651).]

urine volume can be estimated indirectly by determining the specific gravity and calculating the urine volume. It has been reported (287) that the correlation coefficient relating urine specific gravity and volume in rats is 0.85 (Fig. 13A).

2. Bladder Centesis

While this procedure is used extensively in larger laboratory animals, it generally is not a preferred technique in rodents. It has been used somewhat successfully by workers studying water load induced diuresis in newborn rats. Since it may not be possible to empty the bladder of newborn rats completely by exerting suprapubic pressure, the investigators chose to perform bladder centesis at autopsy. Even so, only very small quantities of urine were obtained (292).

3. Urinary Fistula or Cystostomy

Hoy and Adolph (306) developed the technique of establishing a urinary bladder fistula in both newborn and adult rats. A short polyethylene tube with a flange at one end was inserted from inside the bladder through the ventral body wall where it was sutured in place with stainless steel surgical wire. The inner flange was made flush with the mucosa of the bladder. The other end of the cannula was "flanged" by applying heat and was used for attaching collection tubes (Fig. 13B).

4. "Free-Catch"

Rats will frequently spontaneously urinate when handled. Therefore, small quantities of urine can be collected by merely picking up a rat manually and catching the urine in a suitable receptacle.

5. Urethral Catheterization in Female Rats

Cohen and Oliver (122) described a successful method for catheterizing female rats using a No. 4 coude ureteral catheter.* This catheter has a curved tip which the authors claim allows it to be successfully passed by the symphysis pubis. The anesthetized rat is placed in the supine position and the urethral meatus exposed by grasping the fur on either side with forceps. The catheter is inserted and rotated if it meets resistance until it is in the bladder. Repeated catheterizations have been performed successfully with no evidence of trauma or urinary tract infection utilizing this technique (Fig. 13C).

6. External Drainage Catheter in Male Rats

White (651) developed a method of urine collection from anesthetized male rats utilizing a modified polyethylene Leurend Intramedic† catheter which is fitted over the tip of the penis and fixed in place by suturing the catheter to the prepuce. The author reported that following oral "loading" urine flow of 6 to 10 ml/hr was obtained. Residual bladder urine of 0.5 ml was found at autopsy in rats with external drainage catheters (Fig. 13D).

C.   Feces

For certain studies where quantitation is not important, feces may be simply obtained by retrieving voided pellets from beneath suspended cages or from within bedding material. However, contamination with urine, bedding, feed, hair, dander, and other products will occur.

Rodents are naturally coprophagic and will ingest fecal material directly as it is voided from the anus as well as from the floor of the cage. Therefore, special procedures have been devised to prevent coprophagy and collect all feces voided. It has been estimated, for example, that 50 to 65% of feces voided are ingested by rats while housed on screen floors (38).

Once it was recognized that screened floors alone did not prevent coprophagy, early workers described a variety of other caging innovations designed to prevent or at least minimize coprophagy. These were varied from circular "doughnut-

---

*C. M. Bard Company, Summit, New Jersey.
†P.E. 50 or 90, Clay-Adams, New York, New York.

shaped'' cages which allowed forward and backward movement but restricted body flexion, to restraining cages which were basically tubes of metal or plastic similar to devices presently used for short-term restraint during bleeding or other procedures (233).

Because of problems encountered in all caging devices aimed at preventing coprophagy, workers began experimenting with anal cups made from a variety of small plastic laboratory bottles (38, 390, 529, 636). Each succeeding report claimed an improvement and simplification over those previously described. If all fecal and urine excretion must be collected, Frape *et al.* (213) have described a simplified metabolic cage and tail cup to achieve this goal.

In addition to tail cups of various designs, Armstrong and Softley (24) designed a leather jacket which effectively prevented the rat from reaching its anus with its mouth. Jacketed rats were placed in suspended cages with large mesh screened floors designed to permit all (most) of the fecal pellets excreted to drop through. Growth studies were performed in which it was demonstrated that growth was significantly depressed over 6 weeks time in jacketed rats as compared to rats allowed to practice coprophagy. Rats not allowed to practice coprophagy for prolonged periods of time have been reported to develop a ''spectacle-eye'' like that seen in biotin deficiency (233). Others noted the same condition but in rats supplemented with biotin (24). Several investigators did, however, report that rats tended to have a licking tendency and would lick anything they could reach if kept in restraint jackets for prolonged periods (24, 233).

Finally, small amounts of fecal material for diagnostic purposes may be obtained by the use of rectal swabs.

D. Metabolic Cages

The principal use of metabolic cages is to separate and collect urine and feces effectively without contamination by feed, water, hair, dander, or other contaminants. In addition, some few metabolism cages are designed to regulate and collect both inspired and expired gases. Most metabolism cages, however, are merely urino-fecal separators of varying designs.

If collection of all fecal material or urine is required, one or more of the previously described techniques should be considered because metabolism cages do not serve this purpose well.

Lazarow (376) published a detailed review of the evolution of rat metabolism cages and state of the art in 1954. Since that time, a number of new designs have been described in the literature (58, 72, 73, 93, 168, 275, 539).

Two types of modern commercially available rat metabolism cages are depicted in Figs. 14 and 15. The plastic cage (Fig. 14) features a cage compartment into which the rat is placed suspended above the collecting and separating devices on a

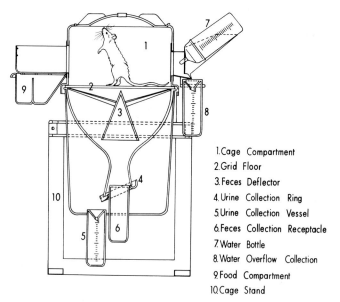

1. Cage Compartment
2. Grid Floor
3. Feces Deflector
4. Urine Collection Ring
5. Urine Collection Vessel
6. Feces Collection Receptacle
7. Water Bottle
8. Water Overflow Collection
9. Food Compartment
10. Cage Stand

Fig. 14. A modern plastic—metal free—standing metabolism cage.

stainless steel grid floor. Water intake measurements are accomplished by the use of a graduated water bottle and overflow catch bottle which collects any spilled water. Food consumption can be measured by the use of the food hopper located outside of the cage proper in a runway or tunnel which is designed to minimize or eliminate food contamination of excreta. When the rat micturates, urine flows through the grid floor and drops directly onto the wall of the collecting chamber or onto the inverted cone-shaped feces deflector, then onto the collecting wall. Ultimately, all urine runs down the wall into the collecting ring and then into the graduated urine collecting vessel. Fecal pellets, on the other hand, are deflected by the inverted cone and dropped off the collection chamber wall and fall vertically into the fecal vessel.

The stainless steel metabolism cage depicted is similar but somewhat less sophisticated in design. It is, however, the most common type of metabolism cage in use today. Neither of the

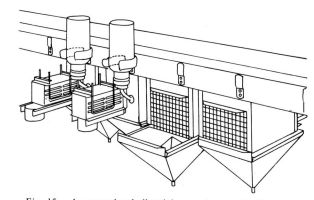

Fig. 15. A conventional all stainless steel rat metabolism cage.

two types of cages is totally satisfactory for regulation and collection of both inspired and expired gases. Cages and equipment designed to measure and collect respiratory gases are described in references by Edwards *et al* (178) and Lazarow (376).

E. Bone Marrow

A number of techniques have been described for the collection and staining of bone marrow from rats and other rodents (86, 285, 490, 623, 627, 628). Most investigators collect the various specimens at autopsy or under deep anesthesia just prior to euthanasia. A commonly used technique consists of dissecting out the femur, cracking it longitudinally or transversely with a small bone cutting forcep and collecting the specimens with a fine (0) brush. Marrow samples are gently painted on microslides for drying, fixation, staining, and examination.

Small biopsy needles can be used for obtaining marrow specimens from the iliac crest of anesthetized rats. Samples have been collected from the femur or tibia by first drilling a hole through the cortex of the bone using a dental drill and then aspirating marrow into a syringe with a fine gauge needle, or dipping a number 00 red sable brush through the orifice into the marrow cavity (86, 601).

The technique for aspirating marrow from the sternebrae of anesthetized mice can also be used for rats. The anesthetized animal is placed in dorsal recumbency and the area over the sternum clipped and prepared with a suitable antiseptic. A small (0.5 to 1.0 cm) midline incision is made, the muscles are dissected free from both sides of a given sternebrae, and a hole is drilled using an appropriately sized dental drill followed by aspiration of marrow using either a fine pipette or needle and a syringe (490).

F. Ocular Fluids

Anterior aqueous humor may be withdrawn readily from the rat utilizing paracentesis employing a 1-ml tuberculin syringe and a 5/8-inch 26 gauge needle. Vitreous humor may be aspirated into a syringe without a needle following dissection of the globe of the enucleated eye (232, 441) (8, 170).

G. Lacrimal Secretion

Lacrimal fluid may be collected at the medial canthus following pharmacologic stimulation of lacrimation with parasympathomimetic drugs. Collection into capillary micropipettes generally is used.

H. Peritoneal Cells or Ascitic Fluid

Collection of peritoneal cells or ascitic fluid can be accomplished by three basic methods: (1) at autopsy, (2) by making a small abdominal incision in the anesthetized rat, or (3) by utilizing a recently described "peritoneal cell glass pipette" in the unanesthetized rat.

Collection at autopsy requires no special comment. However, care in making the abdominal incision so as to avoid losing any fluid is essential. Ascitic fluid can be withdrawn by standard laboratory pipettes filled with a bulb-type suction device.

Cooper and Stanworth (129) describe a technique whereby ascitic fluid is collected in recently killed rats through a small abdominal incision by the use of a 12-ml plastic tube, the end of which has been perforated several times with a hot needle. The tube is inserted deep within the abdomen and the rat held vertically by the tail while 20 ml of sterile phosphate buffer is poured into the tube. The rat is then laid horizontally after clamping off the tube so it will not leak. One then gently massages the abdomen for 90 sec, the clamp is reopened, the rat is again picked up and held in the vertical position, and the ascitic fluid and buffer is removed.

Nashed (451) reported on a simple method for the collection of peritoneal cells from unanesthetized rats, Chinese hamsters, and mice using a specially designed and constructed peritoneal cell glass micropipette having a volume of about 50 ml (Fig. 16). The device is designed to be used as a pipette with one end fitted with a mouth piece and the other terminating with a 15 gauge hypodermic needle. The author claimed that the optimal site for collection is located just anterior to the coxofemoral joint of the right rear leg. The area is shaved and swabbed with 70% alcohol and the pipette filled with approximately 30 to 35 ml of warmed Hank's solution (pH 7.0). The needle is carefully inserted into the abdominal cavity in a caudal direction and the Hank's solution instilled. Aspiration of the Hank's solution, now containing suspended peritoneal cells, commences as soon as it is completely instilled. While pipetting can be performed by mouth, it is recommended that a syringe be used for both introducing the salt solution in as well as

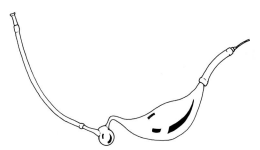

Fig. 16. A specially constructed peritoneal cell glass pipette.

providing suction to remove the specimen. Approximately 70 to 80% of the quantity of solution instilled is recovered using this technique. The author report that 5 to 20 million peritoneal cells are recovered from the average nonstimulated rat.

I. Bile

For a review of surgery of the bile ducts see Lambert (368). Bile can be collected readily for study in the anesthetized rat in short-term acute studies (150). Although more difficult, successful techniques for chronic bile duct cannulation and creation of a biliary fistula have been described. Two anatomical features must be considered in this procedure: (1) the rat has no gallbladder and (2) the bile duct carries both biliary and pancreatic fluids. Consequently, if bile must be collected uncontaminated by pancreatic secretions, the bile duct must be cannulated near the hilum of the liver and before the several pancreatic ducts enter the common duct (Fig. 17A).

The technique of Friedman *et al.* (220) as modified by Nakayama (450) involves the placement of cannulae in the proximal and distal segment of the transected common duct. Anesthetized rats are prepared for surgery, and an incision is made through the right subcostal region. The duct is identified and transected near the hilum of the liver. Polyethylene cannulae (1.22 ml o.d.) are placed in both segments and tunneled from the peritoneal cavity subcutaneously emerging on the dorsal surface of the right hindlimb. Nakayama devised a glass tube used to protect the emerging cannula from trauma while allowing the rat relatively free movement. The two can be connected to one another by a short glass capillary tube to establish normal flow into the rat's intestine.

The "artificial gallbladder" technique of Sawyer and Lipkovsky (535), as modified by Colwell (124), likewise involves the surgical implantation of a cannula. However, it is a specialized cannula containing an 8 to 10 ml capacity bulb with two side arms, one for inflow and the other for the collection of bile. Polyethylene tubing connected to the outflow is tunneled through the abdominal wall and out through a small incision in the right dorsal body wall of the rat. The authors stress appropriate postoperative maintenance regimens as well as the importance of cannulating the common duct close to its origin from the several hepatic ducts if bile uncontaminated by pancreatic secretin is required.

More recently, Knapp, *et al.* (351) have described a much improved technique for collection of bile from the unanesthetized rat. They utilized a PE 50 cannula* for which a 1-cm long jacket made from a piece of PE 250 tubing was permanently affixed. The jacket is placed 1 to 2 cm from the beveled end of the PE 50 cannula which is inserted into the common

*Intramedic Clay-Adams, New York.

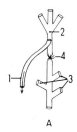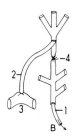

Fig. 17. Technique for collecting bile (A) and pancreatic juice (B). (A) Cannula (1) placed into common bile duct (2), proximal to entrance of pancreatic ducts (3), and ligature (4), distal to cannula and proximal to pancreatic ducts. (B) Cannula (1) into common duct distal to entrance of pancreatic ducts, second cannula (2) ("artificial bile duct") into proximal common duct and also into duodenum (3), ligature distal to "artificial bile duct" and proximal to pancreatic ducts (4).

bile duct. This jacketed cannula serves as the basis for a very secure method of anchoring the cannula to both the mesentary and the abdominal wall as well as ligation in the bile duct. Long-term collection of up to 1 ml of bile per hour has been reported utilizing this technique (351).

In several of the techniques reported above, rats were housed individually in restraining cages similar to those described by Bollman.

J. Pancreatic Juice

It is important to remember that in the rat pancreatic fluid and bile drain into the duodenum by a common duct. A large number of smaller pancreatic ducts empty into the common duct in its terminal portion within the pancreas. Because the bile duct and pancreatic duct radicals lead to a common duct in the rat, it is necessary to stop or divert biliary flow into the lower segment of the common duct in order to collect pancreatic secretions uncontaminated by bile. For a general review of surgery of the pancreatic ducts see Lambert (368).

DeSmul *et al.* (161) recently reviewed several problems associated with profound differences in methodologies used to study pancreatic secretion in the rat. He pointed out that most investigators collect pancreatic juice by the method of Colwell (125) and that many ligate the proximal portion of the common duct to prevent contamination of pancreatic secretion with bile, thus producing obstructive jaundice.

Hotz *et al.* (304) also commented on methods used to study pancreatic secretions in the rat and stated that besides the need to divert bile (but maintain the enterohepatic circulation), it is necessary to ligate the pylorus to prevent endogenous release of gastrointestinal hormones by gastric secretion entering the small intestine. Furthermore, Hotz (304) pointed out that deep narcosis depresses exocrine secretion of the rat pancreas so that results obtained in anesthetized animals may not reflect normal physiologic status.

Pancreatic juice collection in the acute rat preparation, like the collection of pure bile, can be performed by ligating the bile duct below or distal to the point where the several hepatic radicals enter the common duct. D'Amour and Blood (148) described an acute procedure whereby cannulation of the common duct through its orifice into the duodenum (papilla of Vater) is accomplished by using a PE 10 polyethylene tubing and a 32 gauge wire as a stylet to facilitate insertion of catheter.

Junquiera *et al.* (331) described a procedure where pancreatic juices were collected into 10-ml pipettes from a polyethylene cannula placed in the common duct below a ligature to prevent bile contamination. These authors utilized secretin (0.7 units iv) and 1.0 μg of carbamolylcholine (1.0 μg in 0.2 ml of saline po) to stimulate secretion.

Colwell described five different methods for collecting uncontaminated pancreatic juice. All techniques utilized the intra-abdominal glass bulb or "artificial gallbladder" described in Section IV,I. The one method that consistently gave satisfactory results involved dissecting the proximal end of the common bile duct between the pancreas and liver free from hepatic radicals. The duct was then ligated, hemisected, and cannulated with the cannula tied in place with a ligature. The distal portion of the cannula was threaded through the ligated duodenal end of the bile duct into the duodenum acting as an "artificial bile duct" and thus reestablished the enterohepatic circulation. A second cannula, used to collect pancreatic secretion was placed just beyond the openings of the most distal pancreatic ducts (Fig. 17B). This cannula connected to the in-flow port of the so-called "artificial gallbladder." The outflow was connected to a fistula in the flank of the rat. Two sizes of "glass bulb" were used. One device with a volume of 6 to 7 ml was used for smaller rats (250 to 350 gm) and one with a volume of 8 to 10 ml for larger rats (350 to 550 gm). Survival of rats undergoing this procedure average only 9 days. Rats recovered from the surgery within 1 to 2 days and remained in good condition for about a week, although steadily losing weight over the entire period. The volume of pancreatic secretion collected daily ranged from between 1.6 and 3.5 ml per 100 gm of body weight (125, 556). Recently, Hotz *et al.* (304) reported a method of studying pancreatic secretions using a procedure involving collection of duodenal juice by means of a double lumen tube inserted into a chronic gastric fistula. A thin and long (1.0 ml i.d., 60 ml long) polyethylene tube is threaded through a thick and short (4.0 ml i.d., 20 ml long) tube down into the duodenum so that the tip of the duodenal cannula is placed approximately 1 cm distal to the papilla of Vater. Animals are placed in Bollman cages and specimens of gastric and duodenal contents aspirated using a suction system. This procedure clearly is not suitable for collection of pancreatic secretions free from contamination with bile.

K. Female Reproductive Products

1. Ova

Oocytes may be obtained from either normal cycling female rats or following the administration of drugs designed to induce superovulation. In the developing fetus, primordial oocytes and their corresponding populations of M-prospermatogonia may be obtained for studies of gametogenesis prior to sexual differentiation at 14 to 15 days.

Ovarian oocytes may be obtained by first removing the ovaries from postpubertal female rats and placing them in warm normal saline solution. Oocytes are then teased from their preovulatory ovarian follicles with a sharp needle. Oocytes may be picked up and transferred using a Pasteur pipette. Cumulus cells may be removed from the oocyte by use of 0.025% trypsin solution (185, 415) and/or by forcing the cell masses containing the oocytes through a tapered pipette (185, 669).

Tubal oocytes may be collected by flushing the oviducts of female rats on day 1 to 2 of the estrous cycle with a balanced salt solution via the fimbria (669). Alternatively, prepubertal female rats can be induced to superovulate using 20 to 30 I.U. of pregnant mare serum gonadotropin (PMSG) followed in 48 hr by an injection of 20 to 30 I.U. of human chorionic gonadotropin (HCG). Rats are killed 15 to 18 hr afterward and oviducts flushed to collect tubal oocytes (469).

For a description of techniques for egg transfer in the rat consult the work of Bennett and Vickery (51). Clewe and Mastroianni (119) have reported a technique for continuous volumetric collection of oviduct secretions involving chronic cannulization of the oviduct. Adams and Abbot (7) published a bibliography of 451 original articles on egg and embryo transfer studies in 1971.

2. Vaginal Fluid and Cells

The use of exfoliative cytology for diagnosis of human cancer and other uterine and vaginal disorders is now commonplace. Similar techniques are also extensively used to study the morphology of the vaginal mucosa during the estrous cycle in rats. For a thorough description of the principles and practices of cytopreparation see the recent work of Gill (236).

Long and Evans (397) first described and correlated the morphology to ovarian activity in the rat. They described and proposed the separation of the estrous cycle in the rat into five stages based upon the morphology of the external genitalia, the vaginal smear, the uterus, and the ovaries. Nicholas described methods of collecting vaginal fluids and recommended staining with Wright's stain (457). Clarkson and Kalnins (117) similarly reviewed the morphology of the vaginal smear of the rat during the various stages of estrus but recommended the use of the Papanicolaou stain because the cytoplasmic staining reaction accompanying keratinization provided a more reliable indicator of the stage of estrus.

Several techniques to obtain specimens from the vagina can be used including smooth, polished tip glass rods, small saline-wetted cotton tipped swabs or vaginal lavage. Samples for cytology may be spread on a glass microslide, immediately fixed in 95% ethyl alcohol or ether in 95% ethyl alcohol (1:1) and stained with Wright's, Papanicolaou, or other suitable stains.

3. Milk

Cox and Meuller (135) developed an apparatus for milking rats and guinea pigs. Three factors considered to be most important in design and application of such a micromilker include (a) the support of the small easily collapsable teat, (b) the negative pressure applied, and (c) the pulsation rate. Depending on the stage of lactation, 3.0 to 8.0 ml of milk could be obtained.

Temple and Kon (606) claimed an improvement over the apparatus developed by Cox and Meuller. Both groups of investigators used 25 to 40 pulsations per minute and 7 inches of mercury vacuum.

Kahler (333) described a mouse milker which simultaneously milked three to four glands. Pulsation rate used was 72 per minute at 1 to 8 inches Hg vacuum. He was able to obtain 0.7 ml of milk from mice between 10 to 20 days in lactation.

Grosvenor and Turner (263, 264) have studied milk letdown utilizing oxytocin. A dosage of 0.5 units per 1000 gm of body weight have been reported to be effective.

If quantitative milk output is to be measured it is possible to weigh rat pups before and after nursing (439). Morag (440), more recently, attempted to determine the affect of litter size on milk yield and milk removal in the rat and found that litters of 12 to 15 cause maximum effect on both parameters. He recommended that in any bioassay of milk yield, 12 pups be used.

L. Male Reproductive Secretions

While surgical dissection and collection of epididymal, prostatic, or seminal vesicular fluid at autopsy has been used frequently, collection of sperm by electroejaculation allows for repeated sampling over time in the same male.

Blandau and Jordan (57) describe techniques for obtaining specimens of sperm from various portions of the female reproductive tract following breeding. Collection of semen and spermatozoa and the individual secretions of the accessory sexual glands can be performed in the rat by means similar to those used for other rodents. Bennett and Vickery (50) provide

a good review of the many different techniques used in the study of reproductive physiology and in the natural and artificial breeding of laboratory rats and mice.

Early workers such as Scott and Dzuik (546) experienced difficulties when electroejaculating intact rats in that there was entrapment of spermatozoa in the "coagulum" and blockage of the urethra resulting in death from uremia in about 10% of the intact rats within 4 days. They then surgically removed the seminal vesicles and coagulating glands prior to electroejaculation. Specimens obtained were relatively free of accessory gland secretions and the ejaculate was found to be nearly pure sperm. Sperm collected in these operated animals reportedly has no effect on the fertility achievable by artificial insemination (50). Total sperm count obtained by electroejaculation using the techniques described in operated rats were similar to those estimated by Blandau and Jordan in naturally inseminated females.

Birnbaum and Hall (56) also electroejaculated rats with surgically removed coagulating glands. Although they too were able to obtain coagulum-free sperm, some specimens did form a coagulum. They speculated that since they did not remove the seminal vesicles, as did Scott and Dzuik, there might be a coagulant secreted by it which resulted in a certain percentage of coagulated specimens. Then, too, only a very small portion of the coagulating gland which might have been left or which might have regenerated may have resulted in such failures.

Results obtained by Lawson *et al.* (374) with electroejaculation in rats with surgically extirpated coagulating glands were successful. They did not have a problem with coagulating semen. They collected semen on groups of male rats biweekly for up to 18 weeks with good results. Rats were handled gently and restrained in a device developed in their laboratory which placed the rats in a position similar to normal copulation. This report should be read in detail by workers contemplating electroejaculation in rats.

A recently described technique where the ductus deferens is anastomosed to the urinary bladder, permits enumeration of sperm produced each day by counting sperm obtained in 24-hr urine specimens (631). This technique eliminates problems of quantitation of daily sperm output as a result of daily spontaneous ejaculation as described in the rat (471).

M. Pulmonary Cells

Exfoliative cytology of the respiratory tract as a means for diagnosis and as a source of cells for other laboratory purposes is becoming an increasingly important experimental tool.

Pulmonary or alveolar macrophages may be obtained by pulmonary lavage. Brain and Frank (68) reported on a technique whereby following exsanguination of anesthetized rats, lungs were removed, rendered gas-free by evacuating the air space at a pressure of 20 mm Hg for 10 min, and the trachea cannulated with PL-200 polyethylene tubing. One hour later, twelve consecutive 3-min flushes with normal saline were performed. Approximately 5 ml of saline per gram of lung tissue was used for each wash. Saline wash contents were collected after each flushing. A similar procedure more recently has been reported for use in mice (427).

Others recently have described a method for obtaining pulmonary cells by tracheal washings in anesthetized rats and hamsters. Rats were anesthetized using methoxyflurane, placed on a specially designed slanted board, mouth fixed open, and tracheal catheter inserted. The system used for tracheal washing consisted of the catheter itself, an automatic injector, regulator unit, monometer, and vacuum control. Two milliliters of washing fluid were delivered by the injector. For details of the design and operation of this equipment consult the article by Schreiber and Netteshein (544).

N. Lymph

Bollman, *et al.* (62) described a method for collection of lymph from either the liver, intestine, or thoracic duct of the rat. Collection of lymph from the intestine or thoracic duct was successful for up to 10 days, whereas lymph could only be collected from liver for up to 3 days. Cannulation of the hepatic and intestinal lymphatics was aided by the use of Evans blue dye injected into the liver or by feeding a meal rich in fat. Up to 5 ml, 20 ml, and 25 ml of lymph can be collected daily from the hepatic, intestinal, or thoracic duct, respectively. Rats are restrained in a Bollman-type cage during the collection period. A Plexiglas cage reported to have several advantages over the Bollman cage has been used successfully for chronic drainage of lymph or thoracic duct (32).

Gowans (244) described a device for the continuous reinfusion of lymph and lymphocytes in unanesthetized rats with thoracic duct fistulae. The devices allowed quantitation of lymph output at the same time maintaining lymphatic volume since reinfusion was continually in progress.

Reinhardt (513) first described a method of thoracic duct cannulation in the rat.

O. Cerebrospinal Fluid

Techniques for injection of substances into the cerebrospinal space or withdrawal of cerebrospinal fluid will be jointly considered in this section. The techniques described vary from those performed "free hand" (155, 464), to those performed with the rat placed in a stereotaxic instrument (288), or to those performed at autopsy (354).

Noble *et al.* (464) described a simple, rapid, and atraumatic method for introducing substances directly into the lateral ventricles of the brain of the rat. The anesthetized rat is placed in sternal recumbency on an operating board. A midsagittal incision is made from the level of the eyes to the level of the ears exposing the bregma. A small hole, large enough to accept a 27 gauge needle is made through the skull at approximately 1.5 to 2.0 mm lateral and caudal to the intersection of the coronal and sagittal sutures. A 27 gauge needle with only 3.5 to 4.0 mm exposed beyond a stop is inserted into the lateral ventricle while being held in a vertical position (Fig. 18).

This technique undoubtedly could also be used for collection of cerebrospinal fluid. The authors claimed that 15 to 20 rats could be injected per hour.

DeBalbian Verster (155) reported a technique for implanting a chronic cannula in the lateral ventricle without the use of a stereotaxic instrument. This technique has the advantage over that described by Noble in that specimens can be collected (or substances can be injected) from the unanesthetized rat following its recovery from anesthesia. Implantation is performed using a hole drilled according to the coordinates of Noble with a second hole drilled nearby for a stainless steel retaining screw. Construction of a special polyethylene cannula was described.

Details of construction and implantation of a chronic cannula into the lateral ventricle of the rat brain utilizing a stereotaxic instrument have been described by Hayden *et al.* (288) and Hilliard *et al.* (295).

Knigge and Joseph (354) have described a microcannula technique for collecting cerebrospinal fluid directly from the third ventricle of the rat at autopsy immediately following sacrifice by decapitation. Specimen size ranged from 0.3 to 0.5 μl per adult rat. The optimal size of the microcannula used was approximately 0.15 mm O.D. and 60 to 75 mm long.

Techniques for cisternolumbar perfusion or spinal subarachnoid perfusion (171) and ventriculocisterno perfusion (212) have been described recently. Cisternal puncture for the collection of approximately 0.1 ml samples of cerebrospinal fluid in the rat has been described by Clemens and Sawyer (118).

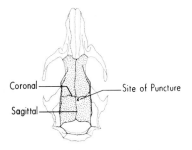

Fig. 18. Dorsal view of rat skull indicating location for insertion of a needle into the lateral ventricle.

P. Saliva

1. General

A recent review of current research on oral fluids by Shannon and Suddick (552) indicates the intense interest by researchers in oral fluid immunology, circadian rhythms in salivary flow, and electrolyte secretions. In addition, the chronically reserpinized rat has been suggested as a possible model for the secretory abnormality of cystic fibrosis based upon the work of Martinez *et al.* (421). Workers in these fields have described numerous techniques for collecting oral fluids.

Collection of saliva from the living rat may be performed by one of three procedures: (a) collection from the oral cavity, (b) cannulation of the duct, and (c) direct collection from the cut end of a surgically isolated duct. In addition, the entire gland can be taken at autopsy for analysis of secretory proteins as reported by Sreebny and Johnson (586). For a detailed review of the anatomy and technique of extirpation of the various glands consult the work by Cheyne (110).

2. Stimulation of Salivation

In many procedures sialogogues are administered to increase the flow of saliva. Pilocarpine at 0.75 to 2 mg per 100 gm of body weight is the drug most commonly employed for this purpose (300, 430, 525). DL-isoproterenol sulfate at a dosage of 8 to 10 mg per 100 gm of body weight has also proved successful (525). Pilocarpine is a parasympathomimetic drug which also stimulates lacrimal, bronchial, and nasal secretions. The extent to which these secretions contribute to the normal "saliva" collected under pilocarpine stimulation has been described by Holloway and Williams (300). These authors noted that the submandibular glands secrete the major portion of pilocarpine-stimulated saliva.

In other procedures, collection of adequate quantities of saliva is facilitated by direct simulation of the secretory fibers of nerves supplying the salivary glands (290, 291, 541, 662). The stimulation of the auriculotemporal nerve supplying the parotid salivary gland (662) and the efferent chordatympany (290) and secretory fibers in the trigeminal part of the lingual nerve (291) supplying the mandibular salivary gland has been described. Schneyer and Hall (541) have compared rat saliva evoked by both auriculotemporal nerve and pilocarpine stimulation. They showed only small differences in electrolyte concentration, but 5 to 20 times higher amylase activity and total protein in pilocarpine evoked saliva when compared to nerve stimulated saliva. The authors concluded that pilocarpine is an adequate substitute for nerve stimulation of the parotid salivary gland if only electrolytes and flow rate are considered but not when protein and amylase activity are considered.

For certain studies, inhibition of salivary secretion is re-

quired and for that purpose hyoscine hydrobromide can be administered (541).

3. Methods

a. Direct Collection from the Oral Cavity. Following anesthetization and pilocarpine administration, Holloway and Williams (300) placed each rat with its head hanging over the lower edge of a sloping board. Saliva was collected into a graduated tube as it dripped from the mouth.

Tatevossian and Wright (605) described a technique whereby resting (unstimulated) saliva was collected from unanesthetized, manually restrained rats, using disks of preweighed filter paper. Saliva was absorbed from the sublingual area.

To avoid contamination of saliva with lacrimal and nasal secretions in pilocarpine-treated rats, some workers prefer collecting saliva directly from the orifice of the salivary duct(s) using microcapillary tubes fitted with a rubber suction bulb (430), or by specially constructed collecting devices as described by Wolf and Kakeheshi (660) or Woodruff and Carpenter (662) for collection of parotid secretions.

b. Cannulation Techniques. Cannulation techniques have been described for all three groups of salivary glands in the rat. All techniques involve the use of either fine polyethylene tubing or glass cannulae placed either into the orifice of the duct or into the proximal end of surgically isolated and transected ducts. Hellekant and Kasahara (291) described the cannulation of the mandibular salivary gland duct using a polyethylene tubing with internal and external diameters of 0.28 and 0.68 mm, respectively.

Techniques for cannulation of the main excretory ducts of the submaxillary salivary glands using polyethylene tubing drawn out to a tip diameter of 150 mm was described by Martinez *et al.* (421). Other techniques using polyethylene tubing (151) or fine glass cannulae have been described.

The parotid salivary gland or, rather Stenson's duct, can be cannulated successfully by using a 30 gauge steel cannula to which short lengths of polyethylene tubing are attached (109, 525). Ulmansky *et al.* (617) have described the cannulation of Stenson's duct via its orifice using a stereoscopic microscope and a polyethylene catheter into which a stylet is placed. A Bardic inside needle catheter* with a 1919 gauge 19 stylet was used. The advantage of the latter technique lies in the fact that collection is through the natural duct orifice and not through a surgically isolated and incised duct. The method is reported to be simple and nontraumatic, and thus can be repeated in the same animal. In addition, it allows instillation of substances into the duct.

Drum (169) described the cannulation of each of the gland

*CR Bard, Inc., Murray Hill, New Jersey.

ducts using cannulae made by inserting the blunt ends of 25 to 26 gauge needle tips into 3- to 4-cm lengths of polyethylene tubing with an internal diameter of 0.023 inches (0.58 mm). Volumes of 0.20 to 0.80 ml were obtained from single sub-maxillary ducts and 0.06 to 0.21 ml from single parotid ducts during 20-min collection periods using 200-gm rats.

c. Collection from Surgically Isolated and Transected Ducts. Saliva may be collected from the proximal end of transected salivary gland ducts. Collection is usually made into tared vessels or graduated test tubes following the administration of a sialogogue.

V. SUBSTANCE ADMINISTRATION

A. General

Routes for administration of drugs, tumors, carcinogens, and other agents must be selected on the basis of intended purpose of such administration. For a full discussion of these principles of drug administration, the reader is encouraged to read Woodard (661).

B. Oral

Numerous methods have been developed to administer materials into or through the oral cavity and ultimately into the stomach of rodents. Marks (417) described a "silk rubber tube" with a syringe needle head for stomach feeding in 1908 (Fig. 19A).

Ferrill (205) described the use and construction of the ball-tipped 15 to 16 gauge hypodermic needle. Only moderate modifications of this simple device have been made over the years (Fig. 19B). Moreland (441) recommends placing a 20° to 30° bend, 2 cm proximal to the ball tip (Figs. 19C and 20). This conventional ball-tip stomach tube is attached to an appropriately sized syringe. The rat is grasped firmly and securely around the shoulders with one hand, the index finger and thumb are placed on either side of the head behind the mandibles and the rat is brought up against the operator's chest for better control. No mouth gag is necessary. The tube is readily passed in the unanesthetized rat between the interdental space, rotated slightly, moved over the tongue, posterior oropharynx and down into the esophagus. Once in the esophagus the material in the syringe is expelled. The tube may also be inserted further for direct injection into the stomach. When properly performed, accidental tracheal infusion is rare. The procedure can be very rapidly performed and an experienced operator can intubate approximately 120 rats per hour (205, 441).

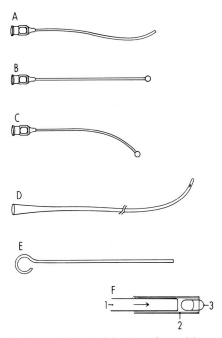

Fig. 19. Devices used for administration of materials po. (A) Original flexible silk tube. (B) Straight ball-tipped needle. (C) Curved ball-tipped needle. (D) Flexible No. 8 French catheter. (E) and (F) "Balling gun" for gelatin capsules: (E) ram portion of device, (F) close up of ram (1) within outer sleeve (2) showing position of gelatin capsule (3).

Shay and Gurenstein (557) described a method of instilling fluids into the stomach of rats utilizing a number 8 French catheter with a depressed eye (Fig. 19D). The rat is held on a flat surface and "the thumb and index fingers so placed at the middle of the distal phalanx of the former is at the angle of the mouth while the latter straddles the base of the skull of the animal." The palm of the hand holds the rat on the table by gentle pressure on its back. The thumb is drawn upward and backward to open the mouth and slightly turn the head of the rat toward the operator. The tube is moistened in normal saline

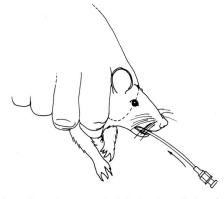

Fig. 20. Restraint and passage of a ball-tipped gastric inoculation needle.

and passed over the tongue and down along either cheek until it slips into the esophagus and then into the stomach. With experience a person can intubate 60 rats per hour.

To eliminate some of the problems that may be encountered using the tubing technique described above, Nelson and Hoar (452) developed a small animal balling gun used with a small gelatin capsule (Fig. 19E). This balling gun eliminates the possibility of irritation of the esophagus, of accidentally "drenching" the trachea and causing a foreign body pneumonia, and minimizes the chance of oral contamination of such innoculates as radioactive or toxic compounds or infectious agents. While this technique was originally described for the guinea pig, it can be used in the rat with minor modifications. The gelatin capsule is deposited in the posterior oral pharynx and the rat's mouth held shut until it is swallowed.

To facilitate the passage of stomach tubes in large numbers of rats, Lehr (381) and Kessel (340) developed simple devices used to fix the rat's mouth in an open position. Lehr's device was merely a keyhole-shaped 3-mm diameter stainless steel wire held in a vise at the narrow portion of the "keyhole." The rat's incisors are hooked over the narrow portion of the keyhole wire and moved toward the expanded portion to open the mouth to any degree necessary. Devices designed to facilitate oral examination of rodents have been described by Johansen (326) and Rizzo (523).

Gibson and Becker (235) described the use of a PE 10 polyethylene catheter fitted to a 30 gauge 3/8-inch blunt tip needle as a means of gastric intubation in newborn rats weighing 5 to 6 gm. Other techniques for intragastric injection of newborn rats have been described also (326, 433, 573).

McPherson (413) reported on the preparation of a smooth, uniform suspension of diet for tube feeding of rats.

Several devices for automatically feeding and regulating diet or water consumption in rats have been reported (326, 433, 573). Other workers have described methods for evaluating the feeding cycles of small rodents (413).

A method was described whereby the rat can ingest fluid through a chronic gastric tube that bypasses the oropharyngeal cavity. Epstein and Teitelbaum (190) developed a water-tight swivel joint permitting injections into moving animals, including into the stomach.

C. Intravenous

1. General

Intravascular injection of substances is generally performed by use of one of the superficial veins described previously (see Section IV or specimen collection). Choice of the most appropriate site for intravenous injections depends on several factors: (a) relative ease of cutaneous venipuncture; (b) age, size,

and sex of the animal; (c) relative risk to the animal; (d) whether anesthesia is required; (e) whether surgical exposure is necessary; (f) whether chronic catheterization is required; and (g) whether the vein is part of the hepatic portal system.

When injecting drugs for pharmaceutical or pharmacokinetic studies, it is most important to know whether or not, for example, the vein is part of the hepatic portal system. Drugs administered orally or intraperitoneally are absorbed into the mesenteric veins which combine to form the portal vein which leads directly to the liver. Drugs metabolized by the liver will undergo much more rapid degradation if injected orally or intraperitoneally than if injected by a route leading directly into the heart. This "first pass" effect can be marked. Nightingale and Mouravieff (460) described the fate of drugs administered by the dorsal vein of penis and by intracardiac injection to determine whether or not blood from the penis entered the portal system before entering the general circulation. They found that it did not, and, therefore, the dorsal vein of the penis was recommended as a site for intravenous injection for certain kinds of studies.

Intravenous injections, in general, may be performed by many of the techniques described in Section IV. The reader is also referred to Table I.

Certain vessels are preferred for intravenous injections while others only allow for injection of substances and not withdrawal of blood.

Injection may be made into the jugular, femoral, caudal (coccygeal), saphenous, lateral marginal, dorsal metatarsal, lingual and dorsal penile vein. Since all of the veins other than the femoral, lingual, and dorsal penile veins have been described amply in Section IV on specimen collection, only the others will be described here. The reader is referred to the bibliography which lists publications that are specifically concerned with the use of vessels for injection and infusion.

2. Femoral Vein

A technique for chronic catheterization of the femoral vein has been described by Eklund and Olin (182). The rat's femoral vein has the distinct disadvantage in that it generally must be exposed via a skin incision prior to catheterization. For this reason, it is used rather infrequently and then only for specialized studies requiring either acute or chronic catheterization.

3. Lingual Vein

While neither commonly used nor described in standard works, the lingual veins of the rats are readily accessible in the anesthetized animal and can be used for intravenous injections. The rat is placed in dorsal recumbency, the tongue blotted dry with a gauze sponge, held between the thumb and forefinger

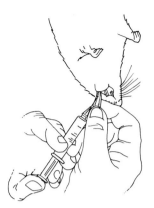

Fig. 21. Intravenous injection utilizing the lingual vein.

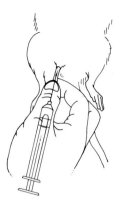

Fig. 22. Intravenous injection utilizing the dorsal vein of the penis. The hand holding the syringe has been removed from the drawing to increase clarity of presentation.

and gently pulled out to one side. The prominent sublingual veins run superficially from the tip posteriorly toward the root of the tongue. A 25 to 27 gauge needle can be inserted into these veins and substances administered. Care must be exercised to exert finger pressure following removal of the needle to prevent hematoma formation (16). This technique is especially easy in larger rats where it is more difficult to penetrate the thickened skin for a caudal vein injection. Greene and Wade (251) suggest the use of a temporary fine silk suture placed through the tip of the tongue to facilitate injection of the sublingual vein (Fig. 21).

4. Dorsal Vein of the Penis

This technique, although described by Salem, *et al.* (530) for the mouse, guinea pig, and hamster, works successfully in the rat. The rat is held by an assistant to control the head and front legs with one hand and the rear legs, feet and tail with the other. The rat is gently stretched to hyperextend its vertebral column. The operator then extrudes and grasps the tip of the penis between the thumb and forefinger. The dorsal vein of the penis is readily visible and can be injected with ease (Fig. 22).

5. Other Intravascular Injection Techniques

Postnikova (500) described a method of intracardiac injection of newborn rats and mice, while Anderson *et al.* (17) described a technique of intravenous injection of the superficial veins lying at the junction of the superior and inferior palpebral vein through the lateral canthus of the unopened eyelids Pinkerton and Webber (491) described the use of opthalmic plexus as a route for intravenous injections.

Many specialized techniques for periodic, pulsed, or continuous long-term intravenous injection or infusion have been described (132, 147, 193, 195, 296, 392, 459, 589, 659).

Steiger, *et al.* (589) developed a technique for long-term (up

to 40 days) intravenous feeding of unrestrained rats using a specially designed harness and fluid infusion assembly. The technique utilizes a jugular catheter with its free end tunneled through the subcutis and exiting in the midscapular region. The catheter is placed through a specially designed harness and connected to a sterile swivel infusion apparatus. Solutions are infused by means of a continuous infusion pump.

D. Intraperitoneal

The rat is restrained in the conventional manner. An assistant usually will be required to control the rear legs and prevent the operator from being scratched. The needle is directed anteriorly and is inserted just lateral to the ventral midline. Care must be taken not to penetrate either a hollow (urinary bladder, uterus, intestine) or solid organ (liver, spleen). Up to 15 to 25 ml fluid may be injected into the peritoneal space. While very commonly used because of its ease and convenience, intraperitoneal injections should be performed only when that route is the route of choice for reasons of scientific merit. Numerous commonly injected substances are irritating to the peritoneum, producing inflammation, adhesions, and sometimes even death.

E. Subcutaneous

Subcutaneous injections can be performed readily using the subcutis of the ventral body wall, flank or shoulders. Rats are manually restrained and an appropriately sized needle inserted through the skin at a 20° to 30° angle. Sometimes it is helpful to hold the point of the needle against the skin and pull the skin onto the needle with the other hand. If more than 1 or 2 ml must be injected, retracting the needle somewhat and redirecting it in the subcutaneous tissue is recommended.

F. Intramuscular

Intramuscular injections of relatively small quantities of material are generally made into the relatively large muscle mass of the posterior aspect of the rear limbs including the semimembranosus, semitendinosus, and biceps femoris muscles. With an assistant holding the rat, the muscle mass is stabilized by grasping the upper leg with the thumb and forefinger is one hand while the other directs the needle into the muscle. Care must be exerted not to accidentally inject intravascularly. While slight withdrawal of the syringe plunger should be performed prior to injection, the smaller needles and vessel sizes encountered may not allow detection of inadvertant vascular puncture in all cases.

G. Miscellaneous Routes

Many other routes of administration are available for specific purposes. Methods of topical application, instillation of the several body orifices, (ears, nose, rectum, vagina) intrathoracic, intracerebral, and inhalation routes of administration are all described in the literature. The reader is referred to the bibliography for description of these methods.

H. Special Techniques in the Neonate

Specialized applications of substance administration techniques in the newborn or neonate can be found in the articles by Gibson and Becker (235) and Bader and Klinger (29).

VI. ANESTHETIC TECHNIQUES

A. General

The appropriate use of anesthetics, analgesics, and tranquilizers to relieve pain and anxiety in laboratory animals is required by Federal law. Although at the present time, rats are not one of the species included under the Animal Welfare Act of 1970 (19) all recognize and accept the professional and moral responsibility to treat this species humanely. The determination of "when" such agents are appropriate and "what" and "how" such agents are used is the professional responsibility of the veterinarian or investigator.

Considerable effort has been made in recent years to better define what is actually meant by "adequate veterinary care" under PL 89-544, which of course includes the proper use of pain-relieving drugs. A symposium held in conjunction with the 112th annual meeting of the American Veterinary Medical Association in 1975 sponsored by the American College of Laboratory Animal Medicine attempted to define the meaning and spirit of the term "adequate veterinary care" as it relates to the topics under consideration (428, 474, 545, 644). Of particular relevance to the subject of anesthesiology was the presentation of Vierck (625). One must assume that "laboratory animal species perceive pain until it can be proved beyond a reasonable doubt that the animals do not experience an intensively adverse feeling state with stimulation that we would label noxious" (625). Consequently, we are obligated to use these classes of pharmacologic agents whenever we know that a particular procedure is painful or when we are not absolutely sure that it is not painful. Other articles concerned with pain perception in animals and humane technique as it relates to pain and suffering have been written by Breazile and Kitchell (70), Brain (69), and Croft (141).

The choice of anesthesia can affect the system under investigation. It is possible, for example, that the carotid occlusion relfex might not have been discovered if barbiturate anesthesia had been used instead of chloralose, since the latter agent greatly augments the reflex. For a further discussion of the various interactions of anesthetics on pharmacologic and clinical biochemical parameters, the reader is referred to an article by Chenoweth and VanDyke (108).

The choice of the most appropriate anesthetic, the dose, and the route of administration depend upon several factors: (1) age and size of the rat; (2) duration and depth of anesthesia required; (3) body system or parameters to be evaluated; (4) facilities, expertise, and equipment available; (5) single or multiple anesthetic procedures required; (6) presence of debility, obesity, pregnancy, lactation, or intercurrent disease; and (7) recovery or acute surgical procedures.

Almost all the anesthetic techniques and pharmacologic agents have been used in rats, including such uncommon methods as hypothermia in the neonatal rat (423, 477) and electrical anesthesia (565). However, only techniques commonly employed will be discussed here. References to anesthetic techniques in addition to those cited are included in the bibliography.

Many recent excellent reviews of the pharmacology of anesthetic agents (108, 596), general techniques of anesthesia (225, 403), anesthesia in laboratory animals (120, 143, 403), and anesthesia in the rat (47, 478, 479) have been published.

B. Inhalation Anesthetics

While ethyl ether has been and probably continues to be the most widely used inhalant anesthetic agent used in rats, halothane and methoxyflurane are becoming increasingly popular. Ether is highly flammable, and the peroxides remaining after evaporation of ether are highly explosive. However, ether is inexpensive and can be used with either an open drop or semiopen method. Use of halothane and/or methoxyflurane in these methods is prohibitively expensive.

In closed systems, halothane or methoxyflurane alone or combined with nitrous oxide and/or oxygen, have proved to be very safe and effective anesthetic agents for rats. Various methods for administering gaseous agents to rats have been described. These include open drop, semiopen drop (e.g., Bell jar), anesthetic mask or cone (328, 422), and endotracheal catheterization techniques. Endotracheal catheterization can be performed either via a tracheostomy (350) or less commonly, by a oroendotracheal route (504).

Chambers designed for controlling the concentration of volatile anesthetics and conserving the anesthetic agent have been described (179, 438, 668). Other systems designed to control and monitor various anesthetic gases also have been described (145, 156, 164, 406, 479). A device has been developed specifically for administration of volatile anesthetics to rats undergoing stereotaxic surgery (392).

When properly used, inhalant anesthetic agents, particularly halothane and methoxyflurane, have significant advantages over injectable anesthetic agents. There is more rapid induction. Recovery period and precise control of anesthetic depth is greater than with most injectable agents.

Chloroform, once commonly used as both an anesthetic and euthanizing agent, should not be used as an anesthetic because of its toxicity.

C. Injectable Anesthetics

1. Barbiturates

Pentobarbital is probably the most commonly used injectable anesthetic in the rat. It may be injected either intravenously or intraperitoneally at a dose of approximately 30–40 mg/kg of body weight. Considerable sex and strain variation exist in the amount of anesthetic required to achieve a desired plane of anesthesia. Females generally require less than males. Strain of rat as well as overall health status, presence of obesity, age, pregnancy, and other factors must be considered as well (55, 446, 492, 510, 559). Such differences also have been noted with other barbiturates, such as amobarbital (39) and hexobarbital (508). Anesthesia ensues in 5 to 15 min if barbiturates are given intraperitoneally and generally lasts 30 to 45 min. Additional small amounts of barbiturates can be given over time to prolong anesthesia. Up to one-quarter of the original dose can be given after the 30 to 45 min time period (438). It is best to dilute the stock pentobarbital solution (60 mg/ml) to 30 mg/ml or less to simplify accurate dosing. Premedication with chlorpromazine (167) and chlorprothixene (528) prolongs the duration of anesthesia.

2. Other Injectable Anesthetic Agents

a. Ketamine. Ketamine hydrochloride is a cataleptoid anesthetic of the phencyclidine group that is used in a wide range of laboratory animal species (408). Intramuscular administration of 44 mg/kg of body weight produces a surgical plane of anesthesia in rats which lasts from 15 to 25 min (645). It is recommended that atropine at a dosage of 0.04 mg/kg of body weight be administered to minimize excessive salivation. Tranquilizing agents also can be combined with ketamine to increase the degree of muscle relaxation for certain types of surgery. Acepromazine or triflupromazine are used most commonly with ketamine.

Youth *et al.* (667) described the use of ketamine in conjunction with pentobarbital. Ketamine and pentobarbital are not chemically compatible and should not be mixed or injected using the same syringe. Ketamine is administered at a dosage of 60 mg/kg body weight intramuscularly followed by pentobarbital 5 min later intraperitoneally at a dose of 21 mg/kg of body weight. Induction is rapid and last at least 1 hr; postoperative recovery is shortened, and mortality is low.

b. Innovar-Vet*. The combination of fentanyl citrate and droperidol is used extensively as an anesthetic agent in laboratory animals including rats (330, 387, 582, 608). Fentanyl citrate is a narcotic with very potent analgesic properties, and droperidol is a tranquilizer. This neuroleptoanalgesic combination is supplied commercially at a concentration of 20 mg/ml droperidol and 0.4 mg/ml fentanyl-citrate. Dosages ranging from 0.01 to 0.05 ml per 100 gm body weight are effective for abdominal surgery (330, 387, 608).

Levallorphan, a narcotic antagonist, tends to reverse the respiratory depression caused by fentanyl citrate when used at a dosage of 0.4 ml per 100 gm of body weight (608).

c. CI 744. Ward *et al.* (638) reported on the use of CI 744, a combination of tiletamine hydrochloride, a phencyclidine derivative, and zolasepan, a diazepinone tranquilizer. This experimental drug is being developed for use in several species including the rat. A dose of 20 and 30 mg/kg of body weight which produced a sleeping time of 68 min and 90 min, respectively, for the two levels administered to rats. Induction time averaged about 3 min.

For dosages and references to other agents which have been used as anesthetic agents in rats, see the works of Barnes and Eltherington (36).

D. Artificial Respiration and Resuscitation

Ventilation parameters for small mammals have been published by Kleinman and Radford (347). Artificial respiration

---

*Innovar-Vet, McNeil Laboratories, Washington, Pennsylvania.

and resuscitation techniques suitable to maintain adequate ventilation have been described by several workers (371, 563). A simple rat resuscitator was described by Ingall and Hasenpusch (318).

VII. EUTHANASIA

A. General

The word euthanasia is derived from the Greek *eu,* meaning good, and *thanatos,* meaning death. The spirit and essence of that meaning is contained in the definition of euthanasia included in the Animal Welfare Act of 1970. "Euthanasia means the humane destruction of an animal accomplished by a method which produces instantaneous unconsciousness and immediate death without visible evidence of pain or distress, or a method that utilizes anesthesia produced by an agent which causes painless loss of consciousness and death following such loss of consciousness" (19). While the above definition may serve as a starting point in one's consideration of and selection of the appropriate method of euthanasia for a given experiment, it does not include all factors. Certainly, one must also include psychological stress, particularly, but not only to the rat but also the personnel. Psychological stress does not have to be "visibly evident" to be real. Studies performed by Ader and Friedman (8) and Carney and Walker (100) among others clearly show changes in plasma corticosterone concentration in the rat subjected to different modes of killing. While decapitation may produce the least psychological stress to the rats, it probably produces the most distress in personnel of all the commonly used and accepted techniques. Guillotining must be properly performed in order to avoid measurable psychological stress to the rats.

One must also know or determine what effects, if any, the method of euthanasia has on experimental data to be collected. Differences in plasma corticosterone levels is only one example of this type of complication (77, 100). Numerous other labile hormones are likewise affected. The well-known effects of pentobarbital on hepatic microsomal enzymes, of chloroform on liver and kidney integrity, and cyanide on the cytochrome system are other examples of these types of interactions. Feldman and Gupta (204) recently described histopathologic changes in rats and other laboratory species which resulted from various means of euthanasia. They recommended the use of overexposure to carbon dioxide or intraperitoneal overdose of sodium pentobarbital as suitable for pulmonary studies. Decapitation or intracardiac injection of sodium pentobarbital was recommended as most suitable for examination for abdominal viscera.

Finally, one must be aware of and control for the effects of

circadian variations and other biorhythms as they related to the method of euthanasia and time of day chosen (172).

Several good reviews of the methods of euthanasia used in laboratory animals are in the literature (71, 204, 547, 576). In addition, the report of the American Veterinary Medical Association's Panel on Euthanasia (408a) should be consulted. Only some of the more commonly used and acceptable methods of euthanasia of rats will be described here. Very brief mention will be made of methods either not commonly used or not recommended for use with rats.

Regardless of the method chosen, great care must be taken to ensure that the animal is dead and not merely deeply narcotized or anesthetized. If any doubt exists, an additional dose of agent should be given until death is certain.

B. Physical Methods

The recommended methods of euthanasia of rats will be divided into (1) physical methods, (2) gaseous or volatile agents, and (3) noninhalant chemical agents.

1. Cervical Dislocation

Cervical dislocation is an acceptable method of euthanizing small rats (less than 250 gm body weight). The operator grasps the neck at the base of the skull with the thumb and forefinger of one hand and the hindlimbs and tail with the other and separates the cervical vertebrae from the base of the skull with a swift but controlled motion. This technique is usually performed on a bench top or other firm surface. Properly performed, this procedure results in instantaneous loss of consciousness and loss of all vital signs within a few minutes. Since the heart continues to beat for a short period of time, cardiac puncture or other blood collection techniques may be successfully employed to obtain a terminal blood specimen. Agonal movements due to spinal reflexes may prove offensive, but it must be remembered that no pain is perceived by the rat since its spinal cord has been severed. Burdizzo emasculators can also be used for bloodlessly and effectively performing cervical separation (432, 547).

2. Decapitation

Guillotines may be used for decapitating small or large rats and are commonly employed for studies in which the use of any pharmacologic agent is contraindicated. Decapitation with a guillotine has been thought to produce instantaneous unconsciousness and a rapid and humane death. However, recent results of studies conducted by Mikeska and Klemm (432) suggest that "decapitation" is "inhumane but is perhaps a necessity for many neurochemical experiments." Electroencephalogrpahic recordings taken before, during and after induction of death by anoxia or decapitation were studied and, in the author's interpretation, the data suggested that neither means of killing rats is humane. These authors went on to state that "the optimal euthanasia method is probably rapid injection of a large dose of barbiturate." (432).

Stimulation of an acute adrenal cortical reaction which results in as much as a fivefold increase in plasma corticosterone is encountered when rats react to others being sacrificed nearby. For this reason, only one rat at a time should be in the room when such a device is being used. It is important to consider the time of day as it relates to adrenocortical cycle in order to minimize circadian effect.

Decapitation is esthetically unpleasant and potentially dangerous to the inexperienced and untrained person. The rat is picked up with the forelimbs retracted back along the body wall and the neck position in the jaws of the opened guillotine. The lever is rapidly thrust downward, cleanly decapitating the rat. Great care must be taken to properly position and restrain the rat and also prevent accidental injury to the operator. The device should be thoroughly cleaned of blood before bringing another animal into the laboratory. Terminal blood collection is accomplished by rapidly thrusting the neck downward into a collection funnel.

3. Concussion

Scott and Ray describe (547) methods of killing laboratory rats by concussion. They describe wrapping the rat firmly about the body in a cloth and striking it very hard behind the ears with a suitable object or striking the dorsal skull very hard against a hard horizontal surface such as a sink or table. While these techniques can be successfully used as a means of euthanizing rats, there is considerable room for technical error, so that unless one is trained and experienced in these methods, other methods should be selected.

C. Gaseous or Volatile Agents

1. Ether

Ether is perhaps the most common volatile agent used for both anesthesia as well as euthanasia of rats. Cotton or soaked gauze pledgets are placed in the bottom of an air-tight chamber such as a desiccator jar. The rat is placed on a floor screen to prevent direct contact with the ether, since it will cause unnecessary irritation. Ether vapors rapidly anesthetize and then kill the rat by depressing the descending reticular formation of the brain which causes respiratory depression and ultimate failure. Ether is inexpensive, moderately rapid in effect, and has the further advantage in that several animals can be

euthanized simultaneously. However, it does have several significant disadvantages. It presents a serious explosion and fire hazard, is irritating to mucous membranes, and may cause excitement (CNS stimulation) and stimulation of the pituitary-adrenal axis during induction. For these reasons, Breazile and Kitchell (71) believe that neither ether nor chloroform fulfill the criteria of euthanasia. While many would agree with this contention regarding chloroform, all other reports reviewed considered ether as a satisfactory euthanizing agent if proper personnel safety precautions are employed.

2. Carbon Dioxide

High concentrations of carbon dioxide produces rapid anesthesia which proceeds to death and is considered to be one of the most humane, safest, and most reliable means of euthanasia. Carbon dioxide is relatively inexpensive, can be used with simple apparatus, is nonflammable and nonexplosive, has no odor, is one and a half times heavier than air, and can be used to euthanize several animals at a time. Scott and Ray (547) describe the construction and use of a chamber as well as techniques of placing a plastic bag over a basket-type cage and euthanizing the rats with carbon dioxide. A CO_2 chamber simply made from a 5 gallon can with CO_2 supplied by a gas cylinder, pressure valve, and laboratory tubing is most effective, inexpensive, and safe to operate (Fig. 23). It is recommended that the chamber be precharged with carbon dioxide for 1 min or so prior to introducing rats to be euthanized. This allows the CO_2 to fill the chamber and dis-

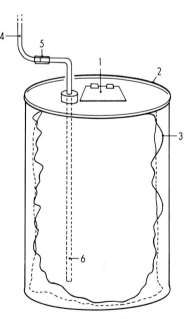

Fig. 23. Carbon dioxide euthanasia chamber. (1) Animal entrance port; (2) 5-gal container; (3) plastic liner; (4) CO_2 entry tubing; (5) rubber junction; (6) inflow tubing reaching nearly to bottom of container.

place the air so that minimal time is necessary to obtain anesthesia and euthanasia. Carbon dioxide administered properly is highly recommended for euthanasia in rats.

3. Other Volatile or Gaseous Agents

Halothane, methoxyflurane, and other gaseous anesthetic agents can be used successfully to produce anesthesia. They are fairly expensive and present explosion hazards or potential toxicity for man and other laboratory animals. While considered acceptable volatile agents for routine anesthesia of rats, their use for routine euthanasia is not recommended. The reader is referred to the recent work of Hughes (310) for further details about the use of these agents for euthanasia.

While chloroform has been used extensively as a euthanizing agent in the past, its routine use is not recommended. Chloroform vapors are very irritating, produce excessive excitement during induction, and present serious potential biohazard to man and other laboratory animals. Likewise, hydrogen cyanide vapors and carbon monoxide, although deadly, present risks far too great to be used routinely for euthanizing rats.

D. Noninhalant Chemical Agents

Of the literally scores of chemical compounds which may be used for euthanizing rats, only the barbiturates and chloral derivitives will be discussed.

1. Barbiturates

Of the many barbituric acid derivatives, sodium pentobarbital and thiopental are most commonly used for euthanasia in rats. Most investigators use anesthetic concentrations of these agents; however, higher concentrations may be either made or purchased commercially. Dosages of 100 mg/kg of body weight, which is two times the LD_{50} dose or three times the anesthetic dosage, have been recommended. The barbiturate is best administered rapidly by intravenous injection resulting in rapid anesthesia, respiratory and cardiac depression, and death. The intraperitoneal route is commonly employed and is generally accepted. However, the barbituric acid derivatives produce unnecessary irritation and may be inadvertently injected into an organ, thus delaying anesthesia and death. Intrathoracic or subcutaneous routes for these agents are not acceptable, since considerable irritation and pain can result from deposition in these sites.

2. Chloral Derivatives

Chloral hydrate and α-chloralose are general anesthetic agents which may be used to produce euthanasia in rats if given intravenously. They are irritating when given by other routes (intraperitoneal, subcutaneous, intrathoracic, etc.)

The American Veterinary Medical Association Panel on Euthanasia (517) does not recommend chloral hydrate as a euthanizing agent but did not comment on α-chloralose.

VIII. SPECIALIZED RESEARCH TECHNIQUES

It is well beyond the scope of this section to review all the literally hundreds of specialized research techniques developed for studies using rats; however, several important categories of techniques will be described: (1) selected surgical techniques, (2) techniques in cardiovascular research, and (3) techniques in respiratory research.

A. Selected Surgical Techniques

While aseptic techniques should be used for all types of surgery in rats, many routine procedures are still performed using "clean conditions." Cesarean derivation of germfree rats, gastrointestinal surgery, and surgery involving the central nervous system are examples of types of surgery which must be done aseptically. On the other hand, most endocrinectomies, splenectomies, and thymectomies may be performed using less than strict aseptic surgical technique.

1. Endocrine Gland Ablation

Endocrinectomies constitute some of the most important and experimentally useful surgical manipulations performed in the rat. Some single source references describing many surgical procedures including endocrinectomies are available (148, 255, 501).

a. Adrenalectomy. Bilateral adrenalectomy is generally performed utilizing a single cutaneous incision located on the dorsal midline commencing at the costal margin and extending caudally 20 to 25 mm. Blunt dissection is used to separate natural muscle—fascial planes between the latissimus dorsi and external abdominal oblique muscles on each side. Identification of the adrenal can be made by visual inspection, as it generally lies on the anterior dorsal aspect of its respective kidney. Care should be exercised to differentiate the gland from lobules of perirenal adipose tissue. Relatively fine, smooth, or toothless curved surgical forceps are used to identify, tease, grasp, and then clamp the connective tissue pedicle thus isolated. A second curved forceps is then placed on the pedicle, and the adrenal gland separated and removed by gentle traction. Ligation of adrenal blood vessels is unnecessary. The opposite adrenal is removed similarly through a separate incision in the muscle fascia on the other side. Muscular layers are routinely closed with a single suture. Skin closure is effected by either nonabsorbable suture material or wound clips. The entire surgical procedure may be performed by experienced personnel in less than 4 min.

Ablation or destruction of accessory adrenals (which vary in size and number but are generally found retroperitoneally over the aorta and inferior vena cava) is more difficult. Techniques for this procedure are described in detail by Poumeau-Delille (501).

b. Hypophysectomy. Four basic surgical approaches to hypophysectomy have been described: (a) temporal (610), (b) parapharyngeal (610), (c) palatal (8), and (d) transauricular (196, 361, 532). Only the transauricular approach of Falconi and Rossi (196) will be described. The position of the hypophysis near the thin wall of the tympanum on an almost flat sella turcica allows for this simplified technique. The transauricular technique utilizes a water suction pump that literally sucks the hypophysis from the floor of the cranium underneath the tectorial membrane. The specially designed needle is introduced into the left auditory canal, kept at a proper angle, and pushed gently forward until the medial wall of the periotic capsule is reached. With additional pressure this thin bone is perforated and the floor of the cranium beneath the tectorial membrane is reached. Suction is applied and the pituitary gland is sucked out in 4 to 7 sec (196). Sato and Yoneda (532) improved on the above technique which "guaranteed complete removal of the gland, as well as a high rate of survival." They utilized an operating board of special designed to facilitate the surgery and a technique of improved exposure involving making a small incision in the tragus of the ear.

c. Pinealectomy. A technique for pinealectomy (346) was first described by Exner and Boese in 1910. Since that time numerous techniques have been developed and improved upon. Recent works should be consulted for details of methodology (60, 298, 307, 507).

d. Thyroidectomy and Parathyroidectomy. The thyroid and associated parathyroid glands lie on the lateral surface of the trachea just distal to the larynx and extend caudally four to five tracheal rings. The two main lobes of the thyroid are joined by an isthmus. The parathyroid glands are closely associated with the thyroids and lie on their anterior and lateral surface. A technique for thyroparathyroidectomy has been described by Ingle and Griffith (319). One may leave the parathyroid behind if only thyroidectomy is required.

e. Pancreatectomy. Total pancreatectomy in the rat is most difficult because of the diffuse nature of this organ and the relative inexcessibility of certain branches of tissue. However, Scow (548) described a technique for removing "at least 99.5% of the pancreas of the rat". Within 18 hr all animals were hyperglycemic, glucosuric, and ketonuric, and all died within 48 hr if not treated with insulin. A detailed consideration of the gross anatomy of the rat pancreas as well as well as the surgical procedure itself is given.

2. Thymectomy

Cloud and Ledny (121) recently published a detailed description of thymectomy in the adult rat. Following the exposure of the thymus, a suction device was used to aspirate the gland. Using this technique, survival was 80%.

Hard (283) described a method for thymectomy in the neonatal rat which also utilizes suction. Pre- and postoperative management, included tranquilization of the dam, warming of the pups to 37°C following surgery, and application of pheromones to the young prior to replacing with the dam. These refinements improved survival over previously described techniques. A description of the mouse thymus as compared to the rat is included in the article. Thymectomy techniques in newborn and adult mice have been published by Sjödin *et al.* (571).

B. Techniques in Cardiovascular Research

1. Blood Pressure Determination

Many techniques, modifications, and improvements of methods used to measure systolic and diastolic blood pressure from anesthetized and unanesthetized rats have been published. The Report of the Committee on the Care and Use of Spontaneously Hypertensive (SHR) Rats of the Institute of Laboratory Animal Resources (518) provides a review of the methods currently believed to be most valid and reproducible. In addition, Chodorowicz *et al.* (112) recently published a review from the literature comparing blood pressure in rats as determined by direct and indirect methods.

a. Direct Techniques. Direct measurement of intraarterial pressure utilizing an intravascular cannula connected to a pressure transducer undoubtedly provides the most accurate results. While direct measurements are most suitable for acute preparations, Stanton (587) described a good method for chronic catheterization of the abdominal aorta. Explicit details of the procedure are presented.

b. Indirect Techniques. Indirect, noninvasive techniques have many advantages for monitoring purposes or for large numbers of animals. All indirect techniques utilize an occluding cuff, similar in principle to that used for routine blood pressure measurement in humans. The cuff is placed over an appendage, most commonly the tail, with a sensing device which usually consists of plethysmographic system placed distal to the cuff. The cuff is inflated and raised to a pressure above the systolic pressure of the rat and then the pressure is reduced slowly. The increase in tail volume when the cuff reaches systolic pressure is observed on the plethysmographic manometer. Other sensing devices, such as a piezoelectric crystal, can be used instead of the plethysmographic device to measure systolic pressure.

The various forms of instrumentation, their calibration, and the very specific and controlled conditions under which the rat's blood pressure must be determined are all extremely important if one is to obtain meaningful and reproducible results. The reader is urged to read Chapter 7 of the ILAR Report (518) for details of the procedure.

2. Determination of Blood Pressure in Minute Vessel

Nakata *et al.* (449) reviewed earlier methods and describe a method of direct blood pressure determination of minute vessels of the liver of the rat.

3. Blood and Plasma Volume Determination

Accepted techniques for determination of blood volume (79) and plasma volume (664) have recently been described. Other selected references are listed in the bibliography.

C. Techniques in Respiratory Physiology

Numerous techniques used to study respiration, respiratory minute volume, oxygen consumption, respiratory exchange, and other respiratory parameters are described in the references.

IX. NECROPSY TECHNIQUES

A necropsy examination is performed following death or euthanasia to assist in arriving at a definitive diagnosis and/or to fulfill the requirements of a given experimental protocol.

A diagnostic necropsy should be performed by or under the direction of a veterinary or comparative pathologist. It should be comprehensive, systematic, and thorough with appropriate notations, organ weights, and measurements made during the procedure as well as specimens properly taken for clinicopathologic or histopathologic studies.

Necropsy examinations of animals euthanized as part of an experimental protocol may frequently only be "partial" in that the prosector may only closely examine the organ or system under study. It is most important, however, that consideration be given to having a few animals from such experimental and control groups examined completely by a pathologist to exclude or identify latent disease.

The reader is referred to general veterinary pathology texts for detailed descriptions of possible gross necropsy protocols and techniques. Texts such as Greene's "The Anatomy of the Rat" (250) and Smith and Calhoun's "The Microscopic Anatomy of the Rat" (577) should be consulted for anatomical description. Finally, the various chapters in this text considering the naturally occurring diseases of rats should be consulted for pathoanatomic and clinicopathologic details.

ACKNOWLEDGMENTS

The author wishes to acknowledge and thank Mrs. Marcia Stefani who provided invaluable service to the author by her diligence in the library and in transcribing this manuscript, and to Miss Laura A. Hayward, Miss Marian I. Eaton, and Ms. Pam Couillard for their preparation of the illustrations.

REFERENCES

1. Abildgaard, C. F. (1964). A simple apparatus for holding rats. *Lab. Anim. Care* **14,** 235–237.*
2. Ablondi, F., SubbaRow, Y., Lipchuck, L., and Personeus, G. (1947). Comparison of blood pressure measurements in the rat as obtained by use of the tail and foot methods and by direct femoral puncture. *J. Lab. Clin. Med.* **32,** 1099–1106.*
3. Abramson, R. G., and Levitt, M. F. (1975). Micropuncture study of uric acid transport in rat kidney. *Am. J. Physiol.* **228,** 1597–1605.*
4. Acheson, R. M., MacIntyre, M. N., and Oldham, E. (1959). Techniques in longitudinal studies of the skeletal development of the rat. *Br. J. Nutr.* **13,** 283–292.*
5. Adamov, A. K., and Makapob, C. B. (1966). A table for immobilization of white mice and rats during intracerebral innoculation. *Lab. Delo.* **12,** 749.
6. Adams, C. E. (1962). Artifician insemination in rodents. *In* "The Semen of Animals and Artificial Insemination" (J. P. Maule, ed.), pp. 316–330. Commonw. Agric. Bur., Farnham Royal.*
7. Adams, C. E., and Abbot, M. (1971). Bibliography No. 45. Reprod. Res. Inf. Serv. Ltd., Cambridge, England.
8. Ader, R., and Friedman, S. B. (1968). Plasma corticosterone response to environmental stimulation and the 24 hour adrenocortical rhythm. *Neuro-endocrinology* **3,** 378–386.
9. Adolph, E. F., Barker, J. P., and Hoy, P. A. (1955). Multiple factors in thirst. *Am. J. Physiol.* **178,** 538–562.
10. Agrelo, C. E., and Miliozzi, J. O. (1974). A technique for repeated

*References followed by an asterisk are listed for the reader's further information.

blood sampling with transfusion in the conscious rat. *J. Pharm. Pharmacol.* **26**, 207-208.

11. Alexander, C. S. (1957). A new simple method for indirect determination of blood pressure in the rat. *Proc. Soc. Exp. Biol. Med.* **94**, 368-372.*

12. Amann, R. P. (1970). Sperm production rates. *In* "The Testes" (A. D. Johnson, W. R. Gomes, and N. L. VanDemark, eds.), pp. 435-441. Academic Press, New York.*

13. Ambrus, J. L., Ambrus, C. M., Harrison, J. W., Leonard, C. A., Moser, C. E., and Cravitz, H. (1951). Comparison of methods for obtaining blood from mice. *Am. J. Pharm.* **123**, 100-104.

14. Anchel, M. (1976). Beyond "adequate veterinary care". *J. Am. Vet. Med. Assoc.* **168**, 513-517.*

15. Anderson, F. F. (1941). A glass-capsule manometer for recording the blood pressure. *J. Lab. Clin. Med.* **26**, 1520-1521.*

16. Anderson, J. M. (1963). Lingual vein injection in the rat. *Science* **140**, 195.

17. Anderson, N. F., Delorme, E. J., Woodruff, M. F. A., and Simpson, D. C. (1959). An improved technique for intravenous injection of newborn rats and mice. *Nature (London)* **184**, 1952-1953.

18. Andrews, R. V., Belknap, R. W., Southard, J., Hess, S., and Strohbehn, R. (1971). Capture and handling of wild Norway rats used for endocrine studies. *J. Mammal.* **52**, 820-823.

19. Animal Welfare Act of 1970, Public Law 89-544 as amended by 91-579.

20. Anonymous (1965). Identification and record keeping for laboratory animals. *Charles River Dig. Four* **4.**

21. Anonymous notes (1967). Prevention of copraphagy in the rat. *Nutr. Rev.* **25**, 192.*

22. Appel, J. B. (1964). The rat: An important subject. *J. Exp. Anal. Behav.* **7**, 355-356.*

23. Arabehety, J. T., Colcini, H., and Gray, S. J. (1959). Sympathetic influences on circulation of the gastric mucosa of the rat. *Am. J. Physiol.* **197**, 915-922.*

24. Armstrong, B. K., and Softly, A. (1966). Prevention of copraphagy in the rat: A new method. *Br. J. Nutr.* **20**, 595-598.

25. Armstrong, G. G., Guyton, A. C., and Crowell, J. W. (1961). Small animal metabolism apparatus. *J. Appl. Physiol.* **16**, 387-389.*

26. Ashe, W. F., Carter, E. T., Hoover, G., Roberts, L. B., Johanson, E., Brown, F., and Largent, E. J. (1961). Some responses of rats to whole body mechanical vibration. Part I. *Arch. Environ. Health* **2**, 369-377.*

27. Ashman, R. B. (1975). A rapid method for skin grafting in germfree rats. *Lab. Anim. Sci.* **25**, 502-504.*

28. Austin, C. R. (1973). Embryo transfer and sensitivity to teratogenesis. *Nature (London)* **244**, 333-334.*

29. Bader, M., and Klinger, W. (1974). Intragastric and intracardiac injections in newborn rats. Methodological investigation. *Z. Versuchstierkd.* **16**, 40-42.

30. Bailey, C. B., Kitts, W. D., and Wood, A. J. (1957). A simple respirometer for small animals. *Can. J. Anim. Sci.* **37**, 68-72.*

31. Baker, R. D. (1967). "Postmortem Examination-Specific Methods and Procedures." Saunders, Philadelphia, Pennsylvania.*

32. Baker, R. D., Guillet, G. C., and Maynes, C. C. (1962). A cage for restraining rats. *J. Appl. Physiol.* **17**, 1020-1021.

33. Ball, L. (1976). Electroejaculation. *In* "Applied Electronics for Veterinary Medical and Animal Physiology" (W. R. Klemm, ed.). Thomas, Springfield, Illinois.*

34. Ballantyne, D. L., and Converse, J. M. (1957). The relation of hair cycles to the survival of suprapannicular and subpannicular skin homografts in rats. *Ann. N.Y. Acad. Sci.* **64**, 958-966.*

35. Ballantyne, D. L., Siegel, W. H., and Kapitchnikov, M. M. (1962). Further observations on massive skin homografts in rats. *Transplant. Bull.* **30**, 143-145.*

36. Barnes, C. D., and Eltherington, L. C. (1973). "Drug Dosages in Laboratory Animals: A Handbook." Univ. of California Press, Berkeley.

37. Barnes, J. M., and Denz, F. A. (1954). Experimental methods used in determining chronic toxicity. *Pharmacol. Rev.* **6**, 191-242.*

38. Barnes, R. H., Fiala, G., McGehee, B., and Brown, A. (1957). Prevention of coprophagy in the rat. *J. Nutr.* **63**, 489-498.

39. Barron, D. H. (1933). Some factors influencing the susceptibility of albino rats to injection of sodium amytal. *Science* **77**, 372-373.

40. Barrow, M. V. (1968). Modified intravenous injection technique in rats. *Lab. Anim. Care* **18**, 570-571.*

41. Baurmash, L., Bryan, F. A., Dickinson, R. W., and Burke, W. C. (1953). A new exposure chamber for inhalation studies. *Am. Ind. Hyg. Assoc., Q.* **14**, 26-30.*

42. Beach, F. A. (1938). A new method for marking small laboratory animals. *Science* **87**, 420.

43. Beauzeville, C. (1968). Catheterization of the renal artery of rats. *Proc. Soc. Exp. Biol. Med.* **129**, 932-936.

44. Bechtol, C. O. (1948). Differential *in toto* staining of bone, cartilage, and soft tissues. *Stain Technol.* **23**, 3-8.*

45. Beinfield, W. H., and Lehr, D. (1956). Advantages of ventral position in recording electrocardiogram of the rat. *J. Appl. Physiol.* **9**, 153-156.*

46. Bellinger, L. L., and Mendel, V. E. (1975). Hormone and glucose responses to serial cardiac puncture in rats. *Proc. Soc. Exp. Biol. Med.* **148**, 5-8.

47. Ben, M., Dixon, R. L., and Adamson, R. H. (1969). Anesthesia in the rat. *Fed. Proc., Fed. Am. Soc. Exp. Biol.* **23**, 1522-1527.

48. Ben-Jonathan, N., and Porter, J. C. (1974). An embolating apparatus for the collection of multiple samples of systemic and hypophysial portal blood. *Endocrinology* **94**, 864-870.

49. Bennett, J. P., Vallance, D. K., and Vickery, B. H. (1967). A method for direct observation of ovulation inhibition in the mature rat. *J. Reprod. Fertil.* **13**, 567-569.*

50. Bennett, J. P., and Vickery, B. H. (1970). Rats and mice. *In* "Reproduction and Breeding Techniques for Laboratory Animals" (E. Halfez, ed.), pp. 299-315. Charles C Thomas, Springfield, Illinois.

51. Bennett, J. P., and Vickery, B. H. (1972). Rats and mice. *In* "Reproduction and Breeding Telchniques for Laboratory Animals" (E. S. E. Hafez, ed.), pp. 299-315. Lea & Febiger, Philadelphia, Pennsylvania.

52. Ben-Ziv, G., Weinman, J., and Sulman, F. G. (1964). A photoplethysmographic method for measurement of systolic blood pressure in the rat. *Arch. Int. Pharmacodyn. Ther.* **149**, 527-535.*

53. Bergström, R. M., Hellström, P. E., and Sendberg, D. (1961). An intrauterine technique for recording the fetal EEG in animals. *Ann. Chir. Gynaecol. Fenn.* **50**, 430-433.*

54. Bernstein, L., and Elrick, H. (1957). The handling of experimental animals as a control factor in animal research-a review. *Metab., Clin. Exp.* **6**, 479-482.

55. Bester, J. F. (1952). A study of the potentiation of pentobarbital anesthesia by glucose and its metabolites. Dissertation, Ohio State University, Columbus.

56. Birnbaum, D., and Hall, T. (1961). An ejaculation technique for rats. *Anat. Rec.* **140**, 49-50.

57. Blandau, R. J., and Jordan, E. S. (1941). A technique of artificial insemination in the white rat. *J. Lab. Clin. Med.* **26**, 1361-1362.

58. Blass, E. M. (1972). An improved rat metabolism cage. *Physiol. Behav.* **9**, 681-683.

59. Bleicher, N. (1965). Care of animals during surgical experiments. *In* "Methods of Animal Experimentation" (W. I. Gay, ed.), Vol. 1, pp. 103-150. Academic Press, New York.*

60. Bliss, D. K., and Bates, P. L. (1973). A rapid and reliable technique for pinealectomizing rats. *Physiol. Behav.* **11**, 111-112.

61. Bollman, J. L. (1948). A cage which limits the activity of rats. *J. Lab. Clin. Med.* **33,** 1348.

62. Bollman, J. L., Cain, J. C., and Drindley, J. H. (1948). Techniques for the collection of lymph from the liver, small intestine, or thoracic duct of the rat. *J. Lab. Clin. Med.* **33,** 1349–1352.

63. Borell, V. (1947). Studies on function of pineal body by means of radioactive phosphorus. *Nord. Med.* **36,** 2137–2141.*

64. Borella, L. E., and Herr, F. (1971). A new method for measuring gastric acid secretion in unanesthetized rats. *Gastroenterology* **61,** 345–356.*

65. Bottiglieri, N. G., and Palmer, E. D. (1963). Technique for gastroscopic examination of rats. *J:. Lab. Clin. Med.* **62,** 842–845.*

66. Boura, A., and Dicker, S. E. (1953). An apparatus for the maintenance of a constant water load and the recording of urine flow in rats. *J. Physiol. (London)* **122,** 144–148.*

67. Bowen, J. M. (1976). Drugs acting on the nervous system. *In* "Handbook of Laboratory Animal Science" (E. Melby and N. Altman, ed.), pp. 65–118. C. R. C. Press, Cleveland, Ohio.*

68. Brain, J. D., and Frank, N. R. (1968). Recovery of free cells from rat lungs by repeated washings. *J. Appl. Physiol.* **25,** 63–69.

69. Brain, L. (1963). Animals and pain. *New Sci.* **19,** 380–381.

70. Breazile, J. E., and Kitchell, R. L. (1969). Pain perception in animals. *Fed. Proc., Fed. Am. Soc. Exp. Biol.* **28,** 1379–1382.

71. Breazile, J. E., and Kitchell, R. L. (1969). Euthanasia for laboratory animals. *Fed. Proc., Fed. Am. Soc. Exp. Biol.* **28,** 1577–1579.

72. Brittain, R. T. (1959). A simple urino-fecal separator for use in experiments on diuresis. *Lab. Pract.* **8,** 279.

73. Brittain, R. T., and Spencer, P. S. J. (1963). An apparatus for the long term collection of urine free from fecal and food contamination. *J. Pharm. Pharmacol.* **15,** 483–485.

74. Brodsky, S. G. (1967). A rapid procedure of i. v. injections in the rat. *Lab. Anim.* **5,** 29.*

75. Brown, C. W., Foldesy, M. E., and Henry, F. M. (1935). Two operative procedures for eliminating hearing in the rat. *J. Gen. Psychol.* **46,** 220–223.*

76. Brown, C. W., and Henry, F. M. (1934). The central nervous mechanisms for emotional responses. II. A technique for destroying the deeper nuclear regions within the cerebrum with a minimal destruction of the intervening cortex. *Proc. Natl. Acad. Sci. U.S.A.* **20,** 310–315.*

77. Brown, G. M., and Martin, J. B. (1974). Corticosterone, prolactin, and growth hormone responses to handling and new environment in the rat. *Psychosom. Med.* **36,** 241–247.

78. Bruckner-Kardoss, E., and Wostmann, B. S. (1967). Cecectomy of germfree rats. *Lab. Anim. Care* **17,** 542–546.*

79. Bruckner-Kardoss, E., and Wostmann, B. S. (1974). Blood volume of adult germ-free and conventional rats. *Lab. Anim. Sci.* **24,** 633–635.

80. Buchsbaum, M., and Buchsbaum, R. (1962). Age and ether anesthesia in mice. *Proc. Soc. Exp. Biol. Med.* **109,** 68–70.*

81. Bunag, R. D. (1973). Validation in awake rats of a tail cuff method for measuring systolic pressure. *J. Appl. Physiol.* **34,** 279–282.*

82. Bunag, R. D., McCubbin, J. W., and Page, I. H. (1971). Lack of correlation between direct and indirect measurements of arterial pressure in unanesthetized rats. *Cardiovasc. Res.* **5,** 24–31.*

83. Bunster, E., and Meyer, R. K. (1933). An improved method of parabiosis. *Anat. Rec.* **57,** 339–343.*

84. Burdette, W. J. (1963). "Methodology in Mammalian Genetics." Holden-Day, San Francisco, California.*

85. Burhoe, S. O. (1940). Methods of securing blood from rats. *J. Hered.* **31,** 445–448.

86. Burke, W. T., Brotherston, G., and Harris, C. (1955). An improved technique for obtaining bone marrow smears from the rat. *Am. J. Clin. Pathol.* **25,** 1226–1228.

87. Butcher, R. L., Collins, W. E., and Fugo, N. W. (1975). Altered secretion of gonadotropins and steroids resulting from delayed ovulation in the rat. *Endocrinology* **96,** 576–586.*

88. Buttle, G. A. H., D'Arcy, P. F., Howard, E. M., and Kellett, D. N. (1957). Plethysmometric measurement of swelling in the feet of small laboratory animals. *Nature (London)* **179,** 629.*

89. Byrne, R. J. (1965). Methods of animal infection with bacteria, fungi, and viruses. *In* "Methods of Animal Experimentation" (W. I. Gay, ed.), Vol. 2, pp. 481–525. Academic Press, New York.*

90. Byrom, F. B., and Wilson, C. A. (1938). A plethysmographic method for measuring blood pressure in the rat. *J. Physiol. (London)* **93,** 301–304.*

91. Calatayud, J. B., Gorman, P. A., and Caceres, C. A. (1965). Electronic monitoring of physiological phenomena in experimental animals. *In* "Methods of Animal Experimentation" (W. I. Gay, ed.), Vol. 2, pp. 528–580. Academic Press, New York.*

92. Calne, R. Y. (1965). Organ transplantation. *In* "Methods of Animal Experimentation" (W. I. Gay, ed.), Vol. 2, pp. 251–272. Academic Press, New York.*

93. Campbell, D. (1957). A compact cage arrangement for study of individual diuresis in large numbers of rats. *Acta Pharmacol. Toxicol.* **14,** 13–19.

94. Carlson, W. D. (1965). Radiography. *In* "Methods of Animal Experimentation" (W. I. Gay, ed.), Vol. 1, pp. 151–165. Academic Press, New York.*

95. Carmichael, E. B. (1938). A system for numbering laboratory animals. *Science* **87,** 557–558.

96. Carmichael, E. B. (1947). The median lethal dose (LD_{50}) of pentothal sodium for both young and old guinea pigs and rats. *Anesthesiology* **8,** 589–593.

97. Carmichael, E. B., McBurney, R., and Cason, L. R. (1946). A trap with holder for handling vicious laboratory animals such as wild rats. *J. Lab. Clin. Med.* **31,** 365–368.

98. Carmichael, E. B., and Posey, L. C. (1932). A convenient pneumograph. *Am. J. Physiol.* **101,** 17.*

99. Carney, F. M. T., and VanDyke, R. A. (1972). Halothane hepatitis: A critical review. *Anesth. Analg. (Cleveland)* **51,** 135–160.*

100. Carney, J. A., and Walker, B. L. (1973). Mode of killing and plasma corticosterone concentrations in the rat. *Lab. Anim. Sci.* **23,** 675–676.

101. Carvalho, J. S., Shapiro, R., Hopper, P., and Page, L. B. (1975). Methods for serial study of renin–angiotensin system in the unanesthetized rat. *Am. J. Physiol.* **228,** 369–375.

102. Case, R. W., Brueckman, C. T., and Lord, C. F. (1958). A new technique for the preparation of sterile carcinogenic pellets for implantation in experimental animals. *Am. J. Hosp. Pharm.* **15,** 565–566.*

103. Cassin, S., and Bogh, B. (1962). An automatic small animal volumeter. *J. Appl. Physiol.* **17,** 150–151.*

104. Castanera, T. J., Kimeldorf, D. J., and Jones, D. C. (1955). An apparatus for measurement of activity in small animals. *J. Lab. Clin. Med.* **45,** 825–832.*

105. Cates, C. C. (1969). A successful method for exsanguinating unanesthetized mice. *Lab. Anim. Care* **19,** 256–258.

106. Chebotar, N. A. (1974). Technic of obtaining oocytes from rats and mice and staining the constituent heterochromatin in meiotic chromosomes. *Tsitologiya* **16,** 1318–1329.*

107. Chen, K. K., and Rose, C. I. (1939). Small animal holders for intravenous injection. *J. Lab. Clin. Med.* **24,** 750–752.*

108. Chenoweth, M. B., and VanDyke, R. A. (1969). Anesthesia in biomedical research. *Fed. Proc., Fed. Am. Soc. Exp. Biol.* **28,** 1383–1385.

109. Chernick, W. S., Bobyock, E., and DiGregorio, G. J. (1971). Collection and quantitation of mouse salivary secretions. *J. Dent. Res.* **50,** 165.

110. Cheyne, V. D. (1939). A description of the salivary gland of the rat and a procedure for their extirpation. *J. Dent. Res.* **18**, 457–468.

111. Cho, M. H. (1961). Quantitation of spontaneous movements of animals given psychotropic drugs. *J. Appl. Physiol.* **16**, 390–391.*

112. Chodorowicz, M., Brzozowski, J., and Bojarski, R. (1973). Modifications of an apparatus for indirect measurement of blood pressure in the rat. *Acta Physiol. Pol.* **24**, 473–477.

113. Chrispens, C. G., and Kaliss, N. (1961). A simple device for facilitating injections in the tail veins of mice. *Am. J. Clin. Pathol.* **35**, 387–388.

114. Christensen, S. (1974). Effects of water deprivation in rats with polydipsia and polyuria due to long term administration of lithium. *Acta Pharmacol. Toxicol.* **35**, 201–211.*

115. Clarke, E. G. C., and Hawkins, A. E. (1957). Appraisement of total activity in laboratory animals. *Nature (London)* **179**, 1361.*

116. Clarke, N. P., Zuidema, G. D., and Smith, G. D. (1957). A method of producing controlled automatic inhalation anesthesia. *J. Am. Vet. Med. Assoc.* **130**, 347–349.*

117. Clarkson, T. B., and Kalnins, P. (1959). The use of the Papanicolaou stain to study cyclic vaginal epithelial changes of the rat. *Proc. Anim. Care Panel* **9**, 35–37.

118. Clemens, J. A., and Sawyer, B. D. (1974). Identification of prolactin in cerebrospinal fluid. *Exp. Brain Res.* **21**, 399–402.

119. Clewe, T. H., and Mastroianni, L. (1960). A method for continuous volumetric collection of oviduct secretions. *J. Reprod. Fertil.* **1**, 146–150.

120. Clifford, D. (1971). Restraint and anesthesia of small laboratory animals. *In* "Textbook of Veterinary Anesthesia" (L. R. Soma, ed.), pp. 369–384. Williams & Wilkins, Baltimore, Maryland.

121. Cloud, C. L., and Ledney, G. D. (1974). A technic for thymectomy in the adult rat. *Lab. Anim. Sci.* **24**, 340–342.

122. Cohen, A. E., and Oliver, H. M. (1964). Urethral catheterization of the rat. *Lab. Anim. Care* **14**, 471–473.

123. Coker, S. T. (1958). A note on a simple small animal respirometer. *J. Am. Pharm. Assoc., Sci. Ed.* **47**, 686.*

124. Colwell, A. R. (1950). The relation of bile loss to water balance in the rat. *Am. J. Dig. Dis.* **17**, 270–276.

125. Colwell, A. R. (1951). Collection of pancreatic juice from rats and consequences of its continued loss. *Am. J. Physiol.* **164**, 812–821.

126. Committee on Physiological Effects of Environmental Factors on Animals (1971). "A Guide to Environmental Research on Animals." Agriculture Board, Nat. Acad. Sci., Washington, D. C.

127. Connoly, J. H. (1958). A simple anaesthetic apparatus for small animals. *J. Clin. Pathol.* **11**, 369–370.*

128. Cook, R., and Dormann, R. G. (1969). Anesthesia of germfree rabbits and rats with halothane. *Lab. Anim.* **3**, 101–106.*

129. Cooper, P. H., and Stanworth, D. R. (1974). A simple and reproducible method of isolating rat peritoneal mast cells in high yield and purity. *Prep. Biochem.* **4**, 105–114.

130. Corbin, J. E. (1965). Controlled exercise. *In* "Methods of Animal Experimentation" (W. I. Gay, ed.), Vol. 2, pp. 451–479. Academic Press, New York.*

131. Cotchin, E., and Roe, F. J. C. (1967). "Pathology of Laboratory Rats and Mice." Blackwell, Oxford.*

132. Cotlove, E. (1961). Simple tail vein infusion method for renal clearance measurements in the rat. *J. Appl. Physiol.* **16**, 764–766.

133. Cottam, S. M., and Halliday, R. (1958). Letter to the editor. *J. Anim. Tech. Assoc.* **9**, 56.*

134. Courtice, F. C., and Morris, B. (1955). The exchange of lipids between plasma and lymph of animals. *Q. J. Exp. Physiol. Cogn. Med. Sci.* **40**, 138–148.*

135. Cox, W. M., and Meuller, A. J. (1937). The composition of milk from stock rats and no apparatus for milking small laboratory animals. *J. Nutr.*, 249–261.

136. Cramer, L. M. (1962). Rapid skin grafting in small animals. *Plast. Reconst. Surg.* **30**, 539–540.*

137. Cramlet, S. H., and Jones, E. F. (1976). Aeromedical review (1–76). Selected topics in laboratory animal medicine. *In* "Anesthesiology," Vol. V. USAF Sch. Aerosp. Med., Aerosp. Med. Divi. (AFSC), Brooks AFB, Texas.*

138. Crary, D. D. (1962). Modified benzyl alcohol clearing of alizarin-stained specimens without loss of flexibility. *Stain Technol.* **37**, 124–125.*

139. Creskoff, A. J., Fitzhugh, T., and Farris, E. J. (1967). Hematology of the rat. *In* "The Rat in Laboratory Investigation" (J. Q. Griffith and E. J. Farris, eds.), pp. 406–420. Hafner, New York.*

140. Croft, P. G. (1953). Electric stunning and electrocution of small animals. *Vet. Rec.* **65**, 259–261.*

141. Croft, P. G. (1957). Aspects of anesthesia. *Lab. Anim. Bur. M. R. C. Lab., Collect. Pap.* **6**, 75–77.

142. Croft, P. G. (1957). The criteria for a humane technique. *Lab. Anim. Bur., M.R.C. Lab.* pp. 19–22.*

143. Croft, P. G. (1960). "An Introduction to the Anesthesia of Laboratory Animals." Univ. Fed. Anim. Welfare, London.

144. Crossland, L. M., and Holloway, P. J. (1971). A technique for tube feeding newborn rats and the effects of administration of various carbohydrate solutions on their subsequent caries susceptibility. *Caries Res.* **5**, 144–150.*

145. Dahlof, L. G., VanDis, H., and Larsson, K. (1970). A simple device for inhalation anesthesia in restrained rats. *Physiol. Behav.* **5**, 1211.

146. Dai, S., and Ogle, C. W. (1972). A new method for the collection of gastric secretion in conscious rats. *Pfluegers Arch.* **336**, 111–120.*

147. Dalton, R. G., Touraine, J. L., and Wilson, T. R. (1969). A simple technique for continuous intravenous infusion in rats. *J. Lab. Clin. Med.* **74**, 813–815.

148. D'Amour, F. E., and Blood, F. R. (1954). "Manual for Laboratory Workin Mammalian Physiology." Univ. of Chicago Press, Chicago, Illinois.

149. D'Amour, F. E., Blood, F. R., and Belden, D. A. (1965). "Manual for Laboratory Work in Mammalian Physiology." Univ. Chicago Press, Chicago, Illinois.*

150. D'Amour, F. E., Blood, F. R., and Belden, D. A. (1965). Effect of decholin on bile flow. *In* "Manual for Laboratory Work in Mammalian Physiology." Univ. Chicago Press, Chicago, Illinois.

151. D'Amour, F. E., Blood, F. R., and Belden, D. A. (1965). The secretion and digestive action of saliva. *In* "Manual for Laboratory Work in Mammalian Physiology." Univ. of Chicago Press, Chicago, Illinois.

152. Davidman, M., Lalone, R. C., Alexander, E. A., and Levinsky, N. G. (1971). Some micropuncture techniques in the rat. *Am. J. Physiol.* **221**, 1110–1114.*

153. Davies, G. E., Evans, D. P., and Horsephal, G. B. (1971). An automatic device for the measurement of edema in the feet of rats and guinea pigs. *Med. Biol. Eng.* **9**, 567–570.*

154. Davies, L., and Grice, H. C. (1962). A device for restricting the movement of rats suitable for a variety of procedures. *Can. J. Comp. Med.* **26**, 62–63.

155. deBalbian Verster, F., Robinson, C. A., Hengeveld, C. A., and Bush, E. S. (1971). Freehand cerebroventricular injection technique for unanesthetized rats. *Life Sci.* **10**, 1395–1402.

156. DeBoer, B. (1948). Effects of thiamine hydrochloride upon pentobarbital sodium ("Nembutal") hypnosis and mortality in normal, castrated, and fasting rats. *J. Am. Pharm. Assoc.* **37**, 302.

157. Delfs, U., and Emmelin, N. (1974). Degeneration secretion of saliva in

the rat following sympathectomy. *J. Physiol. (London)* **239**, 623–630.*

158. Denckla, W. D. (1970). Minimum oxygen consumption in the female rat, some new definitions and measurements. *J. Appl. Physiol.* **29**, 263–274.*

159. Dennis, W., and Bolton, C. (1935). Producing brain lesions in rats without opening the skull. *Science* **81**, 297–298.*

160. Deringer, M. K. (1963). Technique for the transfer of fertilized ova. *In* "Methodology in Mammalian Genetics" (W. J. Burdette, ed.), pp. 563–564. Holden-Day, San Francisco, California.*

161. DeSmul, A., DeWaele, 0. 0., Wissoc, O. P., and Kiekens, R. (1974). Exogenous and endogenous secretin stimulation in the conscious rat. *Digestion* **11**, 39–59.

162. DeToledo, L. (1965). A technique for recording heart rate in moving rats. *J. Exp. Anal. Behav.* **8**, 181–182.*

163. Ditscherlein, G., Spann, M., and Natusch, R. (1968). Kidney puncture studies in the rat. *Z. Gesamte Exp. Med.* **145**, 260–265.*

164. Dobson, C., and Tschirky, H. (1971). Development of an anesthetic apparatus for experimental surgery on rats. *Pharmacology* **5**, 307–313.

165. Dodson, L. F., and Mackaness, G. B. (1957). The estimation of basal blood pressure in the rat by tail oscilloscope. *Brt. J. Exp. Pathol.* **38**, 618–627.*

166. Dohan, J. S. (1940). A numerical system using colors for albino rats and mice. *J. Lab. Clin. Med.* **25**, 872–874.

167. Dolowy, W. C., Thompson, I. D., and Hesse, A. L. (1959). Chlorpromazine premedication with pentobarbital anesthesia in the rat. *Proc. Anim. Care Panel* **9**, 93–96.

168. Draper, H. H., and Robbins, A. F. (1956). A urinofecal separator for laboratory rats. *Proc. Soc. Exp. Biol. Med.* **91**, 174–175.

169. Drum, D. E. (1963). Simple technique for direct cannulation of rat salivary ducts. *J. Dent. Res.* **42**, 892.

170. Dudley, W. R., Soma, L. R., Barnes, C., Smith, T. C., and Marshall, B. E. (1975). An apparatus for anesthetizing small laboratory animals. *Lab. Anim. Sci.* **25**, 481–482.*

171. Dudzinski, D. S., and Cutler, R. W. (1974). Spinal subarachnoid perfusion in the rat–Glycine transport from spinal fluid. *J. Neurochem.* **22**, 355–361.

172. Dunn, J., and Scheving, L. (1971). Plasma corticosterone levels in rats killed sequentially at the "trough" or "peak" of the adrenocortical cycle. *J. Endocrinol.* **49**, 347–348.

173. Dyban, A. P., and Baranov, V. S. (1974). Methods of studying the chromosomes in gametogenesis and embryogenesis in mammals. *Arkh. Anat., Gistol. Embriol.* **66**, 179–189.*

174. East, J., and Parrott, D. M. (1972). Operative techniques for newborn mice using anesthesia by cooling. *J. Endocrinol.* **24**, 249–250.*

175. Eayrs, J. T. (1954). An apparatus for analyzing the pattern of spontaneous activity in laboratory animals. *Brt. J. Anim. Behav.* **2**, 20–24.*

176. Ebmeyer, W. R., and Janssen, P. (1971). Hypophysectomy in rats and mice, a comparison of SPF with conventional animals. *Endokrinologie* **58**, 21–26.*

177. Ebner, H. J., Kuo, E., and Bogden, A. E. (1976). Arcadian patterns of prolactin and growth hormone levels in female rats of two strains. *Lab. Anim. Sci.* **26**, 186–189.*

178. Edwards, G. A., Edwards, C. H., and Gadsden, E. L. (1959). An inexpensive metabolism apparatus for collecting CO_2 excreta and blood from small experimental animals. *Int. J. Appl. Radiat. Isot.* **4**, 264–266.

179. Edwards, J. W., Sloan, R. F., Bleicher, N., and Ashley, F. L. (1959). An ether chamber for small laboratory animals. *Proc. Anim. Care Panel* **9**, 71–73.

180. Ehrenfreund, D. (1968). Use of rodents in behavioral research. *In* "Methods of Animal Experimentation" (W. I. Gay, ed.), Vol. 3, pp. 1–26. Academic Press, New York.*

181. Einstein, E. M., and Woskow, M. H. (1958). Technique for measuring heart potentials in a freely moving rat. *AMA Neurol. Psychiatry* **80**, 397–405.*

182. Eklund, L., and Olin, T. (1970). Catheterization of arteries in rats. *Invest. Radiol.* **5**, 69–74.

183. Emlen, J. T. (1944). Device for holding live wild rats. *J. Wildl. Manage.* **8**, 264–265.

184. Endicott, K. M., and Ott, M. (1945). The normal myelogram in albino rats. *Anat. Rec.* **92**, 61–69.*

185. Engel, W., and Kreutz, R. (1973). Lactate dehydrogenase isoenzymes in the mammalian egg-investigations by micro disc electrophoresis in 15 species of the orders Rodentia, Lagomorpha, Carnivora, Artiodactyla and in man. *Humangenetik* **19**, 253–260.

186. England, S. J. M., and Pasamanick, B. (1961). Radiotelemetry of physiological responses in the laboratory animal. *Science* **133**, 106–107.*

187. Enta, T., Lockey, S. D., and Reed, C. E. (1968). A rapid and safe technique for repeated blood collection from small laboratory animals. *Proc. Soc. Exp. Biol. Med.* **127**, 136–137.

188. Epstein, A. N. (1960). Reciprocal changes in feeding behavior produced by intrahypothalamic chemical injections. *Am. J. Physiol.* **199**, 969–974.*

189. Epstein, A. N. (1960). Water intake without the act of drinking. *Science* **131**, 497–498.*

190. Epstein, A. N., and Teitelbaum, P. (1962). A watertight swivel joint permitting chronic injection into moving animals. *J. Appl. Physiol.* **17**, 171–172.

191. Evans, C. S. Smart, J. L., and Stoddard, R. C. (1968). Handling methods for wild house mice and wild rats. *Lab. Anim.* **2**, 29–34.

192. Evans, E. P., Breckon, G., and Ford, C. E. (1964). An air-drying method for meiotic preparations from mammalian testes. *Cytogenetics* **3**, 289.–294.*

193. Eve, C., and Robinson, S. H. (1963). Apparatus for continuous long term intravenous infusion of small animals. *J. Lab. Clin. Med.* **62**, 169–174.

194. Falabella, F. (1967). Bleeding mice: A successful technique of cardiac puncture. *J. Lab. Clin. Med.* **70**, 981–982.

195. Falck, B., Hillarp, N. A., and Thieme, G. (1959). A simple device for long lasting continuous injections into animals. *Acta Physiol. Scand.* **47**, 259–261.

196. Falconi, G., and Rossi, G. L. (1964). Transauricular hypophysectomy in rats and mice. *Endocrinology* **74**, 301–303.

197. Falk, G. (1955). Maturation of renal function in infant rats. *Am. J. Physiol.* **181**, 157–170.*

198. Fall, M. W. (1974). The use of red light for handling wild rats. *Lab. Anim. Sci.* **24**, 686–687.

199. Fechheimer, N. S. (1960). Mammalian chromosome counts: A simple method for making preparations. *Nature (London)* **188**, 247–248.*

200. Feist, J. H., and Petrie, S. J. (1950). A holding board for small laboratory animals. *Am. J. Clin. Pathol.* **20**, 691–692.*

201. Fekete, A., and Tarjan, E. (1972). Renal hypertension induced by a new procedure in the rat. *Acta Med. Acad. Sci. Hung.* **39**, 177–184.*

202. Fekete, A., and Tarjan, E. (1973). Renal hypertension induced by a new procedure in the rat. *Acta Med. Acad. Sci. Hung.* **30**, 177–184.*

203. Felczak, J. (1968). Technic of dust radiography in experimental rats *J. Pol. Przegl. Radiol. Med. Nukl.* **32**, 87–89.

204. Feldman, D. B., and Gupta, B. N. (1976). Histopathologic changes in laboratory animals resulting from various methods of euthanasia. *Lab. Anim. Sci.* **26**, 218–221.

205. Ferrill, H. W. (1943). A simplified method for feeding rats. *J. Lab. Clin. Med.* **28**, 1624.

206. Festing, M., and Grist, S. (1970). A simple technique for skin grafting

rats. *Lab. Anim.* **4,** 255–258.*

207. Field, L. W., deGraff, W., and Wallis, A. T. (1958). Constant output blood perfusion pump for use in cross transfusion experiments in small laboratory animals. *J. Appl. Physiol.* **12,** 142–144.*

208. Fischer, A. E. (1956). Maternal and sexual behavior induced by intracranial chemical stimulation. *Science* **124,** 228–229.*

209. Ford, C. E., and Hamerton, J. L. (1956). A colchicine, hyptonic citrate, squash sequence for mammalian chromosomes. *Stain Technol.* **31,** 247–251.*

210. Ford, C. E., and Woollam, D. H. M. (1963). A colchicine, hyptonic citrate, air drying sequence for foetal mammalian chromosomes. *Stain Technol.* **38,** 271–274.*

211. Fox, M., and Zeiss, I. M. (1961). Chromosome preparations from fresh and cultured tissues using a modification of the drying technique. *Nature (London)* **192,** 1213–1214.*

212. Franklin, G. M., Dudzinski, D. S., and Cutler, R. W. (1975). Amino acid transport into the cerebrospinal fluid of the rat. *J. Neurochem.* **24,** 367–372.

213. Frape, D. L., Wilkinson, J., and Chubb, L. G. (1970). A simplified metabolism cage and tail cup for young rats. *Lab. Anim.* **4,** 67–73.

214. Fregly, M. J. (1953). Minimal exposures needed to acclimatize rats to cold. *Am. J. Physiol.* **173,** 393–402.*

215. Fregly, M. J. (1959). Extimation of thyroxine output by the thyroid glands of normal and adrenalectomized rats by means of a simple cooling test. *Can. J. Biochem. Physiol.* **37,** 425–432.*

216. Fregly, M. J. (1960). A simple and accurate feeding device for rats. *J. Appl. Physiol.* **15,** 539.*

217. Fregly, M. J. (1963). Factors affecting indirect determination of blood pressure of rats. *J. Lab. Clin. Med.* **62,** 223–230.*

218. Friedel, R., Trautschold, T., Gartner, K., Helle-Feldman, M., and Gaudssuhm, D. (1974). Effects of blood sampling on enzyme activities in the serum of laboratory animals. *Z. Klin. Chem. Klin. Biochem.* **12,** 229.

219. Friedman, M., and Freed, C. (1949). Microphonic manometer for indirect determination of systolic blood pressure in the rat. *Proc. Soc. Exp. Biol. Med.* **70,** 670–672.*

220. Friedman, M., Byers, S. O., and Michaels, F. (1950). Observations concerning production and excretion of cholesterol in mammals II. Excretion of bile in the rat. *Am. J. Physiol.* **162,** 575–578.

221. Friedman, S. M. (1947). The determination of heart rate in the rat. *J. Lab. Clin. Med.* **32,** 1284–1287.*

222. Fuson, R. B., and Eichwald, E. J. (1955). A semi-automatic needle for implanting tissues into experimental animals. *Cancer Res.* **15,** 162–163.*

223. Gaddum, J. H. (1941). A method of recording respiration. *J. Physiol. (London)* **99,** 257–264.*

224. Gafford, R. D., and Craft, C. E. (1959). A photosynthetic gas exchanger capable of providing for the respiratory requirement of small animals. *U.S. Air Force Sch. Aviat. Med. Rep.* **58,** 124.*

225. Galla, S. J. (1969). Techniques of anesthesia. *Fed. Proc., Fed. Am. Soc. Exp. Biol.* **28,** 1404–1409.

226. Gallahgerd, J. A., and Grimwood, L. H. (1953). A simple electrical method for measuring systolic blood pressure in the intact rat. *J. Physiol. (London)* **121,** 163–166.*

227. Gallyas, F., and Merei, F. T. (1965). The procedure of serial withdrawal of 10–200 ml blood samples by small laboratory instrument. *Pfluegers Arch. Gesamte Physiol. Menschen Tiere* **284,** 373–378.*

228. Ganesan, D. (1969). Influence of female sex hormones on pentobarbital sodium anesthesia in rats. *Arch. Int. Pharmaedyn. Ther.* **177,** 88–91.*

229. Ganis, F. M. (1962). Convenient mouse and rat holder for use in various laboratory procedures. *J. Lab. Clin. Med.* **60,** 354–356.

230. Garcia, J. F. (1957). Changesin blood plasma and red cell volume in the male rat as a function of age. *Am. J. Physiol.* **190,** 19–24.*

231. Geller, L. M., and Geller, E. H. (1966). A simple technique for the permanent marking of newborn albino rats. *Psychol. Rep.* **18,** 221–222.

232. Gershbein, L. L., Dan, T. C., and Shurrager, P. S. (1975). Clycolytic enzyme levels of intraocular fluids and lens as compared to the respective sera of various animal species. *Enzyme* **20,** 165–177.

233. Geyer, R. P., Geyer, B. R., Derse, P. H., Zinkin, T., Elvehjem, C. A., and Hart, E. B. (1947). Growth studies with rats kept under conditions which prevent coprophagy. *J. Nutr.* **33,** 129–142.

234. Ghosh, M. N., and Schild, H. O. (1955). A method for the continuous recording of acid gastric secretion in the rat. *J. Physiol. (London)* **128,** 35–36P.*

235. Gibson, J. E., and Becker, B. A. (1967). The administration of drugs to one day old animals. *Lab. Anim. Care* **17,** 524–527.

236. Gill, G. W. (1976). Principles and practice of cytopreparation. *In* "Handbook of Laboratory Animal Science" (E. C. Melby, and N. H. Altman, eds.), pp. 519–552. CRC Press, Cleveland, Ohio.

237. Giovati, A., Piccinin, G. L., and Bertaccini, G. (1970). Recording of respiratory minute volume in the rat. *Farmaco, Ed.* **25,** 512–515.*

238. Girardet, R. (1974). A simple and inexpensive restraining cage for rats. *J. Surg. Res.* **17,** 131–133.

239. Glennon, T. J., Elder, S. T., and Levinson, D. A. (1966). A new method for preparing rats with permanent ECS electrodes. *Psychol. Rep.* **19,** 1125–1126.*

240. Golba, S., Golba, M., and Wilczok, T. (1974). The effect of trauma, in the form of intraperitoneal injections or puncture of the orbital venous plexus on peripheral white blood cell count in rats. *Acta Physiol. Pol.* **25,** 339–345.

241. Gordon, H. A., Wostmann, B. S., and Bruckner-Kardoss, E. (1963). Effect of microbial flora on cardiac output and other elements of blood circulation. *Proc. Soc. Exp. Biol. Med.* **114,** 301–304.*

242. Gottleib, S. F., and Jagodzinski, R. V. (1961). Small animal body plethysmograph. *Fed. Proc. Fed. Am. Soc. Exp. Biol.* **20,** 420B.*

243. Gourevitch, G., Hack, M. H., and Hawkins, J. E. (1960). Auditory thresholds in the rat measured by an operant technique. *Science* **131,** 1046–1047.*

244. Gowans, J. L. (1957). The effect of the continuous reinfusion of lymph and lymphocytes on the output of lymphocytes from the thoracic duct of unanesthetized rats. *Br. J. Exp. Pathol.* **38,** 67–78.

245. Graham, M. M. (1971). Semi-automatic blood samples for the rat. *J. Appl. Physiol.* **30,** 772–773.

246. Grant, L., and Haddrell, D. (1971). An operating table plus head holder for rats. *Med. Biol. Eng.* **9,** 703–704.

247. Gray, R. S., and Axelrod, A. E. (1953). Application of thermistor to measurement of subcutaneous temperatures during hydrothermal injury to the rat. *Proc. Soc. Exp. Biol. Med.* **83,** 269–272.*

248. Grazer, F. M. (1958). Technic for intravascular injection and bleeding of newborn rats and mice. *Proc. Soc. Exp. Biol. Med.* **99,** 407–409.

249. Green, M. C. (1952). A rapid method for clearing and staining specimens for the demonstration of bone. *Ohio J. Sci.* **52,** 31–33.*

250. Greene, E. C. (1955). "The Anatomy of the Rat." Hafner, New York.

251. Greene, F. E., and Wade, A. E. (1967). A technique to facilitate sublingual vein injection in the rat. *Lab. Anim. Care* **17,** 604–606.

252. Grice, H. C. (1964). Methods of obtaining blood and for intravenous injections in laboratory animals. *Lab. Anim. Care.* **14,** 483–493.

253. Grice, H. C., and Middleton, E. J. (1962). Simplified procedure for intestinal anastomosis in the rat. *Can. Vet. J.* **3,** 352–355.*

254. Griffith, J. Q. (1934). Indirect method for determining blood pressure in small animals. *Proc. Soc. Exp. Biol. Med.* **32,** 394–396.*

255. Griffith, J. Q., and Farris, E. J., eds. (1967). "The Rat in Laboratory

Investigation." Hafner, New York.

256. Griffith, J. Q., and Jeffers, W. A. (1967). Method of restraint while working on the tail. *In* "The Rat in Laboratory Investigation." (J. Q. Griffith and E. J. Farris, eds.), pp. 19-23. Hafner, New York.*

257. Griffith, J. Q., and Jeffers, W. A. (1967). General methods. *In* "The Rat in Laboratory Investigation" (J. Q. Griffith and E. J. Farris, eds.), p. 20. Hafner, New York.*

258. Griffith, J. Q., and Jeffers, W. A. (1967). The circulatory system. *In* "The Rat in Laboratory Investigation" (J. Q. Griffith and E. J. Farris, eds.), pp. 278-294. Hafner, New York.*

259. Griffith, J. Q., and Jeffers, W. A. (1967). Methods for obtaining blood. *In* "The Rat in Laboratory Investigation" (J. Q. Griffith and E. J. Farris, eds.), pp. 278-294. Hafner, New York.*

260. Griffith, J. Q., and Jeffers, W. A. (1967). Methods for intravenous injection. *In* "The Rat in Laboratory Investigation" (J. Q. Griffith and E. J. Farris, eds.), pp. 288-289. hafner, New York.*

261. Grossman, M. I. (1958). Pancreatic secretion in the rat. *Am. J. Physiol.* **194**, 535-539.*

262. Grossman, S. P. (1960). Eating or drinking elicited by direct adrenergic or cholinergic stimulation of hypothalamus. *Science* **132**, 301-302.*

263. Grosvenor, C. E., and Turner, C. W. (1957). Estimation of amount of oxytocin released as result of nursing stimuli in the lactating rat. *Proc. Soc. Exp. Biol. Med.* **95**, 131-133.

264. Grosvenor, C. E., and Turner, C. W. (1958). Milk let down activity of synthetic oxytocin (syntocinon) and relaxin in lactating rats. *Proc. Soc. Exp. Biol. Med.* **97**, 189-190.

265. Guest, G. M., Brodsky, W. A., and Nelson, N. (1952). Metabolism cage for rats with feeding device that minimizes food scattering. *Metab., Clin. Exp.* **1**, 89-91.*

266. Gupta, B. N. (1973). Technic for collecting blood from neonatal rats. *Lab. Naim. Sci.* **23**, 559.

267. Gustafson, G., Stelling, E., and Brunius, E. (1968). The use of animals in dental research. *In* "Methods of Animal Experimentation" (W. I. Gay, ed.), Vol. 3, pp. 263-321. Academic Press, New York.*

268. Guyton, A. C. (1947). Measurement of the respiratory volumes of laboratory animals. *Am. J. Physiol.* **150**, 70-77.*

269. Haas, E., and Goldblatt, H. (1948). A simple aid for the cannulation of small blood vessels. *Science* **108**, 166.*

270. Hackl, H. (1954). A simple procedure for narcotizing laboratory animals. *Arch. Hyg. Bakteriol.* **138**, 436-439.*

271. Hadgraft, J. W., and Somers, G. F. (1954). A method for studying percutaneous absorption in the rat. *J. Pharm. Pharmacol.* **6**, 944-948.*

272. Hagen, E. O., and Hagen, J. M. (1964). A method of inhalation anesthesia for laboratory mice. *Lab. Anim. Care.* **14**, 13-15.*

273. Haldi, J., Wynn, W., and Breeding, H. (1961). Apparatus for measuring oxygen consumption of small animals. *J. Appl. Physiol.* **16**, 923-925.*

274. Haley, T. J., and Dickinson, R. W. (1953). A simple apparatus for observation of mesenteric capillary blood flow in rats. *Arch. Int. Pharmacodyn. Ther.* **95**, 243-245.*

275. Haley, T. J., Koste, L., and Duncan, F. (1965). Inexpensive rat metabolism cage. *J. Pharm. Sci.* **54**, 1391.

276. Hall, C. E. (1965). Parabiosis. *In* "Methods of Animal Experimentation" (W. I. Gay, ed.), Vol. 2. pp. 223-250. Academic Press, New York.*

277. Hall, C. E., Hall, O., and Adams, A. R. (1959). An electromechanical apparatus for stressing small animals. *J. Appl. Physiol.* **14**, 869-870.*

278. Halpern, B. N., and Pacaud, A. (1951). Technique de prélèvement d'echantillons de sang chez les petits animaux de laboratoire par ponction du plexus ophthalmique. *Ct. R. Seances Soc. Biol. Ses Fil.* **145**, 1465-1466.

279. Hammond, R. H. (1974). Measurement of growth and resorption of bone in the 7th caudal vertebra of the rat. *Calcif. Tissue Res.* **15**, 11-20.*

280. Hansard, S. L., and Comar, C. L. (1953). Radioisotope procedures with laboratory animals. *Nucleonics* **11**, 44-47.*

281. Hansen, L., and Holm, H. (1971). Apparatus for collecting faeces and urine from rats in metabolic experiments. *Lab. Anim.* **5**, 221-224.*

282. Hanson, H. M., and Brodie, D. A. (1960). Use of the restrained rat technique for study of the antiulcer effect of drugs. *J. Appl. Physiol.* **15**, 291-297.

283. Hard, G. C. (1975). Thymectomy in the neonatal rat. *Lab. Anim.* **9**, 105-110.

284. Harned, B. K., Cunningham, R. W., and Gill, E. R. (1949). A metabolism cage for small animals. *Science* **109**, 489-490.*

285. Harris, C. (1961). The lymphocyte like cell in the marrow of rats. *Blood* **18**, 691-701.

286. Harris, J. M. (1965). Differences in responses between rat strains and colonies. *Food Cosmet. Toxicol.* **3**, 199-202.*

287. Hayashi, S., and Sakaguchi, T. (1975). Capillary tube urinalysis for small animals. *Lab. Anim. Sci.* **25**, 781-782.

288. Hayden, J. F., Johnson, L. R., and Maikel, R. F. (1966). Construction and implantation of a permanent cannula for making injections into the lateral ventricles of the rat brain. *Life Sci.* **5**, 1509-1515.

289. Hecht, K., and Bielecke, F. (1960). A new multipurpose chamber with polygraphic registration for study of conditioned motor nutrition and defense reaction in rats and its methodical use. *Acta Biol. Med. Ger.* **4**, 71-89.*

290. Hellekant, G., and Hagström, E. C. (1974). Efferent chorda tympani activity and salivary secretion in the rat. *Acta Physiol. Scand.* **90**, 533-543.

291. Hellekant, G., and Kasahara, Y. (1973). Secretory fibres in the trigeminal part of the lingual revue to the mandibular salivary gland of the rat. *Acta Physiol. Scand.* **89**, 198-207.

292. Heller, H. (1947). The response of newborn rats to administration of water by the stomach. *J. Physiol. (London)* **106**, 245-255.

293. Hilf, R., Freeman, J. J., Staff, R., Johnson, M. M., and Borman. A. (1963). Effect of steroids upon three transplantable mammary tumors in the Fischer rat. *Cancer Chemother. Rep.* **30**, 1-8.*

294. Hillensj, O. T., Hamberger, L., and Ahr-en, K. (1975). Respiratory activity of oocytes isolated from ovarian follicles of the rat. *Acta Endocrinol. (Copenhagen)* **78**, 751-759.*

295. Hilliard, W. G., Dillistone, E. J., and Oliver, W. T. (1968). A method for cerebroventricular cannulation of the rat. *Can. J. Comp. Med.* **32**, 368-371.

296. Hint, H. C., and Richter, A. Q. (1958). A simple intravenous infusion technique for mice. *Acta Pharmacol. Toxicol.* **14**, 153-157.

297. Hoebel, M., Maroske, D., and Eichler, O. (1971). Measurement of respiratory minute volumes in the rat and guinea pig by a modified spirometer. *Arch. Int. Pharmacodyn. Ther.* **194**, 371-374.*

298. Hoffman, R., and Reiter, R. (1965). Rapid pinealectomy in hamsters and other small rodents. *Anat. Rec.* **153**, 19-22.

299. Holland, L. M., Drake, G. A., London, J. E., and Wilson, J. D. (1971). Intravenous injection with a pulsed dental cleaning device. *Lab. Anim. Sci.* **21**, 913-915.*

300. Holloway, P. J., and Williams, R. A. D. (1965). A study of theoral secretions of rats stimulated by pilocarpine. *Arch. Oral Biol.* **10**, 237-244.

301. Holmquist, D. L., Retiene, K., and Lipscomb, H. S. (1966). Arcadian rhythms in rats: Effect of random lighting. *Science* **152**, 662-664.*

302. Hosko, M. J. (1972). An improved rat head clamp for neurosurgery or electrode implantation. *Physiol. Behav.* **9**, 103-104.

303. Hotz, J., Goberna, R., and Clodi, Ph. H. (1973). Reserve capacity of

the exocrine pancreas, enzyme secretion and fecal fat assimilation in 95% pancreactomized rat. *Digestion* **9**, 212–223.*

304. Hotz, J., Zwicker, M., Minne, H., and Ziegler, R. (1975). Pancreatic enzyme secretion in the conscious rat method and application. *Pfluegers Arch.* **353**, 171–189.

305. Howells, G. R., Wright, C. F., and Harrison, G. E. (1964). A new metabolic cage for rats. *J. Anim. Tech. Assoc.* **14**, 137–144.*

306. Hoy, P. A., and Adolph, E. F. (1956). Diuresis in response to hypoxia and epinephrine in infant rats. *Am. J. Physiol.* **187**, 32–40.

307. Hsieh, K. S., and Ota, M. (1969). Improved procedure of pinealectomy in rats. *Endocrinol. Jpn.* **16**, 477–478.

308. Huang, K., and Bondurant, J. H. (1956). Simultaneous estimation of plasma volume, red cell volume and thioryanate space is unanesthetized normal and splenectomized rats. *Am. J. Physiol.* **185**, 441–445.*

309. Huddleson, I. F., and Carlson, E. R. (1928). A new animal board. *J. Lab. Clin. Med.* **13**, 660–661.*

310. Hughes, H. C. (1976). Euthanasia of laboratory animals. *In* "Handbook of Laboratory Animal Science" (E. C. Melby and N. H. Altman, eds.), Vol. III, pp. 553–560. C.R.C. Press, Cleveland, Ohio.

311. Hughes, H. C., and Lang, C. M. (1972). Hepatic necrosis produced by repeated administration of halothane to guinea pigs. *Anesthesiology* **36**, 466–471.*

312. Hughes, P. C., and Nowak, M. (1973). The effect of the number of animals per cage on the growth of the rat. *Lab. Anim.* **7**, 293–296.*

313. Hulse, S. H. (1960). A precision liquid feeding system controlled by licking behavior. *J. Exp. Anal. Behav.* **3**, 1–3.*

314. Hunt, H. F. (1961). Methods for studying the behavioral effects of drugs. *Annu. Rev. Pharmacol.* 125–144.

315. Hurwitz, A. (1971). A simple method for obtaining blood samples from rats. *J. Lab. Clin. Med.* **78**, 172–174.

316. Huttunen, R. (1975). The proteolytic proenzymes in the peritoneal exudate during acute experimental pancreatitis of the rat. *Acta Chir. Scand.* **141**, 285–288.*

317. Hyman, C., Stone, A., and Sosnow, M. (1953). Application of tissue clearance technics in small animals. *Proc. Soc. Exp. Biol. Med.* **83**, 643–646.*

318. Ingall, J. R., and Hasenpusch, P. H. (1966). A rat resuscitator. *Lab.*

319. Ingle, D. J., and Griffith, J. Q. (1967). Surgery of the rat. *In* "The Rat in Laboratory Investigation" (J. Q. Griffith and E. J. Farris, eds.), pp. 434–452. Hafner, New York.*

320. Irwin, M. H., Steenbock, H., and Templin, V. M. (1936). A technic for determining the rate of absorption of fats. *J. Nutr.* **12**, 85–101.*

321. Jocobs, P., and Adriaenssens, L. (1970). A simple method for repeated blood sampling in small animals. *J. Lab. Clin. Med.* **75**, 1013–1016.

322. Jaffee, R. A., and Free, M. J. (1973). A simple endotracheal intubation technique for inhalation anesthesia of the rat. *Lab. Anim. Sci.* **23**, 266–269.*

323. Jarausch, K. H., and Ullrich, K. J. (1957). Zur technik der entrahme von harnproben aus einzelnen sammelrohren der suagtierniere mittels polyathylen-cappilaren. *Pfluegers Arch. Gesamte Physiol. Menschen Tiere* **264**, 88–94.*

324. Jeffers, W. A., and Griffith, J. Q. (1967). The central nervous system. *In* "The Rat in Laboratory Investigation" (J. Q. Griffith and E. J. Farris, ed.), pp. 196–202. Hafner, New York.*

325. Jemski, J. V., and Phillips, G. B. (1965). Aerosol challenge of animals *In* "Methods of Animal Experimentation" (W. I. Gay, ed.), Vol. I, pp. 273–341. Academic Press, New York.*

326. Johansen, E. (1952). A new technique for oral examination of rodents. *J. Dent. Res.* **31**, 361–365.

327. John, R. (1941). An improved electrically-heated operating board for small animals. *J. Physiol. (London)* **99**, 157–160.

328. Johns, T. N. P., and Olson, B. J. (1954). Experimental myocardial

329. Jokiph, L. (1973). Migration inhibition of peritoneal exudate cells from sensitized and desensitized rats. *Immunology* **25**, 283–295.*

330. Jones, J. B., and Simmons, M. L. (1968). Innovar-Vet as an intramuscular anesthetic for rats. *Lab. Anim. Care* **18**, 642–643.

331. Junqueria, L. C. V., Hirsch, G. C., and Rothschild, H. A. (1955). Glycine uptake by the proteins of the rat pancreatic juice. *Biochem. J.* **61**, 275–278.

332. Kaczmarczyk, G., and Reinhardt, H. (1975). Arterial blood gas tensions and acid base status of Wistar rats during thipental and halothane anesthesia. *Lab. Anim. Sci.* **25**, 184–190.*

333. Kahler, H. (1942). Apparatus for milking mice. *J. Natl. Cancer Inst.* **2**, 257–258.

334. Kassel, R., and Levitan, S. (1953). A jugular technique for the repeated bleeding of small animals. *Science* **118**, 563–564.

335. Keeler, C. E. (1941). Simple floor-net for catching the escaped laboratory rodent. *Science* **93**, 408.

336. Keighley, G. (1966). A device for intravenous injection of mice and rats. *Lab. Anim. Care* **16**, 185–187.*

337. Kellogg, R. H., Burack, W. R., and Isselbacher, K. J. (1954). Comparison of diuresis produced by isotonic saline solutions and by water in rats studied by a "steady state" method. *Am. J. Physiol.* **177**, 27–37.*

338. Kemp, J. G., Hunsaker, W. G., and Palmer, J. (1961). Apparatus for simultaneous infusion and multiple blood sampling. *Proc. Soc. Exp. Biol. Med.* **106**, 498–500.*

339. Kersten, H., Brosene, W. G., Ablondi, F., and SubbaRow, Y. (1947). A new method for the indirect measurement of blood pressure in the rat. *J. Lab. Clin. Med.* **32**, 1090–1098.*

340. Kessel, H. (1964). A simple aid in the intubation of small animals. *Lab. Anim. Care* **14**, 499–500.

341. Khairy, M. (1960). Effects of chronic dieldrin ingestion on muscular efficiency of rats. *Br. J. Ind. Med.* **17**, 146–148.*

342. Kilham, L., and Belcher, J. H. (1954). A device for safe handling of wild animals in the laboratory. *J. Wildl. Manage.* **18**, 402.

343. Kinard, F. W., Danielson, D. N.. and Warren, J. A. (1947). Metabolism cages. *Science* **49**, 105.*

344. King, H. (1966). An innoculation cage for mice and rats. *Z. Med. Labor Tech.* **7**, 40–42.*

345. Kirkpatrick, D., and Silver, D. (1970). A simplified technique for lymphography for the small laboratory animal. *J. Surg. Res.* **10**, 147–150.*

346. Kitay, J. I., and Altschule, M. D. (1954). Chief methods of investigating pineal function. *In* "The Pineal Gland," pp. 8–12. Harvard Univ. Press, Cambridge, Massachusetts.

347. Kleinman, L. I., and Radford, E. P. (1964). Ventilation standards for small mammals. *J. Appl. Physiol.* **19**, 360–362.

348. Klemm, W. R., (1976). "Applied Electronics for Veterinary Medicine and Animal Physiology." Thomas, Springfield, Illinois.*

349. Klinger, W., Kersten, L., and Melhorn, G. (1965). The influence of blood taking technique on the high blood sugar level and the total liver content of glucose in Wistar rats and Agnes-Bluhm mice (Jena) fixation of normal values. *Z. Versuchstierkd.* **6**, 35–47.

350. Kluge, T., and Tveten, L. (1968). Endotracheal anesthesia for intrathoracic surgery in rats. *Acta Pathol. Microbiol. Scand.* **72**, 103–108.

351. Knapp, W. C., Leeson, G. A., and Wright, G. J. (1971). An improved technique for the collection of bile in the unanesthetized rat. *Lab. Anim. Sci.* **21**, 403–405.

352. Kniazuk, M. (1937). A method for determining the heart rate of small animals. *J. Lab. Clin. Med.* **22**, 868–870.*

353. Kniazuk, M., Martin, W. R., and Unna, K. R. (1958). A new method of measuring movement in small animals. *J. Pharmacol. Exp. Ther.*

122, 38A–39A.*

354. Knigge, K. M., and Joseph, S. A. (1974). Thyrotrophin releasing factor (TRF) in cerebrospinal fluid of the third ventricle of rat. *Acta Endocrinol. Copenhagen* **76,** 209–213.

355. Kogan, A. K. (1959). Universal chamber for immobilization of nonanesthetized rats. *Patol. Fiziol. Eks. Ter.* **3,** 69–70.

356. Komarov, S. A., Bralow, S. P., and Boyd, E. (1963). A permanent rat gastric fistula. *Proc. Soc. Exp. Biol. Med.* **112,** 451–453.*

357. Konig, K. G., Schmid, P., and Schmid, R. (1968). An apparatus for frequency controlled feeding of small rodents and its use in dental caries experiments. *Arch. Oral Biol.* **13,** 13–26.*

358. Kontsevoi, V. M. (1968). Comparison of different methods of blood taking in rats. *Lab. Delo.* **2,** 121–122.*

359. Kooyers, W. M., and Pabst, M. L. (1959). Rat blood pressure test for pressor substances and biological assay of epinephrine. *Drug Stand.* **27,** 94–96.*

360. Kowalewski, K., and Chmura, G. (1969). A method permitting prolonged and repeated studies of rat gastric secretion. *Arch. Int. Physiol. Biochim.* **77,** 10–16.*

361. Koyama, R. (1931). Experimentelle untersuchungen uber die hypophysenexstirpation an ratten und die wirkung des vorderlappenextraktes. *Jpn. J. Med. Sci. Pharmacol.* **5,** 41–67.

362. Kozlowski, L. T., and Woods, S. C. (1972). A device for easy administration of volatile anesthetics to rats undergoing stereotoxic surgery. *Behav. Res. Methods & Instrum.* **4,** 197–198.

363. Kraner, K., and Parshall, C. J. (1968). Experimental procedures and surgical techniques performed on intrauterine fetal animals. *In* "Methods of Animal Experimentation" (W. I. Gay, ed.), pp. 211–239. Academic Press, New York.*

364. Kreezer, G. L. (1967). Technics for the investigation of behavioral phenomena in the rat. *In* "The Rat in Laboratory Investigation" (J. Q. Griffith and E. J. Farris, eds.), pp. 203–277. Hafner, New York.*

365. Kuehne, R., Altman, R., and Engel, E. (1973). A simplified method for bleeding animals. *Physiol. Behav.* **10,** 1119.*

366. Laboratory Animals Bureau (1957). "Humane Technique in the Laboratory," Collect. pap. M.R.C. Laboratories, London.*

367. Lambert, R. (1965). "Surgery of the Digestive System in the Rat." Thomas, Springfield, Illinois.*

368. Lambert, R. (1965). Surgery of the bile and pancreatic ducts. *In* "Surgery of the Digestive System in the Rat," pp. 107–170. Thomas, Springfield, Illinois.

369. Lane-Petter, W. (1972). The laboratory rat. *In* "The UFAW Handbook on the Care and Management of Laboratory Animals," 4th ed., pp. 204–211. Churchill-Livingstone, London.*

370. LaPeyrac, F. (1963). Taking of blood from the cavernous sinus. *Rev. Fr. Etud. Clin. Biol.* **8,** 195–196.

371. Lassig, W., and Verron, G. (1974). Technic of infusion in rats. *Z. Versuchstierkd.* **16,** 125–126.

372. Law, T., and Wise, H. W. (1959). A multiple electrode carrier for chronic implantation in small animals. *Electroencephalogr. Clin. Neurophysiol.* **10,** 749–751.*

373. Lawson, R. L., Barranco, S., and Sorensen, A. M. (1966). A device to restrain the mouse, rat, hamster, and chinchilla to facilitate semen collection and other reproductive studies. *Lab. Anim. Car* **16,** 72–79.

374. Lawson, R. L., Krise, G. M., and Sorensen, A. M. (1967). Electroejaculation and evaluation of semen from the albino rat. *J. Appl. Physiol.* **22,** 174–176.

375. Lawson, R. L., and Sorensen, A. M. (1964). Oblation of coagulating gland and subsequent breeding in the albino rat. *J. Reprod. Fertil.* **8,** 415–417.*

376. Lazarow, A. (1954). Rat metabolism cages. *Methods Med. Res.* **6,** 216–242.

377. Leach, L. J., and Spiegl, C. J. (1958). An animal inhalation exposure unit for toxicity screening. *J. Am. Ind. Hyg. Assoc.* **19,** 66–68.*

378. Leavell, G. (1942). Improved procedure for preparation of carcasses of small animals and skeletal parts of larger animals for chemical analysis. *J. Assoc. Off. Agric. Chem.* **25,** 159–163.*

379. Lebeder, K. A. (1962). An apparatus for intravenous administration of solutions to mice and rats. *Lab. Delo.* **8,** 59–61.*

380. LeBouffant, L. (1958). Apparatus for the quantitative study of exposure of animals to dust. Measurement of lung cleaning capacity in the rat. *Staublungenerkrankungen, Ber. Arbeitsmed. Tag., 1956* Vol. 3, pp. 277–284.*

381. Lehr, D. (1945). Stomach tube feeding of small laboratory animals. *J. Lab. Clin. Med.* **30,** 977–980.

382. Lester, D., and Greenberg, L. A. (1948). A simple apparatus for the serial measurement of the respiratory exchange in the rat. *J. Biol. Chem.* **176,** 53–57.*

383. Levine, G., Lewis, L., and Cember, H. (1973). A vacuum-assisted technic for repetitive blood sampling in the rat. *Lab. Anim. Sci.* **26,** 211–213.

384. Levine, S. (1965). Rapid method for obtaining spinal cords from small laboratory animals. *Am. J. Clin. Pathol.* **43,** 272–273.*

385. Levy, B. J., and Fair, W. R. (1973). The location of antibacterial activity in the rat prostatic secretions. *Invest. Urol.* **11,** 173–177.*

386. Lewis, A. C., Rubini, M. E., and Beisel, W. R. (1960). A method for rapid dehydration of rats. *J. Appl. Physiol.* **15,** 525–527.*

387. Lewis, G. E., and Jennings, P. B. (1972). Effective sedation of laboratory animals using Innovar-Vet. *Lab. Anim. Sci.* **22,** 430–432.

388. Lewis, V. J., Thacker, W. L., Mitchell, S. H., and Baer, G. M. (1976). A new technic for obtaining blood from mice. *Lab. Anim. Sci.* **23,** 556.

389. Linder, M. J. (1953). Estimation of growth rate in animals by marking experiments. *U.S. Fish Wildl. Serv., Fish. Bull.* **78,** 65–69.*

390. Linkola, J. (1974). Effects of ethanol on urine sodium and potassium concentrations and osmolality in water-loaded rats. *Acta Physiol. Scand.* **92,** 212–216.

391. Lish, P. M., Clark, B. B., and Robbins, S. I. (1959). Effects of some physiologic substances on gastrointestinal propulsion in the rat. *Am. J. Physiol.* **197,** 22–26.*

392. Little, J. R., Brecher, G., Bradley, T. R., and Rose, S. (1962). Determination of lymphocyte turnover by continuous infusion of H^3 thymidine. *Blood* **19,** 236–242.

393. Littlewood, G., and Platts, T. L. (1953). A simple apparatus for repeated intratracheal injections in rats. *J. Anim. Tech. Assoc.* **4,** 53.*

394. Llaurado, J. G. (1958). A method for rapid adrenalectomy in the rat. *J. Anim. Tech. Assoc.* **8,** 75–78.*

395. Lockhard, R. B. (1968). The albino rat: A defensible choice or a bad habit? *Am. Psychol.* **23,** 734–742.*

396. Long, C. N., Katzin, B., and Fry, E. G. (1940). The adrenal cortex and carbohydrate metabolism. *Endocrinology* **26,** 209–351.*

397. Long, J. A., and Evans, H. M. (1922). The estrus in the rat and its associated phenomena. *Mem. Univ. Calif.* **6,** 1–148.

398. Love, J. W. (1957). A method for the assay of secretin using rats. *Q. J. Exp. Physiol. Cogn. Med. Sci.* **42,** 279–284.*

399. Lovless, B. W., Williams, P., and Heaton, F. W. (1972). A simple automatic feeding apparatus for rats. *Br. J. Nutr.* **28,** 261–262.*

400. Loring, W. E. (1952). A method of positive pressure anesthesia for the rat. *Proc. Soc. Exp. Biol. Med.* **79,** 658–660.*

401. Loring, W. E. (1954). A rapid, simplified method for serial blood volume determinations in the rat. *Proc. Soc. Exp. Biol. Med.* **85,** 350–351.*

402. Ludwig, J. (1972). "Current Methods of Autopsy Practice." Saunders, Philadelphia, Pennsylvania.*

403. Lumb, W., and Jones, E. (1963). Anesthesia of laboratory and zoo animals. *In* "Veterinary Anesthesia" (L. R. Soma, ed.), pp. 427–507. Lee & Febiger, Philadelphia, Pennsylvania.

404. Lundberg, V. (1974). Composite transport systems for iodipamide and iodohippurate out of the cerebrospinal fluid in the rat. *Acta Physiol. Scand.* **92**, 204–211.*

405. Lundy, J. S. (1959). An excellent method for obtaining speedy vasodilation for venapuncture. *Proc. Staff Meet. Mayo Clin.* **34**, 550–551.*

406. Luscheri, E. S., and Mehaffey, J. J. (1967). Small animal anesthesia with halothane. *J. Appl. Physiol.* **22**, 595.

407. Lushbough, C. H., and Moline, S. W. (1961). Improved terminal bleeding method. *Proc. Anim. Care Panel* **11**, 305–308.

408. McCarthy, T. C. (1976). The phencyclidine anesthetics: Their effects on the central nervous, cardiovascular, and respiratory function. *Vet. Anesth.* **3**, 49–56.

408a. McDonald, L. E., Booth, N. H., Lamb, W. V., Redding, N. W., Sawyer, D.C., Stevenson, L., and Wass, W. M. (1978). Report of the AVMA Panel on Euthanasia. *J. Am. Vet. Med. Assoc.* **173**, 59–72.

409. MacEwen, J. D., Urban, E. C., Smith, R. G., and Vorwald, A. J. (1961). A new method for massive dust exposures by inhalation. *J., Am. Ind. Hyg. Assoc.* **22**, 109–113.*

410. Machella, T. E., and Griffith, J. Q. (1967). The digestive system. *In* "The Rat in Laboratory Investigation" (J. Q. Griffith and E. J. Farris, eds.), pp. 166–180. Hafner, New York.*

411. Macnic, G. R., Klose, R. M., and Giebisch, G. (1966). Microperfusion study of distal tubular potassium and sodium transfer in rat kidney. *Am. J. Physiol.* **211**, 529–547.*

412. Macnic, G. R., Klose, R. M., and Giebisch, G. (1966). Micropuncture study of distal tubular potassium and sodium transport in rat nephron. *Am. J. Physiol.* **211**, 548–559.*

413. McPherson, J. C. (1974). The preparation of a smooth, uniform suspension of a diet for tube feeding rats. *Lab. Anim. Sci.* **24**, 90–91.

414. Maistrello, I., and Matscher, R. (1969). Measurement of systolic blood pressure of rats: Comparison of intra-arterial and cuff values. *J. Appl. Physiol.* **26**, 188–193.*

415. Major, J. S., and Heald, P. J. (1974). The effects of ICI 46, 475 on ovum transport and implantation in the rat. *J. Reprod. Fertil.* **36**, 117–124.

416. Malette, W. G., Fitzgerald, J. B., and Eiseman, B. (1960). A rapid decompression changer for small animals. *US Air Force Sch. Avia. Med. Rep.* **60–73.**

417. Marks, L. H. (1908). Stomach tube feeding in mice. *J. Exp. Med.* **10**, 204–205.

418. Marquet, R., Hess, F., Kort, W., and Boeckx, W. (1976). "Microsurgery: Experimental Techniques in the Rat and Clinical Applications." European Press, Ghent, Belgium.*

419. Marsh, D. F. (1948). A convenient apparatus for recording the blood pressure of small animals. *Science* **108**, 393.*

420. Martin, J. M., and Lacy, P. E. (1963). The prediabetic period in partially pancreatectomized rats. *Diabetes,* **12**, 238–242.*

421. Martinez, J. R., Adshead, P. C., Quissell, D. O., and Barbero, G. J. (1975). The chronically reserpinized rat as a possible model for cystic fibrosis. II. Comparison and cilioinhibitory effects of submaxillary saliva. *Pediatr. Res.* **9**, 470–475.*

422. Mauderly, J. L. (1975). An anesthetic system for small laboratory animals. *Lab. Anim. Sci.* **25**, 331–333.

423. Max, B., Dixon, R. L., and Adamson, R. H. (1969). Anesthesia in the rat. *Fed. Proc., Fed. Am. Soc. Exp. Biol.* **28**, 1522–1527.

424. Mazzacca, G., Bianco, A. R., Budillon, G., and Perillo, N. (1975). The source of ascitic fluid in experimental cirrhosis in the rat. *Pathol. Microbiol.* **42**, 66–69.*

425. Mazze, R. I., Cousins, M. J., and Kisek, J. C. (1973). Strain differences in metabolism and susceptibility to the nephrotoxic effects of methyoxyflurane in rats. *J. Pharmacol. Exp. Ther.* **184**, 481–488.*

426. Medd, R. K., and Heywood, R. (1970). A technique for intubation and repeated short duration anesthesia in the rat. *Lab. Anim.* **4**, 75–78.*

427. Medin, N. I., Osebold, J. W., and Zee, Y. C. (1976). A procedure for pulmonary lavage in mice. *Am. J. Vet. Res.* **37**, 237–238.

428. Melby, E. C. (1976). What does "adequate veterinary care" mean to the American College of Laboratory Animal Medicine? *J. Am. Vet. Med. Assoc.* **168**, 508–509.

429. Meltzer, M. R., and Fold, G. E. (1958). A method for correlating bladder emptying and total activity in rodents. *J. Mammal.* **39**, 454–455.*

430. Menaker, L., Sheetz, J. H., Cobb, C. M., and Navia, J. (1974). Gel electrophoresis of whole saliva and associated with histologic changes in submandibular glands of isoproterenol-treated rats. *Lab. Invest.* **30**, 341–349.

431. Metcoff, J., and Favour, C. B. (1944). Determination of blood and plasma volume partitions in the growing rat. *Am. J. Physiol.* **141**, 695–706.*

432. Mikeska, J. A., and Klemm, W. R. (1975). EEG evaluation of humaneness of asphyxia and decapitation euthanasia in the laboratory rat. *Lab. Anim. Sci.* **25**, 175–179.

433. Miller, S. A., and Dymsza, H. A. (1973). Artificial feeding of neonatal rats. *Science* **141**, 517–518.

434. Mitchell, H. H., and Carman, G. C. (1924). Biological value for maintenance and growth of protein of whole wheat, eggs, and pork. *J. Biol. Chem.* **60**, 613–620.*

435. Miyamoto, H., and Chang, M. C. (1973). Fertilization of rat eggs *in vitro. Biol. Reprod.* **9**, 384–393.*

436. Miyamoto, H., Toyoda, Y., and Chang, M. C. (1974). Effect of hydrogen-ion concentration on *in vitro* fertilization of mouse, golden hamster, and rat eggs. *Biol. Reprod.* **10**, 487–493.*

437. Moberg, E. (1938). Eine methode zur unblutigen bestimmung des blutdruckes bei der ratte. *Skand. Arch. Physiol.* **93**, 301.*

438. Molello, J. A., and Hawkins, K. (1968). Methoxyflurane anesthesia of laboratory rats. *Lab. Anim. Care* **18**, 581–583.*

439. Morag, M. (1970). Estimation of milk yield in the rat. *Lab. Anim.* **4**, 259–272.

440. Morag, M. (1975). Effect of litter size on milk yield in the rat. *Lab. Anim.* **9**, 43–47.

441. Moreland, A. F. (1965). Collection and withdrawal of body fluids and infusion techniques. *In* "Methods of Animal Experimentation" (W. I. Gay, ed.), Vol. 1, pp. 1–42. Academic Press, New York.

442. Morrow, W. G. (1975). A method for intratracheal instillation in the rat. *Lab. Anim. Sci.* **25**, 337–340.*

443. Mueller, A. J. (1939). A modified apparatus for milking small laboratory animals. *J. Lab. Clin. Med.* **24**, 426–427.*

444. Mulder, J. B., and Brown, R. V. (1972). An anesthetic unit for small laboratory animals. *Lab. Anim. Sci.* **22**, 422–423.*

445. Munson, E. S. (1970). Effect of hypothermia on anesthetic requirement in rats. *Lab. Anim. Care* **20**, 1109–1113.*

446. Munson, E. S., Martucci, R. W., and Smith, R. E. (1970). Canadian variations in anesthetic requirements and toxicity in rats. *Anesthesiology* **32**, 507–514.

447. Naess, K., and Rasmussen, E. Q. (1958). Approach withdrawal responses and other specific behavior reactions as screening test for tranquilizers. *Acta Pharmacol. Toxicol.* **15**, 99–114.*

448. Nairn, R. C. (1957). Fluid transfer from the peritoneal to the pleural cavities in rodents. *Br. J. Exp. Pathol.* **38**, 62–67.*

449. Nakata, K., Leong, G. F., and Brauer, R. W. (1960). Direct measurement of blood pressures in minute vessels of the liver. *Am. J. Physiol.* **199**, 1181–1188.

450. Nakayama, F. (1959). Metabolic relationship of exogenous phospholipids and phosphate to bilary phospholipids. *Am. J. Physiol.* **196**, 319–324.

451. Nashed, N. (1975). A technic for the collection of peritoneal cells from lab animals. *Lab. Anim. Sci.* **25**, 225–227.

452. Nelson, N. S., and Hoar, R. M. (1969). A small animal balling gun for oral administration of experimental compound. *Lab. Anim. Care* **19**, 871–872.

453. Nelson, N. S., and Rust, J. H. (1965). Use of ionizing radiation for measuring biological phenomena. *In* "Methods of Animal Experimentation" (W. I. Gay, ed.), Vol. 3, pp. 59–170. Academic Press, New York.*

454. Nemec, J., and Wildt, S. (1973). Blood pressure in the Wistar rat. *Z. Versuchstierkd.* **15**, 135–142.*

455. Nerenberg, S. T., and Zedler, P. (1975). Sequential blood samples from the tail vein of rats and mice obtained with modified Liebig condenser jackets and vacuum. *J. Lab. Clin. Med.* **85**, 523–526.

456. Newman, D. L., and Looker, T. (1972). Simultaneous measurement of the systolic blood pressure and heart rate in the rat by transcutaneous method. *Lab. Anim.* **6**, 207–211.*

457. Nicholas, J. S. (1967). Experimental methods and rat embryos. *In* "The Rat in Laboratory Investigation" (J. Q. Griffith and E. J. Farris, eds.), pp. 51–67. Hafner, New York.*

458. Nicholas, J. S., and Barron, D. H. (1932). Use of sodium amytal in the production of anesthesia in the rat. *J. Pharmacol. Exp. Ther.* **46,**, 125–129.*

459. Nicolaidis, S., Rowland, N., Meile, M. J., Marfaing-Jallat, P., and Pesez, A. (1974). A flexible technique for long term infusions in unrestrained rats. *Pharmacol., Biochem. Behav.* **2**, 131–136.

460. Nightingale, C. H., and Mouravieff, M. (1973). Reliable and simple method of intravenous injection into the laboratory rat. *J. Pharm. Sci.* **62**, 860–861.

461. Niwa, K., and Chang, M. C. (1973). Fertilization *in vitro* of rat eggs as affected by the maturity of the females and the sperm concentration. *J. Reprod. Fertil.* **43**, 435–451.*

462. Niwa, K., and Chang, M. C. (1975). Fertilization of rat eggs *in vitro* at various times before and after ovulation with special reference to fertilization of ovarian oocytes matured in culture. *J. Reprod. Fertil.* **43**, 435–451.*

463. Noback, G. J. (1916). The use of the VanWijhe method for the staining of the cartilaginous skeleton. *Anat. Rec.* **11**, 292–294.*

464. Noble, E. S., Wortman, R. J., and Axelrod, J. A. (1967). A simple and rapid method for injecting morepinephrine into the lateral ventricles of the rat brain. *Life Sci.* **6**, 281–291.

465. Nobunaga, T., Nabamura, K., and Imamichi, T. (1966). A method for intravenous injection and collection of blood from rats and mice without restraint and anesthesia. *Lab. Anim. Care* **16**, 40–49.

466. Noller, H. G. (1955). Die blutentnahme aus dem retroorbitalen venenplexus. *Klin. Wochenschr.* **33**, 770–771.

467. Noyes, H. E. (1959). A chronic irradiation source for small animals. *Radiat. Res.* **10**, 400–401.*

468. Oates, H. F., and Stokes, G. S. (1974). Renin stimulation caused by blood collection techniques in the rat. *Clin. Exp. Pharmacol. Physiol.* **6**, 495–501.

469. Oikawa, T., Yanagimachi, R., and Nicolson, G. L. (1975). Species differences in the lectin binding sites on the zona pellucida of rodent eggs. *J. Reprod. Fertil.* **43**, 137–140.

470. Olmsted, F., Concoran, A. C., and Page, I. H. (1951). Blood pressure in the unanesthetized rat I. Spontaneous variations and the effects of heat. *Circulation* **3**, 722–720.*

471. Orbach, J. (1961). Spontaneous ejaculation in rats. *Science* **134**, 1072–1073.*

472. Osborne, B. E. (1973). A restraining device facilitating electrocardiogram recording in rats. *Lab. Anim.* **7**, 185–188.

473. Owen, S. E. (1934). Small animal metabolism cage. *J. Lab. Clin. Med.* **19**, 1135–1137.*

474. Pakes, S. P. (1976). Adequate veterinary care as viewed by the American Association for Accreditation of Laboratory Animal Care. *J. Am. Vet. Med. Assoc.* **168**, 519–521.

475. Panchenko, L. F. (1958). A modified pressure chamber for use in a small laboratory. *Byull. Eksp. Biol. Med.* **45**, 120–121.*

476. Parbrook, G. D. (1966). A halothane vaporizer for small animal anesthesia. *Anesthesia* **21**, 403–405.

477. Parker, G. H. (1939). General anesthesia by cooling *Proc. Soc. Exp. Biol. Med.* **42**, 186–187.

478. Paterson, R. C., and Rowe, A. H. R. (1972). Surgical anesthesia in conventional and gnotobiotic rats. *Lab. Anim.* **6**, 147–154.

479. Payne, J. M., and Chamings, J. (1964). The anesthesia of laboratory rodents. *In* "Small Animal Anesthesia" (O. Graham-Jones, ed.), pp. 103–108. Pergamon, Oxford.

480. Peacock, A. C., and Harris, R. S. (1950). Plastic house for the quantitative separation of urine and feces excited by male rats. *Arch. Biochem.* **27**, 198–201.*

481. Pearce, K. A. (1957). A route for intravenous injection in the albino rat. *Nature (London)* **180**, 709.*

482. Peluso, J. J., and Butcher, R. L. (1974). RNA and protein synthesis in control and follicularly aged rat oocytes. *Proc. Soc. Exp. Biol. Med.* **147**, 350–353.*

483. Pendergrass, E. P., and Griffith, J. Q. (1967). Radiologic considerations. *In* "The Rat in Laboratory Investigation" (J. Q. Griffith and E. J. Farris, eds.), pp. 421–433. Hafner, New York.*

484. Peterson, W. J. (1956). Permanently inserted plastic cannula for direct access to cecal contents of rats. *Exp. Parasitol.* **5**, 427–434.*

485. Pfeffer, J. M., Pfeffer, A., and Frohlich, E. (1971). Validity of an indirect tail cuff method for determining systolic arterial pressure in the unanesthetized normotensive and spontaneously hypertensive rat. *J. Lab. Clin. Med.* **78**.*

486. Pfeiffer, C. A. (1936). Sexual differences of the hypophyses and their determination by the gonads. *Am. J. Anat.* **58**, 195–225.*

487. Phillips, W. A., Stafford, W. W., and Stuut, J. (1973). Jugular vein technique for serial blood sampling and intravenous injection in the rat. *Proc. Soc. Exp. Biol. Med.* **143**, 733–735.

488. Piaskowski, J., and Kacki, J. (1974). Modified treadmill for rats. *Acta Physiol. Pol.* **25**, 179–182.*

489. Piggott, W. R., and Emmons, C. W. (1960). Device for inhalation exposure of animals to spres. *Proc. Soc. Exp. Biol. Med.* **103**, 805–806.*

490. Pilgrim, H. I. (1963). Technic for sampling bone marrow from the living mouse. *Blood* **21**, 241–242.

491. Pinkerton, W., and Webber, M. (1964). A method of injecting small laboratory animals by the ophthalmic plexus route. *Proc. Soc. Exp. Biol. Med.* **116**, 959–961.

492. Pinschmidt, N. W., Ramsey, H., and Haag, H. B. (1945). Studies on the antagonism of sodium succinate to barbiturate depression. *J. Pharmacol. Exp. Ther.* **83**, 45–52.

493. Pissidis, G. A., and Clark, C. G. (1967). An improved technique of perfusion of the stomach for study of gastric secretion in the rat. *Gut* **8**, 196–197.*

494. Pitesky, I., and Last, J. H. (1948). An effective depilatory formula for use on laboratory animals. *Science* **108**, 657.*

495. Pizzuto, J. S., and Perry, H. L. (1960). A new chronic low dose cobalt-60 facility of the radiobiological laboratory. *U.S. Air Force Sch. Aviat. Med., Rep.* **60–22**.*

496. Plaut, S. M., Grota, L. J., Ader, R., and Graham, C. W. (1970).

Effects of handling and the light-dark cycle on time of parturition in the rat. *Lab. Anim.Care* **20,** 447.*

497. Plum, C. M. (1943). A method for collecting large samples of blood from living rats. *Acta Physiol. Scand.* **6,** 289–290.

489. Popovic, V., and Popovic, P. (1960). Permanent cannulation of aorta and vena cava in rats and ground squirrels. *J. Appl. Physiol.* **15,** 727–728.

499. Postnikova, Z. A. (1960). A method of intracardiac injection in newborn rats and mice. *Folia Biol. (Prague)* **6,** 59–60.*

500. Postnikova, Z. A. (1960). Method of intracardiac injection of newborn mice and rats. *Vopr. Virusol.* **5,** 111–112.

501. Poumeau-Delille, G. (1953). ''Techniques biologiques en endocrinologie expérimentale orchez le Rat.'' Masson, Paris.

502. Prehn, R. T. (1953). Tumors and hyperplastic nodules in transplanted mammary glands. *J. Natl. Cancer Inst.* **13,** 859–872.*

503. Princi, F., Church, F., and McGilvray, W. (1949). An improved animal dusting apparatus. *J. Ind. Hyg. Toxicol.* **31,** 106–112.*

504. Proctor, E., and Fernando, A. R. (1973). Oro-endotracheal intubation in the rat. *Fr. J. Anaesth.* **45,** 139–142.

505. Proskauer, G. C., Neumann, C., and Graef, I. (1945). The measurement of the blood pressure in rats with special reference to the effect of changes in temperature. *Am. J. Physiol.* **143,** 290–296.*

506. Quarterman, J., Williams, R. B., and Humphries, W. R. (1970). An apparatus for the regulation of the food supply to rats. *Br. J. Nutr.* **24,** 1049–1051.*

507. Quay, W. B. (1965). Experimental evidence for pineal participation in homeostasis of brain composition. *Prog. Brain Res.* **10,** 646–653.

508. Quinn, G. P., Axelrod, J., and Brodie, B. B. (1958). Species, strain, and sex differences in metabolism of hexobarbitone, amidopyrine and amiline. *Biochem. Pharmacol.* **1,** 152–159.

509. Rapp, K. W., Skiner, J. T., and McHargue, J. S. (1946). A new type of glass cage for metabolism studies. *J. Lab. Clin. Med.* **31,** 598–599.*

510. Raventos, J. (1938). The influence of room temperature on the action of barbiturates. *J. Pharmacol. Exp. Ther.* **64,** 355–363.

511. Raynor, A. C., Yoakum, S. T., and Munster, A. M. (1976). A simple technique for obtaining split-thickness skin grafts in small laboratory animals. *J. Am. Anim. Hosp. Assoc.* **12,** 648–649.*

512. Redfern, R., and Rowe, F. P. (1972). Wild rats and house mice. *In* ''UFAW Handbook on the Care and Management of Laboratory Animals,'' 4th ed., pp. 212–222. Churchill-Livingstone, London.

513. Reinhardt, W. O. (1945). Rate of flow and cell count of rat thoracic duct lymph. *Proc. Soc. Exp. Biol. Med.* **58,** 123–124.

514. Renaud, S. (1959). Improved restraint technique for producing stress and cardiac necrosis in rats. *J. Appl. Physiol.* **14,** 868–869.

515. Renaud, S. (1969). Jugular vein technique for blood collection and intravenous injection in the rat. *Lab. Anim. Care* **19,** 664–665.

516. Renaud, S., and Picard, G. (1964). An improved table for hypophysectomy in rats. *Can. J. Physiol. Pharmacol.* **42,** 870–872.*

517. Report of the American Veterinary Medical Association Panel on Euthanasia (1963). *J. Am. Vet. Med. Assoc.* **142,** 162–170.

518. Report of the Committee on the Care and Use of Spontaneously Hypertensive (SHR) Rats (1900). *Inst. Lab. Anim. Resour. News* **19,** G3-G20.

519. Reynell, P. C., and Spray, G. H. (1956). The simultaneous measurement of absorption and transit in the gastrointestinal tract of the rat. *J. Physiol. (London)* **131,** 452–462.*

520. Richter, C. P., and Schmidt, E. C. H. (1939). Behavior and anatomical changes produced in rats by pancreatectomy. *Endocrinology* **25,** 698–706.*

521. Riley, V. (1960). Adaptation of orbital bleeding technic to rapid serial blood studies. *Proc. Soc. Exp. Biol. Med.* **104,** 751–754.

522. Rizer, R. L., Shaw, L. G., and Orentreich, N. (1970). A method for the intensive repeated plasmaphoresis of the rat. *Lab. Anim. Care* **20,** 105.*

523. Rizzo, A. (1959). A new mouth prop for oral examinations and operative procedures in rodents. *J. Dent. Res.* **38,** 830–832.

524. Roberts, W. H. (1961). Identification of multi-colored laboratory animals using electronic flash photography. *J. Anim. Tech. Assoc.* **12,** 11–14.

525. Robinovitch, M. R., and Sreebny, L. M. (1969). Separation and identification of some of the protein components of rat parotid saliva. *Arch. Oral Biol.* **14,** 935–949.

526. Rubin, H. B., and Brown, H. J. (1972). A Larnes and electrode connector for rats. *J. Exp. Anal. Behav.* **18,** 385–387.*

527. Rusher, D. L., and Birch, R. W. (1975). A new method for rapid collection of blood from rats. *Physiol. Behav.* **14,** 377–378.*

528. Rye, M. M., and Elder, S. T. (1966). A suggestion concerning the anesthetization of the rat. *J. Exp. Anal. Behav.* **9,** 243–244.

529. Ryer, F. H., and Walker, D. H. (1971). An anal cup for rats in metabolic studies involving radioactive materials. *Lab. Anim. Sci.* **21,** 942.

530. Salem, H., Grossman, M. H., and Bilbey, D. J. (1963). Micro method for intravenous injection and blood sampling *J. Pharm. Sci.* **52,** 794–795.

531. Sandiford, M. (1965). Some methods of collecting blood from small animals. *J. Anim. Tech. Assoc.* **16,** 9–14.

532. Sato, M., and Yoneda, S. (1966). An efficient method for transauricular hyphophysectomy in rats. *Acta Endocrinol. (Copenhagen)* **51,** 43–48.

533. Sawyer, C. M., and Everett, J. W. (1956). The small saphenous vein as a route for intravenous injection in the white rat. *Nature (London),* **178,** 268–269.*

534. Sawyer, D. C., Lumb, W. V., and Stone, H. L. (1971). Cardiovascular effects of halothane, methoxyflurane, pentobarbiral, and thiamylal. *J. Appl. Physiol.* **30,** 36.*

535. Sawyer, L., and Lipkovsky, S. (1935). Technic for a bile fistula in the rat and demonstration of the indispensability of the bile. *J. Lab. Clin. Med.* **20,** 958–963.

536. Schaffenburg, C. A. (1959). Device to control constriction of main renal artery for production of hypertension in small animals. *Proc. Soc. Exp. Biol. Med.* **101,** 676–677.*

537. Schaffer, A. (1965). Anesthesia and sedation. *In* ''Methods of Animal Experimentation'' (W. I. Gay, ed.), Vol. 1, pp. 44–102. Academic Press, New York.*

538. Schapiro, S., and Stjarne, L. (1958). A method for collection of intermittent samples of adrenal vein blood. *Proc. Soc. Exp. Biol. Med.* **99,** 414–415.

539. Schiller, K. (1960). A metabolism cage for rats. *Z. Tierphysiol., Tierernahr. Futtermittelkd.* **15,** 305–308.

540. Schlang, H. (1959). A simple cross transfusion and perfusion pump for small animals. *J. Lab. Clin. Med.* **54,** 163–166.*

541. Schneyer, C. A., and Hall, H. D. (1965). Comparison of rat saliva evoked by auriculotemporal and pilocarpine stimulation. *Am. J. Physiol.* **209,** 484–488.

542. Schour, I. (1936). Measurements of bone growth by alizaria injections. *Proc. Soc. Exp. Biol. Med.* **34,** 140–141.*

543. Schour, I., and Massler, M. (1967). The teeth. *In* ''The Rat in Laboratory Investigation'' (J. Q. Griffith and E. J. Ferris, eds.), pp. 104–165. Hafner, New York.*

544. Schreiber, H., and Nettesheim, P. (1972). A new method for pulmonary cytology in rats and hamsters. *Cancer Res.* **32,** 737–745.

545. Schwindaman, D. F. (1976). An evaluation of adequate veterinary care by the United States Department of Agriculture. *J. Am. Vet. Med. Assoc.* **168,** 517–519.

546. Scott, J. V., and Dzuik, P. J. (1959). Evaluation of the electroejaculation technique and the spermatozoa thus obtained from rats, mice, and guinea pigs. *Anat. Rec.* **133**, 655-666.

547. Scott, W. N., and Ray, P. M. (1972). Euthanasia. *In* "UFAW Handbook on the Care and Management of Laboratory Animals," 4th ed., pp. 158-166. Churchill-Livingstone, London.

548. Scow, R. O. (1957). "Total" pancreatectomy in the rat. Operation, effects and post operative care. *Endocrinology* **60**, 359-367.

549. Secord, D. C., Taylor, A. W., and Fielding, W. (1973). Effect of anesthetic agents on exercised atropinized rats. *Lab. Anim. Sci.* **23**, 397-400.*

550. Segaloff, A. (1967). Thymectomy. *In* "The Rat in Laboratory Investigation" (J. Q. Griffith and E. J. Farris, eds.), pp. 434-444. Hafner, New York.*

551. Selye, H. (1953). The use of "granuloma pouch" technic in the study of antiphlogistic corticoids. *Proc. Soc. Exp. Biol. Med.* **82**, 328-333.*

552. Shannon, I. L., and Suddick, R. P. (1975). Current research on oral fluids. *J. Dent. Res.* **54**, B8-B18.

553. Shapiro, R., and Pincus, G. (1936). Pancreatic diabetes and hypophysectomy in the rat. *Proc. Soc. Exp. Biol. Med.* **34**, 416-419.*

554. Sharpe, R. M., Choudhury, S. A. R., and Brown, P. S. (1973). The effect of handling weanling rats on their usefulness in subsequent assays of FSH. *Lab. Anim.* **7**, 31-314.*

555. Shaw, H. M., and Heath, T. (1972). The significance of hormones, bile salts and feeding in the regulation of bile and other digestive secretions in the rat. *Aust. J. Biol. Sci.* **25**, 147-154.*

556. Shaw, H. M., and Heath, T. J. (1973). Basal and postprandial pancreatic secretion in rats. *Q. J. Exp. Physiol. Cogn. Med. Sci.* **58**, 335- 343.

557. Shay, H., and Gruenstein, M. (1946). A simple method for the gastric instillation of fluids in the rat. *J. Lab. Clin. Med.* **31**, 1384-1386.

558. Shay, H., Sun, D. C. H., and Gruenstein, M. (1954). A quantitative method for measuring spontaneous gastric secretion in the rat. *Gastroenterology* **26**, 906-913.*

559. Shearer, D., Creel, D., and Wilson, C. E. (1973). Strain differences in the response of rats to repeated injections of pentobarbital Na. *Lab. Anim. Sci.* **23**, 662-664.

560. Shepeler, V. M. (1963). A holder for mice and rats. *Lab. Delo.* **9**, 55-56.

561. Sholkoff, S. D., Glickman, M. G., and Powell, M. R. (1969). Restraint of small animals for radiopharmaceutical studies. *Lab. Anim. Care* **19**, 662-663.

562. Shuler, R. H., Kupperman, H. S., and Hamilton, W. F. (1944). Comparison of direct and indirect blood pressure measurements in rats. *Am. J. Physiol.* **141**, 625-629.*

563. Siegler, R., and Rich, M. A. (1963). Artificial respiration in mice during thoracic surgery: A simple, inexpensive technique. *Proc. Soc. Exp. Biol. Med.* **114**, 511-513.

564. Sigdestad, C. P., Connor, A. M., Sharp, J. B., and Ackerman, M. R. (1974). Evaluating feeding cycles of small rodents. *Lab. Anim. Sci.* **24**, 919-921.*

565. Silver, M. L. (1939). Electrical anesthesia in rats. *Proc. Soc. Exp. Biol. Med.* **41**, 650-651.

566. Simmonds, W. J. (1957). The relationship between intestinal motility and the flow and rate of fat output in thoracic duct lymph in unanesthetized rats. *Q. J. Exp. Physiol. Cogn. Med. Sci.* **42**, 205-221.*

567. Simmons, M. L., and Brick, J. O. (1970). "Techniques in the Laboratory Mouse." Prentice-Hall, Englewood Cliffs, New Jersey.*

568. Simmons, M. L., and Smith, L. H. (1968). An anesthetic unit for small laboratory animals. *J. Appl. Physiol.* **25**, 324.*

569. Sines, J. O. (1960). Permanent implants for heart rate and body temperature recording in the rat. *AMA Arch. Gen. Psychiatry* **2**, 182-183.*

570. Singer, B., and Stack-Dunne, M. D. (1955). The secretion of aldosterone and corticosterone by the rat adrenal. *J. Endocrinol.* **12**, 130-134.

571. Sjödin K. Dalmasso, A. P., Smith, J. M., and Martinez, C. (1963). Thymectomy of newborn and adult mice. *Transplantation* **1**, 521-525.

572. Skulan, T. W., Brousseau, A. C., and Leonard, K. A. (1974). Accelerated induction to two kidney hypertension in rats and renin angiotensin sensitivity. *Circ. Res.* **35**, 734-741.*

573. Smith, C. J., and Kelleher, P. C. (1973). A method for the intragastric feeding of neonatal rats. *Lab. Anim. Sci.* **23**, 682-684.

575. Smith, D. B., Goddard, K. M., Wilson, R. B., and Newberne, P. M. (1973). An apparatus for anesthetizing small laboratory rodents. *Lab. Anim. Sci.* **23**, 869-871.*

576. Smith, D. C. (1965). Methods of euthanasia and disposal of laboratory animals. *In* "Methods of Animal Experimentation" (W. I. Gay, ed.), Vol. 1, pp. 167-195. Academic Press, New York.

577. Smith, E. M., and Calhoun, M. L. (1968). "The Microscopic Anatomy of the White Rat." Iowa State Univ. Press, Ames.

578. Smith, G. M., Lawrence, A. J., and Colin-Jones, D. G. (1970). The assay of gastrin using the perfused rat stomach. *Br. J. Pharmacol.* **38**, 206-213.*

579. Smith, P. E. (1930). Hypophysectomy and a replacement therapy in the rat. *Am. J. Anat.* **45**, 205-273.*

580. Snell, G. D., and Hummel, K. P. (1966). Bibliography of techniques. *In* "Biology of the Laboratory Mouse" (E. L. Green, ed.), pp. 655-661. McGraw-Hill, New York.*

581. Sobin, S. S. (1946). Accuracy of indirect determination of blood pressure in rats: Relation to temperative of plethysmograph and width of cuff. *Am. J. Physiol.* **146**, 179-186.*

582. Soma, L. R. (1964). Neuroleptoanalgesia produced by fentamyl and droperidol. *J. Am. Vet. Med. Assoc.* **145**, 897-902.

583. Sorg, D. A., and Bruckner, B. A. (1964). A simple method of obtaining venous blood from small laboratory animals. *Proc. Soc. Exp. Biol. Med.* **115**, 1131-1132.

584. Spengler, J. (1960). An apparatus for quantitative continuous recording of food and water consumption of rats. *Helv. Physiol. Pharmacol. Acta* **18**, 50-55.*

585. Spires, R. A. (1959). An apparatus for producing lesions in the hypothalamus of rats and the studies of rats thus treated. *J. Anim. Tech. Assoc.* **10**, 112-115.*

586. Sreebny, L. B., and Johnson, D. A. (1969). Diurnal variation in secretory components of the rat parotid glands. *Arch. Oral Biol.* **14**, 397-405.

587. Stanton, H. G. (1971). Experimental hypertension. *In* "Methods of Pharmacology" (A. Schwartz, ed.), pp. 125-150. Meredith Corp. New York.

588. Steenbock, H., Sell, M. T., and Nelson, E. M. (1923). A modified technique in the use of the rat for determinations of vitamin B. *J. Biol. Chem.* **55**, 399-410.*

589. Steiger, E., Vars, H. M., and Dudrick, S. J. (1972). A technique for long term intravenous feeding of unrestrained rats. *Arch. Surg. (Chicago)* **104**, 330-333.

590. Stern, M. S. (1973). Letter-chimeras obtained by aggregation of mouse eggs with rat eggs. *Nature (London)* **243**, 472-473.*

591. Still, J. W., and Whitcomb, E. R. (1956). Technique for permanent long-term intubation of rat aorta. *J. Lab. Clin. Med.* **48**, 152-154.

592. Stimpfling, J. H., Boyse, E. A., and Mishell, R. (1964). Preparation of isoantisera in laboratory mice. *Methods Med. Res.* **10**, 18-21.*

593. Stitzel, R. E., Furner, R. L., and Gram, T. E. (1968). The influence of sex and drug treatment on microsomal drug metabolism in four rat strains. *Pharmacologist* **10**, 179.*

594. Stockdale, L. G. (1932). Technique for marking rats numerically with

dye. *J:. Comp. Psychol.* **14**, 237–240.

595. Stone, S. H. (1954). Method for obtaining venous blood from the orbital sinus of the rat or mouse. *Science* **119**, 100.

596. Strobel, G. E., and Wollman, H. (1969). Pharmacology of anesthetic agents. *Fed. Proc., Fed. Am. Soc. Exp. Biol.* **28**, 1386–1403.

597. Strong, R. M. (1967). The osseous system. *In* "The Rat in Laboratory Investigation" (J. Q. Griffith and E. J. Farris, eds.), pp. 467–480. Hafner, New York.*

598. Stuhlman, R. A., Packer, J. T., and Rose, S. D. (1972). Repeated blood sampling of *Mystromys albicaudatus. Lab. Anim. Sci.* **22**, 268–270.

599. Sugiura, K. (1965). Tumor transplantation. *In* "Methods of Animal Experimentation" (W. I. Gay, ed.), Vol. 2, pp. 171–222. Academic Press, New York.*

600. Sunahara, F. A. (1961). A simple continuous recording burette. *J. Appl. Physiol.* **16**, 201–202.*

601. Sundberg, R. D., and Hodgson, R. E. (1949). Aspiration of bone marrow in laboratory animals. *Blood* **4**, 557–561.

602. Sunderman, F. W., Kincaid, J. F., Kooch, W., and Birmelin, E. A. (1958). A constant flow chamber for exposure of experimental animals to gases and volatile liquids. *Am. J. Clin. Pathol.* **26**, 1211–1218.*

603. Svorad, D., and Kohout, L. (1959). The use of a light relay for measurement of motility of small laboratory animals. *Pfluegers Arch. Gesamte Physiol. Menschen Tiere* **269**, 308–310.*

604. Syverton, J. T., and Berry, G. P. (1937). An improved instrument for the intracerebral innoculation of experimental animals. *Science* **86**, 567–568.*

605. Tatevossian, A., and Wright, W. G. (1974). The collection and analysis of resting rat saliva. *Arch. Oral Biol.* **19**, 825–827.

606. Temple, P. L., and Kon, S. K. (1937). A simple apparatus for milking small laboratory animals. *Biochem. J.* **31**, 2197–2198.

607. Terkel, L., and Urbach, L. (1974). A chronic intravenous cannulation technique adapted for behavioral studies. *Horm. Behav.* **5**, 141–148.

608. Thayer, C. B., Lowe, S., and Rubright, W. C. (1972). Clinical evaluation of a combination of droperidol and fentanyl as an anesthetic for the rat and hamster. *J. Am. Vet. Med. Assoc.* **161**, 665–668.

609. Thompson, J. H. (1966). An adjustable restraining cage for rats. *Lab. Anim Care* **16**, 520–522.

610. Thompson, K. W. (1932). A technique for hypophysectomy of the rat.

611. Thomson, C., and Hill, A. (1972). The direct innoculation of *Mycoplasma pulmonis* into rat lungs. *J. Comp. Pathol.* **82**, 81–85.*

612. Timiras, P. S. (1965). High altitude studies. *In* "Methods of Animal Experimentation" (W. I. Gay, ed.), Vol. 2, pp. 333–369. Academic Press, New York.*

613. Todd, T. (1959). Animal post-mortem techniques. *J. Anim. Tech. Assoc.* **10**, 94–98.*

614. Toyoda, Y., and Chang, M. C. (1974). Fertilization of rat eggs *in vitro* by epididymal spermatozoa and the development of eggs following transfer. *J. Reprod. Fertil.* **36**, 9–22.*

615. Tsunda, Y., and Chang, M. C. (1975). *In vitro* fertilization of rat and mouse eggs by ejaculated sperm and the effect of energy source on *in vitro* fertilization of rat eggs. *J. Exp. Zool.* **193**, 79–87.*

616. UFAW (1972). "Handbook on the Care and Management of Laboratory Animals," 4th ed. Churchill-Livingstone, London.*

617. Ulmansky, M., Sela, J., Dishon, T., Rosenmann, E., and Boss, J. H. (1971). A technique for the incubation of the parotid duct in rats. *Arch. Oral Biol.* **17**, 609–612.

618. Upton, P. K., and Morgan, D. J. (1975). The effect of sampling technique on some blood parameters in the rat. *Lab. Anim.* **9**, 85–91.*

619. Upton, R. A. (1975). Simple and reliable method for serial sampling of blood from rats. *J. Pharm. Sci.* **64**, 112–114.

620. Vaille, C., Debray, C., Delatour, J., Roze, C., and Souchard, M. (1974). Insulin effect after vagotomy on the external pancreatic and bile secretions in the rat. *Ann. Pharm. Fr.* **32**, 147–153.*

621. Valenstein, E. S. (1961). A note on anesthetizing rats and guinea pigs. *J. Exp. Anal. Behav.* **4**, 6.*

622. Van es, A. A. (1972). A technique for massive skin grafting in rats. *Lab. Anim. Sci.* **22**, 404–406.*

623. VanOye, E. (1953). Note technique concernant l'examen de la moelle ossense chez des petits animaux. *Acta Haematol.* **9**, 388–391.

624. Vermeiden, J. P., and Zeilmaker, G. H. (1974). Relationship between maturation division, ovulation, and leuteinization in the female rat. *Endocrinology* **95**, 341–351.*

625. Vierck, C. J. (1976). Extrapolutions from the pain research literature to problems of adequate veterinary care. *J. Am. Vet. Med. Assoc.* **168**, 510–513.

626. Vlad, A., Winter, D., Sauvard, S., Stanescu, C., and Pal, B. (1970). A new method for tracheal installation without esophogeal obstruction in laboratory animals. *Arzneim.-Forsch.* **20**, 861.*

627. Vogel, M. T. (1946). A simple method for obtaining smears of bone marrow, of the albino rat. *Tech. Bull. Regist. Med. Technol.* 189–190.

628. Volta, A. (1951). Contributo alla technica per lo studio anatomo-funzionale del midollo nel ratto albino. *Haematologica* **35**, 103–112.

629. Vondruska, J. F. (1974). A method for the implantation of solid and liquid materials in the marrow cavity of the femur of small laboratory animals. *Lab. Anim. Sci.* **24**, 672–674.*

630. Vrba, J. (1960). Contribution to the method of determining the toxicity of gases and vapor on small laboratory animals. *Prac. Lek.* **12**, 24–25.*

631. Vreeburg, J. T., VanAndel, M. V., Kort, W. J., and Westbroek, D. L. (1974). The effect of hemicastration on daily sperm output in the rat as measured by a new method. *J. Reprod. Fertil.* **41**, 355–359.

632. Waeldele, G. and Stoclet, J. C. (1973). Permanent catheterization of the thoracic aorta, direct measure of blood pressure, injection of substances and blood sampling in the waking rat. *J. Physiol. (London)* **66**, 357–366.

633. Walker, G. (1935). A method for numbering laboratory rats. *Science* **82**, 397–398.

634. Wallace, M. E., and Hudson, C. A. (1969). Breeding and handling small wild rodents: A method study. *Lab. Anim.* **3**, 107–117.

635. Walton, D. C., and Jones, C. A. (1925–1926). A gassing chamber for short exposures. *J. Lab. Clin. Med.* **11**, 580–583.*

636. Wang, C., and Peters, D. (1963). Modification of anal cup technique for small experimental animals. *Lab. Anim. Care* **13**, 105–108.

637. Wang, L. (1959). Plasma volume, cell volume, total blood volume and F cells factor in the normal and splenectomized Sherman rat. *Am. J. Physiol.* **196**, 188–192.*

638. Ward, G. S., Johnsen, D. O., and Roberts, C. R. (1974). The use of CI 744 as an anesthetic for laboratory animals. *Lab. Anim. Sci.* **24**, 737–742.

639. Wayner, M. (1957). An inexpensive respirator for use with small animals. *Am. J. Physiol.* **70**, 457–458.*

640. Wedeking, P. W., and Babington, R. D. (1974). A head holder for inhalation anesthesia or resuscitation of rats. *Pharmacol., Biochem. Behav.* **2**, 127–129.*

641. Weeks, J. E., and Davis, J. D. (1964). Chronic intravenous cannulas in rats. *J. Appl. Physiol.* **19**, 540–541.

642. Weeks, J. R., and Jones, J. A. (1960). Routine direct measurement of arterial pressure in unanesthetized rats. *Proc. Soc. Exp. Biol. Med.* **104**, 646–648.*

643. Weihe, W. H., Schidlow, J., and Strittmatter, J. (1969). The effect of light intensity on the breeding and development of rats and golden hamsters. *Int. J. Biometeorol.* **13**, 69–79.*

644. Weisbroth, S. H. (1976). Reports from a symposium on adequate veterinary care. *J. Am. Vet. Med. Assoc.* **168,** 507.

645. Weisbroth, S. H., and Fudens, J. H. (1972). The use of ketamine-hydrochloride as an anesthetic in laboratory rabbits, rats, mice, and guinea pigs. *Lab. Anim. Sci.* **22,** 904–906.

646. Weiser, H., and Soliman, M. K. (1960). Thiogenal as a short acting anesthetic in the rat. *Zentralbl. Veterinaermed.* **7,** 519–523.*

647. Weiss, B., and Laties, V. G. (1961). Changes in pain tolerance and other behavior produced by salicylates. *J. Pharmacol. Exp. ther.* **131,** 120–129.*

648. Welshons, W. J., Gibson, B. H., and Scandlyn, B. J. (1962). Slide processing for the examination of male mammalian meiotic chromosomes. *Stain Technol.* **37,** 1–5.*

649. Whitaker, W. L. (1937). A method for studying environmental choices of laboratory animals. *Science* **86,** 314.*

650. White, P. K., and Miller, S. A. (1975). Design of a metabolic cage for infant rats. *Lab. Anim. Sci.* **25,** 344–346.*

651. White, W. A. (1971). A technic for urine collection from anesthetized male rats. *Lab. Anim. Sci.* **21,** 401–402.

652. White, W. C. (1960). A calorimeter for the continuous recording of the energy metabolism of small laboratory animals. *Fed. Proc., Fed. Am. Soc. Exp. Biol.* **19,** 397.*

653. Wiberg, G. S., and Grice, H. C. (1965). Effect of prolonged individual caging in toxicity parameters in rats. *Food Cosmet. Toxicol.* **3,** 597–603.*

654. Williams, J. R., Harrison, T. R., and Grollman, A. (1939). A simple method for determining the systolic blood pressure of the unanesthetized rat. *J. Clin. Invest.* **18,** 373–376.*

655. Wilson, J. T. (1969). A simple restraining apparatus for rats. *Lab. Anim. Care* **19,** 533–534.*

656. Winter, C. A., and Flataker, L. (1951). The effect of antihistaminic drugs upon the performance of trained rats. *J. Pharmacol. Exp. Ther.* **101,** 156–162.*

657. Winter, C. A., and Flataker, L. (1951). The effect of cortisone, deoxycortisone, and adrenocorticotrophic hormone upon the responses of animals to analgesic drugs. *J. Pharmacol. Exp. Ther.* **103,** 93–105.*

658. Wise, R. A. (1974). Strain and supplier differences affecting ethanol intake by rats. *Q. J. Stud. Alcohol* **35,** 667–668.*

659. Wittgenstein, E., and Rowe, K. W. (1965). A technic for prolonged infusion of rats. *Lab. Anim. Care* **15,** 375–378.

660. Wolf, R. O., and Kakehashi, S. (1966). Rat parotid saliva collection technique. *J. Dent. Res.* **45,** 979.

661. Woodard, G. (1965). Principles in drug administration. *In* "Methods of Animal Experimentation" (W. I. Gay, ed.), Vol. 1, pp. 343–359. Academic Press, New York.

662. Woodruff, C. R., and Carpenter, F. G. (1973). Actions of catecholamines and physostigmine on rat parotid salivary secretion. *Am. J. Physiol.* **225,** 1449–1453.

663. Wright, B. A. (1970). A new device for collecting blood from rats. *Lab. Anim. Care* **20,** 274.

664. Wunder, C. C. (1965). Care and growth of animals during chronic centrifugation. *In* "Methods of Animal Experimentation" (W. I. Gay, ed.), Vol. 2. Academic Press, New York.

665. Yale, C. E., and Torhorst, J. B. (1972). Critical bleeding and plasma volumes of the adult germfree rat. *Lab. Anim. Sci.* **22,** 497–502.*

666. Young, L., and Chambers, T. R. (1973). A mouse bleeding technique yielding consistent volume with minimal hemolysis. *Lab. Anim. Sci.* **23,** 428–430.

667. Youth, R. A., Simmerman, S. J., Newell, R., and King, R. A. (1973). Ketamine anesthesia in rats. *Physiol. Behav.* **10,** 633–636.

668. Zauder, H. L., and Orkin, L. R. (1959). Chamber for anesthetization of small animals. *Anesthesiology* **20,** 707–709.

669. Zeilmaker, G. H., and Verhamme, C. M. (1974). Observations on rat oocyte maturation in vitro: Morphology and energy requirements. *Biol. Reprod.* **11,** 145–152.

670. Zeman, W., and Innes, J. R. M. (1963). "Craigie's Neuroanatomy of the Rat." Academic Press, New York.*

671. Ziemnowicz-Radvan, S. (1968). Microsurgery. *In* "Methods of Animal Experimentation" (W. I. Gay, ed.), Vol. 3, pp. 175–210. Academic Press, New York.*

672. Zingg, W., Morgan, C. D., and Anderson, D. E. (1971). Blood viscosity, erythrocyte, sedimentation rate, packed cell volume, osmolality, and plasma viscosity of the Wistar rat. *Lab. Anim. Sci.* **21,** 740–742.*

Chapter 2

Gnotobiology

Henry L. Foster

I. INTRODUCTION

The first report on germfree animals dates back to 1897 (57). The first germfree rat colony was established by Gustafsson in 1948 (34). Since that time there have been major advances in methodology which have facilitated the breeding and utilization of gnotobiotic rats in a continually expanding spectrum of biomedical research. A literature survey for the decade 1966–1976 would yield more than 700 references on germfree rats and their gnotobiotic and disease-free descendants. The vast majority of these research reports are concerned with other experimental uses of these animals rather than their derivation, rearing, and establishment. The recent scientific literature contains reviews specific for the germfree laboratory rat by Pollard (64), Maejima *et al.* (52), and Miyakawa (53).

The quantum jump in the past 20 years in the use of these animals was facilitated in large part by the introduction of the Trexler flexible plastic film isolator system in 1957 (88). This innovation greatly reduced costs while simultaneously increasing design and flexibility. Prior to the late 1950s germfree research was conducted in rigid isolators made of stainless steel or steel and glass. The most commonly used type consisted of a stainless steel cylinder bolted together and gasketed at the joints. Many of these systems contained a steam autoclave as a pass-through lock. These isolator systems were heavy, cumbersome, very costly, and occupied a large amount of floor space, since their weight prohibited placing them in tiers. The advent of the plastic isolator not only reduced the cost to one-tenth to one-fifteenth of the cost of steel systems, but allowed units to be stacked in tiers of two and three in a multitude of sizes and configurations. This marked the birth of a new era permitting vast expansion in the production and availability of gnotobiotic animals. It also paved the way for the development of lightweight, disposable shipping units which made the transport of gnotobiotic animals to distant locations a practical reality.

Terminology

The word *gnotobiotic* is derived from the Greek words *gnotos* and *biota* meaning known flora or fauna. Therefore, when referring to gnotobiotes, one refers to an animal with a known microbial flora. This term also is applicable when a microbial flora does not exist or is not detectable. In other words, gnotobiotic is the broad term encompassing axenic, germfree, and associated (45).

The most recent general review of gnotobiotics by Pleasants in 1974 (62) defines a gnotobiotic animal as follows.

> One of an animal stock or strain derived by aseptic caesarian section or sterile hatching of eggs that is reared and continuously maintained with germfree technics under isolator conditions and in which the composition of an associated fauna and flora, if present, is fully defined by accepted and current methadology.

Axenic animals are gnotobiotes known to be free of all detectable microorganisms. This term is used synonymously with germfree, although the latter is more commonly utilized. The detection of leukemia virus particles by Pollard (68) raises the question whether in fact axenic animals exist at all.

Germfree is the historical term utilized over the longest period of time and is part of the colloquial scientific language. Its definition is the same as axenic, and even though those working within the field prefer axenic as being more accurate, germfree continues to remain the more popular term.

Specific pathogen-free (SPF) animals are those originally cesarean-derived within an axenic environment and subsequently associated with a definable microbial flora. These animals are then transferred from the rigidly controlled isolator system into a controlled larger physical space in which technicians work within protective clothing. It is designed to preclude the entrance of pathogenic organisms. This new environment is most commonly termed a barrier facility (21).

Specific pathogen-free is considered a controversial term in that there does not appear to be a unanimity of opinion as to its true definition. Some have argued that any animal free of one or more specific pathogens, i.e., *Salmonella* or *Mycoplasma*, fits the definition. This controversy is confusing and is gradually being clarified by adherence to the original intent that cesarean derivation and confirmed axenic status is an essential precursor to the true SPF animal.

Pathogen-free is used loosely and interchangeably with SPF, since both categories are implied to be free of pathogens. It is, however, theoretically possible to maintain animals free of pathogens through testing and eradication as well as through the use of broad-spectrum antibiotics (91a).

Cesarean-derived, cesarean-derived barrier-maintained, and *cesarean-originated barrier-maintained* or *sustained* are commonly utilized terms which imply an initial derivation of axenic animals and their subsequent association with a defined microflora followed by the continuing maintenance within a controlled barrier where all materials entering the barrier are subjected to a procedure which removes or destroys pathogenic microorganisms.

Conventional animals are all other animals maintained under accepted husbandry practices but which do not fall within any of the previously described definitions, i.e., axenic, SPF, or pathogen-free. To some working solely in the field of gnotobiotics, animals are either gnotobiotes or conventional. This oversimplification of terms is one of convenience and does not fully consider the largest category of research animals which usually falls in between gnotobiotic and conventional.

For the purposes of this chapter the following abbreviations will be used and reflect the terminology most commonly referred to by the respective authors: GF, germfree; GN, gnotobiotic; SPF, specific pathogen-free; and CV, conventional.

II. THE GERMFREE AND GNOTOBIOTIC LABORATORY RAT

One of the main advantages of using GF and GN laboratory rats in biomedical research is that the nutrition and physiology of many such colonies and strains have been well established. They have been used extensively, for example, in metabolic experiments. These animals are quite prolific in the isolator environment, notwithstanding the greatly enlarged cecum which is thought to impair reproduction.

There are many research areas where the investigator utilizing microbiologically sterile animals can elicit information that cannot be obtained using animals with normal flora. These research areas have included nutrition, immunology, infectious diseases, and dental caries studies. It is probably a lack of training and confidence in GN technology on the part of investigators which limits more extensive use of GN animals. In reality and as a practical matter, the training of GN technicians is a routine procedure and does not require special aptitude or formal education.

The other major uses and importance of GN rats are as nucleus seed stocks for the production of disease-free animals and as diagnostic tools for infectious disease studies, particularly in situations where routinely used culture media are inadequate. In the middle 1950s major laboratory animal breeders reported the use of GN rats as foster nursing stock for the rederivation of breeding colonies (20). It became apparent to the laboratory animal breeding industry that testing, eradication, and selection techniques for the elimination of infectious diseases and parasites were often inconclusive as well as tedious and not totally reliable. Therefore, as a natural evolution of the technology developed for the production of GN animals, the latter became the building blocks and nucleus stock for the production of microbially associated defined flora and disease-free animals. This was accomplished by introducing a known microflora to GF animals, placing them in a barrier, and maintaining them in an environment which precluded the entry of pathogenic organisms.

When a clinical syndrome or a set of pathological findings fail to elicit an etiologic agent by routine microbiologic techniques, the GF rat provides an excellent model for transmission and diagnostic studies. It provides the almost perfect model to establish Koch's postulates, since the use of this definable animal model frequently assures valid results in the determination of specific etiologic agents, i.e., the effect of a single organism can be evaluated in the absence of other microbial forms.

A. Derivation Philosophy

The primary method of deriving GF rats is through surgical intervention of pregnancy at term. Another technique has been reported whereby the gut tract has been rendered sterile through the successive use of a variety of broad-spectrum antibiotics. Van der Waaij (91a) has reported that mice have been rendered GF within sterile isolator systems. The presumed advantages are the rapidity with which this regime can be accomplished as opposed to the traditional and proven method of

cesarean derivation. For certain types of studies of short or medium duration, animals can be freed from microorganisms and maintained within isolator systems free of those organisms utilizing GN techniques. Ultimately the flora will reestablish itself, indicating that total elimination was not accomplished, even though detection could not be made after antibiotic treatments.

Cesarean delivery is the classic method of deriving axenic animals. Early workers delivered GF rats in stainless steel GF tanks where visibility was possible only through small viewing ports. The surgical technician's movements were restricted by the rigid steel isolator, even though sufficient mobility remained to perform the cesarean section. Since the weakest member in any GF system is probably the rubber sleeves and gloves, it was established in the mid-1950s that flexible film polyethylene isolators afforded at least the same degree of microbiological security with the same type of neoprene sleeves and gloves (88) used in the earlier rigid systems.

Therefore, since the middle to late 1950s the most common procedure for cesarean delivery of GF rats is utilization of a 1.5 × 0.6-m polyethylene flexible film isolator fitted with at least two pairs of neoprene sleeves fitted to 22.9-cm glove ports attached to the isolator wall (Fig. 1). In addition, standard surgical gloves are affixed to the wrists of the rubber sleeves to permit maximum tactile sensitivity. There is a 30.5- or 46-cm transfer port in the isolator for the introduction of supplies and instruments. An additional port is installed at one end which is attached to a long tapered sleeve or a rigid clear plastic tube

Fig. 1. Gnotobiotic facility utilizing flexible film isolators.

approximately 7.6 cm in diameter. This sleeve or tube terminates beneath the surface of a liquid germicide trap filled with a warm, 38°C, chlorine solution (Clorox, a product of The Clorox Co., Oakland, California, diluted 1 : 9 with water). A thermostatically controlled electric heating pad is placed between the exterior floor of the isolator and the rigid surface supporting the isolator. This provides warmth to the neonates after delivery. Within the sterile system, in addition to the required surgical instruments, sponges, and water, a plastic cage 35.5 cm long, 30.5 cm wide, and 14.5 cm deep is fitted with a taut Mylar membrane (a product of Dupont Co., East Orange, New Jersey) which provides a work area for the surgeons and which can be replaced after each procedure with a new membrane. This arrangement permits the uterus and fetal membranes to drop to the floor of the plastic cage.

B. Cesarean Methods

Delivery of GF rats can be accomplished by a two-stage hysterectomy technique or by a single-stage hysterotomy procedure (20,64,95). In the latter procedure a plastic Mylar membrane in the floor of the isolator is sealed to the shaved and surgically prepared abdominal wall of the pregnant rat. The surgical technician performs the hysterotomy through the Mylar membrane window in the isolator floor. This method is more tedious than the preferred and more rapidly performed hysterectomy. In addition to speed, the hysterectomy method permits an almost mass production routine, since the two stages can be performed simultaneously by separate surgical teams. One team is responsible for the extraisolator phase which consists of preparation, euthanasia, hysterectomy, and introduction of the uterus into the sterile surgical isolator. Another team of usually two technicians performs the actual cesarean, removing the fetuses from the uterus and its membranes. With the proper planning and coordination, the two surgical teams can perform 6 to 8 cesareans in 1 hr.

C. Derivation Procedure

A vital key to successful cesarean delivery of GF rats is the assurance that the pregnant rat has completed the normal gestation period of 20 to 21 days (20). This is best accomplished by observed or timed matings that are confirmed by the presence of the spermatic plug in the vaginal opening. If the plug is not seen, confirmation can be made by vaginal smear for the presence of spermatozoa. Since a large percentage of natural births occur from 6 PM to 6 AM, normal parturition can be forestalled by the administration of progesterone (58a). This technique is utilized mainly for the convenience of the surgical team and with a high degree of success precludes premature natural par-

turition during the night and allows for better planning for cesareans first thing in the morning.

The timed, gravid female is shaved along the ventral portion of the abdomen from the xiphoid cartilage to the genital opening. The use of a depilatory assures complete removal of hair and a clean incision without the contamination of animal fur. The female is sacrificed usually by cervical fracture outside of but close to the surgical isolator. The surgical site is washed and disinfected. The abdomen is draped with a surgical shroud containing an elliptical opening through which a midline incision is made. Good surgical technique is required to prevent the accidental incision of the intestines with its abundance of microorganisms and parasites. The uterus, with its cervix and cornuae clamped, is lifted from the abdomen onto sterile drapes. After severance from the maternal body, the uterus is lifted and removed to a 38°C primary germicide of 10% chlorine where it remains for 5 sec (Fig. 2). It is then placed into a perforated container which is lowered via the sleeve or

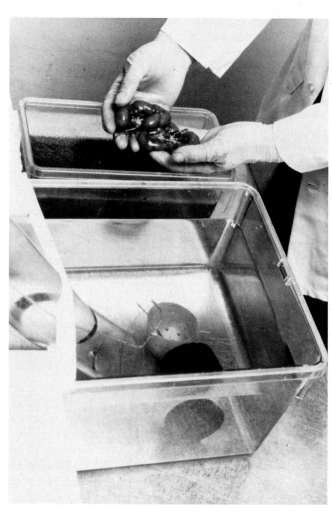

Fig. 2. Gravid uterus being lowered into primary germicidal solution adjacent to germicidal trap.

rigid tube by a nylon cord from inside the sterile isolator to beneath the surface of a 5% iodide germicide (Wescodyne, a product of West Chemical, Long Island City, New York). After an additional 15 to 30 sec, the container still beneath the surface of the germicide is guided into the mouth of a 10.4-cm wide rigid clear plastic tube connected to the isolator wall. A surgical technician, with arms inside the isolator via the surgical sleeves and gloves, raises the uterus within the perforated container to the interior of the sterile isolator allowing the germicide to drain down to the germicidal trap (Fig. 3). Two technicians on opposite sides of the isolator rapidly remove the fetuses from the uterus and separate them from their fetal membranes utilizing a taut Mylar membrane secured to a cage top as a surgical table (Fig. 4). They are quickly washed with surgical sponges and rendered clean from amniotic fluid and blood. This is essential since a foster mother might cannabalize them if body fluids and remnants of the fetal membranes remain. These procedures must be accomplished rapidly, since maternal support is lost upon separation of the placenta from the uterus.

The neonates are dried and massaged to stimulate breathing (Fig. 5). The umbilical cord is separated by clamping and cutting or electric cautery (64). If additional procedures are to be performed, the neonates are loosely wrapped in a small surgical towel and placed on the isolator floor above the warmth of the heating pad resting beneath the isolator floor. The Mylar membrane attached to the plastic cage is punctured, permitting the uterus and membranes to fall below to the floor of the cage. A new membrane is placed across the mouth of the cage and once again held in place by rubber bands. This procedure can now be repeated many times without breaching the integrity of the sterile system.

It is good practice to transfer the neonates to a rearing isolator where they are foster-nursed on lactating GF mothers until weaning age (64). This method has replaced the more cumbersome hand rearing and feeding (33), principally because of the more readily available lactating GF mothers and the increased survival rate as compared to hand-feeding methods. One would only consider hand rearing where minimal antigenic stimulation is desirable, and in these instances highly purified synthetic formulas are utilized. The surgical procedures, equipment, etc., are reviewed by Pleasants (62). There are numerous references to the establishment and maintenance of breeding colonies of GF rats (13,22,28,34,39,40, 44,52,53,75,101).

D. Associating Germfree Rats with a Defined Microflora

The elimination of the rat's normal gut microflora by cesarean derivation results in dramatic changes in the host's physiology, nutrition, tissue morphology, and defense against

Fig. 3. Technician lowered gravid uterus into perforated container attached to cord within rigid tube descending beneath germicidal liquid.

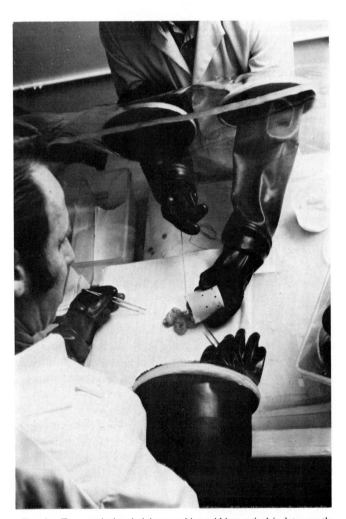

Fig. 4. Two surgical technicians working within surgical isolator gently place uterus in position for cesarean.

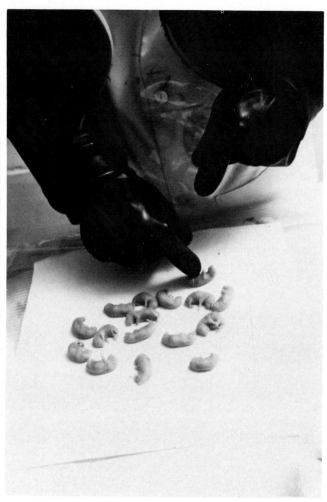

Fig. 5. Surgical technician in isolator manipulates neonates to stimulate respiration.

infectious agents. The most pronounced anomaly of the GF state in rats, as well as other species, is the enlargement of the cecum which can lead to volvulus at the ileo–cecal–colonic junction and eventual death (97b). The content of the cecum and intestines are fluid, and the animal is said to have a chronic mild diarrhea. In addition, aside from low levels of antigenic material in the feed and bedding, the immune system of the GF rat is unstimulated (29). The lamina properia is thin and almost devoid of antibody-producing plasma cells, and lymph nodes are smaller (28, 29a). Also, due to the absence of its vitamin K synthesizing gut flora, the GF rat must have this vitamin added to its food, or it rapidly develops prolonged prothrombin times and hemorrhages (34a). Germfree rats are also much more susceptible to infections than their CV counterparts which is why they often die soon after being introduced into a CV colony (45). This can be prevented if they are first colonized by at least several members of their normal gut microflora.

Gordon and Wostmann demonstrated that GF rats could be normalized by feeding them cecal contents of CV rats (30). However, no attempt was made to determine which member(s) of the microflora was responsible for this phenomenon until Schaedler's and Dubos' classic work describing the bacterial colonization of the gastrointestinal tract of mice, which subsequently became the cornerstone for much of the work which followed (17a,81b,81c). They reported that soon after birth the entire gastrointestinal tract was populated by lactobacilli and a group N streptococcus. During the second week of life, high concentrations of aerobic bacteria, such as enterococci and slow lactose-fermenting coliforms, were observed in the large intestine. Their numbers abruptly dropped during the third week of life when obligately anaerobic bacteria, such as bacteroides, colonized this organ. Throughout the adult life of mice the obligately anaerobic bacteria remained at very high levels, and the aerobic component of the microflora remained suppressed at very low levels. The microflora of the rat has been found to closely resemble that of the mouse (81a,85b).

Schaedler then proceeded to colonize GF animals with a flora consisting of bacteroides, lactobacilli, an anaerobic streptococcus, and a slow lactose-fermenting coliform (81b). This flora was able to drastically reduce the size of the cecum and, therefore, normalize the animals. Consequently this flora has been used extensively to colonize both GF mice and rats prior to their removal from an isolator into a new colony. The process for colonizing GF animals with the Schaedler flora, commonly referred to as "associating animals," merely consists of colonizing an initial isolator of GF rats with pure cultures of each of the individual members of the flora. Additional associations are then achieved by simply introducing an associated animal into a GF isolator and placing fecal pellets from the associated animal into the water bottles of the GF animals on two consecutive days. During the first day the aerobic bacteria colonize the GF animals and lower the oxidation–reduction potential, so that on the second day the bacteroides are able to colonize the animals.

It should be noted that these few bacteria represent a very small fraction of the gut microflora, and many additional members are necessary to normalize a GF animal completely (86b).

E. Microbiological Testing

It is good practice to perform certain examinations on the euthanized dam. Historical data of the health status of the donor female provide excellent reference material should subsequent contaminations occur in the GF or SPF colony. Therefore, prior to losing the identification of a cesarean-delivered litter, examination for *Mycoplasma* and intestinal parasites and serological examination for murine viruses are recommended.

Certainly, at the very least, careful culturing of the ovaries and uterus for *Mycoplasma* should be performed (24,82), since there have been occasional reports of *Mycoplasma pulmonis* isolation from GF rats which may have resulted from *in utero* contamination of the dam (24,39). Careful workers discard neonates as a precaution if the donor female exhibits positive *Mycoplasma* in the reproductive system.

Subclinical *Pasteurella pneumotropica* infections have been reported in GF rats (24,62), and these might be transmitted to GF progeny. Microbiological testing by fecal swabs of the neonates 24 to 48 hr postdelivery assures the asepsis of the surgical procedure as well as the sterility of the rearing isolator. Detailed methods are described in "Gnotobiotes" (95). The methods for gross observation and the detection of bacteria, fungi, and parasites are relatively simple and well standardized. The methods for the presence of exposure to murine viruses are usually accomplished through hemagglutination inhibition (HI) and complement fixation (CF) serological tests for the specific antibodies. Wagner (92) has worked out detailed sterility testing procedures which are utilized as standard procedures in many laboratories. Usually when an isolator becomes contaminated with bacteria, the exhaust air loses its non-animal-like almost sweet odor to the more familiar odor of laboratory rats. In nearly all instances, contaminants can be observed in fecal wet mounts prior to routine culturing. Culturing 24 to 48 hr on appropriate media readily reveals typical contaminants at 37°C. Since 21°, 37°, and 55°C are standard incubation temperatures, molds and thermophilic organisms are also detected in the less common contaminations.

Intrauterine infection represents a potential hazard to the axenic integrity of GF rats (2,4), since vertical transplacental transmission of a leukemia virus (68) and an unidentified virus from the submaxillary gland (5) have been reported. The confirmation of encephalitozoan in GF rabbits (37) suggests the likelihood of such vertical transmission possibilities in rats and other species. With these limited reports as background, the examination of donor stock would be the only means currently available to help reduce the vertical transmission potential in newly derived colonies. Unfortunately the job is tedious to search histologically for leukemia virus and the virus from the submaxillary gland, and negative results would not be conclusive. With regard to encephalitozoan, immunofluorescent, CF, and India dye tests are fairly straightforward procedures and provide a high degree of accuracy (100a,100b).

F. Strains and Stocks

The Lobund Laboratory at the University of Notre Dame has reported on GN colonies of the Fischer strain and the Wistar and Sprague-Dawley rat stocks (52). The Gifu hybrid has been produced in the GF state by Miyakawa in Japan (53). Gustafsson (34) used the Long-Evans hooded stock, and Pleasants (61) reported experiments with the Holtzmann and Lobund stocks. Dajani utilized the GF Agus strain (14).

G. Nutrition

It can be generally stated that the nutritional requirements of animals are inversely proportional to their biosynthetic capacity (45). The need for special diets for GN animals has recently been reviewed by Wostmann (97). Special diets are necessary principally because food sterilization methods usually require compensation for the loss of vitamins and the reduction of nutrient value of proteins resulting from heat sterilization (95). The dietary requirements for microbiologically synthesized vitamins are higher (13,97), because the lack of normal microbial flora affects the absorption, which is greatly enhanced in the GF rat and which leads to the formation of urinary calculi unless dietary levels of calcium are reduced (35,86).

Diets tend to vary according to the specific GF research objectives (45), i.e., antigen-free diets for immunity studies or high sugar content diets in dental caries studies. Some diets have been found nutritionally adequate for short-term experiments even if autoclaved, as long as the diets are supplemented with filter-sterilized heat-labile vitamins (58).

A canned, moist, presterilized (autoclaved) diet of known composition can be provided by spraying it into the isolator system (22,23). Autoclavable diets are also available (40,58,62) as are chemically defined and water-soluble ones. The latter can be filter-sterilized (62) and used as special purified diets for nutritional research (99). Growth of GF rats on these diets is comparable to that of CV animals. Reddy *et al.* (74) grew GF rats from birth to maturity using membrane-filtered, chemically defined, water-soluble diets based on amino acids and glucose. Diets sterilized by gamma irradiation have also been used in rearing GF rats (59). Radiation sterilization using ^{60}Co irradiation is recommended for studies of cholesterol and bile acid metabolism in GN rats (97a). Current diets suitable for studies in nutrition and metabolism of GF rats are listed in Wostmann's review (97).

H. Characteristics of Gnotobiotic Rats

The GF rat is an experimental animal which differs significantly from the CV rat in a number of characteristics. GF rats provide a uniform and relatively stable baseline of morphologic and physiologic activities which, in turn, facilitate studies of superimposed changes. The earliest noted and most conspicuous effect of the GF status on rodents, including rats, is enlargement of the cecum. It becomes voluminous usually five times larger than its CV counterpart on the same diet and may approach 25% of body weight (13). This enlargement sometimes interferes with normal reproduction but can be significantly reduced with dietary manipulation of inorganic ions. The cecal wall is much thinner in GF rats and the cecal contents more liquid than in CV animals. This is due to an excess of water and amnionic-soluble mucins (3,62). Cecectomy of GF rats restores most functional and metabolic parameters within the CV range (97).

The enlarged cecum is associated with altered metabolic functions, particularly slower cholesterol and bile turnover (18), depressed reducing capacity of the cecum contents, and reduced cecal concentrations of chloride and carbonate ions (87). These animals also require less exogenous choline (55), and there is a total absence of metabolism of flavonoid compounds in the gut (33). If the GF rat's cecal contents are replaced with saline, then the water absorption capacity of the cecum becomes normal or greater than in CV rats (27). Germfree rats have been reported to accumulate a compound of unknown composition in their cecum which changes the tone and reactivity of mesenteric vascular smooth muscle to adrenaline (7) and which is normally destroyed by the CV normal flora.

The weight and surface area of the small intestine and the associated lymphoid cells and tissues of GF rats is generally decreased (28). Depending on the diet the rate of peristalsis may be the same (36) or slower (81) in GF than in CV rats. The rate of mucosal sloughing is generally half that of CV controls (28,73), and digestive enzymes, such as proteases, lipase, and amylase, persist longer and farther down the gut under GF conditions (43,71). Urease appears to be absent under GF status (15). The GF rat gut is more efficient in digestion and absorption, in part because the villi are longer and more even (13).

Nutrient requirements in the diet of GF rats are usually higher than CV requirements but vary with experimental conditions (62). In general, there is a higher need for total food and water, for vitamin K, the B vitamins, and for choline to prevent liver cirrhosis. Germfree rats maintained on a diet without supplemental vitamin K rapidly develop a hemorrhagic condition, while CV rats on the same diet do not (13). Antagonism between vitamins A and K occurs only when vitamin A intake is ten times above normal (74,100). On the other hand, GF rat nutrient requirements are less than CV requirements for vitamin A (13,79), lysine, cysteine and vitamin E to prevent liver necrosis, protein (62), calcium, and magnesium and zinc (86). Germfree rats given vitamin A-deficient diets survived much longer than CV rats (13). Assessment of the rat's nutritional requirements are often difficult because of coprophagy.

Germfree rat studies show unequivocally that the CV rat's microbial flora has a significant effect on the basal metabolism, on the response to adrenaline, cardiac output, and vascular distribution (62). The overall metabolic rate of GF rats has been reported to be one-fourth that of CV rats of the same strain (98). This undoubtedly results from reduced (one-third normal) cardiac output and oxygen consumption (98), reduced regional blood flow and distribution (28), decreased heart weight (1,28), decreased total blood volume (11), and decreased pulmonary pCO_2 values (82a). Aortas and portal veins of GF rats have an attenuated reactivity to angiotensin, vasopressin, and epinephrine (2). Reduction of total body fats has been reported (76). In GF rats the lymph nodes and Peyer's patches are small, lack germinal zones, and contain few, if any, plasma cells (8). Serum globulin values are one-third those of CV rats, and GF rats have less total serum proteins (8). The decreased immunological stimulation of GF animals leads to very low titers of agglutinating antibodies for *Streptococcus fecalis, Proteus vulgaris, Pseudomonas aeruginosa, Lactobacillus acidiphilus,* and *Bacillus fragilis* (8). This is accompanied by decreased severity of conjunctival inflammation when infected by bacteria (51). In relation to the above, it has also been reported that the mucosa of the nasal cavity and middle ear have few lymphocytes and no inflammatory infiltrates (25).

Tissue enzyme levels usually tend to be lower in GF rats. There is less mitochondrial succinate oxidase and glycerophosphate dehydrogenase activity in the liver (85). Lower muramidase levels have also been reported (38). However, some tissue enzyme activities are higher in GF rats, namely, peroxidase-mediated antibacterial activity of the salivary glands (54), liver microsomal hydroxylation of steroid hormones (19), and fatty acid synthetase and citrate lyase activity in the liver (72,97).

In general, GF rats eat more, grow better, and are less subject to disease. They absorb saturated and unsaturated fats better, particularly palmitic and stearic acid (16); have greater serum and liver cholesterol concentrations (76); use more total fat in the diet (56); and have higher cholesterol turnover rates (76). A recent report on experimental cholesterol synthesis in GF and CV rats (91) indicates that there is an inverse proportionality between the log phase rate of hepatic cholesterol synthesis and liver cholesterol levels in GF rats. Therefore, the endogenous cholesterol synthesis in GF rats may not be responsible for the high cholesterol levels in plasma or in the liver. Liver cholesterol may play a major role in the regulation of hepatic cholisterogenesis in the GF rat by a mechanism similar to that in the CV rat.

There is total conjugation of bile acids in GF rats compared with almost total lack of conjugation in the cecum of CV rats (47). The bile turnover rates are higher (76) as is the pH of the cecum contents (87). In GF cecal contents the colloid osmotic pressures are approximately 100 mm Hg. This results in a pressure gradient of 60–70 mm Hg between the gut lumen and the blood plasma, in contrast to a smaller gradient in CV rats

(27). Germfree rats have greater reabsorption of bile acids from the gastrointestinal tract and, therefore, have greater recirculation of bile (18). Use of the GF rat in studies of cholesterol metabolism are particularly concerned with the factors which influence the absorption of cholesterol from the gut and its elimination from the body as bile acids via the feces (96,97). The GN rat appears to be unable to decrease the reabsorption of bile acids in the lower gut, a function of normal microbial flora. Recent work (102) indicates that differences in the histochemical nature of mucosaccharides are dependent on whether they are located in areas of normal bacterial flora in CV rats or in areas relatively free of intestinal flora.

Other metabolic parameters which tend to be higher or greater in GF animals as compared to CV animals include pH of cecal contents (87), mean intracolonic oxygen pressure (10), pulmonary arteriovenous oxygen values (82a), plasma levels of some steroids (18), urinary citrate excretion (34), and fasting blood glucose (62). In addition, aortas and portal veins have a higher total calcium content in GF animals (2), and the microvasculature is refractory to catecholamines (7,28).

The GF state has little influence on the functional respiration or oxidative phosphorylation of mitochondria isolated from the liver of adult rats (84). Serum chemistry and hematological values are within the normal range except for the depressed leukocyte level (12). Minimal differences have also been reported in serum β-lysin (38), fasting blood glucose and glucose tolerance (83), metabolism of nicotinamide and nicotinic acid (42), carbon dioxide production in the gut (78), mean pulmonary arterial pO_2 (82a), and activities of hepatic enzymes of urea synthesis (57a). No differences were found in the histology of the eye of GF rats (51).

It is evident that there are many basic physiological and morphological parameters of GF rats which have not yet been studied. Furthermore, one must keep in mind that often reports cannot be reliably compared because of variables of age, sex, and strain of rat as well as environmental conditions, diet, and unknown interactions among these factors. For example, it has been reported that differences in thyroid function and related hepatic enzymes tend to lessen with age of the animals (85) and that the qualitative and quantitative composition of the bile acids varies considerably between male and female GF rats (34b).

III. RESEARCH APPLICATIONS OF GNOTOBIOTIC RATS

A. Immunology

Immunology studies with GF or GN animals enable distinguishing primary mediation lesions from possible microbial infections. From work on the biological effects of radiation, it has been determined that GN rats survive larger doses of total-body X irradiation for a longer time (77). In basic immunological studies, GF or GN rats provide information on the role of the microbial flora in stimulating humoral and cell-mediated immune responses. Immunity, as measured by opsonic activity, is depressed in GF rats infected with *Pseudomonas aeruginosa* (49) whereas the GF rat's responsiveness to H and O antigens of *Escherichia coli* and to sheep erythrocytes is increased (49).

Differences in phagocytosis depend on differences in opsonic activity rather than on functional differences at the cellular level. *In vitro* studies with ^{32}P-labelled *Escherichia coli* opsonized with sera of GF rats indicated that cells from GF rats were slightly more active in ingesting capacity than cells from CV rats. Thus the opsonic activity of CV rat sera is higher than that of GF rat sera when tested *in vitro* (89). Germfree rats have been reported to reject skin allografts more rapidly than CV rats (50), whereas autografts of skin transplants on rat tails heal quickly (6). The latter is postulated to be due to genetic uniformity of histocompatibility factors in GF animals (6).

B. Sensescence, Death, and Postmortem Changes

In general, GF animals tend to live longer than their CV counterparts. Premature death of GF animals may be caused by infection or by environmental factors. Delayed morbidity in 2- to 3-year-old GF rats is a common observation which shows them to be virtually free of age-related kidney, heart, and lung changes (65, 69).

Postmortem differences include a minimum of odoriferous putrefactive changes and autolysis of the intestinal area by digestive enzymes. Dead GF animals undergo drying and mummification if in a dry atmosphere (45).

C. Oral Pathology

Mechanisms of oral pathology have been clarified through GF rat studies. This is particularly true in caries research where *Streptococcus* sp. in association with GN rats fed a cariogenic diet produce carious lesions (31), and in periodontal disease studies where a number of streptococci, actinomycetes, and gram-positive bacilli cause typical periodontal disease under conditions of monoassociation (32).

D. Cancer

Cancer development in GF rats can be related in part to the absence of microbial flora (66,93). Experimental cancer yields are lower in GF rats when the carcinogens tested are of the type necessitating enzymatic metabolic activation (94). In general, the oncogenic potential is the same as in CV rats, but tumor-related changes are more clearly defined in GF animals (70). Germfree rats with either spontaneous or induced tumors have higher numbers of plasma cells but have no germinal zones in their lymph nodes (70).

Gnotobiotic animals are particularly suitable for testing candidate viral carcinogens, since derivation by hysterectomy and GN maintenance have been found to eliminate all known viruses from GF rats (45,62). Nevertheless, GF rats have a very low rate of spontaneous neoplasm development as compared to GF mice (93). The most frequent spontaneous tumors in aged GF rats involve the mammary and pituitary glands (60).

1. Colon Cancer

Cycasin from cycad bean flour is carcinogenic to CV rats whose microflora convert it to a carcinogen, whereas it does not induce tumors in GN rats (41,46). If cycasin is first hydrolyzed to aglycone (MAM), it is then carcinogenic to GF rats (41). Spontaneous colon adenomas are twice as prevalent in GF rats (94). No difference in the incidence of adenocarcinoma have been reported following intracolonic exposure to nitrosoguanadine carcinogens (94), whereas others report greater susceptibility of GF rats to these same direct-acting carcinogens (85a). Results are significantly dependent on route of administration, since oral administration of N-methyl-$\acute{N}$-nitro-N-nitrosoguanadine produced very few adenocarcinomas in GF rats as compared to CV animals (53a). It appears that there is a lower incidence of cancer induction in GF animals when the carcinogens being tested are of the type necessitating enzymatic metabolic activation.

2. Breast Cancer

GN rats are as susceptible to dimethylbenzanthracene (DMBA)-induced breast cancer as CV rats (68) (see also this volume, Chapter 6; Volume I, Chapter 13).

3. Leukemia

A spontaneous, transplantable, lymphatic leukemia called "Nova rat leukemia" has been reported in aged GF Fischer rats (80). Leukemia could not be induced by whole-body irradiation (68), but GF rats were found as susceptible to passage of Gross A leukemia virus as CV rats (68).

4. Urethral Cancer

This type of cancer is rare in CV rats but is relatively frequent in some older GF rat strains (67).

5. Endocrine Cancer

Endocrine-related cancers of the nonleukemic type such as thymomas and mammary neoplasms occur in GF rats (69).

E. Wound Repair

The GF allogenic radiation-induced chimera has a greatly reduced or absent T cell response, whereas the B cell response is almost normal (9). Germfree rats are less sensitive by half to X-irradiation as it affects the rate of wound closure (17).

F. Infectious Diseases

The infectious and chronic diseases of CV rats are described by Tuffery and Innes (90) and may serve as a basis for comparison of monoassociation and experimental infection studies. *Mycoplasma pulmonis* is the primary pathogen in chronic respiratory disease (CRD) (86a). Luckey (46) has provided an extensive bibliography on the effects of bacterial species on the monoassociated rat. No differences were found in GF rats' susceptibility of *Plasmodium berghei* primary infections via mosquito-borne sporozoites, nor were there any differences in the resulting pathology (48). A recent report (78a) indicates that the GF rat can serve as an animal model of nephritis due to *Candida* infections, since the yeasts multiply in these animals' kidneys.

IV. CONCLUSIONS

At the New York Academy of Sciences Conference in 1959, "Germfree Vertebrates: Present Status," there were discussions on the state of the art, recent research uses, and some practical applications of the GF animal in the development of disease-free breeding colonies. Much of these discussions have come to fruition through further study and expanded implementation of the technology.

Prior to the late 1950s equipment for GF animals was fabricated principally of stainless steel with various sizes and shapes of viewing areas. The early pioneers were convinced that the security of GF systems could be best achieved when rigid components were used that could withstand impact and be less vulnerable to breakage. Stainless steel cylinder-shaped GF isolators bolted and gasketed together was the most commonly used system in the United States at the Lobund Institute, National Institutes of Health, and Walter Reed Hospital. In Sweden at the Karolinska Institute, Gustafsson (34) developed rectangular rigid steel isolators with glass tops maximizing the visibility inside. These units could be totally introduced into a steam autoclave for sterilization. In both the Reyniers cylindrical tank and the Gustafsson isolator, neoprene sleeves and gloves were used for manipulation constituting a potential weakness, since they were subject to tear or puncture. Miyakawa (53) in Japan operated remote mechanical arms and hands from outside a small sterile room similar to the equipment sometimes seen in radiation research laboratories. Even though there are very fine research reports from this period, the cost of equipment and its inefficient utilization of space limited the number of workers in the field. Also the complexity of the fabrication and construction added further to discourage interested scientists. It was not until Trexler developed the low cost, lightweight flexible film isolator that GF research came within the budgetary and technical reach of the research community in general. In 1957 a standard flexible film $1.5 \times 0.6 \times 0.6$ m isolator complete with filtration, transfer port, exhaust trap, and flexible sleeves and gloves bore a price of \$300 to \$400 compared to \$5000 for a typical stainless steel tank type isolator. Thus technological advances produced a system that has been proved to be equally or more secure, light in weight to permit use in tiers, and made of clear plastic to allow complete visibility. An extension of the flexible film technology brought forth a lightweight, disposable flexible film shipping unit (88a) weighting 5–6 kg compared to earlier units weighing 70–80 kg and which required a battery-operated blower system. Thus both rearing and research units plus shipping units were readily obtainable to those interested in using and transporting GF and GN animals.

Diets which early workers developed were either mixed and formulated in their laboratories or prepared at significant cost by organizations who specialized in small batch mixing of complicated formulas. From the late 1950s onward, commercial feed manufacturers offered standard laboratory rodent diets prefortified with sufficient thermolabile nutrients to withstand sterilization and still support reproduction and growth. There have, from time to time, appeared in the marketplace canned sterile water and prepackaged sterilized bedding for those with limited sterilization capacity or capability.

In effect, it is entirely feasible to conduct a single GF research project without setting up a vast facility with expensive support, laboratories, and personnel. Also when a single project is completed, the inflated flexible film isolator can be stored flattened in its deflated form for subsequent reuse. The initial set-up costs are nominal as are the operating costs.

Perhaps one of the single most important by-product benefits of the GF and GN rats is the utilization of these animals as seed stock for new colonies. By deriving a strain of outbred CV stock of rats into the GF state, all microorganisms and parasites are eliminated except the few that are thought to be transplacental and thus vertically transmitted. Colonies plagued with external and intestinal parasites, chronic murine pneumonia, and one or more other bacterial or viral diseases can be rendered free of these infections by utilization of the gnotobiotic technology and deriving these animals GF. It is then a routine procedure to associate these rats with a defined "bacterial cocktail" of gut flora prior to removal from isolator systems and introduction into some type of clean barrier facility. Specific pathogen-free rats is the commonly used terminology for rats derived in this manner and is the accepted practice by industry, government, and academia for providing healthy animals for research. Even though testing and eradication techniques can work as does gut sterilization through broad-spectrum antibiotics, the true gnotobiote must be cesarean-derived within a sterile isolator system.

Valuable genetic strains of rats are assured continuity by maintaining them GF, greatly reducing and almost eliminating the possible loss through an epizootic. The National Cancer Institute for many years has maintained Rodent Genetic Centers under contract whereby valuable genetic strains and stocks have been maintained in the GN state. Through the technology herein described, genetically defined CV animals are cesarean-delivered into the axenic state then subsequently associated with a defined flora and maintained in isolators assuring microbial definition and uniformity. These rats can be sent with genetic and microbiological pedigree to laboratories for research utilization or breeding programs.

Like the pure or refined chemical available to researchers, the laboratory rat can be obtained in the purest microbiological sense (axenic) or with an easily described and definable flora (GN) within the isolator. The barrier-reared animal which is an extension of the gnotobiote can be maintained in a controlled environment to preclude contamination by pathogens. Even though isolator systems break down on occasion, usually through human error, their use is a giant step forward toward supplying defined rats as one of the basic tools of biomedical research. Compared to its CV counterpart, the SPF, cesarean-delivered, cesarean-originated, barrier-sustained (COBS) barrier-maintained rat has provided the research community with some point of reference in that at one time during the immediate past they were gnotobiotes. The CV animal usually is not reared in a barrier environment where all materials contacting the animals have undergone decontamination, pasteurization, or sterilization. The probability, therefore, is far greater that microbiological variability does occur in CV animals because of their undefinable origin and their more loosely controlled environment.

The science of breeding and rearing GF, GN, and SPF rats is an attempt by professionals in laboratory animal science to keep pace with the rapidly evolving technology in the instrumentation field. What is the value of highly sophisticated instrumentation designed to make finite measurements of biological materials if the biological tool is uncontrolled and undefinable?

The horizons of GN research applications have expanded to include space travel and its potential and unpredictable effects on man and his biosphere. Other areas of research include oral pathology, cancer, wound repair, infectious diseases, and nutrition. The GF animal, therefore, offers a multitude of research opportunities.

Because of the technology developed through the years, the cost of such research using GF animals is within the scope of most budgets. However, there is still an inherent resistance to undertake research with GF rats principally as a carry over from other eras when the cost of this type of research was excessive and the technology was too highly specialized for the average laboratory setting. The contents of this chapter, in connection with the literature citations, should provide those desirous of conducting research on GN rats the necessary technical information with regard to methodology, characteristics, and utilization.

REFERENCES

1. Albrecht, I., and Souhrada, J. (1971). Defining the laboratory rat for cardiovascular research. *In* "Defining the Laboratory Animal" (H. A. Schneider, ed.), pp. 616–625. Natl. Acad. Sci., Washington, D.C.

2. Altura, B. T., Altura, B. M., and Baez, S. (1975). Reactivity of aorta and portal vein in germfree rats. *Blood Vessels* **12,** 206–218.

3. Asano, T. (1967). Inorganic ions in cecal content of gnotobiotic rats. *Proc. Soc. Exp. Biol. Med.* **124,** 424–430.

4. Asano, T. (1969). Modification of cecal size in germfree rats by long-term feeding of anion exchange resin. *Am. J. Physiol.* **217,** 911–918.

5. Ashe, W. K., Scherp, H. W., and Fitzgerald, R. J. (1965). Previously unrecognized virus from submaxillary glands of gnotobiotic rats. *J. Bacteriol.* **90,** 1719–1729.

6. Ashman, R. B. (1975). A rapid method for skin grafting in germfree rats. *Lab. Anim. Sci.* **25,** 502–504.

7. Baez, S., and Gordan, H. A. (1971). Tone and reactivity of vascular smooth muscle in germfree rat mesentery. *J. Exp. Med.* **134,** 846–856.

8. Balish, E., Yale, C. E., and Hong, R. (1972). Serum proteins of gnotobiotic rats. *Infect. Immun.* **6,** 112–118.

9. Bealmear, P. M., Loughman, B. E., Nordin, A. A., and Wilson, R. (1973). Evidence for graft vs. host reaction in the germfree allogeneic radiation chimera. *In* "Germfree Research" (J. B. Heneghan, ed.), pp. 471–475. Academic Press, New York.

10. Bornside, G. H., Donovan, W. E., and Myers, M. B. (1976). Intracolonic tensions of oxygen and carbon dioxide in germfree, conventional, and gnotobiotic rats. *Proc. Soc. Exp. Biol. Med.* **151,** 437–441.

11. Bruckner-Kardoss, E., and Wostmann, B. S. (1974). Blood volume of adult germfree and conventional rats. *Lab. Anim. Sci.* **24,** 633–635.

12. Burns, K. F., Timmons, E. H., and Poiley, S. M. (1971). Serum chemistry and hematological values for axenic (germfree) and environmentally associated inbred rats. *Lab. Anim. Sci.* **21,** 415–419.

13. Coates, M. E. (1973). Gnotobiotic animals in nutrition research. *Proc. Nutr. Soc.* **32,** 53–58.

14. Dajani, R. M., Gorrod, J. W., and Beckett, A. H. (1975). Reduction *in vivo* of (minus)-nicotine-1-*N*-oxide by germfree and conventional rats. *Biochem. Pharmacol.* **24,** 648–650.

15. Delluva, A. M., Markley, K., and Davies, R. E. (1968). The absence of gastric disease in germfree animals. *Biochim. Biophys. Acta* **151,** 646–650.

16. Demarne, Y., Flanzy, J., and Jacquet, E. (1973). The influence of gastrointestinal flora on digestive utilization of fatty acids in rats. *In* "Germfree Research" (J. B. Heneghan, ed.), pp. 553–560. Academic Press, New York.

17. Donati, R. M., McLaughlin, M. M., and Stromberg, L. R. (1973). Combined surgical and radiation injury—The effect of the gnotobiotic state on wound closure. *Experientia* **29,** 1388–1390.

17a. Dubos, R., Schaedler, R. W., Costello, R., and Hoet, P. (1965). Indigenous, normal, and autochthonous flora of the gastrointestinal tract. *J. Exp. Med.* **122,** 67–76.

18. Einarsson, K., Gustafsson, J. A., and Gustafsson, B. E. (1973). Differences between germfree and conventional rats in liver microsomal metabolism of steroids. *J. Biol. Chem.* **248,** 3623–3630.

19. Einarsson, K., Gustafsson, J. A., and Gustafsson, B. E. (1974). Liver microsomal hydroxylation of steroid hormones after establishing an indigenous microflora in germfree rats. *Proc. Soc. Exp. Biol. Med.* **145,** 48–52.

20. Foster, H. L. (1959). A procedure for obtaining nucleus stock for a pathogen-free animal colony. *Proc. Anim. Care* **9,** 135–142.

21. Foster, H. L. (1959). Housing of disease-free vertebrates. *Ann. N.Y. Acad. Sci.* **78,** 80–88.

22. Foster, H. L., and Pfau, E. S. (1963). Gnotobiotic animal production at The Charles River Breeding Laboratories, Inc. *Lab. Anim. Care* **13,** 629–632.

23. Foster, H. L., Trexler, P. C., and Rumsey, G. L. (1967). A canned sterile shipping diet for small laboratory rodents. *Lab. Anim. Care* **17,** 400–405.

24. Ganaway, J. R., Allen, A. M., Moore, T. D., and Bohner, H. J. (1973). Natural infection of germfree rats with *Mycoplasma pulmonis*. *J. Infect. Dis.* **127,** 529–537.

25. Giddens, W. E., Jr., Whitehair, C. K., and Carter, G. R. (1971). Morphological and microbiologic features of nasal cavity and middle ear in germfree, defined-flora, conventional, and chronic respiratory disease-affected rats. *Am. J. Vet. Res.* **32,** 99–114.

27. Gordon, H. A. (1974). Intestinal water absorption in young and old, germfree and conventional rats. *Experientia* **30,** 214–215.

28. Gordon, H. A., Bruckner-Kardoss, E., Staley, T. E., Wagner, M., and Wostmann, B. S. (1966). Characteristics of the germfree rat. *Acta Anat.* **64,** 367–389.

29. Gordon, H. A., and Pesti, L. (1971). The gnotobiotic animal as a tool in the study of host-microbial relationships. *Bacteriol. Rev.* **35,** 390–429.

29a. Gordon, H. A., and Wostmann, B. S. (1960). Morphological studies on the germfree albino rat. *Anat. Rec.* **137,** 65–70.

30. Gordon, H. A., and Wostmann, B. S. (1959). Responses of the animal host to changes in the bacterial environment: Transition of the albino rat from germfree to the conventional state. *In* "Recent Progress in Microbiology" (G. Tunevall, ed.), pp. 336–339. Almqvist & Wiksell, Stockholm.

31. Green, R. M., Blackmore, D. K., and Drucker, D. B. (1973). The role of gnotobiotic animals in the study of dental caries. *Br. Dent. J.* **134,** 537–540.

32. Green, R. M., Drucker, D. B., and Blackmore, D. K. (1974). The reproducibility of experimental caries studies within and between two inbred strains of gnotobiotic rat. *Arch. Oral Biol.* **19,** 1049–1054.

33. Griffiths, L. A., and Barrow, A. (1972). Metabolism of flavonoid compounds in germfree rats. *Biochem. J.* **130,** 1161–1162.

34. Gustafsson, B. (1948). Germfree rearing of rats. *Acta Pathol. Mic-*

robiol. Scand., **73**, Suppl., 1-130.

34a. Gustafsson, B. E. (1959). Vitamin K deficiency in germfree rats. *Ann. N.Y. Acad. Sci.* **78**, 166-174.

34b. Gustafsson, B. E., Cronholm, T., and Gustafsson, J. A. (1975). Relation of intestinal bile acids in rats to intestianl microflora and sex. *In* "Clinical and Experimental Gnotobiotics" (T. M. Fliedner, H. Heit, D. Neithammer, and H. Pflieger, eds.). Gustav Fischer, New York.

35. Gustafsson, B. E., and Norman, A. (1962). Urinary calculi in germfree rats. *J. Exp. Med.* **116**, 273.

36. Gustafsson, B. E., and Norman, A. (1969). Influence of the diet on the turnover of bile acids in germfree and conventional rats. *Br. J. Nutr.* **23**, 429-442.

37. Hunt, R. D., King, N. W., and Foster, H. L. (1972). Encephalitozoonosis: Evidence for vertical transmission. *J. Infect. Dis.* **126**, 212-224.

38. Ikari, N. S., and Donaldson, D. M. (1970). Serum beta-lysin and muramidase levels in germfree and conventional rats. *Proc. Soc. Exp. Biol. Med.* **133**, 49-52.

39. Kappel, H. K., Kappel, J. P., Weisbroth, S. H., and Kozma, C. K. (1969). Establishment of a hysterectomy-derived, pathogen-free breeding nucleus of Blu-(LE) rats. *Lab. Anim. Care* **19**, 738-741.

40. Kellogg, T. F., and Wostmann, B. S. (1969). Stock diet for colony production of germfree rats and mice. *Lab. Anim. Care* **19**, 812-814.

41. Laqueur, G. L., McDaniel, E. G., and Matsumoto, H. (1967). Tumor induction in germfree rats with methylazoxymethanol (MAM) and synthetic MAM acetate. *J. Natl. Cancer Inst.* **39**, 355-371.

42. Lee, Y. C., McKenzie, R. M., Gholson, R. K., and Raica, N. (1972). A comparative study of nicotinamide and nicotinic acid in normal and germfree rats. *Biochim. Biophys. Acta* **264**, 59-64.

43. Lepkovsky, S., Furuta, F., Ozone, K., and Koike, T. (1966). The proteases, amylase, and lipase of the pancreas and intestinal contents of germfree and conventional rats. *Br. J. Nutr.* **20**, 257-261.

44. Lev, M. (1963). Germfree animals. *In* "Animals for Research" (W. L. Lane-Petter, ed.), pp. 139-175. Academic Press, New York.

45. Luckey, T. D. (1963). "Germfree Life and Gnotobiology." Academic Press, New York.

46. Luckey, T. D. (1969). Gnotobiology and aerospace systems. *In* "Advances in Germfree Research and Gnotobiology" (M. Miyakawa and T. D. Luckey, eds.), pp. 317-353. Iliffe, London.

47. Madsen, D., Beaver, M., Chang, L., Bruckner-Kardoss, E., and Wostmann, B. (1976). Analysis of bile acids in conventional and germfree rats. *J. Lipid Res.* **17**, 107-111.

48. Martin, L. K., Einheber, A., Porro, R. F., Sadun, E. H., and Bauer, H. (1966). *Plasmodium berghei* infections in gnotobiotic mice and rats—Parasitologic, immunologic, and histopathologic observations. *Mil. Med.* **131**, Suppl., 870-889.

49. McClellan, M. A., Hummel, R. P., and Alexander, J. W. (1974). Opsonic activity in germfree and monocontaminated rat sera. *Surg. Forum* **25**, 27-28.

50. McDonald, J. C., Zimmerman, G., Bollinger, R. R., and Pierce, W. A., Jr. (1971). Immune competence of germfree rats. I. Increased responsiveness to transplantations and other antigens. *Proc. Soc. Exp. Biol. Med.* **136**, 987-993.

51. McMaster, P. R., Aronson, S. B., and Bedford, M. J. (1967). Mechanisms of the host response in the eye. IV. The anterior eye in germfree animals. *Arch. Ophthalmol.* **77**, 392-399.

52. Maejima, K., Mitsuoka, T., Namioka, S., Nomura, T., Tajima, Y., and Yoshida, T. (1974). Bibliography of technology for germfree animal research. *Exp. Anim.* **24**, 229-253.

53. Miyakawa, M. (1968). Studies of rearing germfree rats. *In* "Advances in Germfree Research and Gnotobiology" (M. Miyakawa and T. D. Luckey, eds.), pp. 48-62. Iliffe, London.

53a. Miyakawa, M., Sumi, Y., Kanzaki, M., and Imaeda, F. (1975). Tumor induction by oral administration of *N*-methyl-*N*-nitro-*N*-nitrosoguanadine or aflatoxin B1 in germfree and conventional rats. *In* "Clinical and Experimental Gnotobiotics" (T. M. Fliedner, H. Heit, D. Neithammer, and H. Pflieger, eds.). Gustav Fischer, New York.

54. Morioka, T., Saji, S., Inoue, M., and Matsumura, T. (1969). Peroxidase-mediated antibacterial activity in the salivary gland of germfree and conventional rats. *Arch. Oral Biol.* **14**, 549-553.

55. Nagler, A. L., Seifter, E., Geever, E. F., Dettbarn, W. D., and Levenson, S. M. (1969). The nephropathy of acute choline deficiency in germfree conventional, and open animal room rats. *In* "Germfree Biology" (A. Mirand and N. Back, eds.), pp. 317-324. Plenum, New York.

56. Nolen, G. A., and Alexander, J. C. (1965). A comparison of the growth and fat utilization of caesarean-derived and conventional albino rats. *Lab. Anim. Care* **15**, 295-303.

57. Nuttal, G. H. F., and Thierfelder, H. (1897). Thierisches leven ohne bakterien im verdauungskanal. III. Mittheilung versuche an Huhnern. *Hoppe-Seyler's Z. Physiol. Chem.* **23**, 231-235.

57a. Nuzum, C. T. (1975). Activities of hepatic enzymes of urea synthesis in germfree and conventional rats. In "Clinical and Experimental Gnotobiotics" (T. M. Fliedner, H. Heit, D. Neithammer, and H. Pflieger, eds.). Gustav Fischer, New York.

58. Oace, S. M. (1972). A purified soy protein diet for nutrition studies with germfree rats. *Lab. Anim. Sci.* **22**, 528-531.

58a. Otis, A. P. (1975). Charles River Breeding Laboratories, Inc., Wilmington, Massachusetts (personal communication).

59. Paterson, J. S., and Cook R. (1971). Utilization of diets sterilized by gamma irradiation for germfree and specific-pathogen-free laboratory animals. *In* "Defining the Laboratory Animal" (H. A. Schneider, ed.), pp. 586-596. Natl. Acad. Sci., Washington, D.C.

60. Pittermann, W., and Deerberg, F. (1975). Spontaneous tumors and lesions of the lung, kidney and gingiva in aged germfree rats. *In* "Clinical and Experimental Gnotobiotics" (T. M. Fliedner, H. Heit, D. Neithammer, and H. Pflieger, eds.). Gustav Fischer, New York.

61. Pleasants, J. R. (1959). Rearing germfree caesarian-born rats, mice, and rabbits through weaning. *Ann. N.Y. Acad. Sci.* **78**, 116-126.

62. Pleasants, J. R. (1974). Gnotobiotics. *In* "Handbook of Laboratory Animal Science" (E. C. Melby, Jr. and N. H. Altman, eds.), Vol. I, pp. 119-174. CRC Press, Cleveland, Ohio.

64. Pollard, M. (1971). The germfree rat. *Pathobiol. Annu.* **1**, 83-94.

65. Pollard, M. (1971). Senescence in germfree rats. *Gerontologia* **17**, 333-338.

66. Pollard, M. (1972). Carcinogenesis in germfree animals. *Prog. Immunobiol. Stand.* **5**, 226-230.

67. Pollard, M. (1973). Spontaneous prostrate adenocarcinomas in aged germfree Wistar rats. *J. Natl. Cancer Inst.* **51**, 1235-1241.

68. Pollard, M. (1966). Leukemia in germfree rats. *Proc. Soc. Exp. Biol. Med.* **121**, 585-589.

69. Pollard, M., and Kajima, M. (1970). Lesions in aged germfree Wistar rats. *Am. J. Pathol.* **61**, 25-36.

70. Pollard, M., Kajima, J., and Lorans, G. (1968). Tissue changes in germfree rats with primary tumors. *Res. J. Reticuloendothel. Soc.* **5**, 147-160.

71. Reddy, B. S., Pleasants, J. R., and Wostmann, B. S. (1969). Pancreatic enzymes in germfree and conventional rats fed chemically defined, water-soluble diet free from natural substrates. *J. Nutr.* **97**, 327-334.

72. Reddy, B. S., Pleasants, J. R., and Wostmann, B. S. (1973). Metabolic enzymes in liver and kidney of the germfree rat. *Biochim.*

Biophys. Acta. **320,** 1–8.

73. Reddy, B. S., and Wostmann, B. S. (1966). Intestinal disaccharidase activities in the growing germfree and conventional rat. *Arch. Biochem. Biophys.* **113,** 609–616.

74. Reddy, B. S., Wostmann, B. S., and Pleasants, J. R. (1969). Protein metabolism in germfree rats fed chemically defined, water-soluble diet and semi-synthetic diet. *In* "Germfree Biology" (E. A. Mirand and N. Back, eds.), pp. 301–305. Plenum, New York.

75. Reid, L. C., Jr., and Gates, A. S., Jr. (1966). A method of sterilizing supplies for germfree isolators in plastic bags. *Lab. Anim. Care* **16,** 246–254.

76. Reina-Guerra, M., Tennant, B., Harrold, D., and Goldman, M. (1969). The absorption of fat by germfree and conventional rats. *In* "Germfree Biology" (E. A. Mirand and N. Back, eds.), pp. 297–300. Plenum, New York.

77. Reyniers, J. A., Trexler, P. C., Scruggs, W., Wagner, M., and Gordon, H. A. (1956). Observations on germfree and conventional albino rats after total-body x-irradiation. *Radiat. Res.* **5,** 591.

78. Rodkey, F. L., Collison, H. A., and O'Neal, J. D. (1972). Carbon monoxide and methane production in rats, guinea pigs, and germfree rats. *J. Appl. Physiol.* **33,** 256–260.

78a. Rogers, T., and Balish, E. (1976). Experimental *Candida* infection in conventional mice and germfree rats. *Infect. Immun.* **14,** 33–38.

79. Rogers, W. E., Jr., Bieri, J. G., and McDaniel, E. G. (1971). Vitamin A deficiency in the germfree state. *Fed. Proc., Fed. Am. Soc. Exp. Biol.* **30,** 1773–1778.

80. Sacksteder, M., Kasza, L., Palmer, J., and Warren, J. (1973). Cell transformation in germfree Fischer rats. *In* "Germfree Research" (J. B. Heneghan, ed.), pp. 153–157. Academic Press, New York.

81. Sacquet, E., Lachkar, M., Mathis, C., and Raibaud, P. (1973). Cecal reduction in "gnotoxenic" rats. *In* "Germfree Research" (J. B. Heneghan, ed.), pp. 545–552. Academic Press, New York.

81a. Savage, D. C. (1969). Microbial interference between indigenous yeast and lactobacilli in the rodent stomach. *J. Bacteriol.* **98,** 1278.

81b. Schaedler, R. W., Dubos, R., and Costello, R. (1965). Association of germfree mice with bacteria isolated from normal mice. *J. Exp. Med.* **122,** 77–82.

81c. Schaedler, R. W., Dubos, R., and Costello, R. (1965). The development of the bacterial flora in the gastrointestinal tract of mice. *J. Exp. Med.* **122,** 59–66.

82. Schultz, K. D., Appel, K. R., Goeth, H., and Wilk, W. (1974). Experiments to establish a rat stock free of mycoplasma. *Z. Versuchstierk.* **16,** 105–112.

82a. Schwartz, B. F. (1975). Pulmonary gas exchange in germfree and conventional rats. *In* "Clinical and Experimental Gnotobiotics" (T. M. Fliedner, H. Heit, D. Neithammer, and H. Pflieger, eds.). Gustav Fischer, New York.

83. Sewell, D. L., Bruckner-Kardoss, E., Lorenz, L. M., and Wostmann, B. S. (1976). Glucose tolerance, insulin and catecholamine levels in germfree rats. *Proc. Soc. Exp. Biol. Med.* **152,** 16–19.

84. Sewell, D. L., Wostmann, B. S., Gairola, C., and Aleem, M. I. H. (1975). Oxidative energy metabolism in germfree and conventional rat liver mitochondria. *Am. J. Physiol.* **228,** 526–529.

85. Sewell, D. L., and Wostmann, B. S. (1975). Thyroid function and related hepatic enzymes in the germfree rat. *Metab., Clin. Exp.* **24,** 695–701.

85a. Shih, C. N., Balish, E., Lower, G. M., Jr., Yale, C. E., and Bryan, G. T. (1975). Induction of colonic tumors in germfree and conventional rats. *In* "Clinical and Experimental Gnotobiotics" (T. M. Fliedner, H. Heit, D. Neithammer, and H. Pflieger, eds.). Gustav Fischer, New York.

85b. Smith, H. W. (1965). The development of the flora of the alimentary tract in young animals. *J. Pathol. Bacteriol.* **90,** 495–513.

86. Smith, J. C., Jr., McDaniel, E. G., and Doft, F. S. (1973). Urinary calculi in germfree rats alleviated by varying the dietary minerals. *In* "Germfree Research" (J. B. Heneghan, ed.), pp. 285–290. Academic Press, New York.

86a. Sugiyama, T., and Bruckner, G. G. (1975). Mycoplasma–bacteria relationships in the pathogenesis of chronic respiratory disease in conventional rats. *In* "Clinical and Experimental Gnotobiotics" (T. M. Fliedner, H. Heit, D. Neithammer, and H. Pflieger, eds.). Gustav Fischer, New York.

86b. Syed, S. A., Abrams, G. D., and Freter, R. (1970). Efficiency of various intestinal bacteria in assuming normal function of enteric flora after association with germfree mice. *Infect. Immun.* **2,** 376–386.

87. Thompson, G. R., and Trexler, P. C. (1971). Gastrointestinal structure and function in germfree or gnotobiotic animals. *Gut* **12,** 230–235.

88. Trexler, P. C. (1959). The use of plastic in the design of isolator systems. *Ann. N.Y. Acad. Sci.* **78,** 29–36.

88a. Trexler, P. C., and Reynold, L. I. (1957). Flexible film apparatus for the rearing and use of germfree animals. *Appl. Microbiol.* **5,** 6.

89. Trippestad, A., and Midvedt, T. (1971). The phagocytic activity of polymorphonuclear leucocytes from germfree and conventional rats. *Acta Pathol. Microbiol. Scand. Sect. B* **79,** 519–522.

90. Tuffery, A. A., and Innes, J. R. M. (1963). Diseases of laboratory mice and rats. *In* "Animals for Research" (W. Lane-Petter, ed.), pp. 48–109. Academic Press, New York.

91. Ukai, M., Tomura, A., and Ito, M. (1976). Cholesterol synthesis in germfree and conventional rats. *J. Nutr.* **106,** 1175–1183.

91a. van der Waaij, D., Berghus-de Vries, J. M., and Lekkerkerk van der Wees, J. E. C. (1971). Colonization resistance of the digestive tract in conventional and antibiotic-treated mice. J. Hyg. **69,** 405–411.

92. Wagner, M. (1959). Determination of germfree status. *Ann. N.Y. Acad. Sci.* **78,** 89–102.

93. Walburg, H. E., Jr. (1973). Carcinogenesis in gnotobiotic rodents. *In* "Germfree Research" (J. B. Heneghan, ed.), pp. 115–122. Academic Press, New York.

94. Weisburger, J. H., Reddy, B. S. Narisawa, T., and Wynder, E. L. (1975). Germfree status and colon tumor induction by *N*-methyl-*N*-nitro-N-nitrosoguanidine, *Proc. Soc. Exp. Biol. Med.* **148,** 1119–1121.

95. Wostmann, B. S., ed. (1970). "Gnotobiotes: Standards and Guidelines for the Breeding, Care and Management of Laboratory Animals." Natl. Acad. Sci.—Natl. Res. Counc., Washington, D.C.

96. Wostmann, B. S. (1973). Intestinal bile acids and cholesterol absorption in the germfree rat. *J. Nutr.* **103,** 982–990.

97. Wostmann, B. S. (1975). Nutrition and metabolism of the germfree mammal. *World Rev. Nutr. Diet.* **22,** 40–92.

97a. Wostmann, B. S., Beaver, M., and Modsen, D. (1975). Effect of diet sterilization on cholesterol and bile acid values of germfree rats. *In* "Clinical and Experimental Gnotobiotics" (T. M. Fliedner, H. Heit, D. Neithammer, and H. Pflieger, eds.). Gustav Fischer, New York.

97b. Wostmann, B. S., and Bruckner-Kardoss, E. (1959). Development of cecal distention in germfree baby rats. *Am. J. Physiol.* **197,** 1345–1346.

98. Wostmann, B. S., Bruckner-Kardoss, E., and Knight, P. L., Jr. (1968). Cecal enlargement, cardiac output, and O_2 consumption in germfree rats. *Proc. Soc. Exp. Biol. Med.* **128,** 137–1141.

99. Wostmann, B. S., and Kellogg, T. F. (1967). Purified starch-casein diet for nutritional research with germfree rats. *Lab. Anim. Care* **17,** 589–593.

100. Wostmann, B. S., and Knight, P. L. (1965). Antagonism between vitamins A and K in the germfree rat. *J. Nutr.* **87,** 155–160.

100a. Wosu, N. J., Olsen, R., Shadduck, J. A., Koestner, A., and Pakes, S.

P. (1977). Diagnosis of experimental encephalitozoonosis in rabbits by complement fixation. *J. Infect. Dis.* **135,** 944-948.

100b. Wosu, N. J., Shadduck, J. A., Pakes, S. P., Frenkel, J. K., Todd, K. S., Jr., and Conroy, J. D. (1977). Diagnosis of encephalitozoonosis in experimentally infected rabbits by intradermal and immunofluorescence

test. *Lab. Anim. Sci.* **27,** 210-216.

101. Yale, C. E., and Linsley, J. G. (1970). A large efficient isolator for holding germfree rats. *Lab. Anim. Care* **20,** 749-755.

102. Yamada, K., and Ukai, M. (1976). The histochemistry of mucosaccharides in some organs of germfree rats. *Histochemistry* **47,** 219-238.

Chapter 3

Dental Research

Juan M. Navia and Annie Jo Narkates

I. INTRODUCTION

The rat has been used extensively as an animal model to study the biology and pathology of oral tissues. This species has been useful in dental research to obtain valuable biological information, to improve understanding of basic mechanisms of disease processes, and to facilitate the planning of clinical or epidemiological experiments designed to yield information directly applicable to humans.

Dental research involves a large number of scientific disciplines (52) necessary for the study of a variety of oral tissues. Some of these tissues (such as bone, cementum, dentin, and enamel) are mineralized, while others (such as oral mucosa and periodontium) are soft tissues, and still others are either glandular (such as the salivary glands) or sensory (such as taste receptors on the tongue). All these different tissues are within a cavity which is moist and contains a specific flora in various ecological oral niches. These sites are colonized by large numbers of different types of microorganisms, which may include pathogens responsible for oral diseases such as dental caries and periodontal disease.

The broad objectives of dental research involve the understanding of the factors responsible for normal growth and development of oral tissues, as well as for the maintenance of the health of these tissues through elucidation of the etiology of oral diseases and the development of preventive and therapeutic programs.

Because most oral diseases have a multifactorial etiology and are not influenced solely by isolated effects of specific factors but also by their interactions, it has been necessary to make use of an animal, such as the rat, in which some of these different factors and interactions can be experimentally controlled and evaluated. In addition, while some research questions can be answered directly using human subjects, there are many situations where ethical, moral, and legal considerations make such clinical studies impossible. Therefore, it has become critical to develop well defined animal models to be used by dental researchers (60).

The rat fulfills these experimental needs because (a) a large body of valuable information has been published describing its oral anatomy and histology, the biochemistry and metabolism of oral tissues, and the composition and variations of the oral flora at different sites in the oral cavity; (b) it has been found to be susceptible to every major oral disease described in humans; (c) it adapts well to changes in environment (e.g., gnotobiosis), diet or husbandry; (d) it lends itself to a variety of experimental manipulations, such as saliva and blood collection, topical administration of test agents, dietary manipulations; and (e) the costs of procurement and maintenance are low, thus allowing researchers to conduct experiments with suffcient numbers of replications to yield statistically valid results.

Information obtained with animal models usually cannot be directly extrapolated to humans, but the answers obtained with animal models to properly formulated research questions are extremely useful and valid information. The rat has been successfully used to study, for example, the cariostatic effect of fluoride, and a large amount of information is now available on the effectiveness of various fluoride-containing compounds, types of vehicles, and time of administration, yet the effective cariostatic dose of fluoride for the rat is twenty times higher than for humans (9). The same is true of studies done to investigate the microbial aspects of dental caries, where data obtained using the rat model has clarified to a large extent the identification of cariogenic microorganisms, their different modes of implantation and colonization under different dietary conditions, and their susceptibility to antibacterial compounds (72). Still, there are significant differences between microorganisms normally found in the oral cavity of rats versus those of human mouths.

It is important to point out that other animal models have been developed and characterized and have been used in dental research. Examples of these (60) would be the hamster, which has been used in certain types of periodontal studies, and in investigations involving use of the buccal pouch; the dog in periodontal studies (such as gingivitis) in healing of extractions, transplants, and implants; and nonhuman primates in occlusal traumatism and periodontal disease and many others. A fallacy to avoid, however, is to assume that because nonhu-

man primates are phylogenetically closer to humans than rats or hamsters, the results obtained from nonhuman primates can be directly extrapolated to humans, while those obtained with a rodent cannot. There are still sufficient differences between nonhumans primates and man to preclude such direct extrapolation of information. The scientific method demands that a question or hypothesis be formulated and that this question be tested under unbiased conditions in a suitable model which is clearly defined and understood. The answer obtained from the experiment should agree or disagree with the formulated hypothesis, which would then be accepted or redefined to continue the search for truth. Animal models should be used in this manner. Information obtained using animals will help improve the understanding of the biology of health and mechanisms of disease in man, but will never directly answer a research question such as: Will it work under specific stated conditions for humans? Questions of this type can only be answered in clinical experiments using human subjects.

II. ORAL CHARACTERISTICS OF THE RAT

A. Dentition

Rats are monophyodont rodents, having only one set of teeth, with the formula

$$I\frac{1}{1}M_1\frac{1}{1}M_2\frac{1}{1}M_3\frac{1}{1}$$

Schour and Massler (74) have described in great detail the histogenesis and histophysiology of the rat dentition, particularly, the incisor. The incisors erupt continuously throughout the life of the rat, and characteristically are covered by enamel on the labial surface. The incisor enamel is yellow due to the presence in the enamel of an iron-containing pigment, and the enamel becomes increasingly yellow with age starting at about 21 days after birth. The growth of the incisor is rapid, taking approximately 40 to 50 days to renew itself completely. The first appearance of the dental lamina takes place in the fetus at approximately 14 days after conception, and it erupts into the oral cavity at around 6 to 8 days after birth. The eruption of the rat mandibular incisor has been measured by Chiba *et al.* (16) using a photographic method. Measurements taken at 4- or 6-hr intervals indicated the existence of a circadian rhythm of eruption rate characterized by slow eruption during the night when the incisors were in frequent functional occlusion.

Incisors in rodents are maintained at proper length by constant attrition of the occlusal surfaces which maintains beveled and sharp incisal edges. Failure to limit the length of the incisor due to misalignment of the dentition will result in abnormal and uncontrolled development of the incisor as shown in Fig. 1. These incisors are strong and can support high biting

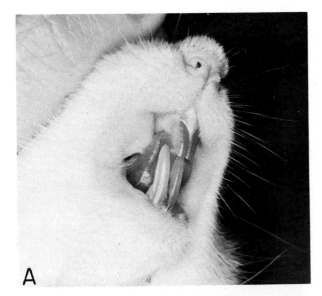

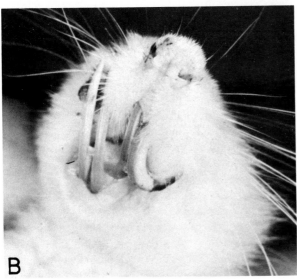

Fig. 1. (A) Normal rat incisors. (B) Misalignment of the rat dentition resulting in abnormal growth of incisor teeth.

loads. Robins (70), using a miniature dynamometer, determined the bitting loads generated by rats and found them to vary between 200 and 4500 gm loads.

The first and second molars develop nearly at the same time, starting approximately the 14th day after conception. Paynter and Hunt (65) have studied the morphogenesis of the rat first molar and found that between 18 and 23 days after conception, the growth of the tooth germ is rapid. This growth rate is reduced with the first appearance of dentine at the tip of the cusps. At this time the spatial relations between the cusps are established, but the depth of the fissure continues to increase for 3 to 4 days more until dentine is formed at the base of the fissure. Completion of the crown is usually achieved by the

tenth to the twelfth day after birth with a 1½ to 2 day difference between the first and the second molars.

Park (64) has studied the morphogenesis of the osteodental fissure formed by the alveolar bony crests surrounding the developing molars. Around the first day after birth, the alveolar bone starts to modify, and, by the sixth day, a completed fissure can be observed obliquely across the long axis of the mandible and oriented directly above the mesiodistal crown fissures of the first and second molars. Approximately 12 days after birth, all major cusps of the first molars are above the level of the alveolar bone crest and root formation progresses rapidly. The growth of the buccal roots produces a thinning of the buccal bone plate, which eventually disappears by the seventeenth day of age. Lester (46) has studied root formation in molar teeth of a strain of Sprague-Dawley rats using thin sections for transmission electron microscopy. He observed that once cellular cementum formation is established, cells of Hertwig's epithelial root sheath are totally embedded between cementum and dentine at the advancing root edge. This developmental effect takes place concomitantly with an increase in the amount and speed of cementum deposition relative to dentine formation at the root edge.

The eruption of the molars in the rat varies depending on the species, the strain, and the nutritional status of the rat. In the Norwegian rat (Sprague-Dawley descended, Crl: COBS, Charles River Laboratory) the first mandibular and maxillary molars usually erupt into the oral cavity around the sixteenth day of age. The second mandibular and maxillary molars erupt 1 or 2 days later, and, by the twentieth day, when nutritionally adequate rats are normally weaned, all four molars have erupted. The third molars erupt sometime around day 32–34 with a 1 day difference between the mandibular third molar and the maxillary molar. Figure 2 illustrates the effect of feeding a low protein diet (8%) to rat dams. The marginal protein-calorie malnutrition imposed on the rat pups due to the limited amount of milk available to them results in a retardation of molar eruption of 1 or 2 days compared to rat pups suckling from rat dams receiving a nutritionally adequate diet containing 25% protein. In some other rodents, such as the cotton rat (*Sigmodon hispidus*), molars erupt much earlier. This rat may wean its young pups as early as 12 to 14 days of age.

Hunt *et al.* (31) have described the morphology and occlusion of rat molar teeth. Rat molars are small in size; three whole molars in a quadrant weigh approximately 40 mg. Molars have a thin layer of enamel, not deeper than 0.1 mm, that does not cover the tip of the cusps. The occlusal surface, therefore, shows exposed dentine, especially after they have been in functional apposition for some days. The mandibular first molar has three fissures (Fig. 3), which are deep, narrow, and extend in a buccolingual direction. The second molar has two major fissures, and the third molar has only one. It is of interest to point out that at eruption rat molars have extensive

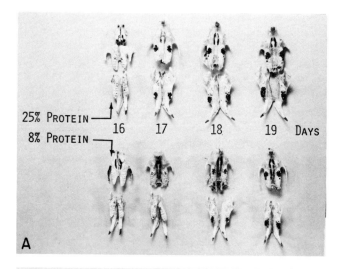

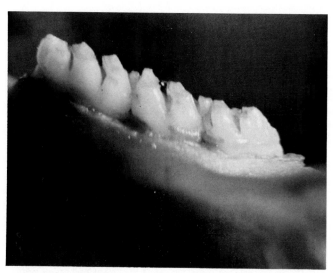

Fig. 3. Mandibular molars showing the sulcal and interproximal areas.

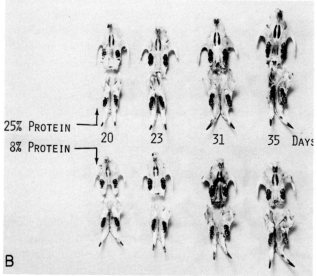

Fig. 2. Eruption pattern for molar teeth of rats (A) 16 to 19 days and (B) 20 to 35 days of age and born to dams fed either a 25% or 8% protein diet. The darkened molars stained with silver nitrate represent the erupted portion of the molar at the stated age (49).

hypomineralized enamel areas on the buccal surfaces and the bases and sides of the fissures, which can be easily disclosed by staining with murexide, alizarin, or similar calcium-binding stains. These hypomineralized surfaces of enamel undergo posteruptive mineralization (maturation) and constitute susceptible areas to dental caries.

B. Salivary Glands and Saliva

The oral environment is to a large extent influenced by saliva, the fluids secreted by several different salivary glands, which constantly bathes tissues in the oral cavity. The washing

and buffering actions of saliva plus its mucins, enzymes, antibodies, and minerals are essential in maintaining the integrity of tissues in the mouth as well as general and oral health. Saliva also has an important role in the formation of the acquired enamel pellicle and the posteruptive mineralization of tooth enamel. Both of these factors have a decisive influence on dental plaque accumulation and calculus formation. Because of this, understanding the biochemistry and physiology of salivary gland function in rats is important to the use of this species as a model.

Salivary glands can be classified according to the type of cells and their secretions into serous, mucous, or combination of both. Rats have three major pairs of glands (16): (a) the parotids (serous) (b) the submaxillary, and (c) the major sublingual (mixed) and a group of minor or lesser sublinguals.

The parotid in the rat has a diffuse appearance, which makes it difficult to differentiate from the fat and lymphoid tissue in intimate contact with the gland. The parotid extends from the masseteric region to the clavicle; a larger proportion of it is found around the ventral and caudal margins of the external ear and extends into the retroauricular area. This gland discharges saliva into the oral cavity through a major duct (Stenson's duct), which is formed by the union of three minor ducts and opens opposite the molar teeth.

The submaxillary and the major sublingual glands in the adult rat are found together forming an almond-shaped capsule situated in the infrahyoid region near the median line. The submaxillary gland is the larger of the two and is clearly visible once the enveloping membrane is removed. The submaxillary gland secretes saliva through Wharton's duct, which opens near the plica sublingualis. The major sublingual gland secretes through an orifice that is usually slightly caudal to that of the submaxillary. The minor sublingual glands are a group of true simple glands usually located on the floor of the mouth with openings on either side of the plica sublingualis.

The parotid in rodents is essentially a serous gland containing a large amount of amylase. Trauma induced by X radiation and the resultant inanition have been found (81) to alter amylase concentration in rat parotid saliva. The cells of this gland will stain with periodic acid–Schiff (PAS), but do not stain with alcian blue. Goblet cells present in the ducts will also stain with PAS. The sublingual glands of rodents are formed mostly by seromucous acini.

Tamarin and Sreebny (86) studied the structure of the rat submaxillary salivary gland and observed five distinct parenchymal zones.

1. Acini. These are the secreting cells, interspaced with myoepithelial cells, which are interconnected by canaliculi to the main lumen of the acinus. A large amount of endoplasmic reticulum Golgi apparatus and secretory material characterizes these secretory cells.

2. Intercalated ducts. These are made up of cuboidal cells and connect the acini to the main ducts of the gland.

3. Granular ducts. These tubules are made up of three types of columnar cells: (a) dark narrow cells, (b) light granular cells, and (c) dark granular cells. The presence of discrete granules characterizes this portion of the gland.

4. Striated ducts. The granular ducts merge into striated ducts lined by tall columnar cells containing little endoplasmic reticulum, few or no granules, and a large number of mitochondria. This portion of the gland is not secretory, but participates actively in the metabolism of water and electrolytes.

5. Excretory ducts. These ducts are formed by tall columnar cells, some of which may be dark and others light. This portion of the gland is probably primarily concerned with water transport.

Jacoby and Leeson (32) have described the postnatal development of the rat submaxillary gland. At the time of birth, the rat gland has no detectable acini, but the intralobular ducts are to some extent differentiated and end in a branching system of terminal tubules with cells containing granules. During the first and also the second weeks, acinar cells start to bud off from the tubules, forming crescents around these terminal tubules. Large numbers of acini and terminal tubules with crescents and buds are seen during the third and fourth weeks. This period is crucial in rat development because they usually are weaned from their dams sometime around the age of 21 days. After 5 weeks the intercalated ducts start to proliferate, and at 6 weeks the phase of acinar development and the regression of the terminal tubules is nearly completed. Differentiation of intralobular striated ducts appears at this time, together with an increase in mitotic activity in the duct cells and a diminished activity in the acini. No sex dimorphism was observed in the glands of rats until the seventh week, when ducts and tubule diameter increase in size more in males than in females. The final stage in the development of the gland is the transformation of the striated ducts into granular tubules. The elaboration of secretion granules continues for a long time, and thus it may still be incomplete even after 4 to 5 months.

Schneyer and Hall (73) studied the growth pattern of the parotid gland of the rat by following the size and weight as well as the DNA and RNA content of the gland. During the first 3 weeks after birth, there is active cellular proliferation, as indicated by a high mitotic index and the concomitant increase in total gland DNA. The parotid, similar in development to the submaxillary, undergoes differentiation of the terminal tubule cells into definite acinar cells during this early period. After weaning, the major activity of the cell is an increase in size with a parallel reduction in mitotic rate, and a slower increase in DNA content. After 10 weeks of age, the parotid weight, cell size, state of cellular differentiation, and cell number do not appear to increase further.

In certain studies, it has been found necessary to extirpate salivary glands either partially or totally. A good three-step surgical procedure has been described by Cheyne (15). An additional technique of interest is the collection of saliva from rats. According to Bernarde et al. (6) large quantities of whole saliva can be collected from rats by anesthetizing them with sodium pentobarbital (20 to 30 mg/kg of body weight), followed by subcutaneous injection of pilocarpine nitrate at a dose level of 3 to 5 mg/kg body weight. The rats are then placed on an inclined board with the mouths protruding over beakers into which saliva is collected. Collection time and time of day must be carefully controlled and standardized to avoid variations in the composition of saliva.

Ericsson (20,21) reported a pH for rat saliva ranging from 8.5 to 9.0, which is in agreement with other investigators (94). Total protein has been found to be higher in rodent compared to human saliva. Whole rat saliva has been reported to contain 1841 mg protein/100 ml (85), but protein concentrations decrease with increased collection time; therefore, shorter collection periods will yield higher concentrations of protein concentration. Concentrations of tyrosine and tryptophan were found to be two and six times higher, respectively, in the rat than in human saliva.

Rat saliva contains a large number of proteins as shown by studies using paper electrophonesis (85) and polyacrylamide gel electrophoresis (67, 68). These include serum proteins as well as salivary gland proteins, such as enzymes (14) and immunoglobulins (30). Enzymes found in rat saliva include, among others, acid and alkaline phosphatases and a Ca^{2+}-stimulated adenosine triphosphatase. Salivary gland secretions contain also glycoproteins, lipids, vitamins, and hormones plus certain minerals and electrolytes including phosphorus, magnesium, iodide, sodium, and potassium.

Feeding rat a liquid diet or a finely powdered ration rather than a pelleted feed eliminates masticatory activity and causes a retention of secretory material in the parotid gland, which is associated with an initial enlargement followed later by atrophy of the gland (33, 82). Recent reports (34) suggest that lysosomes play a major role in the regulation of the gland components and in the subsequent atrophy resulting from lack of mastication.

C. Oral Flora

The definition and description of the oral flora of the rat is extremely important to its use as an animal model in dental research. The type and numbers of microorganisms will vary depending on (a) anatomical site in the oral cavity, (b) colony management, (c) type of diet, (d) rat age, (e) degree of coprophagy, and (f) presence of tissue inflammation or disease.

The oral cavity cannot be considered as a single habitat, but rather as a group of different ecological niches where specific microorganisms will implant and colonize. The dorsum of the tongue, oral mucosa, gingival crevice, smooth surfaces of molars, and deep fissures of molars are all distinct anatomical sites where a characteristic flora will implant and grow. The microbial colonization at these different sites will be determined by the environmental conditions predominating in these areas, but will also be influenced to some extent by bacterial opportunism which could determine which organism will first implant and develop at a site. This initial flora has an ecological advantage through availability of specific dietary residues and environmental conditions and will in turn influence the local environment and the type of bacteria that will continue to colonize the specific anatomical niche. This in part may explain why certain cariogenic streptococci implant successfully on the enamel surfaces of molars, while others fail to do so. It may also explain the preferential growth of strict anaerobes in gingival crevices rather than on the dorsum of the tongue.

Some of the main nutritional and environmental factors that contribute to these differences in microbial colonization include (36) (a) food debris, (b) oxygen availability; (c) volume and composition of saliva; (d) structural characteristics of the site: roughness, fissures and folds, presence of calcium phosphate salts (enamel); (e) gingival or pulp fluids; (f) immunological factors; (g) desquamated epithelial cells; (h) microbial metabolic products (including accummulation of compounds produced by the organisms and byproducts of the metabolism of adjacent colonies); (i) environmental temperature; (j) hydrogen concentration (the pH may vary significantly due to the interaction between the buffering capacity of saliva and the acids and bases produced by fermentation of food residues); and (k) osmotic pressure.

Variations in any of these factors influence colonization.

Changes in the diet, for example, influence the composition of the food residues available to the oral flora, and thus, through specific enrichment, a selection of oral flora will take place. The same is true for gingival inflammation which determines the volume of gingival fluid, and thus modifies the nutritional and environmental conditions at the gingival crevice.

Bacterial sampling of the rat dentition for the purpose of determining bacterial counts and composition are usually done using procedures which allow for collection of bacteria from specific sites, e.g., scrapings, sonication of extracted individual molars or quandrants, or sampling by using cotton swabs rolled over molars and tongue of the rat. These samples are streaked on differential agar or vigorously agitated in a liquid growth medium. The types of bacteria collected using these different approaches may differ significantly. Streptococci are most frequently encountered in bacterial plaque collected from molar surfaces of rats fed a sugar-containing diet. Of these *S. mutans, S. sanguis,* and *S. mitis* are the most common; *S. salivarius* is frequently found, but usually predominates on the dorsum of the rat tongue and in saliva. *Lactobacillus casei* and other *Lactobacillus* also are found in the oral cavity of rats, particularly in animals fed diets containing sugar. *Actinomyces, Neisseria,* and *Enterococcus* also are frequently found in the oral cavity of rats. Enterococci, particularly *S. faecalis,* are characterized by their broad fermentative abilities, and, in some rats, they constitute a large proportion of the bacterial population associated with tooth surfaces. Members of the family Enterobacteriaceae are frequently found when tooth debris samples are grown on eosin-methylene blue (EMB) agar. The presence of these enteric organisms is associated with the practice of coprophagy and seem to be more noticeable in the adult than in the weanling rat. Yeast are found occasionally in tooth scrapings, but in general they are present in low numbers. Staphylococci also are found occasionally in high numbers. In some rats (also, in mice) they seem to implant and colonize readily and may represent a large proportion of the total flora.

III. THE RAT AS AN ANIMAL MODEL FOR THE STUDY OF ORAL DISEASE

A. Experimental Dental Caries

The etiology of dental caries in both humans and rats involves microbial, host (teeth, saliva), and dietary factors. The multifactorial etiology of the disease makes it mandatory that all three factors be present in the appropriate combinations for the disease to be expressed. If, for example, the microbial flora is depressed with an antibiotic or if the intake frequency of a sugar-containing diet is reduced, the severity of caries is proportionally reduced, even though the other etiological factors are present.

Although all rat species are susceptible to the induction of caries, provided they are given a cariogenic challenge, some differences in the severity of lesions have been reported (43) to be dependent on rat strains with different dietary patterns.

1. Types of Carious Lesions

The two types of carious lesions predominating in rat molars are listed below.

a. Smooth Surface Lesions. These are frequently observed on the buccal surface of mandibular and maxillary molars, but can involve other areas too, if the disease is severe. This lesion develops first as a white spot close to the gingival margin, which in the rat is high on the molar crown. It later extends to cover the whole length of the molar and the enamel easily crumbles if probed with a sharp instrument (Fig. 4).

b. Approximal Lesions. These lesions can develop at the enamel walls in the deep fissures of a rat molar (sulcal) or at a contact point between adjacent molars (interproximal). These sites differ from those where smooth surface lesions develop in that they are exposed less to the oral cavity environment, and they constitute more retentive sites that easily trap food debris. The sulci constitute special ecological niches where certain microorganisms can develop in isolation from outside influences (Fig. 5).

2. Microorganisms Used to Induce Caries in Rats

From an etiological point of view, the microbial flora constitutes the factor directly responsible for caries. Even though the

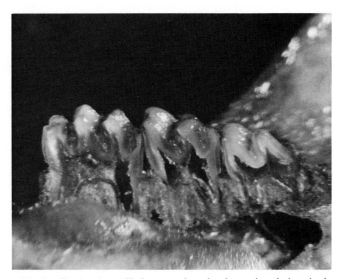

Fig. 5. Sectioned mandibular rat molars showing carious lesions in the aproximal surfaces.

role of bacteria in dental caries had been recognized for a long time (50, 51), their roles were not clearly understood until Orland *et al.* (63) did their classic experiments showing that germfree rats did not develop caries, and that caries develop when rats were inoculated with cariogenic microorganisms. Recently, the microbial aspects of dental caries were discussed in comprehensive reports edited by Stiles *et al.* (84).

The microorganisms most frequently found in dental plaque are members of the genus *Streptococcus*. Fitzgerald *et al.* (23) showed that a single strain of streptococci isolated from rats could produce caries when used to infect germfree rats. This particular strain has been designated as *S. mutans* FA-1. Several other streptococcal species also have been isolated from dental plaque and from other sites in the oral cavity including *S. sanguis, S. mitis,* and *S. faecalis* and its varieties *S. liquefaciens* and *S. zymogenes.* Some of these microorganisms may exhibit special interactions with other microorganisms, and their effects are related to these symbiotic relationships. Such relationships have been described clearly for certain microorganisms found in the rat such as the *Lactobacillus, Veillonella,* and *Neisseria.*

Several organisms in addition to streptococci have been found to be cariogenic (71). Among them a filamentous organism, *Actinomyces viscosus,* is found frequently in rats. Plaque formation and caries result when gnotobiotic rats fed a glucose-containing diet are inoculated with these organisms (88, 89).

The type of cariogenic bacteria predominating on molar surfaces and fissures varies depending on several factors previously discussed, including the trace element composition of the diet. For example, Beighton (5) reported that cadmium significantly depressed the numbers of *S. mutans* in fissure

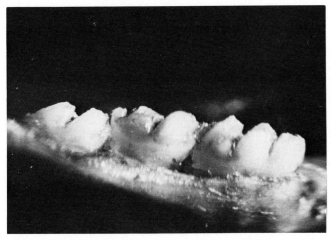

Fig. 4. Carious areas on the smooth surface of mandibular rat molars.

plaque flora and increased the Actynomycetaceae considerably.

Rat pups to be used in caries experiments are usually offered the caries-promoting diet early during lactation. Occasional feeding of this diet at this time modifies the oral environment and either selects out and increases the number of cariogenic organisms already present in the oral cavity of the rat pup or facilitates implantation and colonization of the inoculated cariogenic bacteria. Experimental oral infection with cariogenic bacteria is necessary when working with rats that are not naturally infected. Infection may be accomplished by one of the following procedures:

1. Transmission of cariogenic organisms from a caries-active animal. This method is best accomplished by taking fresh fecal pellets from a rat known to have active carious lesions, and preparing an aqueous suspension that is used to swab the teeth of the recipient rat. A few milliliters of the fecal suspension also can be added to the drinking water, but water contaminated in this manner should be prepared daily. Two or three inoculations on consecutive days may be sufficient to implant a cariogenic flora in the oral cavity. However, this method occasionally is inefficient, and the nature of the cariogenic flora is unknown (61).

2. Caging of caries-active rats with rats to be infected. This procedure, although less effective than the preceding one has also been used with some measure of success.

3. Inoculation with a cariogenic strain of microorganism. Strains of *S. mutans* (or other cariogenic microorganism of high virulence) are grown in broth media. A cell suspension from a 24-hr culture is used to swab the molars of rats and also is added to the drinking water. This procedure is probably best, as it gives greater control and more complete knowledge of the organism being used to induce caries.

The success of such infective procedures is based on the type of organism and bacterial concentration in the inoculum used, the feeding of a caries-promoting diet, the age of the animal, the resident flora at the time of inoculation, and the frequency with which the inoculating procedure is repeated. Cultures of *S. mutans,* containing 10^7 or 10^8 CFU/ml, generally implant readily when the rat is fed a caries-promoting diet containing more than 5% sucrose. Consecutive daily inoculations for 2 or 3 days at the time when molars are erupting (17 to 20 days of age) ensure the development of more smooth surface lesions than if the inoculation is done a week later (59).

3. Experimental Caries-Promoting Diets

a. General Considerations on Dietary and Nutritional Factors Affecting Caries. The diet fed to rats used in caries experiments must fulfill two important nutritional roles: First, it must provide the necessary nutrients to fulfill the nutritional requirement of the host, and, second, it should supply a substrate for the oral microflora. The nutrient composition of these experimental diets contributes to the nutritional requirements of the developing rat pup through the mother during gestation or lactation or directly after weaning. In the rat, tooth formation takes place during gestation and lactation, and the nutritional requirements of protein and other nutrients are especially high during formation of these oral tissues. These nutritional influences of the diet extend into the molar posteruptive period, when food components may still play a role in the maturation of the enamel surface exposed to the oral environment and mediated through saliva.

Diets also play a major role in providing nutrients that oral bacteria require for growth and metabolic activity. Diet residues can be held in retentive areas of the oral cavity and serve as a nutritious substrate, thus facilitating the implantation and growth of specific cariogenic microorganisms. This enrichment effect will facilitate the increase in numbers of these organisms, and the consequent increase in metabolic activity may help initiate carious lesions, if the cariogenic microorganism is sufficiently virulent. In experimental animals, these effects of diet on the selection and growth of the oral flora can take place on recently erupted molars of weanling pups, or in the oral cavity of the dam that is nursing the offspring, to whom she can transmit a selected flora.

Studies done with animals fed nutritionally inadequate diets are undesirable, because they confound the facts to be studied. In animal studies where nutritionally poor diets were used, and where a reduction of caries was observed, no conclusion can be drawn on whether the low caries severity resulted from lack of a nutrient, a reduced diet intake due to loss of appetite, or the lack of stimulation of salivary glands. Caries research should be conducted using diets that stimulate optimal growth and maintenance of the experimental host and at the same time select and maintain the cariogenic flora responsible for the formation of the carious lesions. Dental caries experiments using animals will be meaningful and effective only to the extent in which they fulfill this condition.

Several investigations in man and experimental animals (27, 41,44,79) have shown that caries activity is related to frequency of food intake. König *et al.* (43) have shown that the Osborne-Mendel rat strain, which has a high caries susceptibility, consumes more food and eats more frequently than the less caries-susceptible NIH black rat. When both strains of rats were fed with the same high-frequency eating pattern in an automatically controlled feeding machine (42) the black rats increased the fissure caries scores tenfold, whereas Osborne-Mendel rats maintained the high caries scores regardless of whether the diet was fed *ad libitum* or automaticlaly controlled in the feeding machine.

Aside from nutritional considerations, there also are organoleptic properies (taste, smell, texture) of caries-promoting diets which may influence the eating and drinking pattern of the animal and thus affect the caries process. Mühlemann and König (53) investigated the cariogenic effect of supplementing a diet fed to rats that contained 66% purified cornstarch with 2% NaCl. The total amount of diet consumed was found to be the same whether supplemented or not, but the eating pattern was different. Rats fed the sodium chloride-containing diet spent considerably more time consuming the diet, thus the time of exposure to the caries-promoting diet was longer and could be responsible, among other factors in the experiment, for a slightly higher severity of smooth surface lesions in this group of rats. Other manipulations, such as the addition of sucrose (5 to 10%) to the drinking water, will induce a two- and threefold increase in the amount and frequency of water consumed and a concomitant increase in caries severity.

Substances such as sodium selenite (24, 28) can be detected by rats and hamsters when added to diets and may drastically affect eating and drinking patterns. Testing of the caries effect of chemical compounds and other substances should be done after evaluating their acceptance or rejection by the experimental animal.

Variations in the texture or particle size of the rat diet ingredients also may influence caries. König (40) studied the effect of particle size of corn and sugar diets on the caries incidence in rats and observed that diets containing finely ground corn were more caries conducive than those containing coarse ground corn. Caries-promoting diets are generally dry and finely powdered. Addition of small amounts of water or preparation of the diet in a gel form, using agar, for example, will eliminate or reduce substantially its caries-inducing potential, particularly on the buccal surfaces of molars.

b. Formulations for Caries-Promoting Diets. The early work on caries-promoting diets has been discussed by Navia (54). After some attempts by early investigators to formulate a caries promoting diet, Stephan (83) developed a low fat–high sugar diet designated No. 580 made up of skim milk powder (32%) sucrose (66%), and liver powder (2%), which produced smooth surface lesions in rats consuming this diet. Diet No. 580 led to the development of several purified caries promoting diets commonly used today, including the following shown in the tabulations below.

Diet 2000 component[a]	Amount (gm)
Whole wheat flour	6
Sucrose	56
Skim milk powder	28
Alfalfa leaf meal	3
Liver, whole	1
Yeast, brewers'	4
Sodium chloride, iodized	2

[a] From Keyes (39).

This diet has been fed to rats and hamsters. The latter seemed to do better if they were also fed a supplement consisting of a mixture of apple, carrot, and kale in the form of a puree.

Diet 2700 component[a]	Amount (gm)
Sugar (granulated)	67
Casein	24
Corn oil	5
Cellulose	15
Liver, whole	4
Salt mix (Phillips-Hart)	4

[a] From Shaw (77).

A variation of this diet (2700S) containing 25 gm of sucrose also has been used. The Phillips-Hart salt mixture is slightly modified by increasing the levels of NaCl, copper, and zinc and by addition of cobalt.

Diet MIT 200 component[a]	Amount (gm)
Sucrose (pulverized 6×)	67
Lactalbumin	20
Cottonseed oil	3
Cellulose	6
Vitamin mixture	1
Salt mixture (MIT)	3

[a] From Navia et al. (56).

This diet can be modified into a low sugar-containing diet (No. 305) by replacing 62% of the sucrose in the diet with cornstarch, leaving a total of 5% sugar in the diet. The cellulose serves a double purpose as a source of fiber in the diet, and also to facilitate additions or supplementation of test substances at the expense of this inert ingredient. The formula for Diet No. 305 is shown in the tabulation below.

Diet MIT 305 component	Amount (gm)
Sucrose	5
Cornstarch	62
Lactalbumin	20
Cottonseed oil	3
Cellulose	6
Vitamin mixture	1
Salt mixture (MIT)	3

The composition for vitamin and salt mixtures of diets No. 200 and No. 305 are listed in Table I. These two diets are found to promote caries on all molar surfaces of rats: buccal, sulcal, and interproximal areas.

Recent studies by Van Houte et al. (90) indicate that dietary sucrose in concentrations varying from 1 to 56% aid *S. mutans*

Table I

Composition of the Salt Mixture (MIT No. 1) and vitamin Mixture Used
in the Purified Caries Promoting Diet MIT 200[a]

Salt mixture MIT No. 1	Mixture (gm/kg)	Vitamin mix	Diet (gm/kg)
$CaHPO_4$	491.06	Thiamin · HCl	0.008
KCl	141.86	Riboflavin	0.006
NaCl	232.85	Pyridoxine · HCl	0.003
K_2SO_4	61.83	Vitamin B_{12}	0.030
		(0.1% in mannitol)	
$MgSO_4$	19.82	Niacin	0.020
$MgCO_3$	26.27	Ca pantothenate	0.030
Ferric citrate	18.00	p-Aminobenzoic acid	0.200
$MnCO_3$	3.63	Inositol	2.000
$CuCO_3$	0.53	Biotin	0.0001
$ZnCO_3$	2.70	Menadione	0.003
KIO_3	0.05	Folic acid	0.002
$AlCl_3 · 6 H_2O$	0.29	α-Tocopherol acetate	0.500
$CoCl_2 · 6 H_2O$	0.09	Ascorbic acid	0.100
		Vitamin A acetate	0.06
		(500,000 IU/gm)	
		Vitamin D_2	0.006
		(500,000 IU/gm)	
		Choline HCl	3.00

[a] From Navia et al. (56).

in its colonization of rat teeth. At lower sucrose concentrations or when glucose was exchanged in the diet for sucrose, implantation was impaired. Once the *S. mutans* was implanted, feeding a diet containing no sucrose to the rats did not result in elimination of the bacteria from the molars. Thus, sucrose is important, although not absolutely essential for colonization of *S. mutans*.

4. General Procedures for Induction of Experimental Caries

Several strains of rats have been used including: NIH black, NIH Sprague-Dawley, Osborne-Mendel, Charles River SD-COBS, and Wistar rats among others (61). Because adult rats do not readily develop carious lesions on all molar surfaces, most experiments use weanling rats, and in some experiments even younger animals are used initially in the experiment.

Rats may be housed in either suspended cages or solid-bottom cages. Bedding such as sterilized wood chips or shredded cellulose has been used in the latter cages. It is recommended that the bedding used contain no bactericidal or antibacterial compounds, since they may affect the oral flora and decrease the number of cariogenic bacteria. The practice of coprophagy is somewhat limited when rats are housed in wire-bottom cages which are suspended over trays containing sawdust and other absorbing materials to trap urine and feces. Housing in plastic cages has been found to increase the inci-

dence of rat caries compared to housing of rats in suspended wire-bottom cages (29). This effect could not be explained by the ingestion of zinc from the cage, because supplementation of the diet with this element did not significantly affect caries incidence (55). Recent studies (75) indicated that caging of rats (two per cage) in stainless steel, wire bottom cages resulted in a higher caries severity compared with rats housed in plastic shoe box-style cages with hardwood chip bedding. Reasons for these discrepencies have not yet been resolved, but serve to illustrate the importance of apparently minor environmental factors on experimental results.

Cages should be clearly labeled and randomly positioned on the rack to avoid having all animals in one group clustered at one location. Light, temperature, air drafts, and dust concentration may vary at different sites, and nonrandomized location of experimental subjects may introduce undesirable biases. Color-coded tapes on cages, diet containers, or water bottles help in identifying treatment groups and avoiding costly mistakes.

If the experiment involves the use of young rat pups, they should be housed as litters in plastic cages with bedding. If weanling rats are used, three can be housed during the first 2 weeks of the experiment in a cage with the following dimensions: 17⅞ × 12⅞ × 6⅝ inches. Subsequently, groups of two are found to be better than rats caged singly.

Environmental lighting has been found to affect caries incidence in the cotton rat (22). Artificial illumination providing 12 hr of light and 12 hr of darkness is highly recommended.

Rat pups should be handled using disposable gloves, to minimize possible transmission of undesirable oral bacteria and to prevent imparting odors to the young which could upset the dams and induce cannibalism. Manipulations involved in weighing litters and changing the bedding contributes to making the animals more tractable and easier to handle. Monitoring body weight gains is also important to follow and evaluate the health and development of the litter.

Rats are offered the caries-promoting diet in glass or stainless steel jars and distilled water in glass bottles with stainless steel drinking tubes. In some laboratories tap water is used, but reproducibility of results is improved when distilled water is used.

Early weaning of rat pups tends to increase the severity of caries (59). Although rats are usually weaned at 21 days of age, early weaning at 18 to 19 days forces the pup to consume the caries-promoting diet and facilitates the implantation of cariogenic bacteria which are used at this time to inoculate the erupting molar surfaces. To facilitate early weaning, it is best to remove the dam and maintain the litter together for at least 8 to 10 days. This also tends to cross-infect pups with the cariogenic microorganism and yield a more uniform caries response.

Treatments applied to experimental groups may consist of

addition of test supplements to diets or to the drinking water, topical applications to the molars, or delivery of substances by gavage. Whatever the procedure, care should be taken to ensure that the treatments are applied uniformly to the rats and that their health and development is not impaired.

During treatment, rats should be weighed weekly to monitor their growth and development. This handling also offers an opportunity to check them individually for signs of illness or toxicity due to the treatment. Activity, condition of hair coat, and body weights are external factors useful in following the response of the animals to treatments.

Diet and water should be checked daily. Water bottles and diet remaining in the cup should be discarded and changed completely every 2 days. The night before rats are to be euthanatized, diet should be withdrawn leaving only water, to avoid accumulation of food debris in molar fissures. At termination of the experiment, body weights should be recorded.

Final body weights are influenced by the treatment and also by the number of rat pups in a litter, which affects the nutrition of the pups. Usually the number of rats left in a litter is 8 or 9. Litters of 10 or 12 usually contain rat pups with low body weights that may respond to the treatment abnormally. Studies done in our laboratories (58) have shown clearly the detrimental effect of limiting milk supply early in development. Low protein–calorie intake affects not only body weight and molar size but also caries susceptibility. It is essential to maintain equal sized litters if variability of results is to be kept low.

The experimental period may extend from 5 to 50 days after weaning, depending on experimental objectives. Normally, not more than 10 to 15 days are required to induce definite lesions on the buccal, sulcal, and proximal areas. If the experimental period is short, i.e., less than 20 days, then the third molar is not included in the caries estimation, as it erupts in most rats around 32 to 34 days of age. Longer experimental periods allow the third molar to be exposed to caries attack, but usually the caries attack is so severe on the first and second molars, that carious lesions merge on the tooth surfaces, and the ability to evaluate where the lesion originated is lost. Short experimental periods are usually recommended, except when experimental designs make it mandatory to have long periods.

At the termination of the experimental period, rats should be euthanatized by a humane procedure such as an overdose of nembutal (4) and the heads and other organs dissected for evaluation as required by the experiment. Procedures that may be used to prepare the molars for caries scoring include the following:

1. Steam autoclaving without pressure (212°F) for 3 to 5 min, depending on age of the animals. This allows easy removal of soft tissues from mandibles and maxillae. The defleshed skulls are than air-dried prior to staining.

2. The heads are skinned and immersed in either 90% alcohol or 10% buffered formaldehyde. This procedure fixes the tissue allowing for inspection of molars *in situ*. The use of formaldehyde is essential if the jaws are to be stained with PAS stain.

3. Heads are skinned, placed into trays with compartments and placed in special racks under a hood, where they are left to be cleaned by dermestid beetles (*Dermestes vulpinus*). The jaws are completely free of soft tissue in about 3 to 5 days.

Several procedures have been described to score the carious lesions developing on the buccal, sulcal, and interproximal surfaces of rat molars (91). Possibly, one of the most useful and widely used procedures is the one described by Keyes (37, 38). Some investigators use only one quadrant rather than all four molar quadrants, which Briner (7) reported to be the best experimental unit to use in order to avoid false conclusions, since the ranking of other subunit combinations may vary significantly.

5. Special Experimental Procedures Used for Testing the Cariogenicity of Foods

Awareness of the diet–caries relationship is essential to the proper manipulation of the animal model. If cariostatic or cariogenic properties of snack foods, for example, are to be evaluated in animal models, then it is essential to define the following experimental conditions (57): (1) feeding frequency (eating and drinking pattern); (2) salivary function (flow and composition); (3) structural, physical, and chemical properties of tooth enamel; (4) type of carious lesion (location and degree of involvement); (5) relative proportion of fermentable carbohydrates in the basic diet (avoiding substitutions at the expense of this component); and (6) cariogenic tooth flora (if the inhibitory or stimulatory properties of foods on specific microorganisms are being investigated).

Few studies have been made in which all of these experimental criteria are followed, and, therefore, many of the results reported are confusing and inconsistent.

Nutrients in foods may stimulate or inhibit caries in rats posteruptively, depending on their physical, chemical, or organoleptic properties (57). Although all these properties of foods can influence the severity of caries in experimental animals, they do not necessarily affect the caries process in humans in the same way. Thus, it is obvious that the response of the animal model may be different from the human, which it is suppose to imitate, even though the carious lesions and the microorganisms associated with them have been shown to be identical in many instances. Tests of the cariogenicity of foods and nutrients represent special cases. If they are to be relevant to the human situation, they must be designed in such a way that the caries effect is not due to a physical or organoleptic side effect, such as a change in the diet intake pattern of the rat rather than to the actual biochemical properties of the food or compound being tested.

These tests should not (if possible) be made exclusively with one animal model or with one cariogenic microorganism with specific biochemical properties, because the results will be biased by the specific conditions used in the test. The effects of test foods being evaluated could be different for humans because the food is consumed in a different manner and it will exert its effect on a broader cariogenic flora with different requirements. It is also important to use more than one type of carious lesion to evaluate the effect of the food because different lesion sites will respond differently to the food that is being tested, depending on the flora that has colonized that particular niche.

Caries tests, therefore, should be performed in different animals models, with different types of conventional or gnotobiotic flora, evaluating different surfaces for carious lesions, and with some control of the feeding and eating pattern of the animal. Testing of foods as a between-meal snack in a carefully programmed diet schedule provided by automatic equipment such as the König feeder has been found useful (61). Microbiological analysis should be done frequently during the experimental period to monitor the flora associated with the carious process. Because few studies have been done considering these factors, there is a dearth of reliable information on the cariogenicity of foods evaluated in animal models.

6. Other Types of Caries Studies Using Rats

The rat also has been used in immunization experiments and has been found to be an adequate model to study the effect of immunization on the pathogenesis of *S. mutans* (49,80). Immunization was performed by stimulating the production of salivary secretory antibodies. Because some of these studies challenge the salivary glands directly, Tanzer (87) has pointed out some possible pitfalls relative to the pattern of food and water consumption by these stressed rats, which may bias the results.

B. Experimental Calculogenesis

The rat has been used in studies involving the production of calculus deposits on molar surfaces. Baer (3) has discussed in detail the use of rats in these types of experiments. In a series of papers (2), many of the factors that may enhance or retard the formation of calculus in the laboratory rat have been investigated. As with caries, interactions between salivary, microbial, and dietary factors determine the presence of calculus formation. Therefore, most approaches to stimulate calculogenesis are based on modifications of these factors (62).

Briner *et al.* (8) has described an experimental approach using rats to induce the formation of calculus and to measure it in a reproducible manner. The approach involved feeding albino rats (Rolfsmeyer strain) a calculogenic diet RC16 containing cornstarch (50 gm); nonfat milk solids (32 gm); liver powder (3 gm); cellulose (5 gm); cottonseed oil (1 gm); sucrose (5 gm); $CaCl_2 \cdot H_2O$ (1 gm); $NaH_2PO_4 \cdot H_2O$ (2.7 gm); and $MgSO_4$ (0.3 gm). The diet is, therefore, a high carbohydrate diet containing a large amount of phosphate, calcium, and magnesium. Body weight gains observed with this diet were comparable to those obtained in rats fed a commercial stock diet.

Scoring of the amount and distribution of calculus was performed on jaws which had been autoclaved and stripped of soft tissue. The quandrant was divided into 11 areas and a severity value assigned, such that 0 represented no calculus, 1, barely detectable deposit with up to 25% of the area thinly covered; 2, up to 50% covered; and 3, 50 to 100% of the molar surface covered with calculus. The maximum score for 11 areas in 4 quadrants would be 132. Using this approach, rats fed the RC16 diet developed calculus accumulations in 3 weeks which scored between 50 and 60.

The formation of calculus deposits on molar surfaces of rats is the result of shifting the dynamic equilibrium that exists in the oral cavity between those agents that tend to dissolve the enamel and those that stimulate mineral deposition to the latter. The type, amount, and composition of saliva is probably a major determinant, but other factors such as the presence of mineral elements, (e.g., magnesium) or specific microorganisms (*B. matruchotii*) also may contribute to the mineralization process by stimulating the formation of hydroxyapatite intra- and extracellularly (19, 92).

C. Growth and Development Studies of Oral Tissues

The rat has been found to be useful for the study of growth and development of oral tissues because of its short gestation time (21 days) and also because rat pups are quite immature at birth. Therefore, teratological effects of drugs administered to pregnant rats can be easily studied in the offspring, and, also, direct intervention can be done in the very early stages of development to study the developmental or clinical consequences of such direct treatments on the rat.

It is well known now that the growth of organs usually proceeds in a definite pattern (93), characterized by three principal stages: (a) an increase in the total number of cells, (b) an increase in number and size of individual cells, and (c) an increase in cell size only (hypertrophy alone). In the rat, the growth and development of oral tissues undergoes these three phases during the perinatal periods. Adequate nutrition is essential during these early stages of development.

If nutritional stresses (deficiences, excesses, or imbalance of nutrients) are enforced early in the development of organs when they are in the hyperplastic stage, the effect is largely irreversible. The effect of such stresses enforced in the final stages of development, however, can be corrected and do not have the same significance in terms of overall development.

Paynter and Hunt (65) have performed careful studies of molar tooth development in the rat using model reconstruction of unerupted upper first molars and autoradiographs of teeth from rats injected with tritium-labeled thymidine. The initial dental laminae appeared around the twelfth to the fifteenth day after conception and continued to develop until dentine appeared at the tip of the cusps. Dentine was histologically visible in the distal and central cusps by day 22 and in the mesial cusp 24 hr later. Formation of enamel followed an identical pattern to that of dentine with a 48-hr lag behind dentinogenesis. In the rat, the crowns of molars are completed usually by the eleventh day of age, but the mineralization (maturation) of enamel continues on for several days after eruption which starts at the sixteenth day of age and finishes at 19 to 20 days of age. Nutritional (48, 49) or pharmacological (69) interventions can be done at different times to affect the development of the molars.

A genetic mutation in the Osborne-Mendel rat has given rise to an autosomal recessive animal with osteopetrosis *tl* characterized by unerupted incisors and molars, delayed growth, normal longevity, periorbital encrustation, and increased bone deposition (17, 18). The *tl* strain of rat is similar to the *ia* rat described by Greep (25) with the exception that the osteopetrosis of the latter seems to be transitory and spontaneously regresses by 100–120 days of age. The *tl* rats (designated TLOM/Ndri) may not only be of interest for the study of tooth eruption but also for bone studies of osteoblastic activity. Leonard and Cotton (45) could not identify osteoclasts in the osteopetrotic tissues of the adult *tl* rat, and, therefore, this may be a specially useful model for calcified tissue research due to the longevity of the rat and its apparent lack of osteoclasis.

The rat parotid gland development has been studied by Schneyer and Hall (73) and the submaxillary gland by Jacoby and Leeson (32). These studies indicate that at birth these glands are quite underdeveloped and that full morphogenesis of structures takes several weeks. It is not until about 10 weeks of age that parotid weight, cell size, state of cellular differentiation, and total cell numbers seem to stabilize.

Redman and Sreebny (66) reported that the anlage of the parotid gland was first observed around 14 days after conception; by the seventeenth day lumen formation in a rudimentary Stenson's duct was seen; and just before birth (18–20 days of gestation) the system of ducts and terminal clusters resembling a tubuloacinar complex was first recognized. Thus, secretions from the parotid gland may not reach the oral cavity of rat pups until 1 or 2 days after birth.

It is obvious from this discussion on the growth and development of calcified tissues, such as bone and molars, and of salivary glands of rats that this experimental animal is useful in the study of biological and pathological factors effecting development of oral tissues.

D. Experimental Periodontal Disease

The rat has not found extensive use in the study of periodontal disease. The rice rat (*Oryzomys palustris*) (see vol 2, Chapter 12) has been used extensively by Shaw and co-workers (1,26,78) to study periodontal disease. Most of the experiments done with rats to study this disease have been concerned with the microbiological aspects of periodontal disease using gnotobiotic rats (35). Some studies also have been conducted to study the effect of experimentally induced malocclusion. Occlusal forces (axial, lateral or rotational) applied to teeth have a profound effect on the supporting periodontium, particularly the bone component. The experimental approach used to apply forces involves either the use of a rubber band (a No. 5 orthodontic intermaxillary elastic) or a wire between the first and second maxillary molars which are the most accessible in the case of the rat. Within the first 24 hr, pressure areas in periodontal membrane usually show distinct narrowing and localized vasoconstriction. A few days later this narrowing of the periodontal membrane is found to be associated with alveolar bone resorption. A similar lesion is observed in tension areas where the spaces increase, and vasodilatation and bone deposition develops. The rat has been used to study repair and regeneration of the periodontal ligament after it has been disrupted using scalers fashioned into elevators (10). The regeneration process was studied histologically at various time intervals. This process was characterized by rapid healing of collagen fibers during the first few days. Fifteen days after wounding, the regeneration of the periodontal ligament was nearly complete, but failed to maintain itself. After 20 days, the ligament became less well organized, and a lack of fibers was detected in the periodontal space.

Some of the characteristics of the rat which limit its use in periodontal research are (a) a keratinized gingival sulcus which differs from that of humans, (b) difficulty in establishing and/or defining a subgingival microbial plaque similar to that of humans, and (c) size of the molars and their location within the oral cavity of the rat, which makes it difficult to manipulate and adapt mechanical systems to induce occlusal trauma.

E. Studies in Oral Neoplasia in Rats

Rats to not normally show spontaneously occurring neoplasms of the oral cavity. However, they have been used in studies of salivary tumors as well as transplantable or chemically induced tumors. The most commonly induced salivary gland tumor reported in the literature (60) is the epidermoid carcinoma (squamous cell carcinoma).

The rat has been used to study the induction of experimental tumors in salivary glands of rats in which a pellet of 9,10-dimethyl-1,2-benzanthracene (DMBA) had been implanted into the parenchyma of the gland (13). The technique involves the use of 15 gauge syringe needle fitted with a plunger of the proper diameter to inject a 3 to 4 mm pellet of DMBA weighing approximately 5 mg into the parenchyma of the gland. Using this procedure on young (3–4 months) rats or hamsters, Cataldo and Shklar (12) reported that in 3 weeks there were palpable nodules and at 14 weeks there were large firm tumor masses. All tumors in hamsters were diagnosed as fibrosarcomas, while in another study (13) where rats were used, squamous cell carcinomas were the most frequent.

Recently a transplantable salivary gland tumor model has been developed using the rat. Twenty highly inbred adult male rats (Kx strain) were implanted with DMBA in the left submandibular gland. Individual primary tumors began to appear between 11 and 32 weeks. Ten permanent subcutaneous lines out of fourteen trials were obtained after transplanting portions of the tumors into 30-day-old Kx rats. After some initial changes in degree of anaplasticity, two squamous cell carcinomas and two sarcomas were successfully established as permanent reproducible models.

IV.   CONCLUSION

We have attempted to include in this chapter the most important uses of the rat in investigations dealing with the tissues in the oral cavity. In general, the use of the laboratory rat is subject to the same conditions as any other animal model, namely, that it should be used under highly controlled and standardized experimental conditions to answer a specific research question posed by the investigator. Laboratory results obtained using the rat as a model cannot be directly extrapolated to humans. This is also true of mice, hamsters, and even nonhuman primates. Research questions, such as is compound X useful to control a disease process in humans, can only be answered directly by obtaining the information in clinical experiments. However, despite this general limitation of all animal models, experiments using rats have predictive value and have proved extremely useful in studies relating to growth and development, wound healing, dental caries, and calculus production. The rat has been of limited use in research dealing with periodontal disease and experimental oral neoplasia.

ACKNOWLEDGMENT

Supported by National Institutes of Health Grants DE-02670 and RR00463-10.

REFERENCES

1. Auskaps, A. M., Gupta, O. P., and Shaw, J. H. (1957). Periodontal disease in the rice rat. III. Survey of dietary influences. *J. Nutr.* **63,** 325–343.
2. Baer, P. N., Stephan, R. M., and White, C. L. (1961). Studies on experimental calculus formation in the rat. I. Effect of age, sex, strain, high carbohydrate, high protein diets. *J. Periodont.* **32,** 190–196.
3. Baer, P. N. (1968). Use of laboratory animals for calculus studies. *Ann. N.Y. Acad. Sci.* **153,** 230–239.
4. Baker, H. J. (1977). Experimental methods applied to animals used in dental research. *In* "Animal Models in Dental Research" (J. M. Navia, ed.), pp. 92–122. Univ. of Alabama Press, University, Alabama.
5. Beighton, D. (1977). The effect of cadmium on the normal tooth fissure plaque flora of Sprague-Dawley rats. *Arch. Oral Biol.* **22,** 99–102.
6. Bernarde, M. A., Fabian, F. W., Rosen, S., Hoppert, C. A., and Hunt, H. R. (1956). A method for the collection of large quantities of rat saliva. *J. Dent. Res.* **35,** 326–327.
7. Briner, W. W. (1976). Selection of an experimental unit in rat caries studies. *J. Dent. Res.* **55,** 476–480.
8. Briner, W. W., Francis, M. D., and Widder, J. S. (1971). The control of dental calculus in experimental animals. *Int. Dent. J.* **21,** 61–73.
9. Brown, W. E., and König, G. K. (1977). Cariostatic mechanisms of fluorides. *Caries Res.* **11,** 1–325.
10. Burkland, G. A., Heeley, J. D., and Irving, J. T. (1976). A histological study of regeneration of the completely disrupted periodontal ligament in the rat. *Arch. Oral Biol.* **21,** 349–354.
11. Capuccino, C., and Shklar, G. (1976). A transplantable salivary gland tumor model in the rat. *Arch. Oral Biol.* **21,** 147–152.
12. Cataldo, E., and Shklar, G. (1964). Chemical carcinogenesis in the hamster submaxillary glands. *J. Dent. Res.* **43,** 568–579.
13. Cataldo, E., Shklar, G., and Chauncey, H. (1964). Experimental submaxillary gland tumors in rats. *Arch. Pathol.* **77,** 305–316.
14. Chauncey, H. H., Henriques, B. L., and Tanzer, J. M. (1963). Comparative enzyme activity of saliva from the sheep, hog, dog, rabbit, rat and human. *Arch. Oral Biol.* **8,** 615–627.
15. Cheyne, V. D. (1939). A description of the salivary glands of the rat and a procedure for their extirpation. *J. Dent. Res.* **18,** 457–468.
16. Chiba, M., Tashiro, T., Tsuruta, M., and Eto, K. (1976). Acceleration and Circadian rhythm of eruption rates in the rat incisor. *Arch. Oral Biol.* **21,** 269–297.
17. Cotton, W. R., and Gaines, J. F. (1947). Osteopetrosis with associated unerupted dentition in the albino rat. *J. Dent. Res.* **53,** 935.
18. Cotton, W. R., and Gaines, J. F. (1974). Unerupted dentition secondary to congenital osteopetrosis in the Osborne-Mendel rat. *Proc. Soc. Exp. Biol. Med.* **146,** 554–561.
19. Ennever, J., and Creamer, H. (1967). Microbiologic calcification: Bone mineral and bacteria. *Calcif. Tissue Res.* **1,** 87–93.
20. Ericsson, Y. (1959). Clinical investigations of the salivary buffering action. *Acta Odontol. Scand.* **17,** 131–165.
21. Ericsson, Y. (1964). The role of some salivary constituents in oral pathology with special regard to caries experiments with rodents. *In* "Salivary Glands and Their Secretions" (L. M. Sreebny and J. Meyer, ed.), pp. 281–297. Pergamon, Oxford.

22. Feller, R. P., Edmonds, E. J., Shannon, I. L., and Madsen, K. O. (1974). Significant effect of environmental lighting on caries incidence in the cotton rat. *Proc. Soc. Exp. Biol. Med.* **145,** 1065–1068.

23. Fitzgerald, R. J., Jordan, H. V., and Stanley, H. R. (1960). Experimental caries and gingival pathologic changes in the gnotobiotic rat. *J. Dent. Res.* **39,** 923–935.

24. Franke, K. W., and Patter, V. R. (1936). The ability of rats to discriminate between diets of varying degrees of toxicity. *Science* **83,** 330–332.

25. Greep, R. O. (1941). An hereditary absence of the incisor teeth. *J. Hered.* **32,** 397–398.

26. Gupta, O. O., and Shaw, J. H. (1956). Periodontal disease in the rice rat. I. Anatomic and histopathologic findings. *Oral Surg., Oral Med. Oral Pathol.* **9,** 592–603.

27. Gustafsson, B. E., Quensel, C. E., Swenander-Lanke, L., Lundquist, C., Grahnen, H., Bonow, B. E., and Krasse, B. (1954). The effect of different levels of carbohydrate intake on caries activity in 436 individuals observed for five years. *Acta Odontol. Scand.* **11,** 232–364.

28. Hadjimarkos, D. M. (1966). Effect of selenium on food and water intake in the rat. *Experientia.* **22,** 117–120.

29. Harris, R. S., Navia, J. M., and Cagnone, L. D. (1968). Influence of caging on development of rat caries in studies involving minerals. *Int. Assoc. Dent. Res., Prog. Abstr.* Abstract 362, p. 128.

30. Heremans, J. F., and Vaerman, F. P. (1971). Biological significance of IgA antibodies in serum and secretions. *In* "Progress in Immunology" (B. Amos, ed.), pp. 875–890. Academic Press, New York.

31. Hunt, H. R., Rosen, S., and Hoppert, C. A. (1970). The morphology of molar teeth and occlusion in young rats. *J. Dent. Res.* **49,** 508–514.

32. Jacoby, F., and Leeson, C. R. (1959). The post-natal development of the rat submaxillary gland. *J. Anat.* **93,** 201–216.

33. Johnson, D. A., and Sreebny, L. M. (1971). Effect of food consistency and starvation on the diurnal cycle of the rat parotid gland. *Arch. Oral Biol.* **16,** 177–185.

34. Johnson, D. A., and Sreebny, L. M. (1977). Biochemical changes in rat parotid gland lysosomal enzyme activities after isoprenaline or starvation. *Arch. Oral Biol.* **22,** 291–297.

35. Jordan, H. V., Fitzgerald, R. J., and Stanley, H. R. (1965). Plaque formation and periodontal pathology in gnotobiotic rats infected with an oral actinomycete. *Am. J. Pathol.* **47,** 1157–1167.

36. Keele, B. B., Jr. (1977). The oral environment in experimental animals: The microflora. *In* "Animals Models in Dental Research" (J. M. Navia, ed.), pp. 197–224. Univ. of Alabama Press, University, Alabama.

37. Keyes, P. H. (1958a). Dental caries in the molar teeth of rats. I. Distribution of lesions induced by high-carbohydrate low-fat diets. *J. Dent. Res.* **37,** 1077–1087.

38. Keyes, P. H. (1958). Dental caries in the molar teeth of rats. II. A method for diagnosing and scoring several types of lesions simultaneously. *J. Dent. Res.* **37,** 1088–1099.

39. Keyes, P. H. (1959). Dental caries in the Syrian hamster. VIII. The induction of rampant caries activity in albino and golden animals. *J. Dent. Res.* **38,** 525–533.

40. König, K. G. (1961). Effect of mastication and particle size of corn and sugar diets in caries-incidence in rats. *Arch. Oral Biol.* **6,** 214–220.

41. König, K. G., Savdir, S., Marthaler, T. M., Schmid, R., and Mühleman, H. R. (1964). The influence of experimental variables on eating and drinking habits of rats in caries tests. *Helv. Odontol. Acta* **8,** 92–96.

42. König, K. G., Schmid, P., and Schmid, R. (1968). An apparatus for frequency controlled feeding of small rodents and its use in dental caries experiments. *Arch. Oral Biol.* **13,** 13–26.

43. König, K. G., Larson, R. H., and Guggenheim, B. (1969). A strain-specific eating pattern as a factor limiting the transmissibility of caries activity in rats. *Arch. Oral Biol.* **14,** 91–103.

44. Larson, R. H., Rubin, M., and Zipkin, I. (1962). Frequency of eating as

45. Leonard, E. P., and Cotton, W. R. (1974). "Morphological observations on the lack of osteoclasis in the "tl" strain of rat. *Proc. Soc. Exp. Biol. Med.* **141,** 596–598.

46. Lester, K. S. (1969). The unusual nature of root formation in molar teeth of the laboratory rat. *J. Ultrastruct. Res.* **28,** 481–506.

47. McGhee, J. R., Michalek, S. M., Navia, J. M., and Narkates, A. J. (1976). Effective immunity to dental caries: Studies of active and passive immunity to *Streptococcus mutans* in Malnourished rats. *J. Dent. Res.* **55,** C206-214.

48. Menaker, L., and Navia, J. M. (1973). Effect of undernutrition during the perinatal period on caries development in the rat. III. Effect of undernutrition on biochemical parameters in the developing submandibular salivary gland. *J. Dent. Res.* **52,** 688–691.

49. Menaker, L., and Navia, J. M. (1973). Effect of undernutrition during the perinatal period on caries development in the rat. IV. The effect of differential tooth eruption and exposure to a cariogenic diet on subsequent dental caries incidence. *J. Dent. Res.* **52,** 692–697.

50. Miller, W. D. (1890). "The Microorganisms of the Human Mouth" (original work) (K. G. Konig, ed., Karger, Basel, 1973).

51. Miller, W. D. (1904). A study of certain questions relating to the pathology of the teeth. *Dent. Cosmos* **46,** 981.

52. Morris, A. L., and Greulich, R. C. (1968). Dental research: The past two decades. *Science* **160,** 1081–1088.

53. Mühleman, H. R., and König, K. G. (1967). The effect of addition of amylase, an amylase inhibitor and sodium chloride on the cariogenicity of a purified cornstarch in Osborn-Mendal rats. *Helv. Odontol. Acta* **11,** 152–156.

54. Navia, J. M. (1968). Caries producing diets. *In* "the Art and Science of Dental Caries Research" (R. S. Harris, ed.), pp. 245–255. Academic Press, New York.

55. Navia, J. M., Cagnone, L. D., Lopez, H., and Harris, R. W. (1968). The effect of $MgCl_2$ $ZnCl_2$, $MnCl_2$ and Na trimetaphosphate on dental caries when fed alone or in combination. *Int. Assoc. Dent. Res., Prog. Abst.* 361, p. 128.

56. Navia, J. M., Lopez, H., and Harris, R. S. (1969). Purified diet for dental caries research with rat. *J. Nutr.* **97,** 1933–140.

57. Navia, J. M. (1970). The evaluation of nutritional and dietary factors which modify animal caries. *J. Dent. Res.* **49,** 1213–1227.

58. Navia, J. M., DiOrio, L. P., Menaker, L., and Miller, S. A. (1970). Effect of undernutrition during the perinatal period on caries development in the rat. *J. Dent. Res.* **49,** 1091–1098.

59. Navia, J. M., and Lopez, H. (1977). Sources of variability in rat caries studies: Weaning age and diet fed during tooth eruption. *J. Dent. Res.* **56,** 222–227.

60. Navia, J. M., ed. (1977). "Animal Models in Dental Research." Univ. of Alabama Press, Unviersity, Alabama.

61. Navia, J. M. (1977). Experimental dental caries. *In* "Animal Models in Dental Research" (J. M. Navia, ed.), Chapter 13, pp. 257–297. Univ. of Alabama Press, University, Alabama.

62. Navia, J. M. (1977). Experimental oral calculus. *In* "Animal Models in Dental Research" (J. M. Navia, ed.), Chapter 14. Univ. of Alabama Press, University, Alabama.

63. Orland, F. J., Blayney, J. R., Harrison, R. W., Reyneirs, J. A., Trexler, P. C., Wayner, M., Gordon, H. A., and Luckey, T. D. (1954). Use of germfree animal technic in the study of experimental dental caries. I. Basic observations on rats reared free of all microorganisms. *J. Dent. Res.* **33,** 147–174.

64. Park, A. W. (1969). Morphogenesis and regression of the osteodental fissure in the rat. *Acta Anat.* **73,** 442–462.

65. Paynter, K. J., and Hunt, A. M. (1964). Morphogenesis of the rat first molar. *Arch. Oral Biol.* **9,** 611–626.

66. Redman, R. S., and Sreebny, L. M. (1970). The prenatal phase of the morphosis of the rat parotid gland. *Anat. Rec.* **168,** 127–137.

67. Robinovitch, M. R., and Sreebny, L. M. (1969). Separation and identification of some of the protein components of rat parotid saliva. *Arch. Oral Biol.* **14,** 935–949.

68. Robinovitch, M. R., and Sreebny, 1. M. (1970). Separation of rat parotid isoamylases by preparative polyacrylamide gel electrophoresis. *Arch. Oral Biol.* **15,** 1381–1384.

69. Robinovitch, M. R., Keller, P. J., Johnson, D. A., Tversen, J. M., and Kauffman, D. L. (1977). Changes in rat parotid salivary proteins induced by chronic isoproterenol administration. *J. Dent. Res.* **56,** 290–303.

70. Robins, M. W. (1977). Biting loads generated by the laboratory rat. *Arch. Oral Biol.* **22,** 43–47.

71. Rosen, S. (1973). Laboratory animals and their contributions to oral microbiology. *In* "Oral Microbiology" (W. A. Nolte, ed.), 2nd ed., Mosby, St. Louis, Missouri. pp. 371–411.

72. Scherp, H. (1971). Dental caries: Prospects for prevention. *Science* **173,** 1199–1205.

73. Schneyer, C. A., and Hall, H. D. (1969). Growth pattern of postnatally developing rat parotid gland. *Proc. Soc. Exp. Biol. Med.* **130,** 603–607.

74. Schour, I., and Massler, M. (1949). The teeth. In "The Rat in Laboratory Investigation" (E. J. Farris and J. Q. Griffith, eds.). Hafner, New York.

75. Schuster, G. S., Navia, J. M., Amsbaugh, S., and Larson, R. (1978). Sources of variability in rat caries studies: Microbial infection and caging procedures. *J. Dent. Res.* **57,** 355–360.

76. Shackleford, J. M., and Wilborn, W. H. (1968). Structural and histochemical diversity in mammalian salivary glands. *Ala. J. Med. Sci.* **5,** 180–203.

77. Shaw, J. H. (1947). Carious lesions in cotton rat molars. I. The role of mechanical factors studied by the extraction of antagonistic molars. *J. Dent. Res.* **26,** 47–51.

78. Shaw, J. H., and Dick, D. S. (1966). Influence of route of administration of penicillin on periodontal syndrome in the rice rat. *Arch. Oral Biol.* **11,** 359–371.

79. Shaw, J. H. (1968). An evaluation in rats of the relationship between the frequency of providing food and the caries-producing ability of diets. *Arch. Oral Biol.* **13,** 1003–1013.

80. Smith, D. J., and Taubman, M. A. (1976). Immunization experiments using the rodent caries model. *J. Dent. Res.* **55,** C193–C205.

81. Sodicoff, M., Pratt, N. E., Trepper, P., Sholley, M. M., and Hoffenberg, S. (1977). Effects of X-irradiation and the resultant inanition on amylase content of the rat parotid gland. *Arch. Oral Biol.* **22,** 261–267.

82. Sreebny, L. M., Morasch, J., and Johnson, D. A. (1969). The similar effect of total starvation, powdered chow and liquid diet on the diurnal cycle of the rat parotid gland. *J. Dent. Res.* **48,** 192.

83. Stephan, R. M. (1951). The development of caries on the buccal and lingual tooth surfaces of rats as well as proximal and fissure caries. *J. Dent. Res.* **30,** 484.

84. Stiles, H. M., Loesch, W. J., and O'Brien, T. C. (1976). "Microbial Aspects of Dental Caries," Vols. 1, 2, and 3, Spec. Suppl. Microbiology Abstract Information Retrival Inc., Washington, D.C.

85. Sweeney, E. A., Shaw, J. H., Childs, E. L., and Weisberger, D. (1962). Studies on the protein composition of rodent saliva. I. Application of methods of paper electrophoresis to two strains of laboratory rats. *Arch. Oral Biol.* **7,** 621–631.

86. Tamarin, A., and Sreebny, L. M. (1965). The rat submaxillary salivary gland. A correlative study by light and electron microscopy. *J. Morphol.* **117,** 295–352.

87. Tanzer, J. M. (1976). Some important considerations with respect to food and water consumption in animal caries studies. *J. Dent. Res.* **55,** C215–C220.

88. Van Der Hoeven, J. S. (1974). A slime-producing microorganisms in dental plaque of rats, selected by glucose feeding. *Caries Res.* **8,** 193–210.

89. Van Der Hoeven, J. S., Mikx, F. H. M., Konig, K. G., and Plasschaert, A. J. M. (1974). Plaque formation and dental caries in gnotobiotic and SPF Osborne-Mendel rats associated with *Actinomyces viscosus. Caries Res.* **8,** 211–223.

90. Van Houte, J., Upeslacis, V. N., Jordan, H. V., Skoke, Z., and Green, D. B. (1976). Role of sucrose in colonization of *Streptococcus mutans* in conventional Sprague-Dawley rats. *J. Dent. Res.* **55,** 202–215.

91. Van Reen, R., and Cotton, W. R. (1968). Methods for scoring dental caries in experimental animals. *In* "Art and Science of Dental Caries Research" (R. S. Harris, ed.), pp. 277–312. Academic Press, New York.

92. Vogel, J. J., and Smith, W. N. (1976). Calcification of membranes isolated from *Bacterionema matruchotii. J. Dent. Res.* **55,** 1080–1083.

93. Winick, M., and Noble, A. (1965). Quantitative changes in DNA, RNA, and protein during prenatal and postnatal growth in the rat. *Dev. Biol.* **79,** 451.

94. Zipkin, I., and Hawkins, G. R. (1964). The human tooth and its environment: Formal discussion. *J. Dent. Res.* **43,** 1040–1044.

Chapter 4

Embryology and Teratology

Allan R. Beaudoin

I. INTRODUCTION

The purpose of this chapter is to introduce the fledgling teratologist to the normal embryology of the rat and to the concepts and techniques of experimental teratology. Equally important is an introduction to original articles, reviews, and books that contain in-depth discussions of many of the topics covered in this chapter.

The literature in teratology is voluminous and, therefore, it was impossible to include references to all the investigations dealing with the topics in this chapter. References were not always used to establish priority, but rather to give the reader a source for additional or more detailed information. There are several good source books for the beginning teratologist that include publications from teratology workshops (216, 304), textbooks (245, 284, 300), several volumes of review articles (312), a catalogue of teratogenic agents (256), and a book evaluating drugs as teratogens (246).

II. EMBRYOLOGY

A. Preimplantation Stages

The early investigations of Blandau and his co-workers make it possible to time pregnancy in the rat rather accurately. The onset of behavioral estrus in the female can be determined by the display of lordosis in response to manual stimulation (46). This display is a sign of willingness to copulate and marks the onset of heat lasting an average of 13 hrs. Ovulation is reported to occur approximately 10 hr after the onset of heat (47). On a light/dark cycle of 14/10 hr, rats are reported to ovulate about 2 AM of the proestrus to estrus night (115). The time of ovulation can be altered by changing the lighting schedule.

Spermatozoa begin appearing in the oviduct 15 min after copulation, and at 1 hr they are found throughout the oviduct (45). Fertilization takes place in the ampulla of the oviduct; it is reported that 90% of the ova are fertilized within 3 hr after ovulation, with the appearance of the first segmentation spindle and cleavage between 20 and 25 hr after sperm penetration (220). Thus, if need be, the onset of embryonic development can be estimated with considerable accuracy. The reader is referred to several good reviews of the morphological, biochemical, and immunological aspects of fertilization by Austin and Bishop (13), Metz and Monroy (188), and Monroy (193). Useful information about techniques used to study fertilization is given by Bedford (34).

Increasing the interval between ovulation and fertilization results in a decrease in the number of eggs fertilized and an increase in the number of dead and resorbed embryos in those animals successfully impregnated (44). Overripeness of the rat egg, induced by delayed ovulation following sodium pentobarbital administration, results in abnormal blastocysts, failure of implantation, chromosomal anomalies, embryonic death, and fetal malformation (61,62,126).

After fertilization, the successful future development of the mammalian embryo depends upon a series of complex interactions between maternal factors and the genetic program of the embryo, and, in this respect, mammals differ from nonmammalian forms. During the preimplantation period of embryonic development, the embryo is bathed by the secretions of the genital tract. The composition of these fluids (e.g., protein, enzymes, glycogen) differs with location in the genital tract and with the hormonal conditions of the mother (136). The importance of these secretions to the embryo is apparent, for early cleavage stages will not survive and implant when transferred prematurely to the uterus (219). Furthermore, only blastocysts obtained from the uterus will, when transplanted to extrauterine sites, develop into embryos, whereas blastocysts forced to develop in oviducts will not, but tube-lock blastocysts will develop normally in hormonally receptive uteri (170). Additional aspects of the influence of maternal factors

on the preimplantation embryo are discussed by Beier et al. (38), Hafez (135), and Kirby (171). The physiology and biochemistry of the preimplantation embryo have been reviewed recently by Biggers and Borland (40) and by Wales (281).

Shortly after fertilization, the zygote loses the follicular cells forming the corona radiata and is surrounded only by a homogenous noncellular membrane, the zona pellucida. At the end of the first 24 hr after fertilization, the embryo has cleaved or is about to cleave into the 2-cell stage. During the next 74 hr, cleavage (mitosis) continues with no overall increase in cytoplasmic mass but with the production of progressively smaller cells (blastomeres) within the confines of the zona pellucida. Cleavage is not synchronous in all eggs ovulated, but rather the cleavage rates are variable, and some eggs are reported not to begin cleavage until 48 hr after ovulation (214). During the stay in the oviduct, immature and nonfertilized eggs degenerate. Ninety-six hours after fertilization, the embryo is a morula containing 12–16 cells and located near the oviductal-uterine junction (146). Schlafke and Enders (248) question the production of a true morula in the rat, for they found signs of blastocyst formation at the 8-cell stage. The embryos begin to enter the uterus around 96 hr, and probably all enter during the next 12 hr (146). During the short time the morula is free-floating in the uterine cavity, it completes its transformation into a blastocyst. The formation of the mammalian blastocyst is said to take place by cavitation with an accumulation of fluid between the inner blastocyst cells (90). The source of this fluid appears to be the secretions of the oviduct and uterus that are transported into the developing blastocyst. The fluid is retained in the blastocoel by the presence of junctional complexes between the outer cells of the transforming morula (111). The process of morula to blastocyst transformation may be influenced by the production of steroids by the embryo (101).

Throughout cleavage and up to the morula stage, the blastomeres are totipotent, but in the morula the position of a cell determines its fate (141). The outer cells go on to form trophoblast (contributes to future placenta), and the inner cells go on to form inner cell mass (forms the future embryo). This first unequivocal differentiation of cells is correlated with the first appearance of zonular tight junctions between the cells in the outer layer of the morula (105). These tight junctions effectively isolate the forming inner cell mass from the maternal environment.

The zona pellucida begins to be shed from the blastocyst during day 5 (96–120 hr after fertilization) of pregnancy, and 95% are gone by 8 PM of day 5 (99). The mechanism of blastocyst release from the zona pellucida is not settled. There is evidence in the rat for "hatching" of the embryo out of the zona (248) and for lysis of the zona by lysins from the trophoblast (99) and by estrogen-dependent lysins from the uterus (268).

The rat embryo is resistant to the effects of teratogens during cleavage and blastocyst stages. This resistance is presumed due to the pluripotentiality of embryonic cells during the preimplantation stages and the regulation that can take place following teratogen-induced cell death. Whether the embryo dies or survives and develops depends on the number of cells killed, the ability of the living cells to take the place of the lost cells, and the time remaining prior to the onset of organogenesis during which reorganization must take place. When malformations are induced by treatment during the preimplantation stages (25), it may not be known whether the teratogen has been cleared from the maternal and fetal systems prior to the onset of organogenesis or whether some lingering effect has been produced by the teratogen.

B. Implantation

Implantation involves a direct interaction between embryonic trophoblast and maternal uterine tissue. This interaction is complex and involves progressive changes in the relationship. Blastocyst transplantation experiments have demonstrated that implantation in the rat is dependent upon hormonal conditioning of the uterus. The uterus is in a receptive state for about 12 hr during the late fourth to fifth day of pregnancy (sperm day = day 1). Prior to and following this time, the blastocyst will not implant (235). To be receptive, the uterine endometrium must be exposed to progesterone for a minimum of 48 hr, and estrogen must be present at the end of this period (235). The rat is unique in this requirement (200). Not only is the maternal hormonal state important to the blastocyst but steroid hormones produced by the embryo may play a role in transformation of the morula to the blastocyst and in implantation of the blastocyst (101). Histamine (120), uterine fluid proteinases (231), and surface charge of the blastocyst (79) have also been implicated in implantation. The role of cell surface materials on the trophoblast and uterine epithelium has not been adequately investigated. Apparently extensive lysis of the maternal tissue is not a component of rat implantation (113), and the early rat trophoblast lacks significant proteolytic activity (222).

Implantation begins when the blastocyst, without a zona pellucida, settles into a decidual crypt (43) or, as described by Enders and Schlafke (113), becomes clasped by the endometrium, perhaps due to estrogen-induced edema. This usually takes place on the evening of day 5 (sperm day = day 1). The crypts are shallow depressions in the antimesometrial lining of the uterine lumen. The decidualization of the uterine stroma commences about the same time with the transformation of fibroblasts into decidual cells. In the rat, there is unequal distribution of implantation sites in the uterine cornua, apparently due to a random scatter phenomenon (174). Not all eggs enter-

ing the uterine horns will implant. Harper (137) has shown, in a Wistar rat, the mean number of eggs ovulated to be 12.2 and the mean number of eggs implanted to be 10.9, for a preimplantation loss of 11.1%. Of those that do implant, approximately 93% will be alive at birth.

At implantation, the blastocyst is oriented with the inner cell mass facing the underlying maternal tissue. This constant orientation is postulated to be due to the ability of the inner cell mass to move around inside the trophoblast (171). The following brief description of the initial stages of rat implantation is based principally on a detailed analysis by Enders and Schlafke (113). There is no attachment between blastocyst and maternal uterine epithelium when the blastocyst first settles into the decidual crypt, for the blastocyst can still be flushed from the uterus. Twenty-four hours later the blastocyst has elongated and adheres to the uterine epithelium on one side of the implantation chamber. The earliest evidence for adhesion is the presence of interdigitating microvilli of the two epithelia (182). Elongation of the implantation site continues, and at day 7 portions of the uterine epithelium are lost, and the trophoblast makes its initial contact with the uterine stroma (Fig. 1). The inner cell mass undergoes entypy (a portion of the embryo invaginates into its own yolk sac) and is converted into an egg cylinder with the endoderm outside of the ectoderm (Fig. 2). Subsequently the trophoblast penetrates beyond the basal

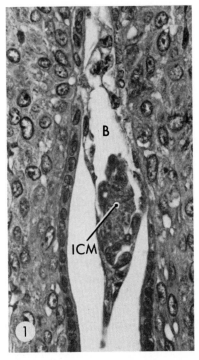

Fig. 1. Implanting rat embryo. Cells of the trophoblast surround the blastocoel (B) and the inner cell mass (ICM). The uterine epithelium has begun to degenerate.

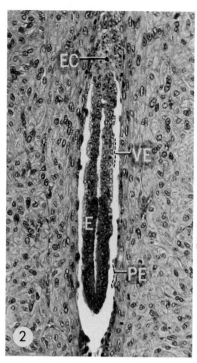

Fig. 2. Egg cylinder stage. E, ectoderm; EC, ectoplacental cone; PE, parietal endoderm and trophoblast; VE, visceral endoderm.

lamina of the uterine epithelium into the uterine stroma. There is evidence that the trophoblast contains lysosomal enzymes and possesses the ability for endocytosis and phagocytosis (249). Implantation is usually completed 2 days after it commences.

At implantation, and throughout gestation, the tissue of one individual is in intimate contact with the tissue of another individual. In most other sites in the body such intimate contact between genetically incompatible tissue would result in rejection of the foreign tissue, yet the embryonic graft is a success. For many years the successful growth of the embryonic graft was ascribed to the uterus being an immunologically privileged site, such as the anterior chamber of the eye. There is now ample evidence that not only is the uterus immunologically competent (36, 37), but that histocompatibility genes are active and transplantation antigens are produced as early as the 2-celled embryo (108). Antigenic material from intrauterine skin homografts does reach regional lymph nodes and does induce a response that causes rejection of the graft (42). Thus transplantation antigens, from within the uterus, are detectable and a transplantation immunity is expressed. The embryo, however, remains unaffected by the maternal immune response, although experimentally this immunity can be breached, as in runt disease (35).

The most likely site of interference with the maternal immune response is the trophoblast. This view is strengthened by the occurrence of ectopic implantations without decidual reac-

tions and by the evidence that enzyme digestion of the mucoprotein coat of the trophoblast allows expression of transplantation antigens (87). In the human, chorionic gonadotropin, elaborated by the trophoblast, is reported to inhibit certain responses of immunocompetent human lymphocytes (1). Such evidence strongly supports the contention that the embryonic trophoblast permits survival of this specialized type of homograft. Why it is eventually rejected (parturition) is not known.

C. The Placentas

In the rat, there are two placentas serving as organs for maternal embryo exchange, a discoidal chorioallantoic placenta and a villiary highly vascularized yolk sac placenta. The two are present together throughout most of gestation. The following synopsis of placental development is based on articles by Amoroso (6), Bridgman (53), Davies and Glasser (89), Everett (114), and Mossman (196).

1. The Chorioallantoic Placenta

The fetal portion of the rat chorioallantoic placenta has two rudiments, tissue derived from the ectoplacental cone and tissue from the allantois. Shortly after implantation, the ectoplacental cone is formed by a proliferation of the trophoblast overlying the inner cell mass. At the same time that the ectoplacental cone is forming, a layer of endoderm forms under the inner cell mass and grows out along the inner surface of the trophoblast. The cells of the inner cell mass grow down into their covering of endoderm (in a direction away from the ectoplacental cone) elongating the inner cell mass and creating a cylinder (the egg cylinder) projecting into the blastocoel with ectodermal cells on the inside covered by endodermal cells on the outside (inversion of the germ layers). Three cavities eventually form within the ectoderm of the egg cylinder (Fig. 3). The most ventral of these is the amniotic cavity lined with ectoderm. The middle cavity is the extraembryonic cavity (exocoelom) lined with extraembryonic mesoderm. The dorsal cavity, just under the ectoplacental cone, is the ectoplacental cavity lined with extraembryonic ectoderm. The membrane separating the exocoelom from the ectoplacental cavity is called the chorion, and the membrane between the exocoelom and the amniotic cavity is the amnion. Both membranes have two layers, one of mesoderm and the other of ectoderm.

During the period 9–10 days after fertilization, the allantois appears as an outgrowth from the posterior region of the embryo and grows into the exocoelom. The ectoplacental cavity is reduced to a cleft and finally is obliterated as the tissue of the chorion fuses with the base of the ectoplacental cone. Subsequently, the vascular mesoderm of the allantois fuses with the mesoderm of the chorion, and the allantoic (umbilical)

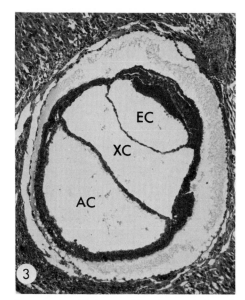

Fig. 3. Egg cylinder with three cavities. Ac, amniotic cavity; EC, ecto-placental cavity; XC, extraembryonic cavity.

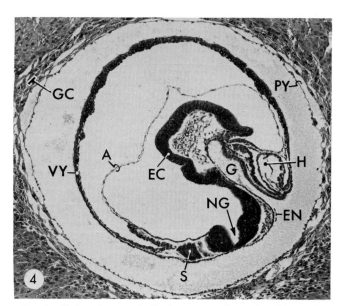

Fig. 4. Embryo in reversal of body curvature. A, amnion; EC, ectoderm; EN, endoderm; G, primature gut; GC, giant cell reticulum; H, heart primordia; NG, neural groove; S, somite, PY, parietal layer of yolk sac and Reichert's membrane; VY, visceral layer of yolk sac.

capillaries thereby have access to the developing fetal placenta (ectoplacental cone). By day 12, the placenta is well vascularized by allantoic vessels except at its lateral margin, which is composed of maternal blood sinuses surrounded by trophoblast (158).

An outgrowth of the mesometrial uterine lining obliterates the uterine lumen and encapsulates the implanting embryo. This encapsulating uterine tissue underlies the ectoplacental cone and becomes the maternal part of the definitive placenta, the dedicua basalis. About day 14, the uterine cavity is reestablished ventral (antimesometrial) to the embryo, creating a capsule of decidua over the implantation site, the decidua capsularis. All the remainder of the uterine lining is called the decidua parietalis.

The ectoplacental trophoblast invades the uterine mucosa eroding maternal tissue, glands, and blood vessels. The maternal secretions, blood, and tissue debris nourish the embryo and collectively are called the embryotroph. The developing placenta expands laterally becoming discoid in shape. Lacunae appear in the trophoblast and form a labyrinth which fills with maternal blood. About day 9, a maternal placental circulation is established. By day 12–13, the chorioallantoic placenta is composed of an inner zone of anastomosing cords or plates of trophoblast bathed in maternal blood, the labyrinth. The labyrinth is capped by a layer of trophoblast (spongiotrophoblast, basal zone) containing maternal blood sinuses but lacking vascularization by embryonic blood vessels. External to this trophoblastic layer are giant cells in contact with the maternal decidua and arranged in a loose reticulum whose interstices are filled with maternal blood. The reticulum of giant cells ensheaths the implantation vesicle, and over the parietal yolk sac

lies in contact with Reichert's membrane (Fig. 4). The giant cells are reported to migrate, phagocytize (5, 116), secrete steroid hormones (92), and have a limited life span, i.e., programmed cell death (104).

The study of the fine structure of the labyrinthine zone reveals the placental barrier to be composed of the embryonic capillary endothelium and a trilaminar trophoblast, at all stages of pregnancy (110,157,247,308). The outer layer of trophoblast (bathed by maternal blood) is cellular; the middle and inner layers are syncytial (110). In late pregnancy, fenestrations appear in the outer layer of trophoblast and in the capillary endothelium (157). The rat chorioallantoic placenta is discoidal, labyrinthine, and hemotrichorial.

2. The Yolk Sac Placenta

The inversion of the germ layers results in an egg cylinder projecting into the yolk sac cavity (old blastocoel) with a single layer of cuboidal and columnar endodermal cells overlying the embryonic and extraembryonic ectoderm. The layer of endoderm (the visceral layer of the yolk sac) is reflected off the base of the chorioallantoic placenta onto the inside of the trophoblast as a layer of flattened endodermal cells (the parietal layer of the yolk sac) that completely encloses the yolk sac cavity. Between the parietal endodermal cells and the trophoblast a homogenous membrane appears called Reichert's membrane. (Fig. 5) Reichert's membrane is attached to the lateral margin of the chorioallantoic placenta and encircles the yolk sac cavity. This membrane has all the morphological and

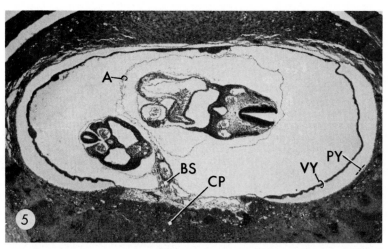

Fig. 5. Rat embryo after reversal of body curvature. A, amnion; BS, body stalk; CP, chorioallantoic placenta; PY, parietal layer yolk sac and Reichert's membrane; VY, visceral layer of yolk sac.

biochemical properties of a basement membrane (78,177,309). External to the layer of trophoblast, with its numerous giant cells and maternal blood sinuses, is the decidua capsularis. Material entering the yolk sac cavity from maternal blood vessels in the decidua capsularis, or from the uterine lumen, must traverse the tissue of the decidua capsularis, the trophoblast, Reichert's membrane, and the cells of the parietal endoderm. About day 12, the visceral endodermal cells form villi, which project into the yolk sac cavity, composed of a simple columnar epithelium with a core of mesenchyme containing branches of the vitelline vessels. The fine structure and function of the visceral layer of the yolk sac has been described by Padykula *et al*. (224).

With continued growth of the embryo, the parietal wall of the yolk sac, including the decidua capsularis, is stretched thin, until at day 15 or 16 the wall ruptures exposing the visceral endodermal cells to the contents of the uterine lumen. The fine structure of these changes has been described by Clark *et al*. (78) and by Jollie (159).

Apparently Brunschwig (58) was the first to suggest that the yolk sac epithelium of the rat might be physiologically a placenta. His conclusion was confirmed by Everett (114). The rat, therefore, has two functional placentas during gestation. The relative importance of the two placentas in maternal–embryo transport during the period of organogenesis is not clearly defined. The yolk sac is the route of iron transport, the chorioallantoic placenta the route of calcium transfer (310). The yolk sac, not the chorioallantois, is the route of transfer of passive immunity from mother to fetus (49). Proteins seem unable to pass the chorioallantoic placenta, but can pass the parietal wall of the yolk sac and are taken up by the endodermal cells of the yolk sac's visceral wall to pass to the embryo

via branches of the vitelline veins lying in the mesenchyme under the visceral endoderm (8,117,179, 186, 187, 254, 294). The absorptive ability of the visceral yolk sac endoderm is inherent to the cells, for they continue to absorb when transplanted to an *in vitro* site (119), and they possess the ultrastructural composition characteristic of protein absorbing cells (224,308). Since the chorioallantoic placenta does not become vascularized by allantoic vessels until day 11 or 12 (114), the yolk sac is the only route for major transport between mother and embryo during the critical stages of early embryogenesis. Protein can be taken up by the visceral yolk sac cells, digested intracellularly, and the breakdown products subsequently released from the cells (187,280,293). This digestion process has been proposed to be vital for supplying the embryo with nutrients and has been found to continue in the yolk sac epithelium after the establishment of the chorioallantoic placenta (31).

In vitro studies have shown the visceral yolk sac to be indispensable for survival of the day 10 rat embryo in culture; in its absence the embryo fails to grow and develop (225). The role the placentas play appears to change as gestation advances, for near term disruption of the yolk sac circulation does not affect fetal survival, whereas disruption of the chorioallantoic circulation is fatal to the fetus (217). With advancing gestational age, the visceral yolk sac cells show increased uptake of some substances (163) and decreased uptake of other substances (96). Studies on the placental transfer of material from mother to fetus in the rat have been helpful to indicate the great number of different substances that are transferred, but have been deficient because no distinction is made between chorioallantoic transfer and yolk sac transfer, and most studies are done late in gestation [see review by Schultz (250)].

The relative importance of the two placentas in teratogenesis is not known. Because many critical stages in organogenesis are passed prior to the onset of function of the chorioallantoic placenta, attention has been focused on the role of the yolk sac placenta in teratogenesis. The parietal wall of the yolk sac and Reichert's membrane are not a barrier to most substances. Material that passes through these membranes enters the yolk sac cavity and is exposed to the endodermal cells of the visceral layer of the yolk sac. Long before it was known that trypan blue was a teratogen, it was demonstrated that the cells of the visceral yolk sac avidly absorb the dye (132). One of the early suggestions for a role of the visceral yolk sac in teratogenesis was protection. Since trypan blue was sequestered by the visceral endodermal cells, and since the teratogenic activity of trypan blue ceased at about the same time that the visceral yolk sac completely encircled the embryo, it was postulated that the yolk sac served to protect the embryo against the action of trypan blue (307). A more recent hypothesis is that trypan blue inhibits lysosomal enzymic activity and pinocytosis in the visceral yolk sac cells, thereby reducing their ability to digest protein (32,295). The result is a reduction in the movement of nutrient materials from the visceral yolk sac to the developing embryo. Any such interference with embryonic nutrition could result in delayed or abnormal growth. One criticism of these experiments is that they were performed, of necessity, on day 17 yolk sacs cultured *in vitro* and, therefore, may not truly represent the morphological and functional conditions of the visceral yolk sac during the organogenetic period.

Several functions of the visceral endodermal cells are known to be altered by teratogens, e.g., the absorption and transport of calcium, sodium, and sulfate ions following maternal trypan blue treatment (164), the transport of the amino acid valine after maternal chlorambucil injection (163), and the *in vitro* incorporation of glucose and glycine in the presence of chloramphenicol and puromycin (124). The dysfunction of the rat yolk sac is postulated as playing a role in the production of malformations by tissue antisera (52).

The emphasis placed on the visceral yolk sac in teratologic studies is not to discount the chorioallantoic placenta, for once it becomes functional it offers a route of passage to the embryo. There are some teratogenic agents (e.g., cadmium and mercury) that accumulate in both chorioallantoic and yolk sac placentas, some (e.g., salicylic acid) accumulate in neither, and some (e.g., trypan blue, and (2,4,5-trichlorophenoxy)-acetic acid) accumulate in the visceral endoderm but not in the chorioallantois (96).

Histochemical (223) and electrophoretic (155,156) analysis of the rat visceral yolk sac throughout gestation reveals a normal sequence of biochemical differentiation with respect to several enzyme families. The parietal yolk sac cells, however, appear inert. Acid and alkaline phosphatases, lactate and ma-

late dehydrogenases, nonspecific esterases, and glucose-6-phosphate dehydrogenases all undergo sequential changes during gestation. The number of isozymes present on a given day is the result of both additions to and deletions from the number of isozymes present on the preceeding day (155,156). In the visceral yolk sac cells, teratogenic folic acid deficiency results in the delayed appearance of isozymes, the premature disappearance of isozymes, or the persistence of isozymes beyond the time they would normally disappear (155,156). In spite of the day-to-day variability in isozyme patterns, there was often recovery by term, and the isozyme patterns became the same in experimental and control animals.

The relationship between the embryo and the yolk sac placenta is unique and complex. Much remains to be learned of how teratogens affect the structural and functional integrity of the visceral endodermal cells. These effects may be a mechanism of teratogenesis in rats.

D. Postimplantation Stages

1. The Organogenetic Period

With the completion of implantation and the appearance of the primitive streak and the associated embryonic mesoderm, the rat embryo enters the postimplantation stages of organogenesis and maturation. The period of organogenesis begins with the primitive streak and ends 8 or 9 days later (Figs. 4–6). During this period, the organ anlagen appear, undergo differentiation, and attain their adult morphology. This is the period of the greatest susceptibility to the action of teratogens, and each organ passes through one or more critical periods. In general, treatment early in the organogenetic period affects principally nervous system, eye, and ear, whereas treatment late in the organogenetic period affects palate, skeleton, and urogenital structures (Fig. 7). It is possible, therefore, to choose the time of treatment to produce malformations in selected organ systems. Not only does organ sensitivity to teratogens change during organogenesis but certain drugs appear to selectively affect the development of certain organs (176).

A synopsis of organ system development is given in Tables I–V. The information for the tables was obtained from many sources. References to specific organs or structures are cited with the tables; for multiple organs and structures see references (9,12,56,77,109,140,145,146,191,214,265,283,300, 301,305,311). Only the major developmental events are listed; no attempt was made to be complete. The events are grouped into 24-hr periods dated from a presumed fertilization during the night of mating. The times given must be considered approximate, but probably vary no more than plus or minus one-half day.

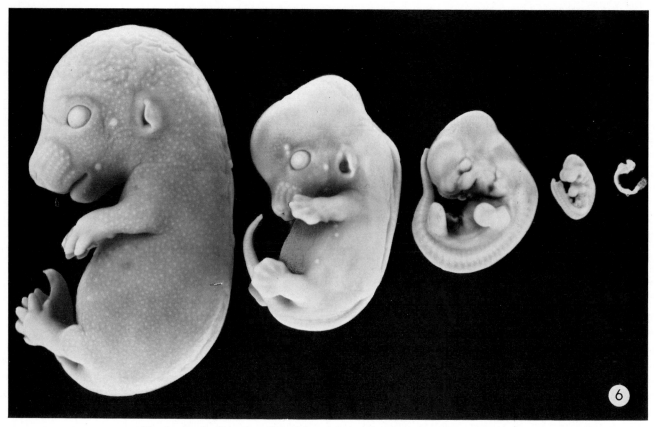

Fig. 6.   Rat development at (from right to left) gestation days 10, 11, 13, 15, and 17.

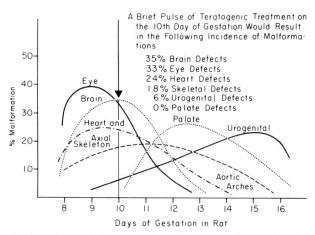

Fig. 7.   Hypothetical representation of how the syndrome of malformations produced by a given agent might be expected to change when treatment is given at different times. The percentage of animals affected as well as the incidence rank of the various types of malformations would be somewhat different from that shown for the tenth day if treatment were given instead on the twelfth or the fourteenth day, for example. [From Wilson (297); with permission.]

Table I

Normal Development of External Body Form and Face[a]

Day 9–10	Somites 1–4 are identifiable; the embryo is bent dorsally; the first branchial arch appears
Day 10–11	Somites 5–20 appear in the cervical region; the second branchial arch appears; the reversal of the body curvature takes place; the dorsal surface of the embryo becomes convex; the tail fold and the lateral body wall folds form; the maxillary processes appear
Day 11–12	Somites 21–33 appear in the thoracic and lumbar regions; the anterior limb buds appear followed approximately one-half day later by the posterior limb buds; the third and fourth branchial arches appear; the medial and lateral nasal processes are recognizable; the olfactory pits are forming; the cervical sinus begins to develop
Day 12–13	Somites 34–45 form in the caudal region; the mandibular, maxillary, and fontonasal processes are prominent; the maxillary processes have grown beneath the optic vesicles to meet the lateral nasal processes; the nasolacrimal grooves form; the lateral and medial nasal processes come into contact and begin to fuse ventral to the nasal pits; the mandibular arches begin to merge to form the lower jaw; the cervical sinus is closing; the auricular hillocks appear on the mandibular arches

(continued)

Table I (*Continued*)

Day 13–14 Somites 46–51 develop; the cervical sinus is closed; the external ear ducts develop from the first visceral cleft; the mouth is narrowing; the auricular hillocks appear on the hyoid arches; the anterior limb buds enter the paddle stage followed shortly by the posterior limb buds; the normal umbilical herniation begins; the upper eyelids appear; the maxillary processes fuse with the medial nasal processes to form the upper lip

Day 14–15 Somites 52–60 are present but their identification becomes difficult; the first vibrissary papillae appear on the maxillary processes; digital condensations begin in the forelimbs followed within the day in the hindlimbs; the lower eyelids form; the mandibular and hyoid arches merge beneath the external auditory canal; the external ear flaps begin to form; the external nares are fully differentiated; the body uncoils from its C shape; the first skeletal cartilage appears in the ribs; the first ossification center appears in the mandible

Day 15–16 Somites 61–65 are present; trunk hair papillae appear; the majority of bones in the body become identifiable

Day 16–17 The digits are completely separated in the forepaws

Day 17–18 The umbilical hernia begins to withdraw; the pinnae cover the ear ducts; the digits are completely separated in the hindpaws

Day 18–19 The eyelids fuse; the umbilical hernia is completely withdrawn

ᵃ From references 60, 103, 109, 125, 183, 189, 199, 267, 296, 313.

Table II

Normal Development of the Nervous System[a]

Day 8–9 The neural plate appears

Day 9–10 The neural folds form and begin to fuse in the region of somites 2–6; the neural crest is formed at the junction of the neural folds and surface ectoderm; the optic evaginations appear from the prosencephalon

Day 10–11 Three brain vesicles are present, the prosencephalon, the mesencephalon, and the rhombencephalon; the infundibulum appears as a depression in the floor of the prosencephalon; the otic placodes form and begin to sink below the surface; the cranial ganglia begin their formation; the optic vesicles contact the surface ectoderm; the neural tube is closed except at the anterior and posterior neuropores; Rathke's pouch begins its outgrowth from the roof of the future mouth

Day 11–12 Five brain vesicles are present, the telencephalon, the diencephalon, the mesencephalon, the metencephalon, and the myelencephalon; the anterior neuropore closes followed shortly thereafter by the posterior neuropore; the optic cups begin to form two layers; the lens placodes are formed; the otic vesicles close and the endolymphatic ducts appear; the olfactory placodes appear and begin their transformation into the olfactory pits; Rathke's pouch makes contact with the infundibulum, forming the primordia of the hypophysis; in the diencephalon the hypothalamus is distinct from the dorsal thalamic area

Day 12–13 The cerebral hemispheres are present; the choroid fissues are identifiable; the auditory portion of the inner ear is developing, and the primordia of the utricle, the saccule, and the cochlea are recognizable; the endolymphatic ducts separate from the otic vesicles

Day 13–14 The primordia of the semicircular canals are forming; the primordium of the cerebellum is present as a thickened plate in the metencephalon; choroid plexuses appear in the first,

Table II (*Continued*)

second, and fourth ventricles; the lens vesicles complete their formation and begin to detach from the epidermis

Day 15–15 A choroid plexus appears in the third ventricle

Day 16–17 The choroidal fissure is completely closed

Day 19–20 In the mesencephalon, the corpora quadrigemina are recognizable

ᵃ From references 2, 10, 57, 81, 252, 266.

Table III

Normal Development of the Circulatory System[a]

Day 8–9 The cardiogenic primordium appears in the mesoderm in front of the oral plate

Day 9–10 Fusion of the bilateral cardiogenic primordia begins ventral to the foregut; aortic arch I is forming; the simple tubular heart forms and begins sporadic contractions

Day 10–11 An S-shaped tubular heart is present; aortic arches II and III appear; the cardinal, the umbilical, and the vitelline veins can be recognized; the atrium, the ventricle, and the sinus venosus can be identified; the embryonic circulation is established

Day 11–12 Aortic arches III and IV are well developed, arches I and II are regressing; the interventricular septum begins to form; the septum primum appears in the atrium

Day 12–13 Aortic arch VI develops; the dorsal and ventral atrioventricular cushions form; the truncoconal septal ridges appear and begin to fuse

Day 13–14 The septum primum is completed and the foramen secundum forms; the atrioventricular cushions are fusing

Day 14–15 The aorta is separated from the pulmonary artery; the atrioventricular canal is divided

Day 15–16 The aortic arches are transformed into the adult configuration; the heart is completely partitioned, except for the foramen ovale

ᵃ From references 59, 190, 303.

Table IV

Normal Development of the Urogenital System[a]

Day 8–9 The intermediate mesoderm forms; it is the precursor of the urogenital system

Day 9–10 The germ cells are identifiable in the yolk sac epithelium

Day 10–11 The nephrogenic ridges appear as condensations in the intermediate mesoderm; the (pro)nephric ducts begin their formation on the dorsal aspect of the nephrogenic ridges; the germ cells begin to migrate from the yolk sac, entering the hindgut and mesenteries

Day 11–12 The first mesonephric tubules develop; the (meso)nephric ducts reach the cloaca, but their lumina are not open the entire length; the gonadal ridges appear medial to the nephrogenic ridges; the first germ cells enter the gonadal ridges

Day 12–13 The lumina of the mesonephric ducts become confluent with the lumen of the cloaca; the ureteric diverticula appear and each becomes surrounded by a cap of mesenchyme; the metanephric blastema; the mesonephroi reach their greatest development and begin to degenerate; the germ cells have finished their

(*continued*)

Table IV (*Continued*)

	migration into the indifferent gonad; the urorectal fold begins to partition the cloaca
Day 13–14	The gonads begin sexual differentiation; the paramesonephric ducts begin to form as infoldings of the coelomic epithelium; the genital tubercle is forming
Day 14–15	The gonads are identifiable as ovaries or testes; the kidneys begin their ascent
Day 15–16	The paramesonephric ducts reach the urogenital sinus; the testes begin their descent; the cloaca is divided into the urogenital sinus and the rectum
Day 17–18	The renal glomeruli and tubules are recognizable; fluid accumulation begins in the kidney; the renal pelves and calyces are dilated; the kidneys are in their final position; the paramesonephric ducts unite to form the utero-vaginal canal
Day 18–19	All of the mesonephroi have degenerated, except for portions appropriated by the male gonad
Day 19–20	The testes lie lateral to the bladder and near the floor of the pelvis; the heterologus set of genital ducts have mostly disappeared; the seminal vesicles appear on the lateral walls of the ductus deferens; the prostatic buds grow out from the urethra

From references 122, 234, 272, 273, 274, 302.

Table V

Normal Development of Digestive System, Pharynx, and Oral Cavity*

Day 9–10	The foregut begins to form
Day 10–11	The oral membrane ruptures; the septum transversum is recognizable ventral to the foregut; the hepatic diverticulum is formed; the first pharyngeal pouch is present; the thyroid primordium appears as a pit in the floor of the pharynx; the hindgut begins its formation
Day 11–12	The laryngotracheal groove appears; the primordium of the dorsal pancreas is present; the stomach is recognizable; the second and third pharyngeal pouches are identifiable; the paired lung buds form; the liver cords begin to develop
Day 12–13	The pleuroperitoneal folds appear; the fourth and sixth pharyngeal pouches form; the ventral pancreatic bud is recognizable; the primordia of the parathyroids, the thymus, and the ultimobranchial bodies are present; the division of the cloaca begins
Day 13–14	The normal umbilical herniation begins; the dorsal and ventral pancreatic primordia fuse; the palatine processes appear on the maxillary processes; the primative primary palate forms on the frontonasal process and begins to fuse with the palatine shelves; the posterior nares become visible; the floor of the mouth is formed by merging of the mandibular processes; the primordia of the tongue appear
Day 14–15	The labiodental grooves appear; the liver has the adult configuration; the pancreatic islet tissue develops
Day 15–16	The cloaca is divided into the urogenital sinus and the rectum; the diaphragm is completed with the closure of the pleuroperitoneal canals; the anal membrane ruptures
Day 16–17	Papillae appear on the dorsum of the tongue; the palatal shelves are elevating and fusion begins
Day 17–18	The umbilical hernia is withdrawing; the fusion of the palate is completed
Day 18–19	The umbilical hernia is withdrawn

From references 11, 39, 77, 83, 102, 107, 142, 180, 238, 240, 315.

2. The Maturation Period

The final stage in *in utero* development is the period of maturation that follows completion of organogenesis. During the maturation period, the organs undergo growth with concomitant histological, physiological, and biochemical changes that result in functional competency. Since the organs have completed their morphological development prior to the period of maturation, they are resistant to the induction of morphological defects. It is clear, however, that teratogens are not without action during this period.

During the past 15 years, it has become increasingly documented that prenatal insult to the embryo can have other effects than the production of overt malformations. Often normal appearing offspring from teratogen-treated mothers exhibit deficiencies in postnatal development and maturation. The study of changes in behavior after prenatal treatment has received the name behavioral teratology. It is beyond the scope of this chapter to analyze the many reports dealing with how agents given during pregnancy affect the subsequent postnatal behavior. Many of the reports suffer from methodological deficiencies, and the results are often contradictory. Extensive reviews of the field have been published by Coyle *et al.* (85) and by Joffe (152).

Behavioral teratology is beset with problems. There are not yet established definitive and universally accepted procedures to be used for analyzing altered behavior following prenatal insult. There is good evidence that genetic differences among strains of rats play an important role in determining not only normal behavior but also behavior following experimental treatment. Ray and Barrett (236) have shown significant differences in shock avoidance among rat strains. When shock avoidance behavior patterns were tested between a high performance strain and a low performance strain, it was found that purebred rats performed as their parents, whereas the offspring of cross-breeding were intermediate in their performance. The use of foster mothers had no bearing on the outcome; the real parent strain completely determined the behavior. Avoidance behavior in shock situations is believed due to interactions between the adrenergic system (related to behavioral activity) and the serotonergic system (related to behavioral suppression). Biochemical analyses revealed differences in brain chemistry consistent with this interpretation (236).

In addition to genetic differences among rat strains, there are other variables that must be defined properly before meaningful interpretations can be made of the data from experiments in behavioral teratology. The effects of the teratogen must be separable from any other influence on behavior. Behavior in adult rats can be influenced by the experiences their mothers had as infants (97), by the handling their mothers received while pregnant (3), and by the use of foster mothers (152). Litter size not only affects adult behavior (255) but also affects the development of the brain as well (314). Rats raised in an enriched environment (increased environmental complexity and training) have distinct changes in brain chemistry and anatomy when compared with animals reared in isolation (98,241). Pre- and postnatal nutrition can also affect the anatomy and biochemistry of the brain (4,259,314).

Three behavioral tests are in common use, open field behavior, maze running, and susceptibility to audiogenic seizures (85). More information about additional types of behavior is desirable. Problems exist to distinguish between transient and permanent effects, to determine the influence of nursing from the teratogen-treated mother, to investigate the role of age and sex of the offspring, and to investigate interactions between teratogen-induced behavior and factors known to modify behavior (*supra vide*). The tenets of teratology may apply to behavioral teratology, but information is lacking to determine their precise role (e.g., critical periods, dose).

It has been suggested by Joffe (152) that three steps are prerequisite for behavioral testing: (a) standardize the environment of the mother prior to and during pregnancy, (b) standardize the postnatal environment, and (c) separate the prenatal and postnatal maternally mediated effects by fostering the offspring at birth.

III. EXPERIMENTAL TERATOLOGY

A. Introduction

The rat has several advantages for experimental teratology.* It is relatively disease resistant, withstands operative procedures well, has a short reproductive cycle, has a high breeding rate with good litter size, has a low spontaneous malformation rate, and has fetuses large enough to work with conveniently. The use of the rat in experimental teratology appears to have begun with the study of radiation effects by Bragg (48) and by Job *et al.* (151) followed by the nutritional experiments of Warkany and co-workers (285). Since then the rat has been used in innumerable investigations in experimental teratology.

To use the rat for teratological investigations, a continued and reliable source of pregnant females is necessary. Depending on the requirements of the research program and the facilities available, a breeding colony can be established or the rats can be purchased from a commercial supplier, either nonpregnant for breeding in the laboratory or pregnant for immediate use. Care must be taken to select a supplier able to

---

*A survey of the field of teratology has been published recently. It is an excellent sourcebook for the beginning teratologist: "Handbook of Teratology" (J. G. Wilson and F. C. Fraser, eds.), Vols. 1, 2, 3 (1977); Vol. 4 (1978). Plenum Press, New York.

provide fast reliable delivery of disease-free rats. Sick animals are not suitable for teratological research.

B. Breeding

A breeding colony can be maintained without difficulty; it has the advantage of providing healthy rats of known life history, and subjected to uniform conditions prior to and during experimentation (see cautions about inbreeding in Volume I Chapter 7). In my laboratory, female rats are bred for only 1 year, or less if the litter size falls consistently below ten. Male rats are also bred only 1 year, or less if they fail to mate in three consecutive exposures. Since only females are used in teratological experiments, the males are culled from the litter during the first week after birth, unless one is needed for breeding. We find female rats weighing 200–225 gm best for breeding for teratological experiments.

Knowledge of the reproductive cycle is essential (Volume I Chapter 7). Two methods are commonly used to obtain pregnant rats, both with success. The first method relies on the reflection of the stage of the estrous cycle in the epithelium of the vagina. Examination of cells from the vaginal epithelium makes it possible to assign the female rat to one of the stages of the cycle, proestrus, estrus, metestrus, or diestrus. Since ovulation in the rat occurs spontaneously near the end of the estrous period, and since the female is receptive to the male during estrus, it is important to be able to identify the proestrous period. Vaginal cells may be obtained by lavage or by a gentle scraping of the vaginal wall with a wire loop, a spatula, or a moist cotton swab. The rat can be held by the tail while in an upright position with all four paws on a surface, or the rat can be held in the hand in an inverted position. In the lavage method, a small amount of water in a medicine dropper is expressed into the vagina and immediately withdrawn and placed on a glass slide for microscopic observation. A cover slip is not necessary. The best viewing is in a dimly lit field. A few grains of methylene blue in the water enhances cell visibility. In the scraping method, the instrument is inserted a short distance into the vagina and during withdrawal it is gently drawn over the vaginal epithelium. The instrument is then stirred in a drop of water on a glass slide. Excessive stimulation of the vagina during the taking of the smear may upset the estrous cycle and produce a pseudopregnancy with a duration of 12–16 days before resumption of the normal cycle (214). The proestrous smear consists of a preponderance of nucleated epithelial cells, singly or in clusters. Females in proestrus are placed overnight with a fertile male. The following morning the vaginal smear is examined for spermatozoa. The copulation plug (a coagulum of semen) is not always a reliable sign of pregnancy in rats.

The second method is to cage four or five sexually mature

females overnight with a single fertile male. Vaginal smears are examined for spermatozoa the following morning. This method requires less time and skill, for the estrous stages do not have to be identified. After a successful insemination, males are rested for 2 or 3 days before being used again.

Overnight mating does permit variation in the time of copulation. Analysis of short mating time (2 hr) versus long mating time (15 hr) revealed no significant difference between the two groups with respect to embryonic weights at day 12 or fetal weights at day 20 (17,143). The adoption of a short mating period protocol does not seem warranted as a means of reducing interlitter weight variability for teratological research. It is more appropriate to use females of the same age and weight for breeding (137).

Meticulous records must be kept for teratological experiments. All animals should be numbered, and the number of each pregnant female together with the number of the impregnating male entered in the record book. Daily observations should be made of each animal, and any change in appearance, behavior, eating habits, and conditions in the animal quarters recorded. It is often advantageous to have a daily record of weight changes, particularly following treatment. The teratogenic agent, its form, route of administration, dose, and time of administration should be noted in the record book. The final entry will be the results of the fetal examinations. Only with such meticulous records is it possible to detect the occurrence of subtle changes with time, season, environment, etc., that may have an effect on the outcome of an experiment.

It is essential that rats used for teratological experiments receive the proper care and diet; healthy animals are a necessity. The animal quarters must have an appropriate light and dark cycle and be kept clean and quiet, well ventilated, and at the proper temperature and humidity. Clean cages and bedding should be provided at frequent intervals, and food and water available *ad libitum*. Sick animals cannot be used, nor can any prophylactic treatment be given just prior to or during the course of the experiment. The animal quarters should not be sprayed for vermin control while the experiments are in progress. Every effort must be made to eliminate all factors that might influence the outcome of the teratological experiment.

In recent years, much discussion has taken place about the selection of the most appropriate method for timing pregnancy (161). There are those who advocate calling the day of finding sperm in the vaginal smear day zero, and there are those who advocate calling it day one. Neither side appears to have convinced the other side to change. A few other investigators time pregnancy from midnight or 2 AM on the night of breeding. Thus, it becomes very important that the method of timing pregnancy is clearly stated in all written reports. Only then is it possible to compare published articles. Depending on the method of timing, the duration of pregnancy in the rat is stated to last 21 or 22 days.

C. Design of the Teratological Experiment

There are several routes for administering a teratogen. If the goal of the experiment is to mimic a human condition, the route of administration should be same as the route in the human, and preferably the metabolism of the agent should be similar in both human and rat. The route of administration not only affects the rate of absorption, and thus the rapidity with which the compound reaches the embryo, but the route may also affect the metabolic fate of the compound. In rats, trypan blue is an effective teratogen when given parenterally, but not when given orally, for trypan blue passes through the gastrointestinal tract with little or no absorption.

Topical application of teratogenic agents can be used, although this is not commonly employed. A major problem is to keep the rat from licking or rubbing off the compound.

Subcutaneous injections are administered easily with little risk to the rat, and they can be given without anesthesia. The dorsum of the neck is a convenient site that can accommodate rather large volumes, or multiple sites can be used. Absorption of substances from subcutaneous sites is rather slow. Intramuscular injection offers slightly more rapid absorption than subcutaneous injection, but only small amounts can be injected at any one site.

Intraperitoneal injection is frequently used in teratological experiments. It requires restraint or anesthesia, and care must be taken not to puncture a viscus. Solutions of extreme pH and tonicity should be avoided lest peritonitis with possible perforation of the intestine ensue. Several milliliters of solution can be injected via this route. To avoid puncturing a viscus, I use enough ether anesthesia to relax the abdominal muscles. The skin and underlying muscles in a lower quandrant of the abdomen are pinched between the thumb and index finger and lifted, creating a tent into which the needle (25 gauge) is inserted. Absorption is very rapid from the peritoneal cavity.

Intravenous or intracardiac injections offer the most rapid access of teratogen to the embryo. Nonphysiologic material may cause thrombosis and should be avoided. A tail vein is the most common route for injection; however, in my laboratory intracardiac injection has been found to be more reliable and less time consuming. A 1-inch 22 gauge needle is placed on the skin just below the xiphoid process. The skin is depressed and the needle inserted beneath and almost parallel to the sternum. The needle is pushed craniad until the pulsations of the heart are felt. Gentle pressure will pierce the heart wall. A slight pull exerted on the syringe plunger will cause blood to appear in the syringe barrel when a heart chamber is successfully entered. The solution is then expressed slowly. We have injected up to 2 ml without difficulty, although smaller volumes are preferable.

Feeding the teratogen in the diet is not often a satisfactory method of administration. Adequate mixing with the food may be difficult; the food may be made less palatable; and accurate measurement of food consumption is troublesome because of the rat's feeding habits. Gavage is a better choice. I have found a soft but firm plastic tube, of the kind found on plastic wash bottles, to make an effective gavage tube. A 2½-inch long tube with a tapered tip and an outside diameter of 3 mm fitted to a syringe works well. Gavage needles are available commercially. The rat may be lightly anesthetized and placed on its side to facilitate passage of the tube to the stomach. Gavage is a good method to use with insoluble compounds.

Substances may also be administered directly into the uterine lumen, the amniotic sac, the umbilical vein, and the fetus. There is even one report of teratogenic effects following administration of the teratogen to the male for several weeks prior to mating (149).

In order to produce morphological defects, the teratogen is administered during the period of organogenesis to expose the formative stages of all organs to the treatment. Two choices exist, to treat with a single dose on a given day (acute exposure), or to treat with multiple doses over several days (chronic exposure). It is difficult to have a truly acute experiment, since once an agent is administered it may not be known how long the embryo is exposed to the agent. It is possible to limit the duration of exposure by treating with an antiteratogen. Antiteratogens are substances that remove or inactivate the teratogen. By controlling the interval between teratogen and antiteratogen administration the period of exposure of the embryo to the teratogen is presumably limited (232). A single dose given on a single day of gestation does have the advantage that it allows the profile of teratogenesis to be determined for the compound administered (25). In general, acute treatments have greater teratogenic affect than chronic treatments. The chronic administration of some compounds stimulate microsomal enzymes that enhance their own metabolism and sometimes the metabolism of other drugs as well (84). Pregnant rats treated with chlorcyclizine over a 4-day period produced significantly more malformed young, than when treated with the same dose over a 16-day period (167). It was found that chlorcylizine stimulates its own metabolism, thus reducing the teratogenic response (168). A single injection of actinomycin D induced three times as many malformed survivors as ten daily injections of a slightly higher dose (299).

It is also possible for substances to exert an inhibiting effect on drug metabolizing enzymes. The compound 2-diethylamino ethyl-2,2-diphenylvalerate hydrochloride (SKF-525A) inhibits the conversion of chlorcyclizine to its active teratogenic form, norchlorcyclizine. A dose of chlorcyclizine given alone caused cleft palate in 85% of the surviving fetuses, but when the same dose was given together with SKF-525A, the fetal concentration of norchlorcyclizine fell and the number of cleft palates was reduced to 35% (233).

Large-scale drug testing programs can save time and money

by using a compromise method between the single treatment and the continuous treatment. Wilson (300) suggests teratogenic treatment be divided into 3-day time spans that cover the period of organogenesis; rats would be treated with three successive doses either on days 9–11, or 12–14, or 15–17.

Since the teratogenic effect produced is due to the concentration of the teratogen reaching the embryo, the dosage administered is critical. The teratogenic concentration of an agent may be a narrow zone between a concentration eliciting no effect and a concentration eliciting a lethal effect. There are several factors that influence the concentration of the teratogen reaching the embryo. The chemical nature and physical form of the agent and the route of administration determine the speed with which the agent reaches the maternal bloodstream. Once in the maternal plasma, the teratogenic substance is subject to a variety of factors tending to reduce its concentration. Exogenous drugs and chemicals may be stored in the body tissues, they may be excreted unchanged, they may be metabolized and detoxified, or they may bind with protein (Fig. 8). These pro-

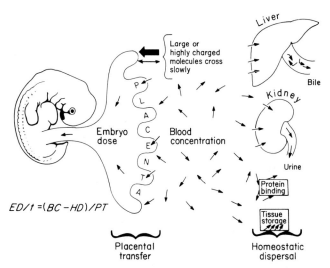

$$ED/t = (BC - HD)/PT$$

Fig. 8. Diagram of the factors that influence embryo dose of a foreign chemical present in the maternal bloodstream. The maternal blood concentration under usual conditions of exposure is subject to considerable change, representing as it does the differential between maternal absorption and the several routes for dispersal, including the maternal homeostatic processes of metabolism, excretion, protein binding, and tissue storage, as well as passage across the placenta to the embryo. These various biologic processes are all functions of time; therefore, the embryo dose depends on the duration as well as the level of a chemical in maternal blood. Embryo dose is proportional to maternal blood concentration, however, only for substances that traverse the placenta by simple diffusion; otherwise the placenta imposes various rates of transfer depending on the nature of the substance being transferred. The formulation $ED/t = (BC - HD)/PT$ is in no sense a precise mathematical one, but does attempt to state the basic fact that embryo dose (*ED*) over a given period of time (*t*) depends on the maternal blood concentration (*BC*) which tends constantly to be reduced by homeostatic dispersal (*HD*) and by placental transfer (*PT*). [From Wilson (300); with permission.]

cesses constantly act to clear the plasma of foreign material (300). The rapidity with which these processes function determines the duration of potential teratogenic exposure.

Teratogenic substances in the maternal blood plasma must traverse the chorioallantoic and yolk sac placentas to reach the embryo. Passage of substances across the placentas is determined by molecular weight, degree of ionization, lipid solubility, concentration gradient, and surface area and thickness of the placental membrane (197). The chorioallantoic placenta of the rat is reported to be capable of drug biotransformations (160), and the yolk sac placenta is reported to concentrate and excrete substances into the uterine lumen (280).

The biotransformation of drugs usually results in the production of less active or inactive compounds. In some instances, however, the metabolite may be teratogenically more active than the parent compound, e.g., salicylic acid is the causative agent in aspirin teratogenesis (166) and norchlorcylizine is the teratogenic metabolite of chlorcylizine (168).

Little is known of the fate of teratogens once they reach the embryo or of the ability of the embryo to protect itself against deleterious agents. Detoxifying enzymes are not present early in embryogenesis. The different tissues of the embryo have different capabilities to react to the same concentration of an agent (threshold of response), and these capabilities change with time.

The problem of how to select the initial dose in teratological experiments is perhaps best met by using serial half-decrements of the adult LD_{50} (300); a one-half, one-quarter, and one-eighth dose most often encompasses the embryo-toxic range. This method is just a guide, and consideration should be given to the known pharmacological and toxicological properties of the agent to be tested. A strictly empirical approach may be necessary. If the goal is to study the genesis of malformations, the dose yielding the maximum number of malformations with the minimum number of resorptions is best.

All teratological experiments must be adequately and properly controlled. The different strains of rats have different rates of occurrence of spontaneous malformations (16,86,127,129, 194,226,246). There is evidence that the somites and organ anlage appear earlier in the development of a Sprague-Dawley rat than they appear in the development of a Wistar rat (221). It is best in teratological investigations, therefore, to continue to use the same strain of rat from the same source and constantly to gather data on the spontaneous incidence of defects. In addition to gathering cumulative control data, it is essential to run a control with each experiment. Both a vehicle control and a noninjected control are necessary. The vehicle or solvent may alter the outcome of the experiment (7) or may itself be teratogenic (118).

The normal prenatal growth patterns of organs and tissue have been studied in rats (18,131,230), and various factors are known to influence prenatal growth, e.g., maternal nutrition,

length of gestation, litter size, sex of fetus, and intrauterine position (20,169,264). Mouse embryos developing nearest the ovary have a significantly higher incidence of spontaneous cleft lip than mouse embryos developing at other uterine sites (275), and embryos near the ovary sometimes have a higher incidence of teratogen-induced malformations (29,33).

Since teratogen-induced malformations are the result of complex interactions between the applied teratogen, the environment, and the genetic constitution of the embryo, it is clear that strain differences can be expected to influence the outcome of teratologic experiments. The same teratogenic treatment given to two or more rat strains often produces different responses (93,94,218,242,243). It has been demonstrated, furthermore, that rats of the same strain, but obtained from different suppliers, may react markedly different to the same dose of the same teratogen administered at the same time in gestation (19).

The goal of each experiment should be to have only one variable, the teratogen administered. All other factors, such as rat strain, food, water, light, and handling should be identical in both the control and the experimental rat.

D.   Recovery and Examination of Fetuses

The usual practice is to recover rat fetuses approximately 24 hr prior to expected parturition. This is necessary because the mother rat will often eat malformed offspring, especially if there is blood associated with the defect. One advantage to recover fetuses near term is the size of the fetus makes handling easier; one disadvantage is some malformed embryos are lost through early death and resorption. Fetuses can be recovered earlier in gestation if the objectives of the experiment so dictate.

An overdose of ether is a simple and effective way of killing the mother. The abdominal hair is wetted with alcohol and a midline incision is made through the abdominal wall, long enough to expose both uterine horns. Each horn is carefully examined for the presence of resorptions. These may vary in form from partially macerated fetuses to complete absence of the fetus, with only the metrial gland remaining. The metrial gland appears as a yellowish nodule at the base of the old placental attachment. Metrial glands form at the site of implantation and persist after the embryo and placenta have resorbed; polytocous animals do not abort. After the number and uterine position of the resorption sites are recorded, the uterine horns are cut open and the fetuses released from their amniotic sacs. Spontaneous or elicited movements reveal the living fetuses. A brief clamping of the umbilical cord, close to the fetus, prior to severing the cord will usually be sufficient to prevent fetal exsanguination (bloodless embryos weigh less than embryos with blood). The placentas are then peeled from their site of attachment. Fetuses and placentas are blotted dry prior to weighing.

After the fetuses are weighed, they are sexed and examined for gross malformations. A surface inspection will reveal such anomalies as the dome-shaped head of hydrocephalus, exencephaly, meningocele, the absence of eye bulges, ear abnormalities, umbilical herniation, gastroschisis, spina bifida, malformations of the paws and the tail, edema, and subcutaneous hemorrhages. Some investigators remove the ovaries and count corpora lutea to indicate the number of eggs available for fertilization and implantation. The number of corpora lutea, however, is not a reliable index of the number implanted (137).

The fetuses are next treated according to future handling. They are placed in Bouin's fluid for subsequent soft tissue examination, or placed in 95% alcohol for subsequent skeletal examination. Fetuses from a single litter may be apportioned between the fixatives or the entire litter placed in one or the other fixative. In any case, both methods of examination should be employed in all teratological investigations. Bouin's fluid is suggested for soft tissue analysis because it is acid enough to decalcify bone and makes future razor blade sectioning easier, and hemotoxylin and eosin staining follows Bouin's fixation very well. Fetuses are left in Bouin's fluid for 7 to 10 days. If the fetus weighs in excess of 4 gm, it is advisable to cut through the abdominal wall to allow better fixation of the internal organs. After fixation, the fetuses are transferred to 70% alcohol overnight to reduce the irritating vapors of Bouin's fluid.

There are two principal methods used to examine fetuses for soft tissue abnormalities, a dissection method (192,215) and the free-hand razor blade slice technique of Wilson (298), or one of its modifications (22). A technique should be selected suitable to the goals of the investigation. The razor blade slice method is easily mastered. It is reliable and repeatable. Much remains to be learned from the examination of histological sections of the fetus, but few investigators are devoting time (and it is considerable) to this pursuit.

After a thorough external examination of the fetus (limbs, digits, tail, and body surface), the initial razor blade cut is made through the mouth, with the fetus placed on its back, and cuts off the upper part of the head. The roof of the mouth is examined for cleft palate (Fig. 9). The upper part of the head is placed mouth side down, and several slices 3–4 mm in thickness are made beginning just in front of the eye and continuing to the ear. These slices reveal the nasal cavities, the eyes, and the fore- and midbrain region together with their ventricles (Figs. 10 and 11). Next, the forelimbs are removed and the trunk sectioned from the neck region (malformations are rarely recognized here) to the kidney. The heart, aortic arches, lungs, diaphragm, and kidney can be identified (Figs. 12–14). The pelvic viscera are removed, and ureters, gonads, and bladder are examined (Figs. 15 and 16).

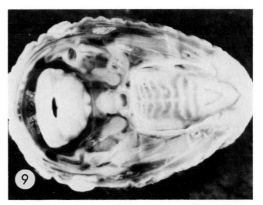

Fig. 9. Section through the mouth to show the palate.

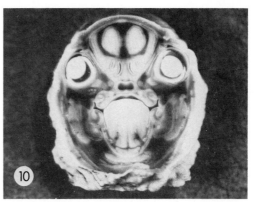

Fig. 10. Section through the eye and the olfactory bulbs. The nasal cavities, palate, and tongue can be identified.

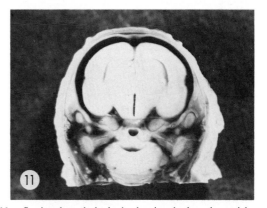

Fig. 11. Section through the brain showing the lateral ventricles and the third ventricle (perpendicular slit in midline).

Alizarin red S is the most commonly used stain for bone. The method for its use can be found in any histological techniques book. Sometimes it is advantageous to stain both bone and cartilage in the same specimen. A good method uses alcian blue with alizarin red S and stains cartilage blue and bone red

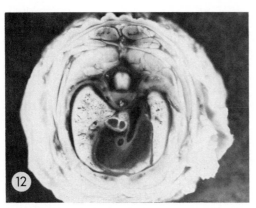

Fig. 12. Section through the thorax. The lungs, ascending aorta (in cross section), and the ductus arteriosus, passing to the left of the bronchi and esophagus, can be identified. The atria are filled with clotted blood and appear dark.

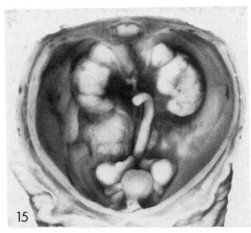

Fig. 15. Section of the male pelvis. The testes lie on each side of the bladder. The adrenals lie adjacent to the cephalic pole of the kidney. The hook-shaped structure is the rectum.

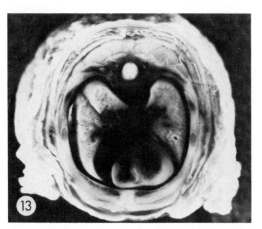

Fig. 13. Section through the thorax showing the lungs, and the right and left ventricles of the heart. The atria are not clearly discernible because of the presence of clotted blood.

(150). There is much confusion about skeletal malformations, and much has been claimed under the guise of malformation when in reality only normal variation existed or, at best, a delay in skeletal maturation. The article by Kimmel and Wilson (165) should be read before analysis of the skeleton is undertaken.

One of the problems in reporting teratological data has been the selection of the experimental unit. Much has been written about the choice of the appropriate sampling unit (embryo versus litter). Haseman and Hogan (138) review the problem in detail and present strong arguments for the litter (pregnant female) as the experimental unit upon which to base statistical

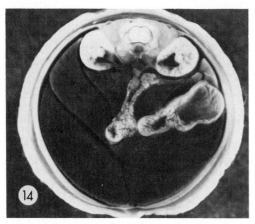

Fig. 14. Section through the renal pelves. The saccular structure on the left side of the embryo is the stomach. Most of the section is occupied by the liver.

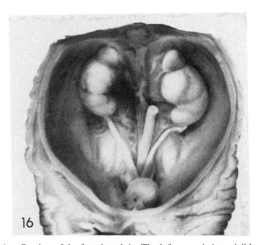

Fig. 16. Section of the female pelvis. The left ovary is just visible beneath the caudal pole of the left kidney. The right ovary is obscured by the right kidney. The two uterine horns are clearly visible running along the dorsal body wall.

calculations. On the other hand, Holson *et al.* (143) believe undue emphasis has been placed on the so-called litter effect, and they propose the embryo as the appropriate sampling unit. Usually the treatment of fifteen to twenty females will give sufficient data for statistical analysis.

E. Useful Techniques

There are several methods of experimental embryology and developmental biology that are adaptable to the study of teratogenesis (88,204).

The preimplantation embryo has been used extensively for the biochemical characterization of the embryo (41), and for the analysis of the nutrients required for embryonic growth during the preimplantation period (55). The *in vitro* treatment of preimplantation stages has not been a fruitful area for teratological investigations because the preimplantation embryo is not sensitive to the induction of malformations (121). There are, nonetheless, well developed embryo transfer techniques that permit *in vitro* treatment of preimplantation embryos with subsequent intrauterine development in foster mothers (100).

In vitro culture of whole embryos and embryo organs is gaining use in experimental teratology. *In vitro* culture allows the direct observation of embryonic growth, the isolation of the embryo from maternal influences, and the possibility to grow the embryo under defined conditions. In teratological investigations, the teratogen may be administered to the mother with subsequent culture of the embryo, or the teratogen may be applied directly to the embryo in culture. The modern use of *in vitro* culture of postimplantation embryos begins with New and Stein (211), and significant advances have been made in the technique since then (206,207,212,239). Whole embryos can be cultured successfully for 1 to 3 days at any stage during the period of organogenesis. At each stage, the embryo has its own limitations and special requirements for growth, e.g., older embryos require increased oxygen concentrations (209,210). The fact that the yolk sac grows well in culture and the chorioallantoic placenta does not may be the limiting factor for the duration of whole embryo culture. During *in vitro* culture, the rat embryo undergoes considerable growth and development, comparable in many respects to that which occurs *in vivo* (80,213). The culture of postimplantation embryos has been used to study energy metabolism (203,257) and to investigate the direct effect of teratogens on the embryo, e.g., trypan blue (276), 6-aminonicotinamide (277), antibodies (208), vitamin A (195), and 6-diazo-5-oxonorleucine (172).

Because of the short duration of whole embryo culture, organ culture is used to observe teratogenic action over a longer period (204). Several embryonic organs have been grown successfully in culture, but the most used is the limb bud. The limb bud is an excellent model system for use in the investigation of the action of teratogens on embryonic tissue, and methods devised for limb bud culture are applicable to many other organs as well (172).

Another technique potentially useful for teratologic investigation is to create allophenic (chimeric) individuals. The technique was developed in mice by Tarkowski (270) and first used in rats by Mayer and Fritz (185). They aggregated 8-cell embryos of Sprague-Dawley albino rats with 8-cell embryos of Long-Evans (hooded) rats. The resultant blastocysts were transferred to the uteri of recipient rats for gestation and normal delivery. Of twenty-five aggregate blastocysts transferred, seven were born, and two exhibited some degree of parental phenotypic mosaicism. Further refinements in this technique may make it possible to study the teratogenic response of tissues from a teratogen-susceptible strain and tissues from a teratogen-resistant strain within the same allophenic rat.

Other methods that show promise for use in experimental teratology include the use of the scanning electron microscope (64,288), the transmission electron microscope (112), electrophoretic analysis of abnormal development (153), autoradiographic analysis of abnormal development (263), surgical lesions (95), biochemical analysis of abnormal development (54,133), antiteratogens (27), placental transfer studies (251), and antenatal diagnosis of developmental aberrations through the analysis of amniotic fluid (184).

In order to identify and characterize mechanisms of action of teratogenic agents, more information must be obtained about the biochemical lesions by which teratogens produce anomalies during prenatal development. The field of biochemical teratology is slowly evolving. Considerable information is available about the biochemistry of the normal embryonic development of the lower vertebrates (289), but too little is known about mammalian embryos.

The cells and tissues of the embryo are different in many respects from the cells and tissues of the adult organism. Embryonic cells and tissues undergo rapid rates of growth and differentiation and consequently may have unique energy requirements, metabolism, synthesis of macromolecules, enzyme ontogeny, and enzymic pathways; and teratogens are reported to affect all of these. The biochemistry of the embryo is in a constant state of flux, with some components increasing, others decreasing, while some remain relatively constant. It is not surprising that embryonic cells have a different degree of sensitivity and response to drugs and chemicals than cells in the adult organism (205).

Some progress is being made in the analysis of the biochemical aspects of teratology (153,154,205). Teratogens produce a variety of biochemical lesions in the embryo. The interested reader is referred to publications by Bass *et al.* (23) and by Neubert and Merker (204), which contain pertinent research and references.

F.   Malformations

The literature is replete with articles listing a myriad of malformations that follow treatment with a multitude of teratogenic agents. Schardein (246) has recently published a text on the teratogenic testing of over 1200 drugs, and Shepard (256) has catalogued more than 800 reported teratogenic agents. There is no need for another list of teratogens.

Tables VI–XI have been constructed to assist the investigator interested in the genesis of malformations. In order to be able to study the embryology of a malformation profitably, that malformation should occur in at least one-half of the surviving fetuses. Thus, every other embryo examined is expected to develop abnormally. The tables list only those references in which the malformation is reported in at least 50% of the surviving offspring. Selecting such a high incidence automatically excludes many articles dealing with many other teratogens producing malformations at lower rates. Malformations that are poorly defined, suspected to be transitory, or represent apparent growth retardation are not listed in the tables, e.g., edema, hydronephrosis, hydroureter, lung defects, malposition of organs, and absence of ossification centers. The problem of intrauterine growth retardation has been reviewed by Brent and Jensh (51). The emphasis of the tables, however, seems justified. The descriptive phase of experimental teratology has passed. New advances in our understanding of abnormal development must come from studies of the genesis of malformations and from studies of how teratogens act to disrupt normal embryonic development, particularly the development of one organ system while sparing another organ system.

Even though the strains of rats used are listed in the tables, it must be remembered that the same strain of rat from different suppliers may respond differently to the same treatment (19).

Table VI

Malformations of the Nervous System

Malformation[a]	Strain	Agent	Reference
Anencephaly/	Wistar	Cyclophosphamide	73
Exencephaly	Wistar	Ethylenethiourea	244
	Wistar	Hydroxyurea	69
	Wistar	Procarbazine	71
	Wistar	Trypan blue + ATP	28
	Hybrid	Vitamin A (excess)	14
	Wistar	Vitamin A (excess)	82, 130
Encephalocele	Wistar	Chlorambucil	198
	?	Dithiocarbamate	229
	Wistar	Hydroxyurea	69
	?	Hypoglycin A	228
	CD	Trypan blue	175
	BR 46	Trypan blue	175
Hydrocephalus	Wistar	Aminothiadiazole	25
	Wistar	Antiserum	50

Table VI (*Continued*)

Malformation[a]	Strain	Agent	Reference
	Wistar	Cadmium	19
	Wistar	Cobalt radiation	262
	Sprague-Dawley	1-(2-Chlorethyl)-3-Cyclo-hexyl-1-nitrosourea	271
	Wistar	Ethylenethiourea	244
	Wistar	Tellurium	106
	Long-Evans	Tellurium	128
	Wistar	Trypan blue	307
	Sprague-Dawley	Zinc deficiency	147
Microcephalus	Fischer	Methylazoxymethanol	134
	Sprague-Dawley	Methylnitrosourea	173
Hydranencephaly	Wistar	Ethylenethiourea	244
Hypoplastic cerebellum	Wistar	Ethylenethiourea	244
Anophthalmia/ microphthalmia	Wistar	Aminothiadiazole	25
	Wistar	Antiserum	50
	Long-Evans	6-Aminonicotinamide	65
	Wistar	Arsenate	26
	Wistar	Cadmium	19
	Wistar	Cobalt radiation	262
	Sprague-Dawley	Ethylenediaminetetraacetic acid	269
	Wistar	Niagara sky blue 6B	24
	Wistar	Radiophosphorus	258
	Long-Evans	Streptonigrin	287
	Wistar	Triparanol	243
	?	Trypan blue	30
	Wistar	Trypan blue + ATP	28
	Long-Evans	Vincristine	94
	Wistar	Vitamin A (excess)	130
	Sprague-Dawley	Vitamin A (excess)	218
	Wistar	X-irradiation	301, 306
Exophthalmos	Wistar	Cyclophosphamide	261
	?	Dithiocarbamate	229

[a] Fifty percent or more of the surviving fetuses have the malformation.

Table VII

Malformations of the Skeletal System

Malformation[a]	Strain	Agent	Reference
Polydactyly	Wistar	Cytosine arabinoside	253
	Wistar	Hydroxyurea	69
Syndactyly	Wistar	Alkane sulfonates	139
	Wistar	Chlorambucil	198
	Sprague-Dawley	Chlorcyclizine	291
	Wistar	1,4-Dimethanesulfonyloxy-butane	198
	Long-Evans	Folic acid deficiency	202
	?	Hypoglycin A	228
	Wistar	Leucine	227

(continued)

Table VII (*Continued*)

Malformation[a]	Strain	Agent	Reference
	Sprague-Dawley	Rachitogenic diet	286
	Wistar	Triethylenemelamine	198
	Wistar	X-irradiation	198
Ectrodactyly	Sprague-Dawley	Acetazolamide	279
	Wistar	Aminothiadiazole	25
	Wistar	1β-D-Arabinofuranosyl-cytosine	74
	Wistar	5-Chlorodeoxyuridine	68
	Wistar	Cyclophosphamide	260
	?	Dithiocarbamate	229
	Wistar	Ethylenethiourea	244
	Wistar	6-Hydroxylaminopurine	70
	Wistar	Mercaptopurine	178
	Sprague-Dawley	Vitamin B_6 deficiency	91
	Sprague-Dawley	Zinc deficiency	148
Fused ribs	Wistar	Arsenate	26
	Wistar	Ethylenethiourea	244
Limb dysplasia	Sprague-Dawley	Acetazolamide	279
	Wistar	6-Aminonicotinamide	237
	Wistar	Cyclophosphamide	260
	Wistar	5-(3,3-Dimethyl-1-triazene)imidazole-4-carboxamide	66
	Wistar	Ethylenethiourea	244
	Wistar	Procarbazene	71
	Sprague-Dawley	Trypan blue	181
Absent/short tail	Wistar	Aminothiadiazole	25
	Wistar	Chlorambucil	198
	Wistar	1,4-Dimethanesulfonyloxy-butane	198
	Wistar	5-(3,3-Dimethyl-1-triazene)imidazole-4-carboxamide	66
	?	Dithiocarbamate	229
	Wistar	Ethylenethiourea	244
	Wistar	5-Fluorocytosine	72
	Long-Evans	Folic acid deficiency	202
	Wistar	Hydroxyurea	69
	Wistar	X-irradiation	198
Clubbed feet	Wistar	Aspirin	166
	Wistar	5-Chlorodeoxyuridine	68
	?	Dithiocarbamate	229
	Wistar	5-Fluorocytosine	72
	Wistar	6-Hydroxylaminopurine	70
	Wistar	Hydroxyurea	69
	Wistar	Mercaptopurine	178

[a] Fifty percent or more of the surviving fetuses.

Table VIII

Malformations of the Face

Malformation[a]	Strain	Agent	Reference
Cleft palate	Long-Evans	6-Aminonicotinamide	63
	Wistar	6-Aminonicotinamide	237
	Sprague-Dawley	B-Aminoproprionitrile	21,292
	Wistar	1β-DArabinofuranosyl-cytosine	74
	Holtzman	β-Methasone	282
	?	Chlorocyclizine	168
	Holtzman	Dexamethasone	282
	?	Dithiocarbamate	229
	Wistar	Ethylenethiourea	244
	Wistar	5-Fluorocytosine	72
	Long-Evans	Folic acid deficiency	202
	Wistar	Hadacidin	67
	Wistar	Hydroxylaminopurine	70
	Wistar	Mercaptopurine	178
	Wistar	Methophenazine	144
	Sprague-Dawley	Norchlorcyclizine	291
	Wistar	Procarbazine	71
	Wistar	Sulfamonomethoxine	162
	Wistar	Sulfadimethoxine	162
	Holtzman	Triamcinolone	282
	Wistar	Vitamin A (excess)	130,201
Cleft lip	?	Dithiocarbamate	229
	Wistar	Ethylenethiourea	244
Agnathia/micrognathia	CD	Cadmium	76
	?	Dithiocarbamate	229
	Wistar	Ethylenethiourea	244
	Wistar	Hydroxyurea	69
	Wistar	Mercaptopurine	178
	Wistar	Procarbazine	71
	Wistar	Streptonigrin	75
	Sprague-Dawley	Ethylenediaminetetraacetic acid	269

[a] Fifty percent or more of the surviving fetuses have the malformation.

Table IX

Malformations of the Digestive System

Malformation[a]	Strain	Agent	Reference
Diaphragmatic hernia	Wistar	Cadmium	19
Gastroschisis/omphalocoel	Sprague-Dawley	1-(2-Chloroethyl)-3-cyclohexyl-1-nitrosourea	271
	?	Dexamethasone	278
	?	Hypoglycin A	228

[a] Fifty percent or more of the surviving fetuses have the malformation.

Table X

Malformations of the Cardiovascular System

Malformation[a]	Strain	Agent	Reference
Heart	Long-Evans	Folic acid deficiency	15
	Long-Evans	Trypan blue	123
	Sprague-Dawley	Trypan blue	290
	Wistar	Vitamin A deficiency	303
Aortic arches	Wistar	Vitamin A deficiency	303

[a] Fifty percent or more of the surviving fetuses have the malformation.

Table XI

Malformations of the Urinary System

Malformation[a]	Strain	Agent	Reference
Renal agenesis	Wistar	Arsenate	26
	Wistar	Mercaptopurine	178

[a] Fifty percent or more of the surviving fetuses have the malformation.

REFERENCES

1. Adcock, E. W., Teasdale, F., August, C. S., Cox, S., Battaglia, F., and Naughton, M. (1973). Human chorionic gonadotropin: Its possible role in maternal lymphocyte suppression. *Science* **181**, 845–847.
2. Adelmann, H. B. (1925). The development of the neural folds and cranial ganglia of the rat. *J. Comp. Neurol.* **39**, 19–171.
3. Ader, R., and Conklin, P. M. (1963). Handling of pregnant rats: Effects on emotionality of their offspring. *Science* **142**, 411–412.
4. Adlard, B. P. F., and Dobbing, J. (1972). Vulnerability of developing brain. V. Effects of fetal and postnatal undernutrition on regional brain enzyme activities in three-week-old rats. *Pediatr. Res.* **6**, 38–42.
5. Al-Abbass, A. H., and Schultz, R. L. (1966). Phagocytic activity of the rat placenta. *J. Anat.* **100**, 349–359.
6. Amoroso, E. C. (1952). Placentation. *In* "Marshall's Physiology of Reproduction" (A. S. Parker, ed.), 3rd ed., Vol. 2, pp. 127–311. Longmans, Green, New York.
7. Anderson, I., and Morse, L. M. (1966). The influence of solvent on the teratogenic effect of folic acid antagonist in the rat. *Exp. Mol. Pathol.* **5**, 134–145.
8. Anderson, J. W. (1959). The placental barrier to gamma globulins in the rat. *Am. J. Anat.* **104**, 403–420.
9. Angulo Y González, A. W. (1932). The prenatal growth of the albino rat. *Anat. Rec.* **52**, 117–138.
10. Armstrong, R. C., and Monie, I. W. (1966). Congenital eye defects in rats following maternal folic acid deficiency during pregnancy. *J. Embryol. Exp. Morphol.* **16**, 531–542.
11. Asling, C. W., Nelson, M. M., Dougherty, H. L., Wright, H. V., and Evans, H. M. (1960). The development of cleft palate resulting from maternal pteroylglutamic (folic) acid deficiency during the latter half of gestation in rats. *Surg., Gynecol. Obstet.* **111**, 19–28.
12. Astaurov, B. L. (1975). "Experimental Animals in Developmental Biology," pp. 517–542. "Nauka" Publishing House, Moscow (in Russian).
13. Austin, C. R., and Bishop, M. W. H. (1957). Fertilization in mammals. *Biol. Rev. Cambridge Philos. Soc.* **32**, 296–349.
14. Baba, T., and Araki, E. (1959). Morphogenesis of malformation due to excessive vitamin A. I. Morphogenesis of exencephaly. *Osaka City Med. J.* **5**, 9–15.
15. Baird, C. D. C., Nelson, M. M., Monie, I. W., and Evans, H. M. (1954). Congenital cardiovascular anomalies induced by pteroylglutamic acid deficiency during gestation in the rat. *Circ. Res.* **2**, 544–554.
16. Banerjee, B. N., and Durloo, R. S. (1973). Incidence of teratological anomalies in control Charles River C-D strain rats. *Toxicology* **1**, 151–154.
17. Barr, M., Jr. (1971). The role of the time allowed for mating on variability of fetal weight in rats. *Teratology* **4**, 1–6.
18. Barr, M., Jr. (1973). Prenatal growth of Wistar rats: Circadian periodicity of fetal growth late in gestation. *Teratology* **7**, 282–288.
19. Barr, M., Jr. (1973). The teratogenicity of cadmium chloride in two stocks of Wistar rats. *Teratology* **7**, 237–242.
20. Barr, M., Jr., Jensh, R. P., and Brent, R. L. (1970). Prenatal growth in the albino rat: Effects of number, intrauterine position and resorptions. *Am. J. Anat.* **128**, 413–428.
21. Barrow, M. V., and Steffek, A. J. (1974). Teratologic and other embryotoxic effects of β-aminopropionitrile in rats. *Teratology* **10**, 165–172.
22. Barrow, M. V., and Taylor, W. J. (1969). A rapid method for detecting malformations in rat fetuses. *J. Morphol.* **127**, 291–306.
23. Bass, R., Beck, F., Merker, H. -J., Neubert, D., and Randhahn, B., eds. (1970). "Metabolic Pathways in Mammalian Embryos During Organogenesis and its Modification by Drugs." Frei Universität, Berlin.
24. Beaudoin, A. R. (1968). Teratogenic activity of six disazo dyes in the Wistar albino rat. *Proc. Soc. Exp. Biol. Med.* **127**, 215–219.
25. Beaudoin, A. R. (1973). Teratogenic activity of 2-amino-1,3,4-thiadiazole hydrochloride in Wistar rats and the protection afforded by nicotinamide. *Teratology* **7**, 65–72.
26. Beaudoin, A. R. (1974). Teratogenicity of sodium arsenate in rats. *Teratology* **10**, 153–157.
27. Beaudoin, A. R. (1976). NAD precursors as antiteratogens against aminothiadiazole. *Teratology* **13**, 95–100.
28. Beaudoin, A. R. (1976). Effect of adenosine triphosphate and adenosine diphosphate on the teratogenic action of trypan blue in rats. *Biol. Neonate* **28**, 133–139.
29. Beaudoin, A. R., and Kahkonen, D. (1963). The effect of trypan blue on the serum proteins of the fetal rat. *Anat. Rec.* **147**, 387–396.
30. Beck, F., and Lloyd, J. B. (1966). The teratogenic effects of azo dyes. *Adv. Teratol.* **1**, 131–193.
31. Beck, F., Lloyd, J. B., and Griffiths, A. (1967). A histochemical and biochemical study of some aspects of placental function in the rat using maternal injection of horseradish peroxidase. *J. Anat.* **101**, 461–478.
32. Beck, F., Lloyd, J. B., and Griffiths, A. (1967). Lysosomal enzyme inhibition by trypan blue; a theory of teratogenesis. *Science* **157**, 1180–1182.
33. Beck, S. L. (1967). Effects of position in the uterus on fetal mortality and in response to trypan blue. *J. Embryol. Exp. Morphol.* **17**, 617–624.
34. Bedford, J. M. (1971). Techniques and criteria used in the study of fertilization. *In* "Methods in Mammalian Embryology" (J. C. Daniel, Jr., ed.), pp. 37–63. Freeman, San Francisco, California.
35. Beer, A. E., and Billingham, R. E. (1973). Maternally acquired runt disease. *Science* **179**, 240–243.
36. Beer, A. E., Billingham, R. E., and Scott, J. R. (1975). Immunogene-

tic aspects of implantation, placentation, and feto-placental growth rates. *Biol. Reprod.* **12**, 176–189.

37. Behrman, S. J. (1971). Implantation as an immunologic phenomenon. *In* ''The Biology of the Blastocyst'' (R. J. Blandau, ed.), pp. 479–494. Univ. of Chicago Press, Chicago, Illinois.

38. Beier, H. M., Kuhnel, W., and Petry, G. (1971). Uterine secretion proteins as extrinsic factors in preimplantation development. *Adv. Biosci.* **6**, 561–574.

39. Bhaskar, S. N. (1953). Growth pattern of the rat mandible from 13 days insemination age to 30 days after birth. *Am. J. Anat.* **92**, 1–53.

40. Biggers, J. D., and Borland, R. M. (1976). Physiological aspects of growth and development of the preimplantation mammalian embryo. *Annu. Rev. Physiol.* **38**, 95–119.

41. Biggers, J. D., and Stern, S. (1973). Metabolism of the preimplantation mammalian embryo. *Adv. Reprod. Physiol.* **6**, 1–59.

42. Billingham, R. E. (1971). The transplantation biology of mammalian gestation. *Am. J. Obstet. Gynecol.* **111**, 469–483.

43. Blandau, R. J. (1949). Embryo-endometrial interrelationships in the rat and guinea pig. *Anat. Rec.* **104**, 331–351.

44. Blandau, R. J., and Jordan, E. S. (1941). The effect of delayed fertilization on the development of the rat ovum. *Am. J. Anat.* **68**, 275–287.

45. Blandau, R. J., and Money, W. L. (1944). Observations on the rate of transport of spermatozoa in the female genital tract of the rat. *Anat. Rec.* **90**, 255–260.

46. Blandau, R. J., Boling, J. L., and Young, W. C. (1941). The length of heat in the albino rat as determined by the copulatory response. *Anat. Rec.* **79**, 453–463.

47. Boling, J. L., Blandau, R. J., Soderwall, A. L., and Young, W. C. (1941). Growth of the graafian follicle and the time of ovulation in the albino rat. *Anat. Rec.* **79**, 313–331.

48. Bragg, H. (1922). Disturbances in mammalian development produced by radium emanation. *Am. J. Anat.* **30**, 133–161.

49. Brambell, F. W. R., and Halliday, R. (1956). The route by which passive immunity is transferred from mother to fetus in the rat. *Proc. R. Soc. London, Ser. B* **145**, 170–178.

50. Brent, R. L. (1964). The production of congenital malformations using tissue antisera. II. The spectrum and incidence of malformations following the administrations of kidney antiserum to pregnant rats. *Am. J. Anat.* **115**, 525–542.

51. Brent, R. L., and Jensh, R. P. (1967). Intra-uterine growth retardation. *Adv. Teratol.* **2**, 140–228.

52. Brent, R. L., Johnson, A. J., and Jensen, M. (1971). The production of congenital malformations using tissue antisera. VII. Yolk-sac antiserum. *Teratology* **4**, 255–276.

53. Bridgman, J. (1948). A morphological study of the development of the placenta of the rat. I. An outline of the development of the placenta of the white rat. *J. Morphol.* **83**, 61–85.

54. Brinster, R. L. (1971). Measuring embryonic enzyme activity. *In* ''Methods in Mammalian Embryology'' (J. C. Daniel, Jr., ed.), pp. 215–227. Freeman, San Francisco, California.

55. Brinster, R. L. (1974). Embryo development. *J. Anim. Sci.* **38**, 1003–1012.

56. Brock, N., and von Kreybig, T. (1964). Teratogenese als pharmakologischtoxikologisches Problem. I. Allgemeine Grundlagen: Normale Embryonalentwicklung von Ratte und Meerschweinchen. *Arzneim.-Forsch.* **14**, 655–664.

57. Browman, L. G., and Ramsey, F. (1943). Embryology of microphthalmos in *Rattus norvegicus. Arch. Ophthalmol.* **30**, 338–351.

58. Brunschwig, A. E. (1927). Notes on experiments in placental permeability. *Anat. Rec.* **34**, 237–244.

59. Burlingame, P. L., and Long, J. A. (1939). The development of the heart in the rat. *Univ. Calif., Berkeley, Publ. Zool.* **43**, 249–319.

60. Butcher, E. O. (1929). Development of somites in the white rat. *Am. J. Anat.* **44**, 381–439.

61. Butcher, R. L., and Fugo, N. W. (1967). Overripeness and the mammalian ova. II. Delayed ovulation and chromosome anomalies. *Fertil. Steril.* **18**(3), 297–302.

62. Butcher, R. L., Blue, J. D., and Fugo, N. W. (1969). Overripeness and the mammalian ova. III. Fetal development at midgestation and at term. *Fertil. Steril.* **20**, 223–231.

63. Chamberlain, J. G. (1966). Development of cleft palate induced by 6-aminonicotinamide late in rat gestation. *Anat. Rec.* **156**, 31–40.

64. Chamberlain, J. G. (1972). 6-Aminonicotinamide (6-AN)-induced abnormalities of the developing ependyma and choroid plexus as seen with the scanning electron microscope. *Teratology* **6**, 281–286.

65. Chamberlain, J. G., and Nelson, M. M. (1963). Congenital abnormalities in the rat resulting from single injections of 6-aminonicotinamide during pregnancy. *J. Exp. Zool.* **153**, 285–300.

66. Chaube, S. (1973). Protective effects of thymidine, 5-aminoimidazolecarboxamide and riboflavin against fetal abnormalities produced in rats by 5-(3,3-dimethyl-1-triazene)imidazole-4-carboxamide. *Cancer Res.* **33**, 2231–2240.

67. Chaube, S., and Murphy, M. L. (1963). Teratogenic effect of hadacidin (a new growth inhibitory chemical) on the rat fetus. *J. Exp. Zool.* **152**, 67–73.

68. Chaube, S., and Murphy, M. L. (1964). Teratogenic effects of 5-chlorodeoxyuridine on the rat fetus; protection of physiological pyrimidines. *Cancer Res.* **24**, 1986–1989.

69. Chaube, S., and Murphy, M. L. (1966). The effects of hydroxyurea and related compounds on the rat fetus. *Cancer Res.* **26**, 1448–1457.

70. Chaube, S., and Murphy, M. L. (1969). Teratogenic effects of 6-hydroxylaminopurine in the rat—Protection by inosine. *Biochem. Pharmacol.* **18**, 1147–1156.

71. Chaube, S., and Murphy, M. L. (1969). Fetal malformations produced in rats by *N*-isopropyl-*a*-(2-methyl-hydrazino)-*p*-toluamide hydrochloride (procarbazine). *Teratology* **2**, 23–32.

72. Chaube, S., and Murphy, M. L. (1969). The teratogenic effects of 5-flurocytosine in the rat. *Cancer Res.* **29**, 554–557.

73. Chaube, S., Kury, G., and Murphy, M. L. (1967). Teratogenic effect of cyclophosphamide in the rat. *Cancer Chemother. Rep.* **51**, 363–379.

74. Chaube, S., Kreis, W., Uchida, K., and Murphy, M. L. (1968). The teratogenic effect of 1β-D-arabinofuranosylcytosine in the rat. Protection by deoxycytidine. *Biochem. Pharmacol.* **17**, 1213–1226.

75. Chaube, S., Kuffer, F. R., and Murphy, M. L. (1969). Comparative teratogenic effects of streptonigrin (NSC-45383) and its derivatives in the rat. *Cancer Chemother. Rep.* **53**, 23–31.

76. Chernoff, N. (1973). Teratogenic effects of cadmium in rats. *Teratology* **8**, 29–32.

77. Christie, G. A. (1964). Developmental stages in somite and post-somite rat embryos, based on external appearance, and including some features of the macroscopic development of the oral cavity. *J. Morphol.* **114**, 263–286.

78. Clark, C. C., Minor, R. R., Koszalka, T. R., Brent, R. L., and Kefalides, N. A. (1975). The embryonic rat parietal yolk sac: Changes in the morphology and composition of its basement membrane during development. *Dev. Biol.* **46**, 243–261.

79. Clemetson, C. A. B., Moshfeghi, M. M., and Mallikarjuneswara, V. R. (1971). The surface charge on the five-day rat blastocyst. *In* ''The Biology of the Blastocyst'' (R. J. Blandau, ed.), pp. 193–205. Univ. of Chicago Press, Chicago, Illinois.

80. Cockroft, D. L. (1976). Comparison of *in vitro* and *in vivo* development of rat foetuses. *Dev. Biol.* **48**, 163–172.

81. Coggeshall, R. E. (1964). A study of diencephalic development in the albino rat. *J. Comp. Neurol.* **122**, 241–269.

82. Cohlan, S. Q. (1954). Congenital anomalies in the rat produced by the excessive intake of vitamin A during pregnancy. *Pediatrics* **13,** 556–569.

83. Coleman, R. D. (1965). Development of the rat palate. *Anat. Rec.* **151,** 107–118.

84. Conney, A. H., and Burns, J. J. (1972). Metabolic interactions among environmental chemicals and drugs. *Science* **178,** 576–586.

85. Coyle, I., Wayner, M. J., and Singer, G. (1976). Behavioral teratogenesis: A critical review. *Pharmacol., Biochem. Behav.* **4,** 191–200.

86. Cramer, D. V., and Gill, T. J., III (1975). Genetics of urogenital abnormalities in ACI inbred rats. *Teratology* **12,** 27–32.

87. Currie, G. A., Van Doornick, W., and Bagshawe, K. D. (1968). Effect of neuraminidase on the immunogenicity of early mouse trophoblast. *Nature (London)* **219,** 191–192.

88. Daniel, J. C., Jr., ed. (1971). "Methods in Mammalian Embryology." Freeman, San Francisco, California.

89. Davies, J., and Glasser, S. R. (1968). Histological and fine structural observations on the placenta of the rat. *Acta Anat.* **69,** 542–608.

90. Davies, J., and Hesseldahl, H. (1971). Comparative embryology of mammalian blastocysts. *In* "The Biology of the Blastocyst" (R. J. Blandau, ed.), pp. 27–48. Univ. of Chicago Press, Chicago, Illinois.

91. Davis, S. D., Nelson, T., and Shepard, T. H. (1970). Teratogenicity of vitamin B$_6$ deficiency: Omphalocele, skeletal and neural defects, and splenic hypoplasia. *Science* **169,** 1329–1330.

92. Deane, H. W., Rubin, B. L., Driks, E. C., Lobel, B. L., and Leipsner, G. (1962). Trophoblastic giant cells in placentas of rats and mice and their probable role in steroid hormone production. *Endocrinology* **70,** 407–419.

93. DeMyer, W. (1964). Vinblastine-induced malformations of face and nervous system in two rat strains. *Neurology* **14,** 806–808.

94. DeMyer, W. (1965). Cleft lip and jaw induced in fetal rats by vincristine. *Arch. Anat. Histol. Embryol.* **48,** 181–186.

95. DeMyer, W., and Baird, I. (1973). Techniques for prenatal neurosurgical operations on rat fetuses and for obtaining postnatal survivors. *Teratology* **7,** 89–98.

96. Dencker, L. (1976). Tissue localization of some teratogens at early and late gestation related to fetal effects. *Acta Pharmacol Toxicol.* **39**(1), 1–131.

97. Denenberg, V. H., and Whimbey, A. E. (1963). Behavior of adult rats is modified by the experiences their mothers had as infants. *Science* **144,** 1192–1193.

98. Diamond, M. C., Krech, D., and Rosenzweig, M. R. (1964). The effects of an enriched environment on the histology of the rat cerebral cortex. *J. Comp. Neurol.* **123,** 111–120.

99. Dickmann, Z. (1969). Shedding of zona pellucida. *Adv. Reprod. Physiol.* **4,** 187–206.

100. Dickmann, Z. (1971). Egg transfer. *In* "Methods in Mammalian Embryology" (J. C. Daniel, Jr., ed.), pp. 133–145. Freeman, San Francisco, California.

101. Dickmann, Z., and Dey, S. K. (1974). Steroidogenesis in the preimplantation rat embryo and its possible influence on morula-blastocyst transformation and implantation. *J. Reprod. Fertil.* **37,** 91–93.

102. Diewert, V. M. (1976). Graphic reconstructions of craniofacial structures during secondary palate development in rats. *Teratology* **14,** 291–314.

103. Dorenbos, J. (1973). Morphogenesis of the sphenooccipital and the presphenoidal synchondrosis in the cranial base of the fetal Wistar rat. *Acta Morphol. Neerl. Scand.* **11,** 63–74.

104. Dorgan, W. J., and Schultz, R. L. (1971). An *in vitro* study of programmed death in rat placental giant cells. *J. Exp. Zool.* **178,** 497–512.

105. Ducibella, T., Albertini, D. F., Anderson, E., and Biggers, J. D. (1975). The preimplantation mammalian embryo: Characterization of intercellular junctions and their appearance during development. *Dev. Biol.* **45,** 231–150.

106. Duckett, S. (1971). The morphology of tellurium-induced hydrocephalus. *Exp. Neurol.* **31,** 1–16.

107. Dvořak, M. (1962). Morphogenesis of the liver of rats (Morfogenesa Krysich Jater). *Cesk. Morfol.* **10,** 182–196.

108. Edidin, M. (1972). Histocompatibility genes, transplantation antigens and pregnancy. *In* "Transplantation Antigens" (B. D. Kahan and R. A. Reisfeld, eds), pp. 75–114. Academic Press, New York.

109. Edwards, J. A. (1968). The external development of the rabbit and rat embryo. *Adv. Teratol.* **3,** 239–263.

110. Enders, A. C. (1965). A comparative study of the fine structure of the trophoblast in several hemochorial placentas. *Am. J. Anat.* **116,** 29–68.

111. Enders, A. C. (1971). The fine structure of the blastocyst. *In* "The Biology of the Blastocyst" (R. J. Blandau, ed.), pp. 71–94. Univ. of Chicago Press, Chicago, Illinois.

112. Enders, A. C. (1971). Techniques for studying fine structure of cleavage, blastocyst and early implantation stages. *In* "Methods in Mammalian Embryology" (J. C. Daniel, Jr., ed.), pp. 247–259. Freeman, San Francisco, California.

113. Enders, A. C., and Schlafke, S. (1967). A morphological analysis of the early implantation stages in the rat. *Am. J. Anat.* **120,** 185–226.

114. Everett, J. W. (1935). Morphological and physiological studies of the placenta in the albino rat. *J. Exp. Zool.* **70,** 243–285.

115. Everett, J. W. (1956). The time of release of ovulating hormone from the rat hypophysis. *Endocrinology* **59,** 580–585.

116. Fawcett, D. W., Wislocki, G. B., and Waldo, C. M. (1947). The development of mouse ova in the anterior chamber of the eye and in the abdominal cavity. *Am. J. Anat.* **81,** 413–443.

117. Fedinec, A. A. (1962). Observations on the passage of tetanus toxin through the placenta and fetal membranes of the rat. *Lab. Invest.* **11**(7), 536–543.

118. Ferm, V. H. (1966). Congenital malformations induced by dimethyl sulphoxide in the golden hamster. *J. Embryol. Exp. Morphol.* **16,** 49–54.

119. Ferm, V. H., and Beaudoin, A. R. (1960). Absorptive phenomena in the explanted yolk sac placenta in the rat. *Anat. Rec.* **137,** 89–92.

120. Ferrando, G., and Nalbandov, A. V. (1968). Relative importance of histamine and estogen on implantation in rats. *Endocrinology* **83,** 933–937.

121. Fisher, D. L., and Smithberg, M. (1972). Early and late effects of *in vitro* exposure of preimplantation mouse embryos to trypan blue. *Teratology* **6,** 159–166.

122. Forsberg, J. C. (1961). On the development of the cloaca and the perineum and the formation of the urethral plate in female rat embryos. *J. Anat.* **95,** 423–436.

123. Fox, M. H., and Goss, C. M. (1956). Experimental production of a syndrome of congenital cardiovascular defects in rats. *Anat. Rec.* **124,** 189–207.

124. Fridhandler, L., and Zipper, J. (1964). Studies *in vitro* of rat yolk sac: Biosynthetic activities, respiration, and permeability to hemoglobin. *Biochim. Biophys. Acta* **93,** 526–532.

125. Fritz, H., and Hess, R. (1970). Ossification of the rat and mouse skeleton in the perinatal period. *Teratology* **3,** 331–338.

126. Fugo, N. W., and Butcher, R. L. (1966). Overripeness and the mammalian ova. I. Overripeness and early embryonic development. *Fertil. Steril.* **17,** 804–814.

127. Fujikura, T. (1970). Kidney malformations in fetuses of A × C line 9935 rats. *Teratology* **3,** 245–250.

128. Garro, F., and Pentschew, A. (1964). Neonatal hydrocephalus in the

offspring of rats fed during pregnancy non-toxic amounts of tellurium. *Arch. Psychiatr. Nerven kr.* **206,** 272–380.

129. Geber, W. F. (1966). Developmental effect of chronic maternal audiovisual stress on the rat fetus. *J. Embryol. Exp. Morphol.* **16,** 1–16.

130. Giroud, A., and Martinet, M. (1956). Tératogenèse par hautes doses de vitamine A en fonction des stades du dévelopement. *Arch. Anat. Microsc. Morphol. Exp.* **45,** 77–99.

131. Goedbloed, J. F. (1976). Embryonic and postnatal growth of rat and mouse. IV. Prenatal growth of organs and tissues: Age determination, and general growth pattern. *Acta Anat.* **95,** 8–33.

132. Goldman, E. E. (1909). Die äussere und inner Sekretion des gesunden und kranken Organismus im Lichte der "vitalen Farbung." *Beitr. Klin. Chir., Teil I* **64,** 192–266.

133. Greengard, O. (1971). Enzymic differentiation in mammalian tissues. *Essays Biochem.* **7,** 159–205.

134. Haddad, R. K., Rabe, A., Laqueur, G. L., Spatz, M., and Valsamis, M. P. (1969). Intellectual deficit associated with transplacentally induced microcephaly in the rat. *Science* **163,** 88–90.

135. Hafez, E. S. E. (1971). Some maternal factors affecting physiochemical properties of blastocysts. *In* "The Biology of the Blastocyst" (R. J. Blandau, ed.), pp. 139–191. Univ. of Chicago Press, Chicago, Illinois.

136. Hamner, C. E. (1970). Composition of oviductal and uterine fluids, *Adv. Biosci.* **6,** 561–574.

137. Harper, M. J. K. (1964). Observations on amount and distribution of prenatal mortality in a strain of albino rats. *J. Reprod. Fertil.* **7,** 185–209.

138. Haseman, J. K., and Hogan, M. D. (1975). Selection of the experimental unit in teratology studies. *Teratology* **12,** 165–172.

139. Hemsworth, B. N. (1968). Embryopathies in the rat due to alkane sulphonates. *J. Reprod. Fertil.* **17,** 325–334.

140. Henneberg, B. (1937). Normaltafel zur Entwicklungsgeschichte der Wanderratte (*Rattus norvegicus* Erxleben). *In* "Normalfaflen zur Entwicklungsgeschichte der Wibeltiere" (F. Keibel, ed.), p. 15, Fischer, Jena.

141. Herbert, M. C., and Graham, C. F. (1974). Cell determination and biochemical differentiation of the early mammalian embryo. *Curr. Top. Dev. Biol.* **8,** 151–178.

142. Hjortsjö, C.-H. (1946). The earlier pulmonal morphogenesis of the albino rat and the dog during 24 and 60 hours, respectively. An experimental embryologic investigation. *Lunds Univ. Arsskr. Avd. 2:* **42,** 1–24.

143. Holson, J. F., Scott, W. J., Gaylor, D. W., and Wilson, J. G. (1976). Reduced interlitter variability in rats resulting from a restricted mating period and reassessment of the "litter effect." *Teratology* **14,** 135–142.

144. Horvath, C., and Druga, A. (1975). Action of the phenothiazine derivative methophenazine on prenatal development in rats. *Teratology* **11,** 325–330.

145. Hoshino, K. (1968). Strain-differences in the skeletal development in the fetus of rats and mice. *Annu. Rep. Res. Inst. Environ. Med., Nagoya Univ.* **16,** 59–68.

146. Huber, C. G. (1915). The development of the albino rat *Mus Norvegicus albinus.* I. From the pronuclear stage to the stage of mesoderm anlage: End of the first day to the end of the ninth day. *J. Morphol.* **26,** 3–114.

147. Hurley, L. S., and Swenerton, H. (1966). Congenital malformations resulting from zinc deficiency in rats. *Proc. Soc. Exp. Biol. Med.* **123,** 692–696.

148. Hurley, L. S., Gowan, J., and Swenerton, H. (1971). Teratogenic effects of short-term and transitory zinc deficiency in rats. *Teratology* **4,** 199–204.

149. Husain, S. M., Bélanger-Barbeau, M., and Pellerin, M. (1970). Malformation in the progeny of thalidomide treated male rats. *Can. Med. Assoc. J.* **103,** 163–164.

150. Inouye, M. (1976). Differential staining of cartilage and bone in fetal mouse skeleton by alcian blue and alizarin red S. *Congential Anomalies* **16,** 171–173.

151. Job, T. T., Leibold, G. J., and Fitzmaurice, H. A. (1935). Biological effects of roentgen rays. The determination of critical periods in mammalian development with x-rays. *Am. J. Anat.* **56,** 97–117.

152. Joffe, J. M. (1969). "Prenatal Determinents of Behavior." Academic Press, New York.

153. Johnson, E. M. (1965). Electrophoretic analysis of abnormal development. *Proc. Soc. Exp. Biol. Med.* **118,** 9–11.

154. Johnson, E. M. (1971). Subcellular effects of teratogens on vertebrate embryos and their associated membranes. *Adv. Biosc.* **6,** 561–574.

155. Johnson, E. M., and Spinuzzi, R. (1966). Enzymic differentiation of rat yolk-sac placenta as affected by a teratogenic agent. *J. Embryol. Exp. Morphol.* **16,** 271–288.

156. Johnson, E. M., and Spinuzzi, R. (1968). Differentiation of alkaline phosphatase and glucose-6-phosphate dehydrogenase in rat yolk-sac. *J. Embryol. Exp. Morphol.* **19,** 137–143.

157. Jollie, W. P. (1964). Fine structural changes in placental labyrinth of the rat with increasing gestational age. *J. Ultrastruct. Res.* **10,** 27–47.

158. Jollie, W. P. (1965). Fine structural changes in the junctional zone of the rat placenta with increasing gestational age. *J. Ultrastruct. Res.* **12,** 420–438.

159. Jollie, W. P. (1968). Changes in the fine structure of the parietal yolk sac of the rat placenta with increasing gestational age. *Am. J. Anat.* **122,** 513–532.

160. Juchau, M. R. (1972). Mechanisms of drug biotransformation reactions in the placenta. *Fed. Proc., Fed. Am. Soc. Exp. Biol.* **31,** 48–51.

161. Kalter, H. (1968). How should times during pregnancy be called in teratology. *Teratology* **1,** 231–234.

162. Kato, T., and Kitagawa, S. (1973). Production of congenital anomalies in fetuses of rats and mice with various sulfonamides. *Congenital Anomalies* **13,** 7–14 and 17–23.

163. Kernis, M. M. (1971). Abnormal yolk sac function induced by chlorambucil. *Experientia* **27,** 1329–1331.

164. Kernis, M. M., and Johnson, E. M. (1969). Effects of trypan blue and Niagara blue 2B on the *in vitro* absorption of ions by the rat visceral yolk sac. *J. Embryol. Exp. Morphol.* **22,** 115–125.

165. Kimmel, C. A., and Wilson, J. G. (1973). Skeletal deviations in rats: Malformations or variations. *Teratology* **8,** 309–316.

166. Kimmel, C. A., Wilson, J. G., and Schumacher, H. J. (1971). Studies on metabolism and identification of the causative agent in aspirin teratogenesis in rats. *Teratology,* **4,** 15–24.

167. King, C. T. G., Weaver, S. A., and Narrod, S. A. (1965). Antihistamines and teratogenicity in the rat. *J. Pharmacol. Exp. Ther.* **147,** 391–398.

168. King, C. T. G., Horigan, E., and Wilk, A. L. (1972). Fetal outcome from prolonged versus acute drug administration in the pregnant rat. *In* "Drugs and Fetal Development" (M. A. Klingberg, A., A. Abramovici, and J. Chemke, eds.), pp. 61–75. Plenum, New York.

169. King, H. D. (1915). On the weight of the albino rat at birth and the factors that influence it. *Anat. Rec.* **9,** 213–231.

170. Kirby, D. R. S. (1965). The role of the uterus in the early stages of mouse development. *Preimplantation Stages Pregnancy, Ciba Found. Symp., 1965* pp. 325–344.

171. Kirby, D. R. S. (1971). Blastocyst-uterine relationship before and during implantation. *In* "The Biology of the Blastocyst" (R. J. Blandau, ed.), pp. 393–411. Univ. of Chicago Press, Chicago, Illinois.

172. Kochhar, D. M. (1975). The use of *in vitro* procedures in teratology.

Teratology **11**, 273–288.

173. Koyama, T., Handa, J., Handa, H., and Matsumoto, S. (1970). Methylnitrosourea-induced malformations of brain in SD-JCL rat. *Arch. Neurol. (Chicago)* **22**, 342–347.

174. Krehbiel, R. H., and Plagge, J. C. (1962). Distribution of ova in the rat uterus. *Anat. Rec.* **143**, 239–241.

175. Kreybig, T. von (1965). Die teratogene Wirkung von Cyclophosphamid während der embryonalen Entwicklungsphase bei der Ratte. Naunyn-Schmiedebergs *Arch. Exp. Pathol.* **252**, 173–195.

176. Kreybig, T. von (1969). The critical sensitivity of the developmental phase and the organotropic action of different teratogenic agents; receptors of morphogenesis in the mammalian embryo. *In* "Teratology, Proceedings of a Symposium Organized by the Italian Society of Experimental Teratology" (A. Bertelli and L. Donati, eds.), pp. 152–159. Excerpta Med. Found. Amsterdam.

177. Kulay, L., and de Moraes, F. F. (1965). A histochemical study on the Reichert membrane of the rat's placenta. *Acta Histochem.* **22**, 309–321.

178. Kury, G. S., Chaube, S., and Murphy, M. L. (1968). Teratogenic effect of some purine analogues on fetal rats. *Arch. Pathol.* **86**, 395–402.

179. Lambson, R. O. (1966). An electron microscopic visualization of transport across rat visceral yolk sac. *Am. J. Anat.* **118**, 21–52.

180. Lejour-Jeanty, M. (1965). Etude morphologique et cytochimique du développement du palais primaire chez le rat. *Arch. Biol.* **76**, 99–168.

181. Lendon, R. G., and Ralis, Z. (1971). Normal posture and deformities of lower limbs in rat fetuses with experimentally produced spina bifida. *Dev. Med. Child. Neurol., Suppl.* **25**, 50–57.

182. Ljungkvist, I. (1972). Attachment reaction of rat uterine luminal epithelium. IV. The cellular changes in the attachment reaction and its hormonal regulation. *Fertil. Steril.* **23**, 847–865.

183. Long, J. A., and Burlingame, P. L. (1938). The development of the external form of the rat, with some observations on the origin of the extra embryonic coelom and foetal membranes. *Calif. Univ., Berkeley, Publ. Zool.* **43**, 143–183.

184. Macri, J. N., Joshi, M. S., and Evans, M. I. (1973). An antenatal diagnosis of spina bifida in the Lewis rat. *Nature (London), New Biol.* **246**, 89–90.

185. Mayer, J. F., and Fritz, H. I. (1974). The culture of preimplantation rat embryos and the production of allophenic rats. *J. Reprod. Fertil.* **39**, 1–9.

186. Mayersbach, H. (1958). Zur Frage des Proteinüberganges von der Mutter zum Foeten. I. Befunde an Ratten am ende der Schwangerschaft. *Z. Zellforsch. Mikrosk. Anat.* **48**, 479–504.

187. Merker, H. -J., and Villegas, H. (1970). Electronenmikroskopische Untersuchungen zum Problen des Stoffaustausches zwischen Mutter und Keim bei Rattenembryonen des Tages 7–10. *Z. Anat. Entwicklungst gesch.* **131**, 325–346.

188. Metz, C. B., and Monroy, A., eds. (1967). "Fertilization: Comparative Morphology, Biochemistry, and Immunology," Vol. 1. Academic Press, New York; Vol. 2 (1969).

189. Milaire, J. (1956). Contributions à l'étude morphologique et cytochimique des bourgeons de membres chez le rat. *Arch. Biol.* **67**, 297–391.

190. Moffat, D. B. (1959). Developmental changes in the aortic arch system of the rat. *Am. J. Anat.* **105**, 1–36.

191. Monie, I. W. (1965). Comparative development of rat, chick and human embryos. *In* "Supplement to the Teratology Workshop Manual. A Collection of Lectures and Demonstrations from the 2nd Workshop in Teratology, Berkeley, California," pp. 146–162. Pharm. Manuf. Assoc. Washington, D.C..

192. Monie, I. W., Kho, K. G., and Morgan; J. (1965). Dissection proce-

dure for rat fetuses permitting alizarin red staining of skeleton and histological study of viscera. *In* "Supplement to the Teratology Workshop Manual. A Collection of Lectures and Demonstrations from the 2nd Workshop in Teratology, Berkeley, California," pp. 163–173. Pharm. Manuf. Assoc., Washington, D.C.

193. Monroy, A. (1965). Biochemical aspects of fertilization. *In* "The Biochemistry of Animal Development" (R. Weber, ed.), Vol. 1, pp. 73–135. Academic Press, New York.

194. Morgan, W. C. (1953). Inherited congenital kidney absence in an inbred strain of rats. *Anat. Rec.* **115**, 635–639.

195. Moriss, G. M., and Steele, E. C. (1974). The effect of excess vitamin A on the development of rat embryos in culture. *J. Embryol. Exp. Morphol.* **32**, 505–514.

196. Mossman, H. W. (1937). Comparative morphogenesis of the fetal membranes and accessory uterine structure. *Contrib. Embryol. Carnegie Inst.* **26**, 129–246.

197. Moya F., and Thorndike, V. (1962). Passage of drugs across the placenta. *Am. J. Obstet. Gynecol.* **84**, 1778–1798.

198. Murphy, M. L. (1959). A comparison of the teratogenic effects of five polyfunctional alkylating agents on the rat fetus. *Pediatrics* **23**, Suppl. 231–244.

199. Myers, J. A. (1917). Studies on the mammary gland. II. The fetal development of the mammary gland in the female albino rat. *Am. J. Anat.* **22**, 195–223.

200. Nalbandov, A. V. (1971). Endocrine control of implantation. *In* "The Biology of the Blastocyst" (R. J. Blandau, ed.), pp. 383–392. Univ. of Chicago Press, Chicago, Illinois.

201. Nanda, R. (1970). The role of sulfated mucopolysaccharides in cleft palate production. *Teratology* **3**, 237–244.

202. Nelson, M. M., Wright, H. V., Asling, C. W., and Evans, H. M. (1955). Multiple congenital abnormalities resulting from transitory deficiency of pteroylglutamic acid during gestation in the rat. *J. Nutr.* **56**, 349–363.

203. Netzloff, M. L., Chepenik, K. P., Johnson, E. M., and Kaplan, S. (1968). Respiration of rat embryos in culture. *Life Sci.* **7**, 401–405.

204. Neubert, D., and Merker, H. -J., eds. (1975). "New Approaches to the Evaluation of Abnormal Embryologic Development." Thieme, Stuttgart.

205. Neubert, D., Merker, H. -J., Köhler, E., Krowke, R., and Barrach, H. J. (1971). Biochemical aspects of teratology. *Adv. Biosci.* **6**, 575–621.

206. New, D. A. T. (1971). Methods for the culture of post-implantation embryos of rodents. *In* "Methods in Mammalian Embryology" (J. C. Daniel, Jr., ed.), pp. 305–319. Freeman, San Francisco, California.

207. New, D. A. T. (1973). Studies on mammalian fetuses *in vitro* during the period organogenesis. *In* "The Mammalian Fetus *in vitro*" (C. R. Austin, ed.), pp. 15–65. Chapman & Hall, London.

208. New, D. A. T., and Brent, R. L. (1972). Effect of yolk-sac antibody on rat embryos grown in culture. *J. Embryol. Exp. Morphol.* **27**, 543–553.

209. New, D. A. T., and Coppola, P. T. (1970). Effects of different oxygen concentrations on the development of rat embryos in culture. *J. Reprod. Fertil.* **21**, 109–118.

210. New, D. A. T., and Coppola, P. T. (1970). Development of explanted rat fetuses in hyperbaric oxygen. *Teratology* **3**, 153–163.

211. New, D. A. T., and Stein, K. F. (1964). Cultivation of post-implantation mouse and rat embryos on plasma clots. *J. Embryol. Exp. Morphol.* **12**, 101–111.

212. New, D. A. T., Coppola, P. T., and Cockroft, D. L. (1976). Improved development of head-fold rat embryos in culture resulting from low oxygen and modifications on the culture system. *J. Reprod. Fertil.* **48**, 219–222.

213. New, D. A. T., Coppola, P. T., and Cockroft, D. L. (1976). Comparison of growth *in vitro* and *in vivo* of post-implantation rat embryos. *J.*

Embryol. Exp. Morphol. **36,** 133–144.

214. Nicholas, J. S. (1962). Experimental methods and rat embryos. *In* "The Rat in Laboratory Investigation" (E. J. Farris and J. O. Griffith, eds.), pp. 51–67. Hafner, New York.

215. Nishimura, K. (1974). A microdissection method for detecting thoracic visceral malformation in mouse and rat fetuses. *Congenital Anomalies* **14,** 41–46.

216. Nishimura, H., and Miller, J. R., eds. (1969). "Methods for Teratological Studies in Experimental Animals and Man." Igaku Shoin, Tokyo.

217. Noer, H. R., and Mossman, H. W. (1947). Surgical investigation of the function of the inverted yolk sac placenta in the rat. *Anat. Rec.* **98,** 31–37.

218. Nolen, G. A. (1969). Variations in teratogenic response to hypervitaminosis A in three strains of albino rat. *Food Cosmet. Toxicol.* **7,** 209–214.

219. Noyes, R. W., Dickmann, Z., Doyle, L. L., and Gates, A. H. (1963). Ovum transfers, synchronous and asynchronous, in the study of implantation. *In* "Delayed Implantation" (A. C. Enders, ed.), pp. 197–212. Univ. of Chicago Press, Chicago, Illinois.

220. Odor, D. L., and Blandau, R. J. (1951). Observations on fertilization and the first segmentation division in rat ova. *Am. J. Anat.* **89,** 29–62.

221. Ogawa, T. (1968). Strain difference in organogenesis in some strains of mice and rats. *Annu. Rep. Res. Inst. Environ. Med., Nagoya Univ.* **16,** 51–58.

222. Owers, N. O., and Blandau, R. J. (1971). Proteolytic activity of the rat and guinea pig blastocyst in vitro. *In* "The Biology of the Blastocyst" (R. J. Blandau, ed.), pp. 207–223. Univ. of Chicago Press, Chicago, Illinois.

223. Padykula, H. A. (1958). A histochemical and quantitative study of enzymes of the rat placenta. *J. Anat.* **92,** 118–129.

224. Padykula, H. A., Deren, J. J., and Wilson, T. H. (1966). Development of structure and function in the mammalian yolk sac. I. Development morphology and vitamin B_{12} uptake of the rat yolk sac. *Dev. Biol.* **13,** 310–348.

225. Payne, G. S., and Deuchar, E. M. (1972). An in vitro study of functions of embryonic membranes in the rat. *J. Embryol. Exp. Morphol.* **27,** 533–542.

226. Perraud, J. (1976). Levels of spontaneous malformation in the CD rat and the CD-1 mouse. *Lab. Anim. Sci.* **26,** 292–300.

227. Persaud, T. V. N. (1969). Developmental abnormalities in the rat induced by the amino acid leucine. *Naturwissenschaften* **56,** 37–38.

228. Persaud, T. V. N. (1972). Teratogenic effect of hypoglycin-A. *Adv. Teratol.* **5,** 77–92.

229. Petrova-Vergieva, T., and Ivanova-Tchemishanska, L. (1972). Assessment of the teratogenic activity of dithiocarbamate fungicides. *Food Cosmet. Toxicol.* **11,** 239–244.

230. Pineau, H., and Capuano, G. (1971). Correlation between age and length during prenatal growth of the rat (*Rattus norvegicus albinus*). *Arch. Anat. Pathol.* **19,** 441–444.

231. Pinsker, M. C., Sacco, A. G., and Mintz, B. (1974). Implantation-associated proteinase in mouse uterine fluid. *Dev. Biol.* **38,** 285–290.

232. Pinsky, L., and Fraser, F. C. (1960). Congenital malformations after a two-hour inactivation of nicotinamide in pregnant mice. *Br. Med. J.* **2,** 195–197.

233. Posner, H. S., Graves, A., King, C. T. G., and Wilk, A. (1967). Experimental alteration of the metabolism of chlorcyclizine and the incidence of cleft palate in rats. *J. Pharmacol. Exp. Ther.* **155,** 494–505.

234. Price, D. (1936). Normal development of the prostate and seminal vesicles of the rat with a study of experimental postnatal modification. *Am. J. Anat.* **60,** 79–128.

235. Psychoyos, A. (1966). Recent researches on egg implantation. *In* "Egg Implantation" (G. E. W. Wolstenholme and M. L'Connor, eds), pp. 4–28. Little, Brown, Boston, Massachusetts.

236. Ray, O. S., and Barrett, R. J. (1975). Behavioral, pharmacological and biochemical analysis of genetic differences in rats. *Behav. Biol.* **15,** 391–417.

237. Ritter, E. J., Scott, W. J., and Wilson, J. G. (1975). Inhibition of ATP synthesis associated with 6-aminonicotinamide (6-AN) teratogenesis in rat embryos. *Teratology* **12,** 233–238.

238. Robinson, A. (1889). Observations on the earlier stages in the development of the lungs of rats and mice. *J. Anat. Physiol. Norm. Pathol. Homme Anim.* **23,** 224–241.

239. Robkin, M. A., Shepard, T. H., and Tanimura, T. (1972). A new in vitro culture technique for rat embryos. *Teratology* **5,** 367–376.

240. Rogers, W. M. (1929). The development of the pharynx and the pharyngeal derivatives in the white rat (*Mus norvegicus albinus*). *Am. J. Anat.* **44,** 283–317.

241. Rosenzweig, M. (1971). Effects of environment on development of brain and behavior. *In* "Biopsychology of Development" (E. Tobach, L. R. Aronson, and E. Shaw, eds.), pp. 303–342. Academic Press, New York.

242. Roux, C., Taillemite, J. L., Aubry, M., and Dupuis, R. (1972). Effets tératogènes comparés du chlorhydrate du [*trans*-1-4-bis(2-chlorobenzyl aminomethyl)cyclohexane] (AY-9944) chez le Rat Wistar et le rat Sprague-Dawley. *C. R. Seances Soc. Biol. Ses Fil.* **166,** 1233–1236.

243. Roux, C., Aubry, M., Dupuis, R., and Horvath, C. (1973). Effets tératogènes comparés du triparanol chez le Rat Wister et le Rat Sprague-Dawley. *C. R. Seances Soc. Biol. Ses Fil.* **164,** 1523–1526.

244. Ruddick, J. A., and Khera, K. S. (1975). Pattern of anomalies following single oral doses of ethylenethiourea to pregnant rats. *Teratology* **12,** 277–282.

245. Saxén, L., and Rapola, J. (1969). "Congenital Defects." Holt, New York.

246. Schardein, J. L. (1976) "Drugs as Teratogens." CRC Press, Cleveland, Ohio.

247. Schiebler, T. H., and Knoop, A. (1959). Histochemische und electronmikroskopische Untersuchungen an der Ratten plazenta. *Z. Zellforsch. Mikrosk. Anat.* **50,** 494–552.

248. Schlafke, S., and Enders, A. C. (1967). Cytological changes during cleavage and blastocyst formation in the rat. *J. Anat.* **102,** 13–32.

249. Schlafke, S., and Enders, A. C. (1975). Cellular basis of interaction between trophoblast and uterus at implantation. *Biol. Reprod.* **12,** 41–65.

250. Schultz, R. L. (1970). Placental transport: A review. *Obstet., Gynecol. Surv.* **25,** 979–1020.

251. Schultz, R. L., and Schultz, P. W. (1971). Experimental approaches to placental permeability. *In* "Methods in Mammalian Embryology" (J. C. Daniel, Jr., ed.), pp. 431–441. Freeman, San Francisco, California.

252. Schwind, J. L. (1928). The development of the hypophysis cerebri of the albino rat. *Am. J. Anat.* **41,** 295–319.

253. Scott, W. J., Ritter, E. J., and Wilson, J. G. (1975). Studies on induction of polydactyly in rats with cytosine arabinoside. *Dev. Biol.* **45,** 103–112.

254. Seibel, W. (1974). An ultrastructural comparison of the uptake and transport of horseradish peroxidase by the rat visceral yolk-sac placenta during mid- and late gestation. *Am. J. Anat.* **140,** 213–236.

255. Seitz, P. F. D. (1954). The effects of infantile experiences upon adult behavior in animal subjects. I. Effects of litter size during infancy upon adult behavior in the rat. *Am. J. Psychiatry* **110,** 916–927.

256. Shepard, T. H. (1976). "Catalog of Teratogenic Agents." Johns Hopkins Press, Baltimore, Maryland.

257. Shepard, T. H., Tanimura, T., and Robkin, M. A. (1970). Energy metabolism in early mammalian embryos. *Dev. Biol., Suppl.* **4,** 42–58.

258. Sikov, M. R., and Noonan, T. R. (1958). Anomalous development induced in the embryonic rat by the maternal administration of radiophosporus. *Am. J. Anat.* **103**, 137–162.

259. Sima, A., and Persson, L. (1975). The effect of pre- and postnatal undernutrition on the development of the rat cerebellar cortex. I. Morphological observations. *Neurobiology* **5**, 23–34.

260. Singh S., and Sanyal, A. K. (1974). Skeletal malformations of forelimbs of rat fetuses caused by maternal administration of cyclophosphamide during pregnancy. *J. Anat.* **117**, 179–190.

261. Singh, S., and Sanyal, A. K. (1976). Eye anomalies induced by cyclophosphamide in rat fetuses. *Acta Anat.* **94**, 490–496.

262. Skalko, R. G. (1965). The effect of Co-60-radiation on development and DNA synthesis in the 11 day rat embryo. *J. Exp. Zool.* **160**, 171–182.

263. Skalko, R. G. (1971). Methods for histologic and autoradiographic analysis of the early mouse embryo. *In* "Methods in Mammalian Embryology" (J. C. Daniel, Jr., ed.), pp. 238–246. Freeman, San Francisco, California.

264. Smart, J. L., Adlard, B. P. F., and Bobbing, J. (1972). Effect of maternal and other factors on birth weight in the rat. *Biol. Neonate* **20**, 236–244.

265. Stotsenburg, J. M. (1915). The growth of the foetus of the albino rat from the thirteenth to the twenty-second day of gestation. *Anat. Rec.* **9**, 667–682.

266. Ströer, W. E. H. (1956). Studies on the diencephalon. I. The embryology of the diencephalon of the rat. *J. Comp. Neurol.* **105**, 1–24.

267. Strong, R. M. (1925). The order, time, and rate of ossification of the albino rat (*Mus norvegicus albinus*) skeleton. *Am. J. Anat.* **36**, 313–355.

268. Surani, M. A. H. (1975). Zona pellucida denudation, blastocyst proliferation and attachment in the rat. *J. Embryol. Exp. Morphol.* **33**, 343–353.

269. Swenerton, H., and Hurley, L. S. (1971). Teratogenic effects of a chelating agent and their prevention by zinc. *Science* **173**, 62–64.

270. Tarkowski, A. (1961). Mouse chimaeras developed from fused eggs. *Nature (London)* **190**, 857–860.

271. Thompson, D. J., Molello, J. A., Strebing, R. J., and Dyke, I. L. (1975). Reproduction and teratological studies with 1-(2-chloroethyl)-3-cyclohexyl-1-nitrosourea (CCNU) in the rat and rabbit. *Toxicol. Appl. Pharmacol.* **34**, 456–466.

272. Torrey, T. W. (1943). The development of the urinogenital system of the albino rat. I. The kidney and its ducts. *Am. J. Anat.* **72**, 113–147.

273. Torrey, T. W. (1945). The development of the urinogenital system of the albino rat. II. The gonads. *Am. J. Anat.* **76**, 375–397.

274. Torrey, T. W. (1947). The development of the urinogenital system of the albino rat. III. The urinogenital union. *Am. J. Anat.* **81**, 139–158.

275. Trasler, D. G. (1960). Influence of uterine site on occurrence of spontaneous cleft lip in mice. *Science* **132**, 420–421.

276. Turbow, M. M. (1966). Trypan blue induced teratogenesis of rat embryos cultured *in vitro*. *J. Embryol. Exp. Morphol.* **15**, 387–395.

277. Turbow, M. M., and Chamberlain, J. G. (1968). Direct effects of 6-aminonicotinamide on the developing rat embryo *in vitro* and *in vivo*. *Teratology* **1**, 103–108.

278. Vannier, B., Jequier, R., and Jude, A. (1969). Sensibilité du Rat Wistar à l'action tératogène de la dexaméthasone. *C. R. Seances Soc. Biol. Ses Fil.* **163**, 1269–1272.

279. Vickers, T. H. (1972). Acetazolamide dysmelia in rats. *Br. J. Exp. Pathol.* **53**, 5–21.

280. Waddell, W. J. (1972) Localization and metabolism of drugs in the fetus. *Fed. Proc., Fed. Am. Soc. Exp. Biol.* **31**, 52–61.

281. Wales, R. G. (1975). Maturation of the mammalian embryo: Biochemical aspects. *Biol. Reprod.* **12**, 66–81.

282. Walker, B. E. (1971). Induction of cleft palate in rats with anti-inflammatory drugs. *Teratology* **4**, 39–42.

283. Walker, D. G., and Wirtschafter, Z. T. (1957). "The Genesis of the Rat Skeleton. A Laboratory Atlas." Thomas, Springfield, Illinois.

284. Warkany, J. (1971). "Congenital Malformations." Yearbook Publ., Chicago, Illinois.

285. Warkany, J., and Nelson, R. C. (1940). Appearance of skeletal abnormalities in the offspring of rats reared on a deficient diet. *Science* **92**, 383–384.

286. Warkany, J., and Nelson, R. C. (1941). Skeletal abnormalities in the offspring of rats reared on deficient diets. *Anat. Rec.* **79**, 83–100.

287. Warkany, J., and Takacs, E. (1965). Congenital malformations in rats from streptonigrin. *Arch. Pathol.* **79**, 65–79.

288. Waterman, R. E. (1972). Use of the scanning electron microscope for observation of vertebrate embryos. *Dev. Biol.* **27**, 276–281.

289. Weber, R., ed. (1965). "The Biochemistry of Animal Development," Vol. 1. Academic Press, New York; Vol. 2 (1967); Vol. 3 (1975).

290. Wegener, K. (1961). Über die experimentelle Erzeugung von Herzmissbildungen durch Trypanblau. *Arch. Kreislaufforsch.* **34**, 99–144.

291. Wilk, A. L. (1969). Production of fetal rat malformations by norchlorcyclizine and chlorcyclizine after intrauterine application. *Teratology* **2**, 55–66.

292. Wilk, A. L., King, C. T. G., Horigan, E. A., and Steffek, A. J. (1972). Metabolism of β-aminopropionitrile and its teratogenic activity in rats. *Teratology* **5**, 41–48.

293. Williams, K. E., Lloyd, J. B., Davies, M., and Beck, F. (1971). Digestion of an exogenous protein by rat yolk-sac cultured *in vitro*. *Biochem. J.* **125**, 303–308.

294. Williams, K. E., Kidston, E. M., Beck, F., and Lloyd, J. B. (1975). Quantitative studies of pinocytosis. II. Kinetics of protein uptake and digestion by rat yolk-sac cultured *in vitro*. *J. Cell Biol.* **64**, 123–134.

295. Williams, K. E., Roberts, G., Kidston, M. E., Beck, F., and Lloyd, J. B. (1976). Inhibition of pinocytosis in rat yolk-sac by trypan blue. *Teratology* **14**, 343–354.

296. Wilson, J. G. (1954). Differentiation and the reaction of rat embryos to radiation. *J. Cell. Comp. Physiol.* **43**, Suppl. 1, 11–38.

297. Wilson, J. G. (1965). Embryological consideration in teratology. *In* "Teratology, Principles and Techniques" (J. G. Wilson and J. Warkany, eds.), pp. 251–261. Univ. of Chicago Press, Chicago, Illinois.

298. Wilson, J. G. (1965). Methods for administering agents and detecting malformations in experimental animals. *In* "Teratology, Principles and Techniques" (J. G. Wilson and J. Warkany, eds.), pp. 262–277. Univ. of Chicago Press, Chicago, Illinois.

299. Wilson, J. G. (1966). Effects of acute and chronic treatment with actinomycin D on pregnancy and the fetus in the rat. *Harper Hosp., Bull.* **24**, 109–118.

300. Wilson, J. G. (1973). "Environment and Birth Defects." Academic Press, New York.

301. Wilson, J. G., and Karr, J. W. (1951). Effects of irradiation on embryonic development. I. X-rays on the tenth day of gestation in the rat. *Am. J. Anat.* **88**, 1–34.

302. Wilson, J. G., and Warkany, J. (1948). Malformation in the genitourinary tract induced by maternal vitamin A deficiency in the rat. *Am. J. Anat.* **83**, 357–408.

303. Wilson, J. G., and Warkany, J. (1949). Aortic-arch and cardiac anomalies in the offspring of vitamin A deficient rats. *Am. J. Anat.* **85**, 113–156.

304. Wilson, J. G., and Warkany, J., eds. (1965). "Teratology, Principles and Techniques." Univ. of Chicago Press, Chicago, Illinois.

305. Wilson, J. G., Roth, C. B., and Warkany, J. (1953). Analysis of the syndrome of malformations induced by maternal vitamin A deficiency:

Effects of restoration of vitamin A at various times during gestation. *Am. J. Anat.* **92,** 189–217.

306. Wilson, J. G., Jordan, H. C., and Brent, R. L. (1953). Effects of irradiation on embryonic development. II. X-rays on the ninth day of gestation in the rat. *Am. J. Anat.* **92,** 153–187.

307. Wilson, J. G., Beaudoin, A. R., and Free, H. J. (1959). Studies on the mechanism of teratogenic action of trypan blue. *Anat. Rec.* **133,** 115–128.

308. Wislocki, G. B., and Dempsey, E. W. (1955). Electron microscopy of the placenta of the rat. *Anat. Rec.* **123,** 33–63.

309. Wislocki, G. B., and Padykula, H. A. (1953). Reichert's membrane and the yolk sac of the rat investigated by histochemical means. *Am. J. Anat.* **92,** 117–151.

310. Wislocki, G. B., Deane, H. W., and Dempsey, E. W. (1946). The histochemistry of the rodents placenta. *Am. J. Anat.* **78,** 281–345.

311. Witschi, E. (1962). Development: Rat. *In* "Growth, Including Reproduction, and Morphological Development" (P. L. Altman and D. S. Dittmer, eds.), pp. 304–314. Fed. Am. Soc. Exp. Biol., Washington, D.C.

312. Woollam, D. H. M., ed. (1966). "Advances in Teratology," Vol. 1, Academic Press, New York; Vol. 2 (1967); Vol. 3 (1968); Vol. 4 (1970); Vol. 5 (1972).

313. Wright, H. V., Asling, C. W., Dougherty, H. L., Nelson, M. M., and Evans, H. M. (1958). Prenatal development of the skeleton in Long-Evans rats. *Anat. Rec.* **130,** 659–677.

314. Zamenhof, S., and Van Marthens, E. (1971). Study of factors influencing prenatal brain development. *Mol. Cell. Biochem.* **4,** 157–168.

315. Zuckerkandl, E. (1903). Die Entwicklung der Schilddrüse und der Thymus bei der Ratte. *Ergeb. Anat. Entwicklungsgesch.* **21,** 1–28.

Chapter 5

Toxicology

G. Bruce Briggs and Frederick W. Oehme

I. INTRODUCTION

The biological effects of chemicals are determined by the dose of the chemical applied (9). It is the responsibility of the science of toxicology to determine under what conditions and circumstances a certain chemical will be safe, beneficial, or detrimental when applied to one or more biological forms at various dose levels. Federal law requires that new drugs, cosmetics, color additives, food additives, pesticides, household substances, and certain basic chemicals must have efficacy and/or safety data submitted to regulatory agencies prior to marketing (18,45,47,48). Environmental pollutants, such as industrial chemicals, must be evaluated to assess their potential to produce adverse effects if workers come in contact with them or when they are introduced into the air we breathe or the water we drink (11,19,23,43). Medical devices surgically implanted into humans must also be tested for safety and efficacy in laboratory animals before they can be sold (15).

Because of the variabilities involved in experimental technique, and the fluctuating degree of biological response in nonhuman experimental systems, the results of toxicological and pharmacological research are often met with varied interpretations and opinions. The ideal study situation would be to evaluate the safety of a specific chemical directly in human beings. This, however, is impossible; thus, the toxicologist is forced to utilize animal systems designed to elicit a statistically significant measurable biological response to provide an endpoint for evaluation. Therefore, experimental animals have become the customary tool of toxicology.

Most toxicological data have been obtained by animal experimentation, but the practice also has several shortcomings: (a) there are occasional differences in human versus subhuman metabolism of and response to a chemical, and (b) animal subjects are unable to communicate subjective discomfort and/ or the experimenter is not able to detect subtle effects on higher nervous function. A variety of toxicologic studies are therefore performed on large numbers of animals in order to gain an overall assessment of toxicity or pharmacologic effect and to relate that to specific reactions that might occur in humans receiving the same chemical. Because of its small size, ease of handling, and low maintenance costs, the laboratory rat has played an extremely important role in providing the necessary biological data for evaluating the efficacy and safety of chemicals (2,3,5,16,25,35,39,42).

The purpose of this chapter is to examine the role of the laboratory rat in developing these new chemical agents and in evaluating the potential risks from environmental pollutants. Methods for testing of chemicals are presented with special emphasis given to how the proper animal model is selected and used.

II. GENERAL CONSIDERATIONS IN SELECTING ANIMALS FOR TOXICOLOGICAL RESEARCH

Species selection of test animals in toxicologic or pharmacological experiments is often dictated by convenience and precedent. Rodents (rats, mice), rabbits, or guinea pigs are the most commonly used animals. Dogs and cats are used less commonly because of size and cost of conducting a chronic study, especially one of 2 or more year's duration. Large farm animals and primates are used for special purposes.

The objectives of studies using rats must be clearly defined, and the animal model should be selected in accordance with these objectives. Strain and specific inbreeding to produce unique characteristics in certain rats are important experimental criteria which offer the experimenter the opportunity to utilize certain characteristics for a desired objective. The basic purpose of any study which uses the laboratory rat is to develop biological data which are accurately corrected so as to be both reliable and reproducible. The study should be designed so the data can be both biologically and statistically evaluated (46). The conditions of the experiment must be controlled and defined to eliminate variables, so that the data collected can be duplicated by later researchers.

Data collected from studies using the laboratory rat are often applied to extrapolate to the human species coming in contact with a specific chemical agent or a combination of chemicals. This contact may be related to either the normal use of the material, such as when a drug, cosmetic, or food additive is used, or when man is exposed at work or through the environment. The laboratory rat is also useful in providing information concerning what the adverse responses might be when these agents are used in excessive amounts or when unusually high levels are present in the environment. Through the years a series of techniques and procedures have been developed to evaluate toxicity. More recently, a National Academy of Sciences expert panel has generated a document which recommends specific principles and procedures for such laboratory studies (37).

III. CONSIDERATIONS FOR SELECTING THE LABORATORY RAT FOR TOXICOLOGICAL RESEARCH

The laboratory animal in these studies serves as a biological test tube in which test materials are introduced by various routes. The variability of individual parameters and hence the evaluation and reproducibility of the results depend to a large extent on the quality and health of the animals used. Reduced variability not only increases the accuracy of results but also decreases the number of animals needed for each test procedure.

A variety of factors influence the proper selection of the animal model. Usually more than one species is selected when evaluating a specific chemical because there may be many similarities and some differences in toxic responses between species. An attempt is made to select animal models which closely resemble man in the way the test material is absorbed, distributed, produces its effect, and is eliminated. The species employed must also permit the identification of toxic signs which may occur in humans.

In this context, the laboratory rat plays a unique and important role in screening new chemical agents for pharmacological effects and in conducting more sophisticated studies involving drug distribution within the body and mechanisms of drug action and toxicity. Laboratory rats are commonly used as the rodent species of choice, along with the mongrel or Beagle dog as the nonrodent species, to evaluate the safety of new drugs and chemicals.

For screening techniques to determine the pharmacological effects and metabolic state of an experimental compound, rats and mice are most often used. Inbred rats generally respond with greater sensitivity to drugs than do random-bred animals. Screening studies to determine drug effectiveness generally use highly inbred strains to eliminate variability in evaluating the pharmacologic action of a chemical agent. These rodent models are easily housed, are readily available at a minimum of expense, and can be maintained at a large uniform population for each screening procedure. Depending upon the purpose of the experiment, either normal healthy rats or those with genetic or experimentally altered organ functions (which specifically increase the sensitivity to the test drug) are used in pharmacologic screening studies. Specific examples of these rat model systems are listed in Table I. Organs or tissues from normal healthy rats are often assayed for the presence or absence of drugs to evaluate the distribution and biological retention of the compound under study.

Toxicity studies are usually conducted using random-bred stocks of laboratory rats. Inbred species are avoided for this purpose since such a selection may enhance or diminish a reaction which is characteristic of the species. The microbial status of the rat, whether it is monitored, barrier-maintained, defined microbiologically associated, gnotobiotic, or axenic, may also be important depending upon the objectives of the study (33).

At the time of the final decision, selection of the laboratory rat as the rodent model of choice is usually based on the convenience and scientific benefit it provides, and this is balanced against the various nonhuman characteristics present in rats that limit extrapolation of the derived scientific data. One of

Table I

Representative Pharmacological Screening Procedures Used in the Laboratory Rat

Test performed or system tested	Recommended procedure
Screening tests using healthy rats	
Central nervous system screening procedures	Maze studies
	Spontaneous motor activity studies
	Confinement motor activity studies
	Anticonvulsant tests
	Ataxic dose 50 study
Analgesic screening procedures	Eddy hot-plate analgesic method
	Nilsen electroshock analgesic method
	D'Amour-Smith thermal radiation
Antiinflammatory screening methods	Rat foot edema studies
Blood pressure screening procedure	Carotid artery cannulation procedure
Gastrointestinal screening studies	Restraint-induced ulcer study
Anticholinergic drug screen	Mydriasis test
Range-finding studies	14-Day exaggerated dose level studies
Screening studies using genetically or experimentally altered rats	
Gastrointestinal screening studies	Gastric motility study
	Gastric secretion studies
	Gunn rat model
Antihypertensive screening test	Spontaneous hypertensive rat study
Antidiabetes screening procedure	Egyptian sand rat model
	Brattleboro rat model
	Alloxan-induced diabetes study
Rat organ or tissue assay studies	
Cardiovascular screening procedures	Isolated rat auricle study
	Heart norepinephrine assay
Gastrointestinal screening studies	Charcoal traverse study
	Isolated perfused gut studies
Endocrine function study	Adrenal catecholamine assay
Central nervous system screening procedure	Tissue monamine studies

the most convenient aspects of using the laboratory rat is its previous and current widespread use. This has resulted in voluminous and well established scientific literature about the rat. These data provide the baseline background information for selection of the proper species and stock or strain of laboratory rat for the specific purposes of the study.

The relatively small size of the rat permits large numbers of animals to be evaluated; this increases the statistical significance of the data collected. The small size also permits very expensive experimental chemicals to be given in relatively large doses to a significant number of experimental animals. Vivarium and technical labor can be maximally employed with these small-sized animals to generate significant data.

As an animal model, the laboratory rat has considerable uniformity in size and genotype. This permits reliable animal-to-animal and group-to-group comparisons, making data collection much more significant. The uniformity of the laboratory rat physiology and biochemistry, when taken together with its well-documented and reliable developmental characteristics, makes the laboratory rat a highly precise research tool.

Because of its diverse use, supply of laboratory rats is readily available at an economic price. Its disease status results in incoming rats having a relatively short quarantine period, during which it can readily adapt to specific research conditions. These factors permit a rapid start-up in initiating research projects. The rat's reproductive cycle and breeding characteristics makes it easy to breed under research conditions. It has a short gestation and breeding cycle, contributing to its advantageous use in teratology and multigeneration toxicity studies. The relatively short lifetime of the laboratory rat (when compared to larger animal species) permits lifetime evaluations for chronic effects and carcinogenic potential in a relatively short period.

The rat lacks a well-developed vomiting reflex. This permits the oral introduction of drugs that would normally evoke a vomiting response in other animal models and helps to ensure proper dosing and the opportunity for maximal digestive tract absorption. The rapid growth rate of the laboratory rat permits utilization of body weight gain as a sensitive indicator of early toxic effects.

Laboratory rats have a good disposition. They can be easily handled by technicians and animal care personnel following a short training period. Their natural curiosity and moderate cage activity permits rats to be utilized effectively in behavioral studies. In such conditions they perform well and are good indicators of such toxic chemical manifestations as depression, hyperactivity, and interferences with learning.

The natural inclination of rats to groom themselves when in normal health is important. It makes the rats economical to maintain and allows use of the loss of grooming behavior as an indicator of a toxic effect. Rats in a study that are found with an unkempt appearance suggest disruption of one or more biological systems.

As general research tools, the laboratory rat lends itself well to involvement with routine laboratory procedures. Blood pressure, various pharmacological screening procedures, and securing of blood samples are all easily accomplished and well tolerated. Since numbers are important in statistical evaluations, the large litter size generated by most rats permits accumulation of a significant data base. This permits good comparison between control and the various dose level groups. For studies of genetics or birth defects, the ability to generate several "individuals" per pregnant rat is a significant benefit.

IV. VARIABLES TO BE CONTROLLED

The results obtained from laboratory animal studies are dependent primarily upon three general factors: the laboratory animal itself, the environment in which the laboratory animal is being kept, and the characteristics of the drug or chemical being administered. Before a study can be properly executed, all the factors which modify the animal's response must be recognized and controlled so that, as far as possible, the responses are uniform. Each major consideration is dependent upon a number of subordinate factors that must be considered in total.

A. Animal Factors

The species, strain, age, body weight, sex, health status, state of sexual cycle (estrus, pregnancy, lactation, sexual maturity), genetic background (inbred or random-bred), and commercial source must all be specified for the needs of the study being conducted.

Pharmacological screening tests are usually performed in male rats only to eliminate the physiologic variation due to the cycling of female hormones during the estrous cycle. Metabolism, mechanism of action, and toxicity studies are conducted in both sexes to compare differences in efficiency and safety of the drug in the total population. The stage of estrous cycle and whether pregnancy or lactation is present affect the circulatory system and may interfere with normal absorption and distribution of chemicals. The additional stress of pregnancy or lactation usually renders the test material more toxic; studies not designed to evaluate these effects should not use pregnant or lactating rats.

The laboratory rat used in physiological and metabolic studies should be old enough to assure that enzyme systems are sufficiently developed to metabolize the test material. Rats weighing approximately 100 gm are generally used for preclinical safety evaluations because they are rapidly growing, and reduction in rate of body weight gain is one of the most sensitive toxic observations. This age rat is old enough to adapt to cage housing systems and to automatic watering devices. The rats will reach sexual maturity during the dosing period of subchronic toxicity studies; hence effects on the reproductive system may be evaluated. Rats on lifetime bioassay studies for carcinogenicity should be dosed as early in life as possible, with weanling animals usually used (see this volume, Chapter 6).

The intended human use of the new drug should be taken into consideration when selecting the animal size. If the drug is to be used primarily in infants, neonates should be used in the testing program. If the proposed drug is being developed for use in geriatric patients, aged rats should be utilized.

The genetic and biologic characteristics of the laboratory rat must be appropriate for the study objective. Ideally the model used must have a history of use, and the scientist should be familiar with literature references before employing that particular model. The breeder should be able to provide the investigator with background genetic information, such as the number of generations the stock or strain has been in existence. In general, animals with genetic defects or new strains should be avoided unless one of the objectives of the study is to evaluate the drug in an animal model which mimics the human disease under study.

Health status is of paramount importance and is often affected by the source through which the rats are secured. Unless otherwise specified, only totally healthy laboratory rats should be used in drug evaluation. The source of the animals should be reliable and of proven credibility. The incidence of spontaneous diseases in the suppliers stock must be identified and acceptable to the evaluation procedures used in the study being conducted (1,10,30,38). Animals that have been exposed to infectious diseases, received poor nutrition during gestation or the neonatal growth period, or have been in contact with toxic pesticides, disinfectants, or poor housing conditions will perform in an uncontrolled manner during the experiment (33). This will generate unreliable data of no value to the experimenter. A reliable source for the experimental animals will cooperate with the investigator in assuring that the rats received have been conscientiously raised, carefully selected and shipped, and continuously protected from the numerous agents that could detrimentally affect the animals' health status and jeopardize the pending study.

B. Environmental Factors

Environmental factors should be defined and controlled so variability is reduced and will not interfere with data interpretation. The conditions to be followed are those recommended in the "Guide for the Care and Use of Laboratory Animals" (34) and "A Report on Long-Term Holding of Laboratory Rodents" (33). Animal welfare act requirements should be met for animals in quarantine and while on research studies.

The major animal environmental factors which must be controlled include the conditions of shipping and quarantine, the season of the year (as it affects stress and exposure to the shipped animals), room temperature and humidity, barometric pressure if appropriate for the study, chemical and hazardous agents in the air, air circulation, the spectrum and intensity cycle of light, the quantity and quality (constituents, vitamins) of the diet, the quality and method of availability of water, the cage size and material, the type of bedding, the bedding source and frequency of bedding changes, handling procedures (technician handling and socialization, extraneous noise in the experimental procedure, physical contact between animals), and the type and degree of sanitation programs to be used.

An ideal experimental system to house the laboratory rat should permit individual room adjustments within 2°F for any

temperature within a range of 65° to 85°F. The relative humidity should be maintained throughout the year within a range of 30 to 70%. The ideal temperature and humidity for rats is between 65° and 75°F with 45–55% relative humidity. The temperature and humidity should be kept as constant as possible to avoid fluctuation stress. A 24-hr monitoring system should provide a constant record of variations in temperature and humidity. An emergency power system should be available to maintain proper environment in case of energy failure.

Sudden changes in room pressure should be avoided by having a properly balanced air ventilation system. Laboratory animal rooms should have a slightly positive pressure when compared to surrounding areas. This prevents chemical or biological contamination from entering the animal rooms. An even flow of air within the total room area is ideal. Laminar flow mass air displacement systems provide the most uniform distribution of air and offer air exchange rates up to 150 exchanges per hour. Conventional air circulation systems are adequate if they provide air exchanges of 10 to 15 times per hour with 0.815 ft³ of fresh air per minute per rat.

Biological and chemical air contaminants, such as microorganisms, insecticides, pesticides, disinfectants, and air deodorizers, should be avoided unless they have been previously tested for toxicity and found inert. If the vivarium is located close to manufacturing areas, precautions should be taken to assure that aerosol chemicals do not enter the animal room ventilation system. The most common noxious constituent of air in animal rooms is ammonia. This contaminant is toxic and must be removed by a good sanitation program and adequate ventilation.

Lighting at the rate of 100 ftc of light at the cage level is recommended. Lighting with a spectrum as close to natural sunlight as possible is desirable. Animal rooms should be windowless, and a time-controlled lighting system should provide a regular diurnal lighting cycle of 12 hr each of light and dark.

Laboratory animals should be provided with clean water and food at all times unless the protocol indicates otherwise. The manufacturer should provide data as to the constituents and quality control procedures used to ensure stability and uniformity of the diet. In general, it is advisable to use a brand of feed produced specifically for rats by a company with nationwide distribution. Water should be free of significant bacteria and chemical contaminants and should not contain unpalatable compounds which could interfere with its acceptance by the animals.

The recommendations for cage size and materials made in the "Guide for the Care and Use of Laboratory Animals" (34) should be carefully followed. The cage selected should correspond to that required for the largest size to which the rats will grow during the study. Rats on short-term studies may be housed in groups if appropriate for the protocol. Generally, rats on long-term and carcinogenicity studies should be individually housed to assure accurate dosing in the feed and to avoid cannabilization of dead animals. Stainless steel or galvanized wire hanging cages are generally used for short-term and chronic studies because of ease of cleaning and sanitation requirements. Plastic boxes are preferred for reproduction and carcinogenicity studies to keep the young rat pups warm and to keep them from falling through the grid floor. Such cages are also used to keep potential carcinogens in a confined area.

Soft wood bedding should be avoided because it may produce alterations in liver microsomal enzymes (44). Any bedding utilized should be clean, free of contaminants, and dust-free (47).

Once the study is implemented, the handling procedures to which the rats are subjected are extremely important. Experience and trained animal care and technical personnel should handle the rats so as to not cause pain or discomfort to the animals. Improper restraint may produce behavioral changes which interfere with the judging and evaluation of significant chemically produced clinical alterations.

All animals used in the study must be individually identified by tattoo, ear punching, or color markings. In addition, the individual animal's cage should be marked with the same identifying number or symbol. Cage identification must be moved to any new cage to which the animal is changed.

A regular scheduled sanitation program for water bottles, feeders, cages, racks, and room must be conducted. Maintenance of this sanitation system is necessary to reduce room odor and limit microbial populations. (See Volume I, Chapter 7).

C. Drug Factors

Such items as the physical and chemical characteristics (solubility, concentration, stability, pH) of the experimental compound, its route and method of administration, the potential stress of administration, suspending agents used in the formulation, particle size, dosage form, and documented or suspected drug interations are important variables that influence the biological response to be observed. Most of these variables may be predetermined by the inherent characteristics of the chemical under test and by the formulation developed for projected commercial availability. However, the route and method of drug administration may be dependent upon the objective of the study and the necessity for obtaining a specific biological concentration or response.

Laboratory rats may be dosed with test materials by several different routes. Oral dosing is most common and may be accomplished by oral gavage, using a rubber catheter (14 French) or a dosing syringe, or by dietary inclusion in the feed or addition to the drinking water. If the chemical is to be added to the feed or water, stability data must be obtained on the

chemical–diet mixture to ensure proper dosing levels throughout the experimental period.

The experimental protocol may also call for topical or skin exposure. This is a frequent route of study of pesticide or insecticide testing or if carcinogenicity is being evaluated by the development of skin tumors. Administration via rectal infusion may be necessary if that particular route is a possibility for the future commercial preparation or if absorption studies from the rectum are a specific objective. Application to the eye may be employed for ophthalmic and cosmetic preparations and as a test for irritability.

Subcutaneous, intradermal, intramuscular, intraperitoneal, and intravenous administrations may also be used, especially if digestive tract or skin absorption of the new chemical is limited and toxic effects are specifically being evaluated. All injections are made with small gauge (21 to 25 gauge) short (¼ to ½-inch) needles. Intravenous injections are usually given in one of the prominent tail veins of the rat; intraperitoneal injections are made through the posterior ventral abdominal wall; intramuscular and subcutaneous injections may be given in the inside thigh tissue; while other subcutaneous injections and intradermal administration of drugs may also occur at any other readily available skin surface.

Exposure to drugs and chemicals via the respiratory route is a more specialized effort and requires highly sophisticated exposure chambers and dosage regulating techniques. Aerosols and vapors are generated for the experimental compound and introduced into closed systems involving the experimental rat. The animal may be exposed via tracheal intubation, through a face mask or cone, by total head exposure, or through placement in an exposure chamber in which the total animal is placed in the chemical-laden atmosphere.

The specific route of administration to the laboratory rat is usually determined by the proposed exposure mode to be used in humans.

V. USE OF THE LABORATORY RAT IN PHARMACEUTICAL TESTING

The system of new drug discovery is founded primarily on using laboratory animals to determine the pharmacology and metabolism of a new chemical and to assess the safety of the drug by performing a series of investigations using laboratory animals (40). The manufacturer of a new drug is responsible for conducting the animal and human tests which are required to show that the test material is both effective and safe under normal use conditions. Data collected from these studies must be evaluated by industrial scientists and submitted to the Food and Drug Administration before the new drug can be marketed. It frequently takes more than 10 years and an expenditure of more than 10 million dollars to develop a single new drug. As an average, more than 5000 chemicals are synthesized and investigated in order to finally produce one marketed drug (14).

The development of new drug is an undertaking of many individualized and often complex steps. The laboratory rat plays an important role in these studies. The key research and development activities for a new drug are schematically presented in Fig. 1; the major steps in this process are outlined by the following sequence.

1. Formulation of chemicals which demonstrate a desired pharmacological effect.

2. Production of a sample of test material to be used for animal and clinical research studies.

3. Perform analytical chemistry procedures to identify the chemical structure, determine stability, and purity of the chemical.

4. Conduct pharmacological screening tests for efficacy using laboratory animal models.

5. Perform metabolism studies to determine the absorption, distribution, and elimination of the test material.

6. Conduct more sophisticated laboratory animal studies to aid in the evaluation of the mechanism of action of the new drug.

7. Conduct preclinical toxicity tests in laboratory animals and submit these data along with data from the studies indicated above to the Food and Drug Administration (FDA) in an Investigational Exemption for a New Drug (IND).

8. Following approval of the IND and clinical protocol by the FDA, phase I clinical trials in healthy human volunteers is initiated using very low dose levels to evaluate the safety and assess blood levels of the new drug.

9. Extensive studies are conducted in laboratory animals to assess the subchronic and chronic toxic effects as well as to characterize the potential for the drug to produce mutagenic, reproduction, teratogenic, and carcinogenic effects.

10. Phase II and III human clinical trials are conducted in normal and diseased human patients to determine the efficacy of the new drug. All data is scientifically evaluated and submitted to the FDA in the form of a New Drug Application (NDA). Following regulatory agency approval, the new drug is marketed.

In recent years, increasing emphasis has been placed on establishing the efficacy, mechanism of action, and safety for any new drugs or chemicals. The recent passage of the Toxic Substances Control Act has expanded this concern to all manufactured chemicals, including those used in industry or for manufacturing purposes. Data collected during the develop-

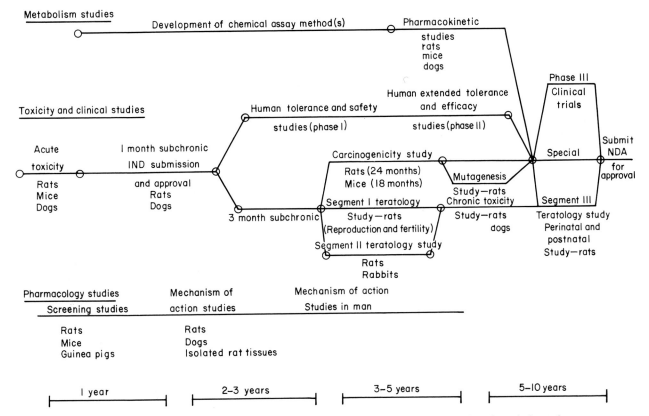

Fig. 1. Key research and development activities in the pharmacologic and toxicologic testing of a typical new drug.

ment of each such chemical must be reported in sequential steps to the FDA. Continual review of this data is performed by regulatory officials at periodic intervals, and the overall progress is critically scrutinized before the new drug or chemical is permitted to be introduced into human volunteers and eventually marketed as a pharmaceutical or commercial product. Included in these evaluations are pharmacologic screening procedures covering a variety of organ systems and tissues and utilizing healthy rats, genetically or experimentally altered rats, or isolated organ or tissue assay preparations (Table I).

The mechanism for the development of drugs includes the submission of a Notice of Claim of Investigational Exemption for a New Drug (IND) to the FDA. This document contains the data accumulated from animal tests for efficacy, metabolism, and acute and subchronic toxicity. Additional animal data may be required, depending upon the type of proposed human administration and whether single, multiple, or unlimited applications of the drug are intended. A summary of the general

FDA guidelines for animal toxicity studies in various phases is provided in Table II. Following a 30-day period after the submission of the IND, clinical investigation in human volunteers may begin.

When human test data are compiled, the complete animal data, including chronic toxicity, carcinogenicity, mutagenicity, and teratogenic potential of the test material as well as all available human information, are submitted to the FDA as a New Drug Application (NDA). Following additional extensive critical review by the regulatory agency and its eventual approval of the documented efficacy and safety, the new drug or chemical may then be marketed by the manufacturer. The fact that this generation of data requires considerable cost and that the formal preparation of documents and applications and their regulatory review necessitates many years of time is readily apparent. The utilization of rodents, such as the laboratory rat, in this overall process serves to reduce costs to an acceptable level, and by using large numbers of rats and because of the

Table II

Food and Drug Administration Guidelines for Animal Toxicity Studies

Category	Duration of human administration	Study phase	Requirements for subacute or chronic toxicity studies
Oral or parenteral	Several days	I, II, III, NDA	2 Species, 2 weeks
	As long as 2 weeks	I	2 Species, 2 weeks
		II	2 Species, as long as 4 weeks
		III, NDA	2 Species, as long as 3 months
	As long as 3 months	I, II	2 Species, 4 weeks
		III	2 Species, 3 months
		NDA	2 Species, as long as 6 months
	6 Months to unlimited	I, II	2 Species, 3 months
		III	2 Species, 6 months or longer
		NDA	2 Species, 12 months (nonrodent), 18 months (rodent)
Inhalation (general anesthetics)		I, II, III, NDA	4 Species, 5 days (3 hr/day)
Dermal	Single application	I	1 Species, single 24-hr exposure followed by 2-week observation
	Single or short-term application	II	1 Species, 20-day repeated exposure (intact and abraded skin)
	Short-term application	III	As above
	Unlimited application	NDA	As above, but intact skin study extended for as long as 6 months
Ophthalmic	Single application	I	Eye irritation tests
	Multiple application	I, II, III	1 Species, 3 weeks daily applications as in clinical use
			1 Species, duration commensurate with period of drug administration
Vaginal or rectal	Single application	I	Local and systemic toxicity in 2 species
	Multiple application	I, II, III, NDA	2 Species, duration and number of applications determined by proposed use
Drug combinations		I	LD_{50} determinations
		II, III, NDA	2 Species, as long as 3 months

rat's relatively short life span, some of the testing procedures are kept to a workable time frame.

With increasing sophistication of laboratory animal studies, larger numbers of laboratory animals are being experimentally treated for longer periods of time. Concern is increasing for the potentially harmful effects of small doses of chemicals exposed to humans for many years. Such considerations require data from lifetime studies in experimental animals. In some instances, studies involving several generations are necessary to provide information on potentially long-term effects in humans. Data from these studies also provide useful insight into the design of human trials and can help to predict what effects might be expected in initial clinical studies in humans. Since rodent and nonrodent species of laboratory animals are used in this evaluation, the benefit from this foundation of knowledge may be negated if improper animal models are used, if animal studies are improperly conducted, or if the data are misinterpreted.

VI. STUDIES USING THE LABORATORY RAT IN EVALUATING THE SAFETY AND EFFICACY OF NEW DRUGS

When using the laboratory rat in studies of biological responses to prospective new drugs, a series of well-defined investigations are undertaken. These fall into the general areas of pharmacological screening investigations, metabolic and pharmacokinetic evaluations, and a variety of toxicological studies investigating acute, subchronic, and chronic exposures, potential effects upon reproduction and the developing individual, and potential for mutagenicity and carcinogenicity.

A. Pharmacological Screening Studies

These evaluations consist of routine biological assay procedures using animal models and organ systems. Intact rats or isolated rat organs suspended in simulated biological fluids are used. Complex pharmacological procedures (employing special rat models) offer insight into the specific mechanism of action for the new drug. The models are selected specifically for the expected therapeutic effect of the compound, and biological response is measured from samples taken from the intact rat or by physiological constituent changes in isolated tissue preparations.

B. Metabolic and Pharmacokinetic Studies

Normal healthy rats are used to evaluate the absorption, dis-

tribution, and elimination of the new drug. How the potential therapeutic agent is altered to produce the desired or undesired effects at receptor sites is studied in this series of investigations (5,6,8). Because the rat absorbs and eliminates many chemicals in a similar manner as man, the laboratory rat is extensively used to evaluate the metabolism of the drug. Following systemic administration of the compound, frequently by intravenous injection, blood levels are determined periodically to establish disappearance rates from the circulating blood. Concentration of the compound at various tissue sites is correlated with the mechanism and duration of action to understand the kinetics of the chemical and its specific tissue distribution, accumulation, and persistence of action. Urine and feces are also quantitated for specific drug and metabolite levels to determine specific routes and rates of elimination.

C. Toxicological Studies

These studies are performed in healthy rats to evaluate potential harmful effects and to determine dose levels at which no toxic effects are detected (50,51). Table III lists the type of toxicological studies employed, together with the numbers of various species, stocks, and strains of laboratory animals commonly used. The use of specific animal models and inbred stock for specific assays is based upon previous experience and published data of their desirability, the biochemical or anatomical characteristics that provide unique advantages for the specific study, the high degree of standardization for rats and other inbred species, the possible presence or absence of physiological response systems similar to those expected in man, and the ease of handling, the suitability of housing, and the advantages of maintaining that particular species under standardized laboratory environmental conditions.

These toxicological studies are designed to produce a variety of effects in a progressive series of experiments leading from the simplest and least complex (acute short-term trials) to longer and more sophisticated studies involving many months or lifetimes and elaborate evaluation techniques and criteria.

Example of a Toxicity Study in Rats

Because exploration of the pharmacologic and toxic effects of new drugs in humans is out of the question, initial clinical evaluations must be performed by laboratory animal experimentation. Rodents, particularly the laboratory rat, are the most common form of laboratory animal used for this type of study.

Following the securing and stabilization of a suitable number of the desired strain of rat, the experimental chemical is administered for a variable period of time, depending upon the specific intent of the study. In most cases, administration is

Table III

Common Species, Stocks, and Strains of Laboratory Animals Used in Various Types of Toxicity Studies

Type of study	Species, stocks, and strains used	Number of animals used
Acute studies		
LD$_{50}$, single dose, 14-day observation	Mice	50–60
	Rats (Sprague-Dawley)	50–60
	Guinea pigs (Hartley)	50–60
	Dogs (Beagles)	6–8
Dose range, 14 days consecutive dosing	Rats (Sprague-Dawley)	5/sex/level, 3 dose levels
	Dogs (Beagles)	2/sex/level, 2–3 dose levels
Subchronic studies		
30-day preclinical	Rats (Sprague-Dawley)	15–20/sex/level, 3 dose levels plus control
	Dogs (Beagles)	2–3/sex/level
90-day subchronic	Rats (Sprague-Dawley)	20–25/sex/level, 3 dose levels plus control
	Dogs (Beagles)	24–30
	Primates (Rhesus)	24–32
Chronic studies		
6 months to 1 year dosing	Rats (Sprague-Dawley)	30–50/sex/level, 3 dose levels plus control
	Dogs (Beagles)	32
Carcinogenicity studies		
18 months to 24 months dosing	Mice (B6C3F1)	50/sex/level, 3–4 dose levels
	Rats (Fischer, Osborne-Mendel, Long-Evans, Sprague-Dawley)	50/sex/level, 3–4 dose levels
Teratology studies	Rats (Sprague-Dawley)	20 females/group
	Rabbits (New Zealand, Dutch Belted)	10 females/group
Mutagenesis studies	Mice (Sprague-Dawley)	Varies with study
	Rats (Sprague-Dawley)	

conducted for 30 days or more using the oral route (via inclusion in the diet or by gavage) or less commonly by intraperitoneal or subcutaneous injection. Animals are then observed daily for alterations in behavior and for any unusual clinical effects resulting from the administration of the drug. Blood samples may be collected from selected individuals in each experimental group on a weekly or biweekly basis. A complete range of blood hematologic parameters, clinical chemistry evaluations, and organ function (enzyme) studies may be performed on the collected specimens depending upon the specific mechanism of action of the drug. Urine samples are often collected by placing the experimental animals in

metabolism cages for 12 or 24-hr periods. Urinalyses are performed, and in some instances metabolite examinations for excretion and breakdown products of the administered chemical may also be conducted.

Since blood samples may be taken by cardiac puncture or, if only small amounts are desired, by skin puncture or bleeding from the medial canthus of the eye, some risk may be involved for the experimental animal. Hence the study protocol may require serial sacrificing of those animals being bled in order to evaluate also any pathologic changes that might be occurring at an early stage in the chemical administration. In those instances, sacrifice is performed, and a gross postmortem examination conducted. Tissues from all major organs and from any portion of the body exhibiting variations from normal are collected and subjected to detailed histopathological study.

Throughout the course of the trial, animals are handled daily and weighed at periodic intervals. Body weight and feed consumption are recorded, together with data on the laboratory evaluations being performed on the blood and urine samples collected. Since this may involve many thousands of data entries, automated data collection and storage techniques have been incorparated as an important part of toxicity studies. Simple automation includes merely entering the collected values (weights, feed consumed, and hematologic or enzyme values) in a computer system that can be retrieved and evaluated in an organized fashion at any time. More sophisticated systems include direct entry of data from balances and analytical instruments into a large computer bank that stores information until the experimenter asks the computer to print out the results in a specified fashion or to manipulate the data for comparative or evaluation purposes.

Even after the period of chemical administration is complete, some individuals in each experimental group may be observed for any effects from chemical withdrawal or as a result of delayed biological response. Sacrifice and gross and microscopic tissue examinations are then performed on all remaining experimental animals.

When all such studies are completed, they are compiled and subjected to statistical evaluation, and the results are evaluated in light of the prospective usefulness of the experimental chemical. Summary reports, together with original detailed documentation of the individual tests performed, are then prepared for submission to and consideration by the FDA as a part of the total safety evaluation for that new proposed commercial compound. Table IV is an outline of a typical subchronic (90 day) oral toxicity study in rats giving the scope of animals used, clinical pathology tests and other measurements performed, pathology criteria including tissues to be evaluated, and physical and chemical data about the new compound and other related chemicals. Specific requirements for the study are also included, as well as a general summary of information already available about the new compound for use in handling this material during the course of the toxicity trial.

The predictive value of tests in laboratory animals for developing drugs to be used in humans is highly variable, and there are many limitations to the extrapolation of data from nonhuman species to man (46). Apart from the obvious differences in anatomical body structure, the differences in organ function and physiology are considerable. Digestive tract differences result in varying absorptive rates following oral dosing; differences in skin characteristics produce absorption variations in topical toxicity trials. Even slight variation in kidney function between the laboratory animal and man produces significant effects on excretion rates and the ability to eliminate

Table IV

Sample Protocol for a Subchronic Toxicity in Rats

Compound Number: XYZ
Test: 90 day oral toxicity study in rats
Starting date: ——————————————
Dose schedule: 7 days a week for 90 consecutive days
Strain: Sprague-Dawley rats
Weight or age: 7–8 weeks

| Number of animals | | Dose level | | Multiple of estimated daily human dose |
Females	Males	(mg/kg)	Route	
15	15	Control (water)	po (gavage)	—
15	15	10		10×
15	15	50		50×
15	15	100		100×

Volume: 5 ml/kg
Clinical pathology tests
    Number of animals: 10 females and 10 males from each group.
    Interval of tests: 1 before dosing, + at 1 week, at 1 month, and at 3 months after initial dose.
        Hematology: Hemoglobin, packed cell volume, total and differential leukocyte count.
        Serum biochemistry: Plasma glucose, urea nitrogen, alkaline phosphatase, glutamic-oxalacetic transaminase.
        Urinalysis: pH, glucose, ketones, bilirubin, occult blood, urine sediment examination.
    Organ weight: Liver, kidneys, adrenals.
General tests: Daily record of general physical health and pharmologic-toxicologic effects will be recorded and maintained.

| | | Testing intervals | |
	Predosing	During drug treatment
Food consumption	2×	Weekly
Water consumption	2×	Weekly
Urine output	2×	During weeks 1, 4, 12
Body weight	2×	Weekly
Clinical ophthalmology	1×	During weeks 4 and 12

Special tests: none
Number of animals to be necropsied: Controls: 6 females 6 males
                                         High dose: 6 females 6 males
Tissues to be fixed and sectioned (in 10% neutral buffered formalin):
    Heart, aorta, lung, thyroid, parathyroid, adrenal, mesenteric lymph node, gonads, pancreas, liver, kidneys, spleen, stomach, intestines (duodenum, ileum, colon), urinary bladder, peripheral nerve and surrounding muscle, bone and bone marrow (sternum), mammary gland, brain with cervical spinal cord, lumbar spinal cord, eye, thymus, prostate, salivary glands, and pituitary from all animals. All gross lesions.
Special study: All gross lesions will be photographed and recorded.
Data pertaining to the new drug:
    Solubility: XYZ is soluble in water.
    Stability: XYZ solutions can be made up weekly. Since XYZ is sensitive to light, the powder and solutions should be stored in the refrigerator.
    Molecular weight: Salt 100; Base 80. Therefore 100 mg of XYZ contains 80

(continued)

Table IV (*Continued*)

mg of the free base. In the report, the dose will be stated as the base (or acid) when applicable, with equivalent amount of salt in parenthesis.

Drug requirement: Approximately 125 gm of XYZ are required from a single lot of material.

Also included in the protocol should be the purpose of the study, a review of the pharmacological and metabolism data, rationale for dose selection, and toxicity data from related chemicals.

doses of the test compounds that otherwise might become toxic due to retention in the body. Even if these physiological differences were minimal, a paramount difference between man and the laboratory animal is that of biochemistry. Enzyme activity levels in liver and other major organs are considerably different between species, and there is no single laboratory animal that mimics man's precise biochemical enzyme pathways for all foreign compounds. Not only is only the quantitative enzymatic difference obvious, but the almost absolute absence of specific biochemical pathways produces total absence of biotransformation ability in some species when dealing with a specific group of chemicals. Cats, for example, have a unique deficiency in the ability to conjugate with glucose, resulting in significant toxicity when that biological pathway is an important excretory mechanism.

For this reason, studies in laboratory animals to evaluate a new drug typically include more than one species, and usually investigate toxicity in at least three laboratory animals. The more uniform the response of a drug among several species, the more likely it is to show the same response in man. Conversely, if the activity of the test agent varies markedly in several species, one should anticipate difficulty in extrapolating that data to humans (28). It is the understanding of these anatomical, physiological, and biochemical differences between species that make utilization of the various laboratory animals for toxicity testing a challenging and always stimulating effort.

D. Acute Toxicity Studies

These evaluate the short-term effects of high doses of the new drug, and in this way provide fundamental data for the design of more complex animal studies (39). The simplest acute toxicity trial is an LD_{50}. This is a single dosing of five or six groups of 10 male and 10 female sexually mature rats, frequently by intravenous administration. The purpose is to determine the dose which kills 50% of a representative rat population. The LD_{50} studies may also be conducted using other routes of administration, with oral dosing and intraperitoneal injection being commonly used. A specific time period is allowed to elapse following the single dose administration, at which time the percentage of live animals in each

dosage group is determined. If one of the dosage groups has an exact 50% mortality, the LD_{50} is readily approximated. In most instances, no one group has this desired 50% reaction, in which case the dose levels and percentage mortality may be plotted on graph paper and the 50% lethal dose determined by graphing and line intersection. With the advent of increased statistical sophistication and computer techniques, the specific LD_{50} dose level is now often calculated by computer applied formulas with the results being expressed as an estimated LD_{50} and a 95% confidence range (28,35). Such acute toxicity trials are intended only as an estimation of the lethality of that compound under the conditions of the experiment, but they do serve to provide preliminary data for other studies involving more frequent dosing. When used in combination with the LD_{50} values for other compounds administered under similar conditions to the same animal species, these estimates may also give useful clues to the type of toxicity expected in other tests.

A second common acute toxicity study involves dosing rats for up to 14 days to determine biological effects. Various dose ranges are used for these short but repeated dosing studies, and they provide data useful for later longer term trials. Several dose groups are used with the dosage in the highest dose group being increased in a geometric or logarithmic fashion until toxic effects occur.

E. Subchronic Toxicity Studies

These studies are designed to provide data on the toxic effects of the new drug and to determine the dose level and time course required for these effects to be produced (39). Dosing during these studies is usually for 30 to 90 days duration (Table III) and in addition to determining clinical effects also involves a variety of biochemical and pathologic evaluations. A representative protocol for this type of study is presented in Table III. Subchronic studies usually involve oral dosing and provide a useful insight into the potentially toxic effects of repeated administration of the new compound on a variety of body systems. Because of the frequency of administration, the dose level is usually reduced from that used in the acute trials. However, by utilizing the laboratory rat as the animal model, large numbers of animals may be included in each dose group and the effect of sporadic mortality is counterbalanced by the numbers of surviving rats.

F. Studies of Chronic Effects

Studies of chronic effects of new drugs are performed in rats dosed with the material for periods from 6 months to the lifetime of the animals (4,39). Usually three dose groups of 20 to 50 male and female rats and a control group of equal or double this size are used. The purpose is to evaluate the potential toxic effects of the drug when administered over an extended period of time compatible with that for which the drug might be used when commercially available. These tests are specifically required for drugs which will be given to human patients for long periods of time, frequently to stabilize chronic disease states. Specific rat models for the counterpart human disease (diabetes, high cholesterol, genetic deficiencies) are especially useful in these studies.

G. Reproductive, Teratogenic, Perinatal and Postnatal Studies

These types of studies are becoming more significant with the increasing emphasis on developmental pathology and toxicology. The purpose of these investigations is to evaluate the toxic effects of the test chemical on the male and female reproductive systems. At the same time, the drug is studied for any potentially harmful effects on the developing fetus when the dam is dosed during a specific period of pregnancy. In addition the potential effect of the drug on the young animal is assessed when the dam is dosed near the end of her pregnancy. Following the Thalidomide incident, guidelines for testing drugs that might cause birth defects were developed in 1966 by FDA (26,27). These Goldenthal Guidelines specify the number of various animal species to be used, the period of pregnancy during which the pregnant dam is exposed to the drug, the techniques by which the potentially teratogenic changes are to be studied, and the evaluation criteria for declaring a new compound a teratogen.

There is no single test for effects on chromosomes and genes adequate to evaluate the *potential mutagenic effect* of a new drug. A series of *in vitro* and *in vivo* studies are usually conducted. Included in the *in vivo* studies are at least two tests using the laboratory rat. The dominant lethal test involves the administration of the drug to male rats which are then sequentially mated to groups of untreated females over the duration of the spermatogenic cycle. The genetic basis for dominant lethality is the induction of chromosomal damage and rearrangements, such as translocations, resulting in nonviable zygotes which are detected as resorption sites in the uterus (31). By dosing a sexually mature male with the test chemical, any mutagenic effects may be produced in his developing sperm. By mating the treated male to proved fertile, but untreated, females, any reduction in fetal viability must result from the drug treatment. Percentages of resorption sites per pregnant female are compared to similar evaluations from nontreated male–female matings.

The second *in vivo* mutagenic study involving the laboratory rat is the host-mediated assay. This procedure involves the introduction of nonpathogenic strains of bacteria into the abdominal cavity of rats. The rat is then dosed with the test drug.

The bacteria are later removed from the abdominal cavity and studied for the formation of mutant strains. The development of strains with unusual growth characteristics is common in bacteria following chemically induced chromosomal alteration. Since bacteria reproduce themselves many times in a few hours, the exposure of normal bacteria via physiological fluid to a mutagenic compound would result in a high likelihood that several mutations would occur in the exposed bacteria. Altered growth characteristics in the mutant strains are easily detected by routine bacteriologic procedures. Unfortunately, the metabolism and distribution of the test drug in the rat will often vary, and the bacteria in the abdominal cavity may not come in contact with the test compound. As a result, this assay may yield false negative results in the presence of a known mutagen if the chemical is not present in the fluid of the abdominal cavity. The host-mediated assay has thus lost favor as an acceptable test (41).

H. Carcinogenic Studies

A vitally important toxicologic assay that has assumed great significance in the past few years is the carcinogenic study. The National Cancer Institute has recently developed guidelines for the evaluation of potential carcinogenic effects from drugs and other chemicals (41). These test agents are dosed to laboratory rats at a schedule which simulates human exposure. Since the life span of the rat approximates 24 months, rats are dosed for this entire period; with the additional time required for pathologic evaluation, the typical carcinogenic study requires 3 or more years to complete. Groups of 50 rats per sex are dosed orally or by the parenteral route. Any rats dying during the course of study are immediately examined for tumor development, and at the completion of the trial, all surviving animals are sacrificed and subjected to complete gross and microscopic pathologic examination for any tumor foci. In these studies particular attention must be given to environmental control parameters to avoid chronic respiratory disease and to assure that contamination of the animal room air with chemicals that might be low-level carcinogens does not occur (see this volume, Chapter 6). For comparative purposes, 1 month of dosing in the laboratory rat is considered equal to 34 months of drug administration in man.

I. Review of Toxicological Data by the Food and Drug Administration for Approval of a New Drug

Following completion of the acute and subchronic toxicity studies, these data together with chemical information and the results from other animal models are submitted to the FDA as a Notice of Claimed Investigational New Drug Application (IND) (29). If, in the opinion of the reviewing officials, these preclinical data demonstrate that there will not be unreasonable hazard in man, clinical studies can be initiated according to a protocol submitted by the drug developer and approved by the FDA (12,40). Studies would be continuing during this time on the chronic toxicity trials and on those involving reproduction, mutagenicity, and carcinogenicity. Following completion of the long-term toxicity studies, the studies dealing with reproduction, teratogenic, perinatal and postnatal effects, mutagenicity, and carcinogenicity data from these investigations and all human clinical trials are submitted to the FDA in the form of a New Drug Application (NDA). After the FDA reviews all the data and approves it, the drug is licensed and may be marketed. This process is illustrated in timed sequence in Fig. 1.

VII. USE OF THE LABORATORY RAT FOR SAFETY AND EFFICACY TESTING OF NONDRUG CHEMICALS

In addition to the development of compounds for therapeutic use in man and animals, a variety of other chemical agents are continually being generated for use as cosmetics, food colorings and additives, agricultural compounds and pesticides, industrial chemicals used in producing consumer products, and numerous chemical by-products of our civilization which enter the environment. Governmental regulations are enforced dealing with all of these areas, and laboratory rats and other animal models are needed for the laboratory evaluation of the toxicity and biological effects from these materials.

A. Cosmetic Evaluation

The laboratory rat is used to evaluate the effects of cosmetics and skin and hair care products on the basic physiology of the skin (17,18,20). These studies are performed under the FDA's responsibility for regulating the safety and efficacy of these chemicals. Skin pigmentation, percutaneous absorption, moisture, and morphological alteration studies are commonly performed with the rat as the animal model. Acute toxicity studies to evaluate the oral and dermal toxicity of cosmetics and cosmetic ingredients are also conducted. New voluntary guidelines developed by the Cosmetic, Toiletry and Fragrance Association (13) and future legislation by the FDA will increase the use of the laboratory rat in subchronic toxicity testing.

B. Food Additive Evaluation

Acute, subchronic, and chronic feeding studies using rats

and other nonrodent species are commonly employed. Although a list of commonly employed food additives generally recognized as safe (GRAS) was developed during the 1960's with the intent of exempting these materials from extensive toxicity testing, reevaluation of many of these chemicals has raised questions about the total safety of their inclusion in man's diet. As a result, all new food additives, and many of those originally placed on the GRAS list, are now subject to intensive toxicological study(24,32,36,48). Compounds which are suspect as cancer producers are evaluated for carcinogenicity using the laboratory rat. Food, drug and cosmetic coloring agents used in foods and cosmetics must undergo lifetime animal carcinogenicity studies before they can be utilized. The FDA regulates these food and color additives.

C. Pesticides and Agricultural Chemicals

The Federal Insecticide, Fungicide, and Rodenticide Act (19) requires that acute oral LD_{50} studies be performed on all compounds used in agriculture, including pesticides (19). Regulatory authority is invested in the Environmental Protection Agency (EPA). Amendments currently under consideration and the recently implemented Toxic Substances Control Act (43) will additionally require subchronic and chronic testing of these materials. The necessity for specific long-term studies dealing with potential carcinogenic properties will result in the laboratory rat being used to a much greater degree than previously. Since many agricultural chemicals have the potential for being contaminants in meat and milk intended for human consumption these compounds must additionally be tested for toxicity according to the food additive regulations.

D. Industrial Chemical Exposures in the Workplace and Home

Safety of the workplace has become an important objective of federal regulations. The National Institute of Occupational Safety and Health, the scientific arm of the Occupational Safety and Health Administration (OSHA), is currently developing threshold limit values for exposure of workers in the workplace to industrial chemicals. Primary concern is for inhalation exposures and potential teratogenic, mutagenic, and carcinogenic effects. Chemicals shown to produce toxic reactions must undergo extensive testing procedures similar to those performed by the drug industry in evaluating new therapeutic agents. The recently passed Toxic Substances Control Act (43) will require that all current and new commercial chemicals undergo extensive testing in laboratory animals prior to their manufacture and distribution. For compounds already being utilized, a specific time schedule is being developed for the submission of toxicity information to regulatory agencies. This act is promulgated by the EPA and requires essentially the same type of toxicological evaluations currently being called for with other drugs and chemicals. Materials used in the home are also subjected to safety evaluation (21). The Federal Hazardous Substances Act (21) requires that compounds and devices for household use be reasonably tested in laboratory situations. The Consumer Product Safety Commission implements control for these types of studies. Laboratory rats are a standard animal model for the vast majority of these toxicity evaluations.

E. Environmental Contaminants

Recently enacted clean air and clean water legislation (11,23) will require that oral and inhalation toxicity studies in rats be performed on compounds found to be polluting the atmosphere and waters and on industrial effluents generated by industrial plants and being discharged into the environment. These environmental safeguards are under the jurisdiction of the EPA and require initial testing of discharges and pollutants in rats with extensive additional studies required if long-term toxicity or health effects are suspected.

VIII. GOOD LABORATORY PRACTICES IN CONDUCTING TOXICITY STUDIES

As the necessity for more sophisticated toxicity evaluations produced expansions and modifications of existing procedures, control of the quality of testing and the evaluation of results became increasingly complex. In the mid-1960s, the World Health Organization published guidlines for the preclinical testing of therapeutic agents (49). With the recent concern for the adequate conduct and interpretation of toxicity studies in the United States, the FDA has recently proposed Good Laboratory Practices regulations (GLPs) (22) to establish standards for the performance of such studies.

These GLPs will have a significant impact on the methods of operation for the pharmaceutical and chemical industries. The regulations were specifically designed for long-term toxicology, teratogenicity, mutagenicity, and carcinogenicity studies for the development of human drugs, biologics, medical devices, food additives, and color additives. However, the influence of these regulations will spread to all toxicology laboratory testing procedures. Many of these techniques will utilize intact laboratory rats or isolated rat tissues.

The animal care section of the GLPs requires that all animals be handled according to the standards set forth in the Animal

Welfare Act and the "Guide for the Care and Use of Laboratory Animals" (34). This section of the new regulations includes the following basic requirements: All incoming rats must be placed in quarantine. A physical examination on all rats must be performed by a qualified person to evaluate the health status in accordance with acceptable veterinary medical practice. All diseased rats must be isolated and management carried our according to the Guide (34). All animals should be free from any disease that would interfere with the study being conducted. Provisions allow for the treatment of sick animals if the drugs used do not interfere with the study and the treatment employed is properly recorded. Rats frequently removed from their cages must have individual permanent identification numbers. All information concerning a specific rat must be recorded using its permanent identification number. Rats should not be housed with other species, and only one study should be conducted in each animal room; this is to avoid possible contamination or dosing of animals with the wrong test material. No rat can be moved to a new location without approval by the study director. Feed and water must be analyzed periodically to assure that interfering contaminants are not present. Feed for the laboratory rats must be stored in a room separate from the test areas. Adequate veterinary care, as defined by the Guide (34) should be provided all animals in the study.

IX. CONCLUSION

Ever since it has been realized that man is not the only animal that responds in a relatively predictable fashion to foreign compounds (7), laboratory animals have been used as a model for drug and chemical testing. In this role, the laboratory rat is becoming increasingly more important for evaluating the safety of such new compounds. The efficient utilization of the rat animal model has been facilitated by improvements made in breeding, ventilation systems and environmental control, sanitation practices, caging and equipment, and animal techniques such as microanalysis of blood samples for hematologic and biochemical evaluations. The laboratory rat has been improved to decrease biological variability and to reduce the incidence of spontaneous diseases. Defined tumor incidence, improved genetic characteristics, and extensive research experience with the laboratory rat have greatly improved the uniformity of its response and the reliability of data thereby obtained.

REFERENCES

1. American Society for Experimental Biology (1971). "Diseases of Laboratory Animals Complicating Biomedical Research." Am. Soc. Exp.

Biol., Chicago, Illinois.

2. Annals of the New York Academy of Sciences (1965). Evaluation and mechanisms of drug toxicity. *Ann. N.Y. Acad. Sci.* **123**, 1–366.

3. Baker, S.B.D.C. (1971). "The Correlation of Adverse Effects in Man with Observations in Animals," Vol. XII. Exerpta Found., Amsterdam.

4. Barnes, J. M., and Denz, F. A. (1954). Experimental methods used in determining chronic toxicity. *Pharmacol Rev.* **6**, 191–242.

5. Boyd, E. M. (1972). "Predictive Toxicometrics." Williams & Wilkins, Baltimore, Maryland.

6. Brodie, B. B. (1969). The absorption of drugs from the gastrointestinal tract. *In* "The Physiological Equivalence of Drug Dosing Forms," Food and Drug Directorate, Ottawa, Canada.

7. Brodie, B. B., and Reid, W. D. (1969). Is man a unique animal in response to drugs? *Am. J. Pharm.* **141**, 21–26.

8. Burns, J. J., Cucinell, S. A., Koster, R., and Conney, A. H. (1965). Application of drug metabolism to drug toxicity. *Ann. N.Y. Acad. Sci.* **123**, 273–286.

9. Casarett, L. J., and Doull, J. (1975). "Toxicology: The Basic Science of Poisons." Macmillan, New York.

10. Charles River Digest (1971). "Disease Diagnosis in Mice and Rats," Vol. X, No. 3. Charles River Dig., Wilmington, Massachusetts.

11. Clean Air Act, PL 91-604 (1970). 42 U.S. Congress, Washington, D.C.

12. Code of Federal Regulations (1960). Section 8.501 (21 CFR 8.501) and Amendments of 1960. (Title II, Pub. L. 86-618 74 Stat 404-407). Food and Drug Admin., Washington, D.C.

13. Cosmetic, Toiletry and Fragrance Association, Inc. (1976). "Safety Substantiation Guidelines." CTFA., Washington, D.C.

14. DeHaen, P. (1976–1977). Research—driving force in the drug industry. Part I and II. *Drug & Cosmet. Ind.*, December 1976 and January 1977.

15. Device Listing (1976). *Fed. Regist* **41**, No. 250.

16. Durbin, C. G., and Robens, J. F. (1965). "The Use of Laboratory Animals for Drug Testing." U.S. Dept. of Health, Education and Welfare, Food and Drug Admin., Washington, D.C.

17. FDC Reports (1976). CTFA launches massive cosmetic ingredient review, September 27, *FDC Rep.* pp. 4–10.

18. Federal Food, Drug, and Cosmetic Act Section 706, 74 Stat 399–403.

19. Federal Insecticide, Fungicide, and Rodenticide Act. Section 754, P.L. 94-40, 89 Stat 16 USC 688 cc-1.

20. Federal Register (1976). Color additives. *Fed. Regist.*

21. Federal Register (1973). Federal Hazardous Substances Act, September 23, 1973, 27012-27038.

22. Federal Register (1976). Non-clinical laboratories studies, proposed regulations for good laboratory practice, November 19, p. 51206.

23. Federal Water Pollution Act (1972) P.L. 92-500, 33 USC.

24. Food and Drug Directorate, Ottawa, Canada (1970). "Guide for Preparation of Submissions on Food Additives." Dept. of National Health and Welfare, Ottawa, Canada.

25. Food and Drug Officials of the United States (1959). "Appraisal of the Safety of Chemicals in Foods, Drugs and Cosmetics." Food and Drug Officials of U.S., Austin, Texas.

26. Goldenthal, E. I. (1968). Current views on safety evaluation of drugs. *FDA Pap.* **2**, 13–18.

27. Goldenthal Guidelines (1966). "Guidelines for Reproduction Studies for Safety Evaluation of Drugs for Human Use." Food and Drug Admin. Washington, D.C.

28. Goldstein, A., Aronow, L., and Kalman, S. M. (1974). "Principles of Drug Action: The Basis of Pharmacology," 2nd ed. pp. 357–393. Wiley, New York.

29. Gyarfas, W. J., and Welch, A. (1969). The IND procedure: Assuring safe and effective drugs. *FDA Pap.* **3**, 27–31.

30. Habermann, R. T., Fletcher, G. W., Jr., and Thorp, W. T. S. (1954). Identification of some internal parasites of laboratory animals. *U.S., Public Health Serv. Publ.* **343.**

31. Hollaender, A. (1973). "Chemical Mutagens; "Principles and Methods for Their Detection," Vol. 3. Plenum, New York.

32. Hopkins, H. (1976). Countdown on color additives. *FDA Consum.* 5–7.

33. Institute of Laboratory Animal Resources (1976). A report on long-term holding of laboratory rodents. *ILAR News* **19**, No. 4,L1-25.

34. Institute of Laboratory Animal Resources (1974). "Guide for the Care and Use of Laboratory Animals." DHEW Publ. No. 74-23, pp. 25–26. Washington, D.C.

35. Litchfield, J. T. (1962). Evaluation of the safety of new drugs by means of tests in animals. *J. Clin. Pharmacol. Therapeutics,* **3**, 665–672.

36. National Academy of Sciences (1959). "Principles and Procedures for Evaluating the Safety of Food Additives," Publ. No. 750. Natl. Acad. Sci., Washington, D.C.

37. National Academy of Sciences (1977). "Principles and Procedures for Evaluating the Toxicity of Household Substances." Natl. Acad. Sci., Washington, D.C.

38. Newberne, P. M., Salmon, W. D., and Hare, W. V. (1961). Chronic murine pneumonia in an experimental laboratory. *Arch. Pathol.* **72**, 106–115.

39. Paget, G. E. (1970). "Methods in Toxicology," Davis, Philadelphia, Pennsylvania.

40. Pharmaceutical Manufacturers Association (1966). "The Investigation of New Drugs; The Conceptual Relation of Studies in Animals to Studies in Man," PMA Rep. Pharm. Manuf. Assoc., Washington, D.C.

41. Sontag, J. M., Page, N. P., and Saffiotti, V. (1976). "Guidelines for Carcinogen Bioassay in Small Rodents," Tech. Rep. Ser. No. 1. Natl. Inst. Health, Washington, D.C.

42. Stilly, F., (1976). "The $100,000 Rat and Other Animal Heroes for Human Health." Putnam, New York.

43. Toxic Substances Control Act (1976). PL94-469, 90 Stat 2003, 15 USC.

44. Vesell, E. S. (1967). Induction of drug-metabolizing enzymes in liver microsomes of mice and rats by softwood bedding. *Science* **157**, 1057–1058.

45. Vorherr, H. (1971). Drug evaluation for efficacy. *Drug Intell. & Clin. Pharm.* **5**, 4–11.

46. Weil, C. S. (1972). Guidelines for experiments to predict the degree of safety of a material for man. *Toxicol. Appl. Pharmacol.* **21**, 194–199.

47. Williams, F. P., and Habermann, R. T. (1962). A study of bacterial contamination in commercially prepared animal feeds and bedding. *Proc. Anim. Care Panel,* February, pp. 11–14.

48. World Health Organization (1970). "Specifications for the Identity and Purity of Food Additives and their Toxicological Evaluation," Publ. No. 445. World Health Organ., Geneva.

49. World Health Organization (1966). "Principles for Pre-clinical Testing of Drug Safety," Publ. No. 341. World Health Organ., Geneva.

50. Zbinden, G. (1964). The problem of the toxicologic examination of drugs in animals and their safety in man. *Clin. Pharmacol. Ther.* **5**, 537–545.

51. Zbinden, G. (1966). Animal toxicity studies; a critical evaluation. *Appl. Ther.* February, pp. 128–133.

Chapter 6

Experimental Oncology

John C. Peckham

I. INTRODUCTION

Cancer is the second most common cause of human death in the United States (328,384) and comprises a hundred or more related diseases (386) that result from many causative factors with complex interactions (12,48). Unraveling the complex etiology and pathogenesis of cancer poses one of the greatest challenges to modern biomedical science (310). Some of the most important advances in experimental oncology have resulted from animal experimentation; and the laboratory rat has been a principal participant.

As the Norway rat (*Rattus norvegicus*) became a popular laboratory animal in the second half of the nineteenth century

(149), its use in cancer research was inevitable. The reports by Yamagima and Itchikawa in 1914 and 1915 of experiments in which skin cancer was induced in rabbits (386) were the culmination of a series of efforts by many individuals to chemically induce cancer in experimental animals. The less fortunate investigators had chosen the wrong species or the wrong compound, did not persist long enough in their experiments, or failed to treat a sufficiently susceptible area of the skin (333). The rat, which has a longer latent period for skin cancer development than the mouse and rabbit, was used in many of these unsuccessful experiments, some as early as 1889 (386).

The first skin cancer was produced in the rat by Jean Clunet of France in 1910 by application of radium (328). The discovery of the first carcinogenic pure chemical compound, 1,2,5,6-dibenzanthracene (dibenz[a,h]anthracene) by Kennaway in 1929 was soon followed by the identification of 3,4-benzpyrene (benzo[a]pyrene) a potent carcinogen in coal tar (145). At present, more than 1500 known and potential carcinogens have been identified (70). These carcinogens include chemical, viral, and physical agents of both exogenous and endogenous origin (144).

The laboratory rat has played an important role in the identification of many of these carcinogens and continues to be the most common laboratory animal used in carcinogenesis experiments today (258,327).

II. RATS AS RESEARCH ANIMALS IN ONCOLOGY

A. Selection of Rats for Oncologic Research

A perfect laboratory animal or other biological test system for oncologic research does not exist. The sensitivity of animal bioassays, unless thousands of animals are used, seldom can detect the induction of tumor incidences below 10% (307). Species variations in the susceptibility to carcinogens are well known (46,73). A specific carcinogenic agent may induce one or more types of tumors that may vary in number and type among different species, within different strains of the same species, and between different sexes, ages, and routes of exposure within the same strain (98,151,248,338,384). Current chemical bioassay studies recognize this problem and utilize both sexes of at least two species, usually the rat and the mouse, for the initial evaluation of carcinogenic potential (329). There also are situations and carcinogens where nonrodent species, such as the hamster, rabbit, dog, and nonhuman primates, have proved useful.

Some characteristics which should be considered in the choice of animal used in cancer research include (a) alert, active and disease free; (b) uniform small size; (c) docile; (d) economical to house; (e) readily available; (f) easy to breed

with high fertility; (g) adaptable to varied experimental conditions, (h) relatively short life span with uniform longevity; (i) well characterized genetically; (j) have spontaneous tumors very rarely or not at all but respond readily and consistently to all classes of carcinogens; (k) have metabolism and pathogenetic responses similar to man; and (l) have ample published information on its normal and abnormal behavioral, physiological, biochemical, toxicological, and pharmacological responses (248,329,342).

Rats are excellent laboratory animals that fit many of the criteria outlined above. They are hardy, readily available, easy to breed, have few serious diseases, are of convenient size, economical to house and maintain, usually docile, genetically well defined, have a relatively short life span, have few spontaneous tumors with well-known incidences in several strains, are sensitive to many carcinogenic agents including known human carcinogens, and have a large volume of published information on their physiologic, pharmacologic, and pathologic responses (248,329,338,342,358,359). Because of these qualities, rats are used successfully in all areas of oncologic research including causation, biological development, biochemical mechanisms, early diagnosis, prevention, and therapy.

B. Stocks and Strains in Common Use

A large number of strains both random-bred (outbred) and inbred, have been used in oncologic research. Many of these rats are used in pharmacologic and toxicologic evaluations of drugs as well as in experimental oncology. This wide distribution is useful because where the biological activity of a chemical compound is known for a specific strain, appropriate dosage levels can be selected more quickly and efficiently for long-term carcinogenesis experiments. (248). There appears to be a trend toward the use of random-bred stocks in chemical carcinogenesis bioassay studies (248,383). Such stocks more closely duplicate the genetic heterogeneity of man and have a variable, moderately high incidence of a few specific tumors with a low incidence of other types. Inbred strains usually have a more constant and relatively high incidence of specific tumors and a low incidence of other types of tumor. Random-bred Sprague-Dawley rats have a moderate to high number of mammary gland and pituitary tumors and few testicular interstitial cell tumors. In contrast, Fischer rats, an inbred strain, have a moderate number of mammary, pituitary, and hematopoietic-lymphoreticular tumors and a high incidence of testicular interstitial tumors (46,264,316). A strain without spontaneous tumors that responds to known carcinogens would simplify the identification of carcinogenic chemicals in bioassay tests, but no such strain is known.

Inbred strains have proved very useful when a carcinogen is

known to induce a specific tumor in high incidence. If the carcinogen is a chemical, that strain can be used to identify and study other structurally similar compounds as well as the cellular mechanism of carcinogenesis (248). Specific tumors having high incidences in inbred strains are valuable as models in cancer therapy studies for both screening new chemotherapeutic drugs and developing improved treatment techniques.

More than 50 random-bred and inbred strains of rats are used in cancer research laboratories throughout the world. Some of these strains are the AIMS, AXC9935 (ACI/Du), August 990 (A990/Du), BD, BDI, BDII, BDIII, BDIV, BIV, BDVI, BDVIII, BDIX, BDX, BR-46, Buffalo (BUF), Charles River CD, Chester Beatty (CB, hooded; CB, Wistar), Copenhagen, Curtis-Dunning, Debreceu, Donryu, Doure-doure, Fischer (F344/Du), Harlan, Holtzman, Hooded, ICI, Japanese, Kasukabe, King (PA), Lewis (LEW), Lister hooded, Long-Evans, Marshall 520 (M520/Du), Moriyama, MRC, NIH Black, New Zealand (NZR and NZBW hooded), Oosawa, Osborne-Mendel (OM), Porton, R-Amsterdam, Rudenstam, Sabra, Sherman, Slonaker, Sprague-Dawley, Thompson, Wistar (WAG/g, WF, WN), Zimmerman 61 and Zygar (17,18,25,26,29,87,149,279).

The most frequently used stocks and strains in the United States are the Sprague-Dawley, Wistar, Osborne-Mendel, Fischer, New Zealand, Holtzman, Long-Evans, and BR-46. The National Cancer Institute Carcinogenesis Bioassay Program currently uses primarily Fischer and Osborne-Mendel strains (342).

C. Experimental Protocols

Because of the diversity of oncologic research, no single protocol or set of procedures will meet the objectives of all cancer studies. Even with recognized chemical carcinogens, such as benzo[a]pyrene, 3-methylcholanthrene, and 2-naphthylamine, oncologists had difficulty reproducing cancer in animals until appropriate experimental conditions were discovered and applied. With weaker carcinogenic agents, the problem of causation becomes more difficult. Past studies in determining carcinogenic activity were frequently inadequate in one or more aspects which appear important by modern standards and sometimes raises serious questions about their results. Some of these deficiencies include genetically poorly defined strains of experimental animals; inadequate details on the amounts of compound administered; unknown chemical purity of the compound tested; insufficient numbers of animals; lack of data on the starting age of animals, the length and frequency of treatment, the latent period to first tumor appearance, and average time of tumor appearance; lack of description of dietary conditions, food intake, and weight gains; inadequate histopathology; and no or imperfect

controls (383). Having recognized that many problems exist and having identified some of the requirements for carcinogen testing, several individuals and groups of investigators, including the Division of Cancer Cause and Prevention of the National Cancer Institute, have published recommendations and guidelines for chemical carcinogenesis bioassay studies (5–8, 14,32,43,44,132,248,335,342,383).

The time of onset for tumor induction is apparently proportional to the duration of exposure and the total dosage received over that period (95,248). In order to detect weak carcinogens, the duration of the test must be long; at least half and preferably more than half of the life span of the animal (303). The average life span of the laboratory rat is 2 to 3 years, with an occasional animal reaching 4 years or more. A period of long finite duration is more practical than total life span studies because a few animals may greatly exceed the mean life span of the species which unnecessarily prolongs the length of an experiment. It has been recommended that no bioassay test be accepted as negative unless a statistically significant proportion of the test animals survives for 30 months, the effective length for rat experiments (363).

1. Carcinogenesis Bioassay Protocols for Young Adult Rats

Reliable demonstration of the carcinogenicity of a chemical hinges not only on characteristics of the compound but also on the experimental animal chosen and the total environment of the test animals (383). In general, the experimental design for carcinogenic testing in a life span study involves the selection of a responsive strain of animals and use of enough young adult males and females in treated and control groups to provide an adequate number of survivors at the termination to permit pathologic and statistical evaluations (248). Two or three dosage levels of the chemical are utilized. The high dosage is the highest level that can be tolerated without toxicity (125). The low dosage should be near the therapeutic dose or threshold exposure level for man (248). The route of administration is usually the same as the anticipated route of exposure for man. If the chemical is added to the diet, food consumption is recorded. Body weight records are maintained for determining weight gains and calculating chemical intake or individual animal dosage. Animals are observed frequently for signs of illness and tumor development. Animals dying or killed during the study are examined by a thorough necropsy to determine the cause of death and to identify the presence of tumors, suspicious lesions, and possible intercurrent disease. Systematic histopathologic evaluations of all tumors, other lesions, and selected organs are done routinely. At termination all animals are examined grossly. Depending upon the protocol, all or a selected number of treated and control animals are examined histopathologically. In some tests, only selected high dosage and control animals are evaluated histopathologically. When specific target organs can be identified, these organs from all animals on the study are evaluated (248).

2. Carcinogenesis Bioassay Protocols for Combined Life Span Studies of Young Adult Rats and Their Offspring

Neonatal and prenatal exposure has been shown to increase the sensitivity of rats to many chemicals, and protocols have been designed utilizing this principle (363). These protocols involve mating at 8 to 9 weeks of age and exposure to the chemical under test during the second half of pregnancy (day 12 for rats). Exposure continues for the desired interval, and these animals constitute a group of young adults under conventional long term tests. Their offspring, which may have been exposed *in utero* and postnatally through maternal milk and excreta, should then be exposed directly for as long as desired after weaning. This procedure results in two different experimental groups, one composed of the parent generation and a second composed of progeny (363).

D. Tumor Transplantation

The ultimate criterion for identifying a carcinogenic agent is the demonstration of malignant lesions, which spread by invasion in the organ of origin, extension into nearby organs, or metastasize to other organs by the bloodstream or lymphatics (253,271). In the absence of this information, the second most reliable method of assessing malignancy is subcutaneous transplantation of a fragment of living suspect tissue to an untreated isologous host. The recipient should be 4 weeks of age and of the same sex as the donor. A malignant transplant must not only survive but grow progressively more than one transplant generation. It should kill the animal by reaching a large size or by metastasizing to other organs (271). Transplantation studies also are a means of correlating the biological behavior of a neoplastic lesion with its histologic pattern. The results of these studies can be used to predict the behavior of tumors of the same tissue and morphology in random-bred animals (271). A further use of tumor transplantation has been made in chemotherapy where a consistent number of tumors having approximately the same size and stage of development are needed for a single experiment. Transplantable tumors were important in the realization that immunity and genetic homogenity played a role in host resistance (140). These tumors also have contributed greatly to the understanding of neoplastic progression, metastasis, growth, other biological behavior, and cellular metabolism (185).

In the "Atlas of Tumor Pathology" fascicle on transplantable and transmissible tumors of animals, ten transplantable tumors of rats were described (352). The "Laboratory Animal Handbook 3" listed 254 transplanted tumors of rats including 174 epithelial tumors, 133 hepatomas, 2 carcinosarcomas, 24 reticuloendothelial tumors, and 54 connective tissue tumors (279). Detailed bibliographies of transplantation experiments have been complied by Roberts and associates (274–278,280–286).

III. RAT TUMORS AS MODELS OF HUMAN NEOPLASTIC DISEASE

A. Validity of Rat Models

Both rat and man have the essential prerequisites for cancer: they are multicellular and possess a population of actively reproducing and growing cells (328). In addition, as vertebrates and mammals, they share many similarities in cellular growth and differentiation (386). Rats have a wide array of neoplasms, both spontaneous and induced, that are nearly all comparable in morphology and biological behavior to those observed in man (61,114,393,394). Some tumors can be demonstrated to undergo progression from hyperplasia to benign lesions to malignancy similar to those of man (80,325,373,386). Spontaneous malignant tumors are rare in rats, however (329).

Man has two peak periods in life when the risk of developing neoplasms is greatest; in childhood and in old age (48). The rat has a similar pattern with appropriate exposure to carcinogens. The probability of developing a neoplasm during childhood may be related to maternal exposure and transplacental transfer of carcinogens. Experimentally this situation can be reproduced with several carcinogens (85,363).

The second peak of cancer risk in both rats and man is observed as a general increase in tumors that is directly associated with increased longevity. In man, 60% of fatal carcinomas occur in people over 60 years of age (14% of the population) and 84% occur in people over 50 (38% of the population) (386). In the rat, the majority of spontaneous tumors occurs toward the end of the life span (383). The frequency of spontaneous mammary gland tumors in rats increases with age. The average latent period in one report was 12 months, with a range of 4.5 to 18 months. The cumulative probability of mammary tumors at the end of 18 months was 70% for females and 49% for males (393). In another report the incidence of mammary tumors at 21 months was 50%, and, at 25 to 30 months, reached 85% (356). Other spontaneous tumors in rats occur with a similar pattern (383).

Since the latency period for tumor induction is related and proportional to total life span, the shorter life span of the rat is an asset in carcinogenesis experiments. The ratio of rat to human life span is 1 : 30 (337). An experiment of 2.5 years in the rat is roughly equivalent to 75 years in man, over 90% of the human life span. The shorter life span of the rat thereby shortens the length of carcinogenesis experiments and makes the results available sooner. In addition, many generations of rats can be studied within a single human generation time. The latent period for human cancer is usually very long depending upon such variables as the carcinogen, the degree of exposure, and individual susceptibility (154,300). For some forms of cancer associated with occupational exposures, latent periods of 1 to 75 years have been reported. Most, however, average 15 to 30 years (154). Positive results from experiments in animals are sometimes available in 1 year or less (248).

Rats and man both respond to a variety of carcinogenic stimuli including chemical, physical, and viral agents, although viral induction of malignant lesions in man is still controversial (65,118,144,231,299,331). In fact, with a few notable exceptions, all chemical carcinogens which are known to induce tumors in man are also active in animals. Arsenicals are an exception still under experimental study (120).

In past decades (as well as presently), carcinogenic agents were suspected first by clinical observations of excess cancers in man in specific occupational groups (65). Subsequently, they were confirmed experimentally in laboratory animals (305,386). In the past few years especially, chemicals that were found first to be carcinogenic in animal tests have in some cases been found subsequently to be carcinogenic by direct observations in human populations (305,380). Examples include diethylstilbestrol, aflatoxin B_1, bis(chloromethyl) ether, and vinyl chloride (180,182,195,305).

A well established carcinogenic effect in any one mammalian species is considered as an indication of the potential carcinogenic effect in humans (44,150,305,364). The value of animal data in the prediction of carcinogenicity for humans has been amply demonstrated and consequently has been repeatedly endorsed by expert national and international committees (114,120,146). This principle has served as a basis for regulations promulgated by the Occupational Safety and Health Administration, the Environmental Protection Agency, and the Food and Drug Administration. Regulatory actions based upon these rulings have been upheld by the United States Judiciary (120).

B. Spontaneous Neoplasia

A spontaneous neoplasm is a tumor observed in an untreated animal having no known cause (61,288). Since spontaneous tumors have an equal opportunity to occur in animals exposed to chemicals as well as in nontreated animals, the oncologist must know the type and incidence of spontaneously occurring tumors and related hyperplasias for the strain of rat being used and the usual time of occurrence for these tumors (248,383). The use of strains with a low spontaneous tumor incidence reduces the problems of interpretation (248).

Spontaneous tumor incidences for several rat strains are well known (46,248,383). These incidences are subject to variation (248,383), which is important since criteria used in determining whether or not a drug is carcinogenic depend upon the presence of an increased tumor incidence, a decrease in the latency of tumor appearance, or the appearance of tumors in organs where spontaneous tumors did not occur (46,89). Assessment of each criterion requires comparison of results obtained in treated animals to those obtained in concurrent controls. In addition, knowledge of spontaneous tumor incidences for the strain used in past experiments aids in evaluating experimental variation (46). This information is essential when evaluating weak carcinogens on a lifetime test where the percentage of induced tumors differs less than 20% from those occurring spontaneously in untreated control groups (248).

The incidence of spontaneous tumors is influenced by the same factors as carcinogenesis in general. These include age, sex, nutrition, endocrine status, degree of immunologic competence, and environment (46,291,383). Some tumors may result from weakly carcinogenic contaminants in food, water, bedding, and cages (46,291,383). Therefore, the use of a standardized protocol with adequate numbers of animals in both treated and untreated groups is essential in the design of meaningful experiments (248,342,383).

Spontaneous neoplasms of high frequency are valuable as models for basic and applied research in cellular mechanisms of carcinogenesis and other areas of oncology. A description of spontaneous neoplasms of rats is presented in detail in Chapter 13, Volume I of this text.

C. Experimentally Induced Neoplasia

Although a carcinogen may cause the appearance of tumors in organs where spontaneous tumors do not usually occur in a given strain, the major effect of experimental carcinogens is to increase the types of tumors usually seen spontaneously and to shorten the period of latency. Another effect is to increase the progression to malignancy in organs where spontaneous hyperplasia and benign tumors are usually seen. The basic morphology of the neoplasms is the same whether spontaneous or experimentally induced.

The tissue specificity of carcinogens varies. Some carcinogens exert effects on a single organ or a few tissues (97). Other carcinogens have effects against a wide variety of tissues (223,355,375,379,383). In experimental oncology this can be very useful since it may be possible to select an organ that is rarely affected by spontaneous tumors and, by selection of the proper dosage and route, to induce a high incidence of a specific tumor. This principle has been used in development of transplantable tumor lines for tumor models which rarely occur spontaneously in animals but which are of great importance in man. One example is the colon adenocarcinoma in rats (83,84,136).

The tissue specificity and the type of tumors induced are greatly affected by the route of exposure. The nature of the carcinogen has a marked influence on this response. Some chemical carcinogens act directly, and others must be activated by cellular enzymes. The absorption from the skin and gastrointestinal tract also varies markedly.

In studies that have as their object the assessment of risk for man, the preferred route of administration should simulate the expected route of human exposure. The pharmacokinetic characteristics of the drug following different routes of exposure and the effect of the route on metabolic activity and detoxication also must be considered in selecting the route of administration (125,363).

The experimental induction of tumors by carcinogens is greatly influenced by age, sex, diet, hormone status, immune state, and other unknown inborn and acquired endogenous tissue predispositions (128,144,255,291,383).

Age has proved a very important factor and has received considerable attention during the past 20 years (98,363). Following results obtained in experimental transplacental carcinogenesis and in epidemiological studies, prenatal exposure became a popular concept (363). Prenatal exposure to a chemical increases the susceptibility of the offspring to the carcinogenic effect of the same chemical or of another chemical later in life, although the carcinogenic effect of the chemical is expressed in the same target organ.

Prenatal exposure of rats to certain nitrosamines and nitrosamides is followed by a high incidence of tumors in the offspring. The tumors in the offspring, especially tumors of the central and peripheral nervous system, appear following treatment of pregnant rats at a dose that does not produce tumors in the mothers. When genetically resistant animals are exposed *in utero* to a carcinogen, not only do many more tumors develop than when exposure occurs during adult life but a high proportion of these grow rapidly and behave as malignant tumors (as in genetically susceptible strains). Such neoplasms are highly antigenic in syngeneic recipients. Therefore, transplacental exposure can circumvent genetically determined resistance. Diethylstilbestrol, a chemical for which carcinogenicity had been demonstrated in five animal species, was shown to be carcinogenic to man following prenatal exposure (124, 305,363).

So far, tumors of offspring have been reported in five animal species (mouse, rat, hamster, guinea pig, and pig) following prenatal exposure to at least 30 substances of various chemical structures, such as urethane, polycyclic hydrocarbons, *N*-nitroso compounds, cycasin, and aflatoxin administered by different routes (363,365). Fetal nervous tissue, when exposed to a variety of carcinogens during intrauterine life, generally appears to have a higher susceptibility as compared to that of adults (363).

While the chemical structure and route of administration seems unrelated to the effect revealed by prenatal exposure, the timing of treatment during pregnancy is often critical. Exposure during the first half of pregnancy tends to result initially in an embryo-toxic effect and subsequently in a teratogenic effect. Exposure in the second half of pregnancy results in a carcinogenic effect. The probable explanation is the sequential change in metabolic competence acquired by different fetal tissues during their development (363). Although prenatal exposure may reveal a carcinogenic effect at unusual target sites and at very low levels of exposure (258), it is not recommended as a routine procedure to replace conventional testing in young adults (303). A more sensitive procedure would be to integrate prenatal exposure with long-term postnatal exposure in order to maximize the sensitivity of the bioassay system (363).

D. Rat Tumor Models

A variety of rat tumors are presented (Table I) as examples of models for human neoplastic diseases. There tumors provide, with varying degrees of analogy, models for the majority of human cancers. In recent years as more carcinogenic chemicals have been identified, the oncologist has been given a greater selection of chemicals to induce neoplastic lesions and a greater selection of target tissues (97,99). Thus, there currently are few tumors in man that have no rat counterpart. Where the incidence of spontaneous or induced tumors is low, transplantation has been used to augment the models available. One of the rat's greatest assets in experimental oncology has been its ability to provide uniform neoplastic tissues in sufficient volume for chemical determinations as needed in chemotherapy, pharmacology, and biochemistry.

Several excellent models of epithelial and mesenchymal cancer now exist in the rat and are discussed below.

1. Breast Cancer

Breast cancer, a significant human cancer, has been studied extensively in the rat. The rat provides an effective model for human breast cancer (137,138,176,370,393). Mammary gland tumors, both benign and malignant, occur as spontaneous and induced lesions in both male and female rats. The incidence of mammary tumors varies greatly within and among different strains. Female Sprague-Dawley rats have been reported to develop from 14 to 57% spontaneous mammary tumors (92,111). Most of these appear after the age of 540 days. Eighty-eight percent of them are benign fibroadenomas. In

Table I

Selected Rat Tumor Models of Human Neoplatic Disease

Organ or tissue tumor	Strain[a]	Carcinogen[b]	Incidence (%)	References
Skin (394)				
Basal cell carcinomas	—	AA	90	394
Squamous cell carcinomas	—	DMBA	70	394
Subcutis and soft tissues (59, 66,147,292,302,326, 372)				
Sarcomas (fibromas, rhabdomyosarcomas, and others)	SD	Radiation	—	66
	—	Chemicals	—	35,46,386
	—	Plastics	35	243,244,246
	—	Metals	32	245
Sarcoma	BUF	NQO	88	217
Malignant fibrous histiocytomas	—	Tannin	100	262
Skeletal muscle				
Rhabdomyosarcoma	OM	MSV	—	250
Breast (1,3,4,68,69,297,393)				
Mammary gland tumors	SD	Unknown (spontaneous)	60	92,111,356
Mammary gland tumors	—	FAA	64	383
Mammary gland adenocarcinomas	SD	DMBA	100	138,176,383,392
Hematopoietic and lymphoreticular system (31,90,100,101,103, 165,167,193,212,213, 313,340,346,352,353)				
Stem cell or monocytic leukemias	F	Unknown	25	90,212,273,313
Stem cell erythroblastic leukemias	LE	TMBA	82	156
Granulocytic leukemia	W	Untreated	1	213
Leukemias and sarcomas	—	DMBA	100	353
Plasmacytomas	SD and OM	MSV	20	139
Lymphocytic leukemia	CB	MSV	80	139
Leukemia	—	NBU	100	31,152,241
Hodgkins disease	W	TB	40	130
Thymoma	BUF	Unknown	54	390
Lymphosarcomas	—	ONCV	70	139,218
Bone (189)				
Osteosarcomas	—	Radiation	90	189
Osteosarcomas	—	Radiation (local)	28	78
Osteosarcomas	NZ	MSV	80	242
Nasal turbinates (208a,258a)				
Esthesioneuroepithelioma and carcinomas	SD	BCME	100	293
	SD	BCME	10	175,177,179,233
Lungs and trachea (174,258a)				
Squamous cell carcinoma	SD	BCME	7	175,233
Bronchioalveolar squamous cell carcinomas	—	MCA	40	81,178,324

(*continued*)

Table I (*Continued*)

Organ or tissue tumor	Strain[a]	Carcinogen[b]	Incidence (%)	References
Heart (163a)				
Neurosarcomas (Anitschkow-cell sarcomas	BD	PDET	36	164
Blood vessels				
Hemangioendothelial sarcomas	SD	ANFOD	100	117
Salivary glands (131)				
Carcinomas and sarcomas	—	DMBA (local)	85	292,320
Liver (10,41a,63,100,183, 316a,347,379,381,391)				
Hepatocellular carcinomas	—	FAA	95	383
Hepatocellular carcinomas	—	Afla. B_1	100	234,236,387
Benign and malignant tumors	—	Urethane	42	374
Pancreas (290,294,295)				
Acinar tumors	SD	HAQ	71	142,143
Acinar tumors	W	AS	17	191
Acinar hyperplasia	W	AS	100	191
Esophagus (97,99,260,296)				
Squamous cell carcinomas	—	EBN	100	260
Squamous cell carcinomas	W	NP	82	162
Squamous cell carcinomas	OM	DHS	75	190
Stomach, squamous (fore-stomach) (57,97,226, 296)				
Squamous cell carcinomas	—	PCH	—	226
Squamous cell carcinomas	BD	MNUTH	90	97
Stomach, glandular (57, 97, 226, 296)				
Adenocarcinomas	—	2,7-FAA	—	223,226
Adenocarcinomas	—	AcMNU	90	97
Adenocarcinomas	W	MNNG	80	58
Small intestine (97,259,266, 296)				
Adenocarcinomas	BD	DMH	51	259
Adenocarcinomas	W	DMADP	87	296
Adenocarcinomas	LIS	B. fern	100	119
Adenomatous polyps	SD	B. fern	90	315
Large intestine (75,76,97,259 261,296,344)				
Adenocarcinomas	BD	DMH	100	62,98,235
Adenocarcinomas	BUF	DMH	60	204
Adenocarcinomas	W	ADP	—	259
Adenocarcinomas	BD	AOM	50	62
Kidney (2,139a,184,339)				
Lipomatous tumors	P and OM	Pyrrol.	17	322
Wilm's tumors	W	Transp.	—	361,362
Renal cell carcinomas	—	DMN	20	225
Stromal and renal cell tumors	W	DMN	—	64
Renal cell carcinomas	LEW and SD	Stpt.	31	201,202
Ureter (139a)				
Carcinomas	BN	Unknown	20	54

(*continued*)

Table I (*Continued*)

Organ or tissue tumor	Strain[a]	Carcinogen[b]	Incidence (%)	References
Urinary bladder (71,126, 144a,197,265,360,373)				
Carcinomas	SD and F	FANTF	100	80,115,116,360
Carcinomas	ACI and W	BHBN	100	126,161,265
Prostate (104,107,123,123a, 256,257,297)				
Squamous cell carcinomas and adenocarcinomas	W	MCA (local)	—	107
Seminal vesicles (123,123a,				
Carcinoma (Flexner-Jobling tumor)	—	Transp.	100	100, 351
Testes (50,79,166,223a,270, 297,298,313,325,369)				
Interstitial cell tumor	F	Unknown	90	79,90,313
Interstitial cell tumor	W	EGH	58	368
Interstitial cell hyperplasia	ACI	Unknown	80	338
Uterus (34a,82,123,166,217, 252,297)				
Uterine tumors	M520	Unknown	33	338
Uterine tumors	F	Unknown	67	338
Adenocarcinomas	BUF	NQO	28	217
Polypoid tumors	—	Unknown	21	166
Deciduomata	—	Unknown	—	82
Leiomyomas	—	Estrogens	25	252
Ovary (49,66a,113,148,158, 215,249,297,298,354)				
Granulosa cell tumors	OM	Unknown	33	338
Granulosa cell carcinoma	AXC	Transp.	67	158
Mucinous adenocarcinomas	OM	Transp.	100	354
Pituitary gland (34,47,127a, 297,298,338)				
Chromophobe tumors	W	Unknown	60	53,160
Chromophobe adenomas	BUF	Unknown	75	338
Chromophobe adenomas	W	Unknown	90	338
Chromophobe tumors	—	Unknown	30	169,264,338
Chromophobe tumors	BUF	F or A	60	220,222
Adrenal glands (105,108, 151a,220,297,298,338)				
Pheochromocytomas	W	Unknown	86	129
Pheochromocytomas	M250	Unknown	67	338
Pheochromocytomas	LE	GH	—	216
Cortical tumors	OM	Unknown	90	338,341
Cortical tumors	W	Unknown	29	53
Cortical tumors	OM	LP-DMAB	72	224
Thyroids (51,52,219,227, 227a,267,297,298)				
Parafollicular tumors (C cell alveolar or medullary tumors)	SD	Unknown	46	318,319,358,359
	LE	Unknown	45	186–188
	W	Unknown	40	53
Follicular tumors	S	I. def.	100	34,47,159,169
Follicular tumors	SD	ThU	100	214
	SD	^{131}I	62	336
Parathyroids (298)				
Adenomas or hyperplasia	LE	Unknown	40	188

(*continued*)

Table I (*Continued*)

Organ or tissue tumor	Strain[a]	Carcinogen[b]	Incidence (%)	References
Islets of Langerhans (290, 297,298)				
Islet cell tumors	SD	Unknown	17	297,298
Islet cell tumors	SD	DEAM-HAQ	59	142
Islet cell tumors	—	Stpt-NA	—	290
Brain (208a)				
Gliomas, ependymomas, and neuriomas	SD	ENU	100	163,171,172
Glial tumors	F	ONCV	—	77,157,385
Peripheral nerve (208a)				
Neurofibromas	OM	Ergot	61	232
External acoustic duct (254)				
Sebaceous gland tumors	—	FAA	80	254,383
Pleural cavity				
Mesotheliomas	—	Asbestos	5	91,376,377
Peritoneal cavity				
Sarcomas	—	*C. fasc.*	91	60,88,106,109, 110,208

[a] Key to abbreviations of strains: ACI or AXC strains; BD, BD strain; BN, Brown Norway; BUF, Buffalo; CB, Chester Beatty; F, Fischer; LEW, Lewis; LE, Long-Evans; LIS, Lister; M520, Marshall 520; NZ, New Zealand; OM, Osborne-Mendel; P, Porton; S, Sherman; SD, Sprague-Dawley; W, Wistar.

[b] Key to abbreviations of carcinogens: AA, 2-aminoanthracene; AcMNU, N^1-acetylmethylnitrosourea; ADP, 4-aminodiphenyl; Afla. B_1, aflatoxin B_1; ANFOD, 5-acetamido-3-(5-nitro-2-furyl)-6H-1,2,4-oxadiazine; AOM, azoxymethane; AS, azaserine; B. fern, bracken fern; BCME, bis(chloromethyl) ether; BHBN, N-butyl-N-(4-hydroxybutyl)nitrosamine; *C. fasc.*, *Cysticercus fasciolaris*; DEAM-HAQ, 6-diethylaminomethyl-4-hydroxy-aminoquinoline 1-oxide; DHS, dihydrosafrole; DMADP, 3,2′-dimethyl-4-aminodiphenyl; DMBA, 7,12-dimethylbenz(a)anthracene; DMN, dimethylnitrosamine; DMH, 1,2-dimethylhydrazine; EGH, endogenous gonadotropic hormone; EBN, ethyl-N-butylnitrosamine; ENU, ethylnitrosourea; FAA, N-2-fluorenylacetamide; 2,7-FAA, N,N^1-2,7-fluorenylenebisacetamide; FANTF, N[4-(5-nitro-2-furyl)-2-thiazolyl]formamide; F or A, fluorene or anilide compounds; GH, growth hormone; HAQ, 4-hydroxyaminoquinoline 1-oxide; I. def., iodine deficiency; LP-DMAB, low protein–p-dimethylaminobenzene; MCA, 3-methylcholanthrene; MNNG, N-methyl-N^1-nitro-N-nitrosoguanidine; MNU, N-methyl-N-nitrosourea; MNUTH, methylnitrosourethane; MSV, Moloney sarcoma virus; NBU, N-nitroso-N-butylurea; NQO, 4-nitroquinoline 1-oxide; NP, N-nitrosopiperidine; ONCV, oncogenic viruses; PCH, polycyclic hydrocarbons; PDET, 1-pyridyl-3,3-diethyltriazene; Pyrrol, pyrrolizidine alkaloids and aldehydes; Stpt, streptozotocin; Stpt-NA, streptozotocin and nicotinamide; TB, trypan blue; ThU, thiouracil; TMBA, 7,8,12-trimethylcholanthrene; Transp., transplantable tumor.

Albany rats, 40 to 50% of females over 14 months of age develop mammary tumors. The majority are fibroadenomas, but adenomas, fibromas, and adenocarcinomas may occur. In certain other strains, spontaneous fibroadenomas are rarely seen, and carcinomas have not been reported (297).

A rat mammary tumor usually begins with patchy lobular hyperplasia, involving both stroma and epithelium. This progresses to fibroadenoma, fibroma, or adenoma. These lesions, in turn, may give rise to adenocarcinomas or to tumors with sarcomatous elements. Rats developing mammary tumors frequently exhibit low fertility, abnormalities of estrus, endometrial polyps, uterine infections, myomas, and ovarian cysts suggestive of a generalized endocrine abnormality (297).

Active carcinogens increase the number of tumors per animal, increase the degree of malignancy, and shorten the induc-tion period. The daily intragastric instillation of 10 mg of 3-methylcholanthrene into Sprague-Dawley females has produced a 100% incidence of mammary tumors within 60 days, some tumors appearing as early as 1 month after treatment (297).

An effective model for mammary cancer being used in chemotherapy by many investigators utilizes 9,10-dimethyl-1,2-dibenzanthracene (7,12-dimethylbenz[a]anthracene or DMBA), which is considered a more effective carcinogen than methylcholanthrene (155,323,393). In a typical experiment, groups of female Sprague-Dawley rats 50–55 days of age are given a single dose at 20 mg/rat of DMBA in 1.0 mg of sesame or corn oil by stomach tube (138,176,393). The tumor incidence in the groups at 120 days post-DMBA treatment ranges from 63 to 100%, with a mean of 75%. At 6 months post-

DMBA treatment, 90% of 69 survivors have palpable tumors. The majority of these tumors are carcinomas (138). Hormone dependency of this induced tumor closely parallels its human counterpart (127,138).

A transplantable mammary tumor which has been used extensively is the Walker carcinosarcoma 256. This tumor arose spontaneously in the region of the mammary gland of a pregnant albino rat in 1928 (352). After limited success during the first two transplant generations, the Walker tumor now grows successfully in 80 to 100% of a variety of rat strains. Regression is rare, about 2% (100). The tumor becomes palpable in size at 1 week following intramuscular or subcutaneous inoculation and kills the host in 14 to 70 days. It invades local tissues and regional lymph nodes with rare pulmonary metastases (100). Microscopically, the Walker carcinoma 256 is composed of three types of cells: carcinomatous epithelial cells, undifferentiated epithelial cells, and sarcomatous spindle cells. These cells may form three patterns: a carcinoma of the first two cell types, a sarcoma of the latter cells, or a mixed carcinosarcoma of all three.

Another transplantable mammary tumor of the rat is the Mammary fibroadenoma R2737. This is a slower growing fibroadenoma that arose spontaneously in the mammary gland of a female rat of the August strain in 1946. Implanted subcutaneously, it grows without regression in 100% of female rats of the August strain, becomes palpable in 4 to 6 weeks after inoculation, and kills the host in 6 to 10 months. It does not metastasize and is not invasive (352).

2. Liver Cancer

Spontaneously occurring hepatocellular carcinomas which normally have a low incidence in the rat can be readily induced by carcinogens (316a,347). These lesions frequently have been termed "hepatomas," an imprecise term which has led to semantic problems in communications (10,347). These liver tumors have been studied extensively by oncologists, especially in biochemical experiments concerning the mechanisms of carcinogenesis. Both induced and transplanted hepatic neoplasms are used in these studies. Over 133 "hepatomas" are available for transplantation and more can be developed readily. These vary from well-organized, slow-growing neoplasms, such as Morris hepatoma 7787, to fast-growing tumors such as hepatoma 3683 (63).

Among the most frequently used hepatocarcinogens are *N*-acetyl-2-aminofluorene (AAF), and 4-dimethylaminoazobenzene (DMAAB) (379). In a typical experiment AAF is fed in the diet at concentrations of 0.025 to 0.06% which results in an incidence of hepatocellular carcinomas approaching 100% in 7 to 9 months. A concentration of 0.125% in the diet for 95 days has produced multiple tumors in 135 days. In addition to

liver tumors, AAF produces tumors of mammary gland, ear, urinary bladder, lung, eyelid, skin, brain, thyroid, kidney and other tissues, including leukemia (379). The compound DMAAB is a powerful hepatocarcinogen for the male rat when fed at a level of 0.06% in the diet for several months.

Another hepatic cancer model utilizes low dietary concentrations of aflatoxin B_1 (0.1 to 1.0 ppm) or 15 daily intragastric doses of 25 μg of aflatoxin B_1 in 0.1 ml of dimethyl sulfoxide (DMSO) per rat. Weanling rats of the Charles River CD, Fischer, or Wistar strains are used. The technique results in hepatocellular carcinomas after 12 to 18 months in 100% of animals (234,236,387).

All the transplantable hepatic tumors of rats arose as chemically induced lesions (100). Among the most commonly used hepatic tumors are the Novikoff hepatoma and hepatoma 3683. The Novikoff hepatoma arose in 1951 in the liver of a Sprague-Dawley rat (Holtzman subline) treated with DMAAB. Hepatoma 3683 arose in 1951 in the liver of an AXC 9935 rat treated with 2-diacetylaminofluorene. Hepatoma L-C18 is an hepatic tumor which arose in 1947 in a Fischer rat treated with AAF (100).

Hepatoma 3683 grows in 100% of subcutaneous transplanted rats with a latent period of 10 to 14 days and kills the host in 14 to 28 days. Novikoff hepatoma grows in 100% of intraperitoneally transplanted hosts with a latent period of 3 to 10 days and kills the host in 7 to 20 days. Hepatoma L-C18 grows in 100% of subcutaneously transplanted rats with a latent period of 14 days and death of the host in 45 to 50 days. In some series 10% of the lesions have regressed. The L-C18 invades local tissues and regional lymph nodes with metastases to the lungs (100).

Nineteen transplantable lines of ascites hepatoma were developed by Yoshida during the period 1951 to 1957 (391). These tumors were developed from 156 solid hepatic tumors induced in Japanese albino rats by feeding *p*-dimethylaminoazobenzene or *p*-dimethylazobenzene. The more frequently used of these tumors are AH13, AH66, AH130 and AH7974. The transplantation success rate is 80 to almost 100% when nondiluted ascitic fluid is administered intraperitoneally to Donryu rats. The median day of death is 7 to 31 days (range 5 to 139 days).

Another series of transplantable hepatomas was developed by Morris. These tumors were induced by low doses of carcinogenic aromatic compounds (*N*-2-fluorenyldiacetamide, *N*-2-fluorenylphthalamic acid, and 2,4,6-trimethylaniline) in inbred ACI and Buffalo strains (221). Two hepatomas, 3683 and 3924A, grow rapidly with deaths occurring in 1 to 3 weeks. Hepatoma 7288 is intermediate with deaths in 4 to 6 weeks. Hepatomas 7800 and 7787 are slow growing with deaths occurring in 3 to 12 months (381). The slowly growing tumors are well differentiated with few mitotic figures. They become less differentiated as the growth rate increases.

Hepatoma 3683 is a poorly differentiated anaplastic carcinoma (221).

3. Colon Cancer

Colon cancer currently is the second leading cause of cancer deaths in the United States. Spontaneous colon tumors are rare in rats and usually occur with a frequency of 1% or less (211). With the discovery of an efficient colon carcinogen, 1,2-dimethylhydrazine (DMH), the incidence has been increased to 100%. Other tumors produced by DMH occur in the small intestine, liver, and kidney at lower frequencies. Administration of ten weekly doses of DMH (30 mg/kg body weight) dissolved in physiologic saline and given by gastric intubation to weanling rats of several strains results in virtually 100% incidence of colon carcinomas within 5 to 7 months (235).

In another experiment, BDIX strain of rats given weekly subcutaneous injections of DMH for 12 weeks, (total dosage of 180 mg/kg of body weight) had a colon tumor frequency of 100%. The mean latent period for males was 162 days, and females was 195 days. When azoxymethane, another colic carcinogen, was given in weekly subcutaneous injections for a total dosage of 60 mg/kg to the BDIX strain rats, 48% had colon tumors and mean latent periods of 170 (males) and 255 (females) days. BDII strain rats in the same experiment were less susceptible to DMH (63% males and 44% females) and azoxymethane (25% males and 13% females) and had longer mean latent periods. Both strains developed multiple colonic adenocarcinomas. Benign polyps also were present. The tumors are most frequently localized in the colon and rectum. These tumors have morphologic and biological characteristics similar to the same disease of man (62,228,229,235,261,343,344).

The ability to induce these tumors consistently has permitted the development of transplantable colon tumor lines as has been done with mouse tumors (83,84,136). Since 1971, several transplantable colon tumors for rats have been reported (133,204,348,350,351).

4. Hematopoietic and Lymphoreticular Cancer

Neoplasms of the hematopoietic and lymphoreticular tissues occur spontaneously in many rat strains. Several of these are comparable to human neoplasms in morphology and behavior, although some differences exist (101,213,340,346,353). The tumors of the rat hematopoietic system and lymphoid tissues have been described well by Swaen and Van Heerde (353).

Models for oncologic research include a stem cell or atypical monocytic leukemia that occurs spontaneously in Fischer rats at a frequency of about 25% (90,212,313). A similar leukemia is observed in about 20% of Wistar/Furth rats. The leukemia seldom develops before 18 months of age, with average age at

death of about 24 months (353). Recently, this leukemia was observed in young Fischer rats at ages of less than 4 months (R. B. Thompson, personal communication).

Spontaneous reticulum cell sarcomas (histiocytic malignant lymphomas) occur in rats of several strains at differing frequencies. One Wistar colony had a frequency of almost 30% in animals more than 24 months old. In other Wistar colonies the frequencies were 1 and 5% (353).

Several chemical carcinogens induce tumors of the hematopoietic system in the rat. One compound which has been used extensively is 7,12-dimethylbenz[a]anthracene (DMBA). A great variety of tumors has been reported after DMBA treatment. They include reticulum cell sarcomas, granulocytic leukemias, lymphosarcomas, lymphocytic leukemias, stem cell leukemias, and erythroleukemias. The yield of tumors ranges from few to 100% and is dose and strain dependent (353).

The development of lymphosarcomas and lymphocytic leukemias has been observed by several investigators following inoculation of rats with Gross, Friend, Rauscher, Moloney, Graffi, and murine erythroblastosis viruses (353). Although the viral etiology of leukemia is firmly established for the mouse, it is not in the rat. Cell-free extracts of spontaneous hematopoietic tumors consistently have resulted in negative findings (353).

Several transplantable hematopoietic tumors are available in rats. Leukemia L5222, a monocytic-myeloid leukemia, was induced by ethylnitrosourea in a BDIX rat in 1967 and serves as a model for chemotherapy. It is transplanted intraperitoneally by diluted heart blood (165). Another monocytic transplantable tumor is an acute leukemia IRC741 which occurred spontaneously in a Fischer rat. It is transplantable by subcutaneous implantation or intraperitoneal injections of cellular suspensions and grows progressively in 100% of inoculated Fischer rats (103,167).

Granulocytic leukemias are a less frequent tumor in rats, but one transplantable line has been used extensively. This is the chloroleukemia 123, Shay (Shay chloroma). It is characterized by leukemic blood and local tumors of a green color. It arose in a Wistar rat in 1949 after intragastric instillation of 20-methylcholanthrene. Chloroleukemia 123 is transplanted by whole blood given intraperitoneally to rats less than 7 days of age. Wistar, Long-Evans, and Sherman rats have been used as hosts. The transplantation rate is 80 to 100%, with a latent period of 42 to 70 days and death in 60 to 90 days. The tumor disseminates widely (100).

Leukemia DBLA-6 is used in chemotherapy screening. It is a myelogenous leukemia that was induced in 1970 in Donryu rats by oral administration of N-nitrosobutylurea. This leukemia is transplanted intraperitoneally by nondiluted ascitic fluid containing 10^4 to 10^5 cells in 7 to 9 week old Donryu rats. The mean survival time is 10 to 16 days. It also can be administered intravenously with 10^3 cells, resulting in a mean survival time of 14 to 21 days (31).

The Murphy-Sturm lymphosarcoma arose in 1938 in a Wistar rat given 1,2,5,6-dibenzanthracene and has characteristics of a lymphosarcoma, or lymphocytic leukemia, or both depending upon the technique of transplantation. It can be transplanted by leukemic blood or affected lymph nodes in Wistar and Sprague-Dawley rats with progressive growth in 75 to 100%. The latent period is 4 to 8 days, and death occurs in 8 to 35 days. The neoplastic tissues disseminate widely. About 20% of the tumors regress (100,352).

Lymphosarcoma R2788 arose in 1947 as a spontaneous reticulum cell or lymphoblastic neoplasm from the mesenteric lymph nodes of an AXC/9935 rat. It grows progressively in 100% of AXC/9935 rats inoculated subcutaneously, with death at 25 to 35 days after inoculation. The latent period is 8 to 10 days. The tumor invades the regional lymph nodes and metastasizes to the deep lymph nodes and lungs (103).

The Yoshida sarcoma is a frequently used transplantable ascitic tumor which is sometimes classified with the monocytic leukemias and sometimes with the reticulum cell sarcomas (353). It arose in 1943 in an albino Japanese rat that was fed *0*-aminoazotoluol and painted with potassium arsenite. The tumor involved the peritoneal and retroperitoneal tissues that were attached to the capsule and omentum of the testes. The peritoneal cavity contained thick milky ascitic fluid composed of tumor cells. It can be transplanted in an ascitic form intraperitoneally or a solid form subcutaneously or intramuscularly (100,103,352). The Yoshida tumor has been tranferred in the Donryu, albino Japanese, Wistar, and Marshall 520 strains with a 90 to 100% success rate. The ascites form has a latent period of 3 to 4 days with death in 6 to 141 days. The solid form has a latent period of 10 to 13 days, and death occurs from 14 to 20 days after inoculation. It infiltrates the skin and muscle and metastasizes to lung, mesentery, colon, and heart (103,352).

5. Urinary Bladder Cancer

N-[4-(5-Nitro-2-furyl)-2-thiazolyl]formamide (FANTF) is a nitrofuran derivative that is almost 100% effective in producing carcinomas of the urinary bladder in rats. The tumors are malignant, can be transplanted, and metastasize. In most cases, only bladder carcinomas are produced in large numbers. In an experiment where 0.188% of FANTF was fed to female Sprague-Dawley rats for 46 weeks all 29 rats that survived for 34 weeks developed carcinomas of the bladder. Histologically, these tumors were identified as eight squamous cell carcinomas, seven transitional cell carcinomas, one transitional cell papillary carcinoma, and thirteen mixed carcinomas (116). Male Sprague-Dawley and Fischer rats have been fed 0.188% levels with similar results (115,360). Male and female Fischer

rats fed 0.2% had progressive lesions beginning with mild hyperplasia at 2 to 4 weeks, to moderate hyperplasia at 6 to 8 weeks, severe nodular and papillary hyperplasia at 10 to 14 weeks and microinvasive carcinomas by 25 weeks (80).

Another potent selective urinary bladder carcinogen in the rat is *N*-butyl-*N*-(4-hydroxybutyl)nitrosamine (BHBN) (126,161, 265). Male Wistar rats given 0.05% BHBN in their water for 12 weeks had 100% bladder cancer, whereas those treated for 8 weeks had 90% bladder cancer at the end of 40 weeks. The histologic types of urinary bladder cancers induced by BHBN were transitional cell carcinomas (95.1%), squamous cell carcinomas (3.3%), undifferentiated carcinomas (2.5%), and carcinosarcomas (0.3%). The morphologic characteristics of these tumors are similar to those seen in human patients (126,144a,373).

6. Brain Cancer

The discovery that nitrosoureas are neurotrophic carcinogens has provided an excellent model for research into the mechanisms of primary tumors of the central and peripheral nervous system.

The resorptive carcinogens, methylnitrosourea (MNU) and ethylnitrosourea (ENU), induce tumors of the nervous system in Sprague-Dawley and Fischer rats. When young adult rats were given weekly intravenous doses of 5 mg MNU up to a total of 180 mg/kg body weight, 97% of the animals developed tumors of the nervous system. Of a total of 71 neoplasms, 58 were of the central nervous system and 13 of the peripheral nervous system. The average survival time was 339 days (172,348).

When ENU was administered intravenously to pregnant rats at day 20 of gestation in single doses of 1, 5, 20, or 50 mg/kg, all offspring of the dams receiving 50 mg/kg developed tumors of the nervous system. The average survival time was 211 days. With decreasing doses, the tumor incidence decreased and the survival time increased (172). Nonneural tumors comprised 6.4% of the total number of neoplasms (171). The classification of the tumors were gliomas (oligodendrogliomas, astrocytomas, mixed gliomas, gliosarcomas, and anaplastic gliomas), ependymomas (including glioendymomas), a meningoma, and neurinomas (171,172). Gliomas and mixed gliomas are the most frequent of brain tumors induced by the nitrosourea derivatives (317). Neurofibromas of the ear were reported in 61% of Osborne-Mendel rats fed 5% crude ergot for 2 years (232). Glial tumors or sarcomas have been also produced by intracerebral inoculation of several oncogenic viruses including Moloney, Rous, and an avian sarcoma virus (77,157,385).

7. Soft Tissue Cancer

In contrast to man where the incidence of sarcomas is low

(311), the soft stromal tissues of rodents and especially those of the rat are particularly sensitive to the induction of sarcomas (59,66,263). These tumors are predominantly spindle cell in type and occur in both the subcutis and body cavities. Considerable research has been directed to define the mechanisms of soft tissue carcinogenesis, especially in the area of foreign body-induced sarcomas and solid state or smooth surface carcinogenesis. The increased use of prosthetic and cosmetic implants made of plastics and other synthetic materials may lead to an increased incidence of foreign body sarcomas in man (59,96,147).

A conspicuous characteristic of foreign body-induced sarcomas in man and animals is that they appear in a great variety of histological types. In man, these include fibrosarcomas, chondrosarcomas, meningiomas, as well as asbestos-induced mesotheliomas. In the rat, the findings are similar. The lesions include rhabdomyosarcomas, leiomyosarcomas, mesenchymosas, liposarcomas, reticulosarcomas, plasmacytomas, histiocytomas, and unclassified sarcomas (59). Similar stromal lesions of rats occur as a result of many other carcinogens, such as chemical compounds, X-irradiation, and neonatal injections of certain viruses (35,56,66,145,217). The majority of soft tissue tumors are fibromas, fibrosarcomas, and lipomas. Those involving serosal surfaces frequently have been termed simply peritoneal and pleural sarcomas. Similar lesions of the pleural cavity, especially those associated with asbestos, have been classified as mesotheliomas (91,376,377).

Evidence has been presented that the cause of the foreign body sarcomas is physical and not chemical, since it can be produced by essentially inert, insoluble materials such as glass, metals, and certain polymers. When implant surfaces were roughened, the tumor incidence decreased. Also, size of the implant is a factor. Materials such as cellophane that were carcinogenic when implanted as film sheets did not cause tumors when the same material was implanted in powdered, finely shredded, or perforated forms. Examples of carcinogenic materials in addition to cellophane are polyethylene, polyvinyl chloride, teflon, nylon, dacron, polymethylmethacrylate, as well as foils of metals, such as tantalum, silver, gold, platinum, and even ivory (326). The incidence with cellophane has been reported to reach 51% with a range of 30 to 40% in rats. The latent period is 12 to 15 months (59).

Other soft tissue carcinogens include iron–carbohydrate complexes [such as iron-dextran (289)], tannin extracts (262), asbestos, chromates, nickel, X-irradiation, many chemicals (35,145,154), and the peritoneal and hepatic larval tapeworm cysts of *Taenia taeniaeformis* (109). The peritoneal sarcomas caused by *Cysticercus fasciolaris* have been investigated extensively (88,106,109,110,208). Most of the sarcomas produced were spindle cell, or polymorphic sarcomas, or a mixture of the two. Some contained bone or cartilage. The inci-

dence of the sarcomas ranged from 31 to 100% in experimental groups depending upon the number of parasites present (88,106).

8. Bone Cancer

The induction of bone tumors in rats has resulted from many studies on the effects of ionizing radiation on skeletal tissue (370). Spontaneous tumors of rat bone occur rarely (189). The induced osteosarcomas of rats are an adequate model for the human tumor (78).

Methods for the induction of tumors in rat bone are based mainly on the effect of chemical or physical agents on physiological osteogenesis. Bone tumors are readily induced by radioactive isotopes which have an affinity for bone tissue ^{32}P, ^{45}Ca, ^{89}Sr or ^{90}Sr, ^{91}Y, ^{140}La, ^{144}Ce, ^{149}Ba, ^{147}Pm, ^{226}Ra, ^{238}U, ^{239}Pu, and isotope mixtures). All these isotopes, in the form of salt solutions, may be administered intravenously, intraperitoneally, or orally (usually in the drinking water). Radioisotopes are considered the most suitable and reliable bone selective carcinogen in rats (77,189).

Intraosseous administration of chemical carcinogens such as 9,10-dimethyl-1,2-benzanthracene and 3,4-benzpyrene results in 75% or more of osteogenic sarcomas (189).

Osteosarcomas have been induced in New Zealand black rats by intratibial inoculation of Moloney sarcoma virus at 1 and 4 days of age. The tumor incidence in litters of inoculated rats averaged over 80%, with radiographic evidence of osteosarcoma in 10 to 15 days after inoculation. Lesions in rats inoculated at 4 days of age were more osteoproliferative, were slower growing, and had a consistent tumor-associated cachexia. Metastases to lungs, sublumbar lymph nodes, or both were observed in over 93% of tumor-bearing rats (242). Osteosarcomas also have been produced by neonatal injection of polyoma virus (170).

Metaplastic and sometimes neoplastic bone lesions have been associated with the spindle cell sarcomas in soft tissues. Among 7500 neoplasms induced in the rat's liver by *Cysticercus fasciolaris*, 49 (0.7%) were bone-forming tumors. In subcutaneous neoplasms, predominantly fibrosarcomas and spindle cell sarcomas, induced by 3,4-benzpyrene in paraffin, 66 of 2351 (2.8%) were bone tumors (106).

A transplantable osteogenic sarcoma 344 arose in the wall of the glandular stomach of a Marshall 520 rat after intramural injection of 20-methylcholanthrene in 1950. The tumor is transplanted by subcutaneous implants with progressive growth in 100 percent of the animals. It has a latent period of 7 to 10 days and kills the host in 60 to 90 days. There is little bone formation, but the transplanted tumors contain large amounts of acid phosphatase and have a fibrillar matrix (100,352).

9. Testicular Cancer

The male Fischer rat has been found to have a spontaneous frequency of interstitial cell tumors of the testis ranging from 63 to 90%. These tumors are similar in appearance to those of human patients and occur in the absence of other testicular tumors (79,90,223a,313). Interstitial cell hyperplasia and tumors have been observed also in a high percentage (80%) of ACI strain rats (338).

10. Other Cancers

Additional rat models are available which mimic other cancers of man, including tumors of the salivary glands, esophagus, stomach, small intestine, pancreas, blood vessels, heart, trachea, lungs, kidney, ovary, uterus, prostate, seminal vesicles, adrenals, thyroid, parathyroid, pituitary, skin, skeletal muscle, nasal turbinates, and external acoustic duct (see Table I). More detailed descriptions of these tumor systems can be found in the references cited in the table. The effectiveness of these tumors as models of human disease varies greatly with some having been studied extensively, whereas others received little attention by oncologists.

Some of the tumors of rats have little application to the study of human cancer. Tumors of the eye are very rare in rats, and no transplantable models have been developed (314). Preputial and clitoral gland tumors which resemble modified sebaceous gland tumors, lymphatic tumors, and a few other tumors that occur in rats have not been developed as experimental models.

IV. USE IN CARCINOGENESIS

A. Definitions and Terminology in Oncology

In reviewing cancer literature, one finds considerable confusion as a result of discrepant terminology. The inconsistency of terminology has resulted in part from the integration of the many disciplines of science that are involved in the field of cancer research and are compounded by the ever changing terminology and new research developments in these disciplines. Scientists who are actively involved in oncology throughout the world have diverse backgrounds of language, experience, and training which further increases the difficulties in communicating the rapidly increasing body of information evolving from cancer research.

The World Health Organization through the International Agency for Research on Cancer (IARC) of Lyon, France and the earlier International Union against Cancer (Union Internationale Contre le Cancer, UICC) of Geneva, Switzerland have recognized the communication problem and in cooperation have proposed a current internationally standardized terminology (10,13,144). "Carcinogenesis" is defined as the generation of benign and malignant neoplasia and includes all phases of the process of neoplastic generation including the invasive phases The terms "tumorigenesis," "oncogenesis," and "blastomogenesis" are less specific and do not necessarily include the invasive phase. Also, carcinogenesis is considered the most adequate term for use in connection with the more complex, multifactorial patterns of exposure of the host or target tissue to causative agents. When "carcinogenesis" is used to describe induction of neoplasia in the broadest possible sense it will, by definition and by convention, imply induction of sarcomas and leukemias as well as carcinomas (144).

"Carcinogens" are defined as agents which induce carcinogenesis and whose origin, nature, and identity are unequivocally clear. They may be of exogenous or endogenous origin and of physical, chemical, or viral nature. They induce carcinogenesis in one or more tissues of the host. For consistency the terms "tumorigen," "oncogen," and "blastomogen" should not be used. The term "tissue specificity" is proposed in preference to the term "organotropy."

"Carcinogenic factors" are causative agents whose origin, nature, and identity are not unequivocally clarified (144).

In chemical carcinogenesis, the strict meaning that the causative agent must induce neoplasia has been broadened to indicate both induction and enhancement of neoplasia by chemicals. For practical purposes, no distinction is made between the induction of tumors and the enhancement of tumor incidence, although it is noted that there may be fundamental differences in these processes.

The response to a carcinogen in experimental animals may be observed as (a) a significant increase in the frequency of one or several types of neoplasms, as compared with other than zero frequency in control animals; (b) the occurrence of neoplasms not observed in control animals; (c) a decreased latent period as compared with control animals; or (d) both a decreased latent period and increased frequency of neoplasms (13,46,89).

B. Carcinogens and Carcinogenic Factors

The majority of all known causative agents for cancer can be placed into three primary groups, physical, chemical, and viral carcinogens. These have both exogenous and endogenous origin and have multiple patterns of exposure and predispositions. The rat responds to all these agents and has been the most frequently used laboratory animal in carcinogenesis studies (327,329).

The statement has been made that there is no cancer without

a carcinogen, just as there is no infectious disease without a specific microorganism (154). In a broad sense this statement is true. Spontaneous intracellular changes that induce cancer, such as somatic mutations, will be difficult to identify specifically as carcinogens just as some infectious agents have been difficult to establish as pathogens.

1. Physical Agents

Sources of physical carcinogens include the sun, irradiation, and nuclear fission. Radiation is believed to be the cause of less than 5% of the cancer occurring in man (120). Rats are particularily sensitive to ionizing radiation and have been subjects of extensive research in this field (329,370). Wistar rats are more sensitive to X rays than the Sprague-Dawley (366). Rats also are susceptible to ultraviolet radiation of the skin. This susceptibility has been attributed to the relative thinness and lack of keratinization of rat skin (386). Hairless rats are less responsive to action of ultraviolet light. This may be due to the greater thickness of the stratum corneum or a decrease in susceptible hair follicle cells in these animals (394).

Another type of seemingly physical carcinogenesis is represented by foreign body or solid state carcinogenesis associated with soft tissue cancer (59,66).

2. Chemical Agents

Chemical carcinogens have both endogenous and exogenous origin and represent the largest group of known carcinogens. They are believed to be the cause of up to 90% of cancer in man (120).

Exogenous sources of chemical carcinogens are industrial chemicals at all stages of production from raw materials to final products, organic and inorganic chemicals, pharmaceuticals, food additives, food contaminants, and plant products including natural oils and fungal toxins. Over three million distinct chemical compounds including three-quarters of a million organic chemicals are known (33,43). Of these chemicals more than 1300 were included in the 1974 Toxic Substances List of the National Institute for Occupational Safety and Health as carcinogens or carcinogenic factors (181). This list has been expanded to over 1500 chemicals (70).

In addition to chemicals of exogenous origin, several endogenous carcinogens have been identified. The carcinogens are formed during anabolic and catabolic cellular metabolism and have been suggested to cause some of the spontaneously occurring neoplasms. Some examples are 3-hydroxyanthranilic acid, estrogens, androgens, and bile acids (144).

Some examples of chemical classes containing carcinogens are listed in Table II (35,72,94,145,153,168,209,210,351, 389).

Because of the immensity of the field of chemical carcino-

Table II
Selected Chemical Carcinogens of Man and Animals[a]

Type	Groups	Chemical
Alkylating agents	Nitrogen mustard	Methyl bis(2-chloroethyl) amine
	Epoxides	1-Ethyleneoxy-3,4-epoxycyclohexane
	Aziridines	N-acetylethyleneimine
	Strained ring lactones	β-Propiolactone
	Alkanesulfonates	1,4-Dimethanesulfonoxybutane
	N-nitroso compounds	Dimethylnitrosamine
		Methylnitrosourethane
		N-nitrosomethylurea
		N-methylnitroso-N-nitroguanidine
		Methylnitrosoacetamide
	Azoxy compounds	Cycasin
		Methylazoxymethanol
	Triazene	1-Pyridyl-3,3-diethyl-triazene
Polycyclic aromatic hydrocarbons		3,4-Benzopyrene (benzo[a]pyrene)
		20-Methylcholanthrene (3-methylcholanthrene)
		1,2,5,6-Dibenzanthracene (Dibenz [a,h]anthracene)
		9,10-Dimethyl-1,2-benzanthracene (7,12-dimethyl-benz[a]anthracene)
Aromatic amines and amides		2-Naphthylamine
		2-Acetylaminofluorene
		4-Acetylaminobiphenyl
		4-Acetylaminostibene
		2-Acetylaminophenanthrene
		3-Hydroxyanthranilic acid
Aminoazo compounds		4-Dimethylaminoazobenzene
Chlorocarbons		Carbon tetrachloride
		Chloroform
Heterocyclic and other compounds		Auramine
		4-Nitroquinoline N-oxide
		Aflatoxin B_1
		Ethionine
		Griseofulvin
		Pyrrolizidine alkaloids
		Procarbazine
		Safrole
		Tannic acid
		Urethane

(continued)

Table II (*Continued*)

Type	Groups	Chemical
Hormones and steroids		Esterone
		Diethylstilbestrol
		Cholesterol
Metals and inorganic chemicals		Arsenic
		Asbestos
		Beryllium
		Chromates
		Cobalt
		Nickel
		Nickel carbonyl
		Iron–dextran complexes
Radiochemicals		Radium-224
		Plutonium-239
		Americium-241
		Phosphorus-32
		Strontium-90
		Thorium-228
Plastics		Cellophane
		Polyethylene
		Polyvinylchoride
		Polystyrene
		Polymethylmethacrylate
		Polytetrafluoroethylene

*a*From Badger (1962), Casarett (1972), Clayson (1974), Druckrey (1972), Magee (1971), Miller and Miller (1967), and Warwick (1971).

gens and carcinogenesis, the subject can only be covered briefly in this chapter. For general information, several references can be consulted (32,33,35,145,154,367). For specific compounds, the ''Survey of Compounds which Have Been Tested for Carcinogenic Activity'' (11,17–19,25–29,132,144, 251,334) and ''Monographs on the Evaluation of Carcinogenic Risk of Chemicals to Man'' (13,15,16,20–24) are recommended. These references present data on some of the thousands of experiments in which rats have been used.

3. Viral Agents

Viral carcinogens can be either of exogenous or endogenous origin and may cause no more than 5% of human cancer (120). Some are readily transmitted horizontally by contact from infected hosts, whereas others are transmitted vertically from one generation to another, from the dam to offspring. Several viruses have been identified with rat tumors, such as those associated with mammary gland tumors, adenovirus 9 and R-35 (1,2,4,68,69), and with leukemia (193). Examples of viral carcinogenic agents effective in rats include Moloney sarcoma virus. The Moloney virus is associated with monocy-

tic leukemia but when injected into bone results in osteosarcomas. Newborn rats also are susceptible to polyoma virus which was isolated from mouse tissues (139). Rous sarcoma virus of the chicken has been used to induce cancer in the rat (77,328,385).

C. Host or Target Tissue Predisposition in Carcinogenesis

Normal cells of the host may be conditioned by exposure to predisposing, noncarcinogenic agents prior to the initiation of a carcinogenic process. Such a predisposition to neoplasia may be either an inborn genetic defect or acquired. Some of the acquired influences are partial hepatectomy, irritation, wound healing, age, sex, and diet (144,237,363,383,386).

D. Carcinogenic Processes and Mechanisms

Carcinogenesis may be the result of one or multiple exposures to one or more carcinogens (48,144,145). These various possible patterns of exposure are collectively termed carcinogenic processes and include the exposures to different types of carcinogens at one or more times in different sequences. The carcinogenic process is initiated by exposure to a carcinogen. If neoplasia results from a single or repeated exposure to a single carcinogen, the exposure is termed ''unifactorial.'' The causative agent in such a case is termed a ''solitary carcinogen'' and the process, ''solitary carcinogenesis.'' In the prevention of cancer, solitary carcinogens have the greatest importance and carry the greatest risk to man. As experimental models, solitary carcinogenesis is the simplest for identification and investigation of environmental and industrial hazards. The solitary carcinogen may require metabolic activation to form a proximate or ultimate carcinogen which initiates the carcinogenic process in the susceptible cells (230).

The International Agency for Research on Cancer (IARC) monographs describe 196 of the 1500 potential chemical carcinogens. Of the 196 chemicals, 101 solitary carcinogens have been identified. Ninety-four chemicals are known to be definitely carcinogenic to animals, 15 to man and animals, and 2 to man alone (364).

In most real life situations, carcinogenesis results from exposure to more than one carcinogen, ''multifactorial exposure.'' The neoplastic process is termed ''syncarcinogenesis.'' In syncarcinogenesis, neoplasia may arise from the synergistic action of subcarcinogenic doses of two or more carcinogens. The origin, nature, and number of synergistically acting carcinogenic factors is unimportant. It is also possible for the multifactorial exposures to be anticarcinogenic with a resulting decrease in neoplasia (144).

If the exposure to a solitary carcinogen is simultaneous or followed by at least one other solitary carcinogen and the effects are irreversible, the process is further termed "pluricarcinogenesis."

If the agent in the second exposure is not a solitary carcinogen but is a cocarcinogen, an agent that is not carcinogenic by itself and is not able to initiate neoplasia, the process is termed "cocarcinogenesis." In cocarcinogenesis, the first event or the effect of exposure to the carcinogen is irreversible, whereas the second event, the effect of the cocarcinogen, is reversible. The first exposure has been also termed "initiation," whereas the second has been termed "promotion" (140,144,145). Experimental models using rats have been used extensively to clarify both types of syncarcinogenesis. Although they are suspected to operate in man, clear evidence is lacking. Additional studies into syncarcinogenesis have been recommended (12).

V. USE IN CANCER THERAPY

The role of the rat in cancer therapy studies has been to provide a model of human neoplastic diseases that can be used to identify effectively potential anticancer drugs and improve the therapeutic efficacy of those currently available. Over 360,000 chemicals and other materials have been tested on laboratory animals, and over 300 potential new chemotherapy drugs identified (328). In addition to drugs, cancer therapy also involves surgery, radiation, immunology, and various combinations of the four. In the past, cancer therapy has emphasized the destruction or removal of neoplastic lesions after their presence is detected (203,304,345). Attention is now being directed toward modification of the initiated carcinogenic process to prevent the appearance of the clinical disease (269,305). Development of new chemotherapeutic drugs and vaccines that stop or inhibit the carcinogenic process could be effective for both cure and prevention.

All types of neoplastic lesions have been used in cancer therapy studies: spontaneous, induced, and transplantable. The chief requirements for therapy experiments are that the tumors be uniform in type, available in large numbers, produced by a simple technique, be typical for some aspect of the human disease, be manipulated with relative ease to provide reliable data, be reproducible and give quantitative data, and be economical (135).

Some examples of rat tumors used in cancer therapy studies are the mammary gland tumors induced by 7,12-dimethybenz[a]anthracene (138,323), Yoshida sarcoma (103), lymphosarcoma R2788 (103), leukemia, IRC-741 (102,103, 167), sarcoma R92)102), adenocarcinoma R2426 (102), squamous cell carcinomas M-C961 and M-C972 (102), Walker carcinosarcoma 256 (188,192), and Hepatoma

(hepatocellular carcinoma) L-C18 (102). The human sarcoma, HS1, and the human epidermoid carcinoma, HEp3, have been transplanted successfully to Wistar and Fischer-344 rats following conditioning by X-irradiation and cortisone or hydrocortisone (198,247,357,366,388).

Since curative therapy has its greatest success (up to 95% for most cancer) when the lesions are small, a goal of oncologists has been to find biochemical, immunological, or morphological markers which could be used to identify precancerous changes and to inrease the frequency of early diagnoses (9,30,121,134,140,196,207,301,311). If neoplasia is an intrinsically heritable form of abnormal hyperplasia (263), the problems of differentiating normal actively growing and hyperplastic cells from neoplastic cells can more easily be appreciated. The anticipated value of cancer-specific antigens as a means of detecting neoplastic lesions in early stages has been largely unfulfilled.

Chemically induced cancers in animals have great differences in immunogenicity. Most rat liver tumors induced by the azo dyes are strongly antigenic, whereas N-2-fluorenylacetamide tumors of the liver and mammary gland are rarely antigenic (140). Spontaneous tumors of animals also tend to have weak or no new antigens (140). α-Fetoproteins and other carcinoembryonic antigens have been detected that are associated with malignant lesions, but they appear to lack the specificity needed for effective early diagnosis (36–42,45,112,199,200,349,371).

The National Cancer Institute has been supporting research that is directed toward the identification of early lesions and the mechanisms in the development of cancer, especially epithelial cancer. One objective of this research is to modify the initiated lesion and involves the study of organ cultures using trachea and urinary bladders from rats having preneoplastic lesions initiated by chemical carcinogens. The chemicals being tested include vitamin A (retinol) and related retinoids (203,345).

VI. USE IN CANCER PREVENTION

Since 75 to 90% of cancer in man (195) is related to chemicals in our environment and our agricultural and industrial technology (378), it is thought that a large proportion of these cases could be prevented. In fact, with proper safeguards, it has been claimed that all occupationally related cancers could be prevented (378). The prevention of cancer should be the goal of our society and the responsibility of health scientists, management, and labor. This responsibility must extend beyond the scientific and economic considerations to the moral and social issues involved (378).

The total elimination of exogenous chemical carcinogens is not possible or practical no matter how worthy the goal may be. Other factors must be considered in the cost versus benefit and benefit versus risk evaluation (173,238,239,287,306,312). The identification of all chemical carcinogens probably will not be possible because a relative few are solitary carcinogens and identification of the carcinogens in syncarcinogenesis of man is exceedingly difficult (73,74,144,364). Not all workers exposed to known occupational carcinogens develop cancer. Therefore, other influencing factors, some of which are unknown, are involved.

Cancer is not a disease limited to civilized or modern man, for it has been known for centuries (140,328). Some natural products contain mycotoxins and plant toxins which accidentally or intentionally are added to foods and drinks, or used as medicinals, or taken for pleasure (240,321).

The carcinogenic hazards of our present civilization result from industrial, occupational, medicinal, and accidental exposures and include a long list of industrial chemicals, pesticides, herbicides, insecticides, food additives, drugs, cosmetics, natural occurring substances, and other materials (55,124,194,330,332).

The identification and control of these hazards is a responsibility of the federal government through the concerted action of several federal agencies including the Environmental Protection Agency, Food and Drug Administration, National Institute for Occupational Safety and Health working with legislative authority from Congress and the President (114,120, 181,268,305).

Prevention of cancer is much more effective than treatment. Many forms of cancer reach an incurable state before diagnosis. The cost of prevention is easily repaid in reduced medical expenses, decreased pain and suffering, and increased life span (120,309).

The rat is a key component of the National Cancer Institute's Carcinogenesis Bioassay Program to identify chemical carcinogens which has been recently expanded to use thousands of animals at a cost of millions of dollars. A standard carcinogenicity test in rodents costs about $100,000 per chemical and takes about 2 years to complete (205,308).

VII. CONCLUSIONS

The rat has proved many times its usefulness in experimental oncology. The hundreds of references in the literature attest to this fact. In a review of this length, a complete survey of the available literature is impossible. General references (35,86,93,140,145,154,263,272,316,328) and results of selected specific experiments have been presented as sources for additional information.

As new areas of investigation are realized, the rat will continue to be a valuable laboratory model for human neoplastic disease. In development of new techniques for more rapid carcinogenesis tests, such as the *Salmonella*/microsomal test (74,122,206,307,383), animal experiments still provide the baseline for comparison to man.

REFERENCES

1. Ahmed, M., Korol, W., Larson, D., Molnar, M., and Schidlovsky, G. (1972). Transformation of rat mammary cell cultures by R-35 virus isolated from spontaneous rat mammary adenocarcinoma. *J. Natl. Cancer Inst.* **48**, 1077–1083.

2. Akimova, P. N. (1963). Transplantable carcinoma of rat kidney (Strain RA), *Vopr. Onkol* **9**, 51–55 (Engl. summ.).

3. Ankerst, J., Jonsson, N., Kjellen, L., Norrby, E., and Sjögren, H. O. (1974). Induction of mammary fibroadenomas in rats by adenovirus type 9. *Int. J. Cancer* **13**, 286–290.

4. Ankerst, J., Steele, G., and Sjögren, H. O. (1974). Cross-reacting tumor-associated antigen(s) of adenovirus type 9-induced fibroadenomas and a chemically induced mammary carcinoma in rats. *Cancer Res.* **34**, 1794–1800.

5. Anonymous (1959). "Food Protection Committee Report: Problems in the Evaluation of Carcinogenic Hazard from Use of Food Additives," Publ. No. 749. Nat. Acad. Sci.—Natl. Res. Counc., Washington, D.C.

6. Anonymous (1967). General recommendations. *In* "Potential Carcinogenic Hazards from Drugs" (R. Truhaut, ed.), pp. 245–246. Springer-Verlag, Berlin and New York.

7. Anonymous (1969). Principles for the testing and evaluation of drugs for carcinogenicity. *W.H.O., Tech. Rep. Ser.* **426.**

8. Anonymous (1971). Food and Drug Administration Advisory Committee on Protocols for Safety Evaluation: Panel on Carcinogenesis Report on Cancer Testing in the Safety Evaluation of Food Additives and Pesticides. *Toxicol. Appl. Pharmacol.* **20**, 419–438.

9. Anonymous (1971). Recommendations of subcommittee on biochemical and metabolic studies. *Liver Cancer, Proc. Work, Conf., 1969* IARC Sci. Publ. No. 1, pp. 171–172.

10. Anonymous (1971). Report and recommendations of subcommittee on morphology, epidemiology and pathology. *Liver Cancer, Proc. Work. Conf., 1969* IARC Sci. Pub. No. 1, pp. 173–174.

11. Anonymous (1971). "Survey of Compounds Which Have Been Tested for Carcinogenic Activity," Public Health Serv. Publ. No. 149, 1961–1967 Volume. US Gov. Printing Office, Washington, D.C.

12. Anonymous (1971). Recommendations of subcommittee on priorities in human carcinogenic studies. *Liver Cancer, Proc. Work. Conf., 1969* IARC Sci. Publ. No. 1, pp. 175–176.

13. Anonymous (1972). "IARC Monographs on the Evaluation of Carcinogenic Risk of Chemicals to Man," Vol. 1. Int. Agency Res. Cancer, Lyon, France.

14. Anonymous (1973). "The Testing of Chemicals for Carcinogenicity, Mutagenicity and Teratogenicity." Ministry of Health and Welfare, Ottawa, Canada.

15. Anonymous (1973). "IARC Monographs on the Evaluation of Carcinogenic Risk of Chemicals to Man," Vol. 2. Int. Agency Res. Cancer, Lyon, France.

16. Anonymous (1973). "IARC Monographs on the Evaluation of Carcinogenic Risk of Chemicals to Man," Vol. 3, Int. Agency Res. Cancer, Lyon, France.

17. Anonymous (1973). "Information Bulletin of the Survey of Chemicals

Being Tested for Carcinogenicity," No. 1. Int. Agency Res. Cancer, Lyon, France.

18. Anonymous (1973). "Informational Bulletin of the Survey of Chemicals Being Tested for Carcinogenicity," No. 2. Int. Agency Res. Cancer, Lyon, France.

19. Anonymous (1973). "Survey of Compounds Which Have Been Tested for Carcinogenic Activity," Public Health Serv. Publ. No. 149, Suppl. 2. US Govt. Printing Office, Washington, D.C.

20. Anonymous (1974). "IARC Monographs on the Evaluation of Carcinogenic Risk of Chemicals to Man," Vol. 4. Int. Agency Res. Cancer, Lyon, France.

21. Anonymous (1974). "IARC Monographs on the Evaluation of Carcinogenic Risk of Chemicals to Man," Vol. 5. Int. Agency Res. Cancer, Lyon, France.

22. Anonymous (1974). "IARC Monographs on the Evaluation of Carcinogenic Risk of Chemicals to Man," Vol. 6. Int. Agency Res. Cancer, Lyon, France.

23. Anonymous (1974). "IARC Monographs on the Evaluation of Carcinogenic Risk of Chemicals to Man," Vol. 7. Int. Agency Res. Cancer, Lyon, France.

24. Anonymous (1974). "IARC Monographs on the Evaluation of Carcinogenic Risk of Chemicals to Man," Vol. 8. Int. Agency Res. Cancer, Lyon, France.

25. Anonymous (1974). "Informational Bulletin on the Survey of Chemicals Being Tested for Carcinogenicity," No. 3. Int. Agency Res. Cancer, Lyon, France.

26. Anonymous (1974). "Informational Bulletin on the Survey of Chemicals Being Tested for Carcinogenicity," No. 4. Int. Agency Res. Cancer, Lyon, France.

27. Anonymous (1974), "Survey of Compounds Which Have Been Tested for Carcinogenic Activity," Public Health Serv. Publ. No. 149, 1968-1969 Volume. US Govt. Printing Office, Washington, D.C.

28. Anonymous (1974), "Survey of Compounds Which Have Been Tested for Carcinogenesis Activity," Public Health Serv. Publ. No. 149, 1970-1971 Volume. US Govt. Printing Office, Washington, D.C.

29. Anonymous (1975). "Informational Bulletin on the Survey of Chemicals Being Tested for Carcinogenicity," No. 5. Int. Agency Research Cancer, Lyon, France.

30. Anthony, P. P., Boutwell, R. K., Farber, E., Friedell, G. H., Guillino, P. M., Lipkin, M., Nettesheim, P., Price, J., and Sporn, M. B. (1976). Summary statement of the symposium on early lesions and the development of epithelial cancer. *Cancer Res.* **36**, 2706.

31. Aoshima, M., Ishidate, M., and Sakurai, Y. (1971). Comparative studies on biological characteristics of transplantable rat leukemia, DBLA lines. *Gann* **62**, 95-106.

32. Arcos, J. R., Argus, M. F., and Wolf, G. (1968). "Chemical Induction of Cancer," Vol. 1, pp. 340-463. Academic Press, New York.

33. Arcos, J. R., and Argus, M. F. (1974), "Chemical Induction of Cancer," Vol. 2. Academic Press, New York.

34. Axelrad, A. A., and LeBlond, C. P. (1955). Induction of thyroid tumors in rats by a low iodine diet. *Cancer* **8**, 339-367.

34a. Baba, N., and von Haam, E. (1976). Tumours of the vagina, uterus, placenta and oviduct. *In* "Pathology of Tumours in Laboratory Animals" (V. S. Turusov, ed.), IARC Sci. Publ. No. 6, Vol. I, Part 2, pp. 161-188. Int. Agency Res. Cancer, Lyon, France.

35. Badger, G. M. (1962). "The Chemical Basis of Carcinogenic Activity." Thomas, Springfield, Illinois.

36. Baldwin, R. W., and Embleton, M. J. (1969). Immunology of spontaneous arising rat mammary adenocarcinomas. *Int. J. Cancer* **4**, 430-439.

37. Baldwin, R. W., and Embleton, M. J. (1970). Detection and isolation of tumour-specific antigen associated with a spontaneously arising rat mammary carcinoma. *Int. J. Cancer* **6**, 373-382.

38. Baldwin, R. W., and Embleton, M. J. (1974). Neoantigens on spontaneous and carcinogenic-induced rat tumours defined by *in vitro* lymphocytotoxicity assays. *Int. J. Cancer* **13**, 433-443.

39. Baldwin, R. W., and Pimm, M. V. (1973). BCG immunotherapy of local subcutaneous growths and post-surgical pulmonary metastases of a transplanted rat epithelioma of spontaneous origin. *Int. J. Cancer* **12**, 420-427.

40. Baldwin, R. W., and Pimm, M. V. (1973). BCG immunotherapy of pulmonary growths from intravenously transferred rat tumour cells. *Br. J. Cancer* **27**, 48-54.

41. Baldwin, R. W., and Vose, B. M. (1974). Embryonic antigen expression on 2-acetylaminofluorene induced and spontaneously arising rat tumours. *Br. J. Cancer* **30**, 209-214.

41a. Bannasch, P., and Massner, B. (1977). Fine structure of cholangiofibromas induced in the rat by *N*-nitrosomorpholine. *Virchows Arch. B* **24**, 295-315. (Engl. Summ.).

42. Bansal, S. C., and Sjögren, H. O. (1973). Correlation between changes in antitumor immune parameters and tumor growth *in vivo* in rats. *Fed. Proc., Fed. Am. Soc. Exp. Biol.* **32**, 165-172.

43. Bates, R. R. (1976). Laboratory approaches to the identification of carcinogens. *Ann. N. Y. Acad. Sci.* **271**, 29-38.

44. Berenblum, I. (1967). Principles of methodology relating to the potential carcinogenicity of medical drugs. *In* "Potential Carcinogenic Hazards from Drugs" (R. Truhaut, ed.), pp. 28-32. Springer-Verlag, Berlin and New York.

45. Bersohn, I., Purves, L. R., and Geddes, E. W. (1971). Liver function tests in primary cancer of the liver in the Bantu. *Liver Cancer, Proc. Work. Conf., 1969* IARC Sci. Publ. No. 1, pp. 158-167.

46. Bickerton, R. K. (1974). Spontaneous tumors and related pathologic changes. *In* "Carcinogenesis Testing of Chemicals" (L. Golberg, ed.), pp. 55-70. CRC Press, Cleveland, Ohio.

47. Bielschowsky, F. (1953). Chronic iodine deficiency as cause of neoplasia in thyroid and pituitary of aged rats. *Br. J. Cancer* **7**, 203-213.

48. Bingham, E., Niemeier, R. W., and Reid, J. B. (1976). Multiple factors in carcinogenesis. *Ann. N. Y. Acad. Sci.* **271**, 14-21.

49. Biskind, M. S., and Biskind, G. R. (1944). Development of tumors in the rat ovary after transplantation into the spleen. *Proc. Soc. Exp. Biol. Med.* **55**, 176-179.

50. Biskind, M. S., and Biskind, G. R. (1945). Tumor of rat testis produced by hetero-transplantation of infantile testis to spleen of adult castrates. *Proc. Soc. Exp. Biol. Med.* **59**, 4-8.

51. Bollman, R., and Pearse, A. G. E. (1974). Calcitonin secretion and APUD characteristics of naturally occurring medullary thyroid carcinomas in rats. *Virchows Arch.* **15**, 95-105.

52. Boorman, G. A., van Noord, M. J., and Hollander, C. F. (1972). Naturally occurring medullary thyroid carcinoma in the rat. *Arch. Pathol.* **94**, 35-41.

53. Boorman, G. A., and Hollander, C. F. (1973). Spontaneous lesions in the female WAG/Rij (Wistar) rat. *J. Gerontol.* **28**, 152-159.

54. Boorman, G. A., and Hollander, C. F. (1974). High incidence of spontaneous urinary bladder and ureter tumors in the Brown Norway rat. *J. Natl. Cancer Inst.* **52**, 1005-1008.

55. Boyland, E. (1967). Biochemical aspects of carcinogenesis with special reference to alkylating agents and some antibiotics. *In* "Potential Carcinogenic Hazards from Drugs" (R. Truhaut, ed.), pp. 204-208. Springer-Verlag, Berlin and New York.

56. Boyland, E., Dukes, C. E., and Grover, P. L. (1963). Carcinogenicity of 2-naphthylhydroxylamine and 2-naphthylamine. *Br. J. Cancer* **17**, 79-84.

57. Bralow, S. P. (1972). Experimental gastric carcinogenesis. *Digestion* **5**, 290-310.

58. Bralow, S. P., Gruenstein, M., Meranye, D. R., Bonakdarpour, A., and Shimkin, M. B. (1970). Adenocarcinoma of glandular stomach and

duodenum in Wistar rats ingesting N-methyl-N¹-nitro-N-nitroguinidine, histopathology and associated secretory changes. *Cancer Res.* **30**, 1215–1222.

59. Brand, K. G. (1973). Foreign body induced sarcomas. *In* "Cancer (F. Becker, ed.), Vol. I, pp. 485–511. Plenum, New York.

60. Bullock, F. S., and Curtis, M. R. (1925). Types of cysticercus tumors. *J. Cancer Res.* **9**, 425–433.

61. Bullock, F. D., and Curtis, M. R. (1930). Spontaneous tumors of the rat. *J. Cancer Res.* **14**, 1–115.

62. Burdette, W. J. (1973). Special rat strains provide model for adenocarcinoma of large intestine. *Comp. Pathol. Bull.* **5**, 1-2,4.

63. Butler, W. H. (1971). Pathology of liver cancer in experimental animals. *Liver Cancer, Proc. Work. Conf., 1969* IARC Sci. Publ. No. 1, pp. 30–41.

64. Campbell, J. S., Wiberg, G. S., Grice, H. C., and Lon, P. (1974). Stromal nephromas and renal cell tumors in suckling and weaned rats. *Cancer Res.* **34**, 2399–2404.

65. Carnow, B. W. (1976). Discussion. *Ann. N. Y. Acad. Sci.* **271**, 496.

66. Carter, R. L. (1973). Tumours of soft tissues. *In* "Pathology of Tumours in Laboratory Animals" (V. S. Turusov, ed.), IARC Sci. Publ. No. 5, Vol. I, Part 1, pp. 151–167. Int. Agency Res. Cancer, Lyon, France.

66a. Carter, R. L., and Ird, E. A. (1976). Tumours of the ovary. *In* "Pathology of Tumours in Laboratory Animals" (V. S. Turusov, ed.), IARC Sci. Publ. No. 6, Vol. I, Part 2, pp. 189–200. Int. Agency Res. Cancer, Lyon, France.

67. Casarett, G. W. (1972). Pathogenesis of radionuclide-induced tumors. *AEC Symp. Ser.* **29**, 1–14.

68. Chopra, H. C., and Taylor, D. J. (1970). Virus particles in rat mammary tumors of varying origin. *J. Natl. Cancer Inst.* **44**, 1141–1147.

69. Chopra, H. C., Bogden, A. E., Zelljadt, I., and Jensen, E. M. (1970). Virus particles in a transplantable rat mammary tumor of spontaneous origin. *Eur. J. Cancer* **6**, 287–290.

70. Christensen, H. E. (1976). Discussion. *Ann. N. Y. Acad. Sci.* **271**, 493–494.

71. Clayson, D. B. (1974). Bladder carcinogenesis in rats and mice: Possibility of artifacts. *J. Natl. Cancer Inst.* **52**, 1685–1689.

72. Clayson, D. B. (1974). Historical evolution of short-term tests for carcinogenicity. *In* "Carcinogenesis Testing of Chemicals" (L. Golberg, ed.), pp. 79–89. CRC Press, Cleveland, Ohio.

73. Clayson, D. B. (1976). Benzidine and 2-naphthylamine-voluntary substitution or technological alternatives. *Ann. N. Y. Acad. Sci.* **271**, 170–175.

74. Clayson, D. B. (1976). Discussion. *Ann. N. Y. Acad. Sci.* **271**, 460–461.

75. Cleveland, J. C., and Cole, J. W. (1969). Relationship of experimentally induced intestinal tumors to laxative ingestion. *Cancer* **23**, 1200–1203.

76. Cleveland, J. C., Litvak, S. F., and Cole, J. W. (1967). Identification of the route of action of the carcinogen 3, 2′-dimethyl-4-aminobiphenyl in the induction of intestinal neoplasia. *Cancer Res.* **27**, 708–714.

77. Cloyd, M. W., Burger, P. C., and Bigner, D. D. (1975). R-type virus-like particles in avian sarcoma virus-induced rat central nervous system tumors. *J. Natl. Cancer Inst.* **54**, 1479–1482.

78. Cobb, L. M. (1970). Radiation-induced osteosarcoma in the rat as a model for osteosarcoma in man. *Br. J. Cancer* **24**, 294–299.

79. Cockrell, B. Y., and Garner, F. M. (1976). Interstitial cell tumor of the testis in rats. *Comp. Pathol. Bull.* **8**, 2.

80. Cohen, S. M., Jacobs, J. B., Arai, M., Johansson, S., and Friedell, G. H. (1976). Early lesions in experimental bladder cancer: Experimental design and light microscopic findings. *Cancer Res.* **36**, 2508–2511.

81. Cone, M. V., and Nettesheim, P. (1973). Effects of vitamin A on 3-methylcholanthrene-induced squamous metaplasia and early tumors in the respiratory tract of rats. *J. Natl. Cancer Inst.* **50**, 1599–1606.

82. Coppola, J. A., Ball, J. L., and Brown, H. W. (1966). The incidence of spontaneous deciduomata in pseudopregnant rats. *J. Reprod. Fertil.* **12**, 389–390.

83. Corbett, T. H., Griswold, D. P., Roberts, B. J., Peckham, J. C., and Schabel, F. M. (1975). A mouse colon-tumor model for experimental therapy. *Cancer Chemother. Rep., Part 2* **5**, 169–186.

84. Corbett, T. H., Griswold, D. P., Roberts, B. J., Peckham, J. C., and Schabel, F. M. (1975). Tumor induction relationships in development of transplantable cancers of the colon in mice for chemotherapy assays, with a note on carcinogen structure. *Cancer Res.* **35**, 2434–2439.

85. Corbett, T. H. (1976). Cancer and congenital anomalies associated with anesthetics. *Ann. N. Y. Acad. Sci.* **271**, 58–66.

86. Cotchin, E., and Roe, F. J. C., eds. (1967). "Pathology of Laboratory Rats and Mice." Davis, Philadelphia, Pennsylvania.

87. Curtis, M. R., Bullock, F. D., and Dunning, W. F. (1931). A statistical study of the occurrence of spontaneous tumors in a large colony of rats. *Am. J. Cancer* **15**, 67–121.

88. Curtis, M. R., Dunning, W. F., and Bullock, W. F. (1933). Is malignancy due to a process analogous to somatic mutation? *Science* **77**, 175–176.

89. D'Aguanno, W. (1974). Interpretation of test results in terms of carcinogenic potential to the test animals: The regulatory point of view. *In* "Carcinogenesis Testing of Chemicals" (L. Golberg, ed.), pp. 41–44. CRC Press, Cleveland, Ohio.

90. Davey, F. R., and Moloney, W. C. (1970). Postmortem observations on Fischer rats with leukemia and other disorders. *Lab. Invest.* **23**, 327–334.

91. Davis, J. M. G. (1974). Histogenesis and fine structure of peritoneal tumors produced in animals by injections of asbestos. *J. Natl. Cancer Inst.* **52**, 1823–1837.

92. Davis, R. K., Stevenson, G. T., and Busch, K. A. (1959). Tumor incidence in normal Sprague-Dawley female rats. *Cancer Res.* **16**, 194–197.

93. De Serres, F. J. (1974). Mutagenesis assessment. *In* "Carcinogenesis Testing of Chemicals" (L. Golberg, ed.), pp. 101–107. CRC Press, Cleveland, Ohio.

94. Dickens, F. (1967). Drugs with lactone groups as potential carcinogens. *In* "Potential Carcinogenic Hazards from Drugs" (R. Truhaut, ed.), pp. 144–151. Springer-Verlag, Berlin and New York.

95. Druckrey, H. (1967). Quantitative aspects in chemical carcinogenesis. *In* "Potential Carcinogenic Hazards from Drugs" (R. Truhaut, ed.), pp. 60–78. Springer-Verlag, Berlin and New York.

96. Druckrey, H. (1967). Discussion of papers by Professor Shabad and Dr. van Esch. *In* "Potential Carcinogenic Hazards from Drugs" (R. Truhaut, ed.), pp. 201–203. Springer-Verlag, Berlin and New York.

97. Druckrey, H. (1972). Organospecific carcinogenesis in the digestive tract. *In* "Topics in Chemical Carcinogenesis" (W. Nakahara, S. Takayama, T. Sugimura, and S. Odashima, eds.), pp. 73–103. Univ. Park Press, Baltimore, Maryland.

98. Druckrey, H., and Lange, A. (1972). Carcinogenicity of azomethane dependent on age in BD rats. *Fed. Proc., Fed. Am. Soc. Exp. Biol.* **31**, 1482–1484.

99. Druckrey, H., Preussmann, R., Ivankovics, S., and Schmahl, D. (1967). Organotrope carcinogene wirkung bei 65 verschiedenen N-Nitroso-Verbindungen an BD-ratten. *Z. Krebsforsch.* **69**, 103–201.

100. Dunham, L. J., and Stewart, H. L. (1953). A survey of transplantable and transmissible animal tumors. *J. Natl. Cancer Inst.* **13**, 1299–1377.

101. Dunn, T. B. (1969). Comparative aspects of hematopoietic neoplasms of rodents. *Natl. Cancer Inst., Monogr.* **32**, 43–47.

102. Dunning, W. F. (1958). Response of some isologously transplanted rat neoplasms to steroids. *Ann. N. Y. Acad. Sci.* **76**, 696–704.

103. Dunning, W. F. (1958). Transplantable lymphomas in rats and respon-

siveness to drugs. *Ann. N. Y. Acad. Sci.* **76,** 643–658.

104. Dunning, W. F. (1963). Prostate cancer in the rat. *In* "Biology of the Prostate and Related Tissues." *Natl. Cancer Inst., Monogr.* **12,** 351–369.

105. Dunning, W. F., and Curtis, M. R. (1952). The incidence of DES-induced cancer in reciprocal F_1 hybrids obtained from crosses between rats of inbred lines that are susceptible and resistant to the induction of mammary cancer by this agent. *Cancer Res.* **12,** 702–706.

106. Dunning, W. F., and Curtis, M. R. (1945). The experimental production of extraskeletal bone-forming neoplasms in the rat. *Radiology* **44,** 64–76.

107. Dunning, W. F., Curtis, M. R., and Segaloff, A. (1946). Methylcholanthrene squamous cell carcinoma of the rat prostate with skeletal metastases and failure of the rat liver to respond to the same carcinogen. *Cancer Res.* **6,** 256–262.

108. Dunning, W. F., Curtis, M. R., and Segaloff, A. (1953). Strain differences in response to estrone and the induction of mammary gland, adrenal, and bladder cancer in rats. *Cancer Res.* **13,** 147–152.

109. Dunning, W. F., and Curtis, M. R. (1946). Multiple peritoneal sarcoma in rats from intraperitoneal injection of washed, ground *Taenia* larvae. *Cancer Res.* **6,** 668–670.

110. Dunning, W. F., and Curtis, M. R. (1953). Attempts to isolate the active agent in *Cysticercus fasciolaris*. *Cancer Res.* **13,** 838–842.

111. Durbin, P. W., Williams, M. H., Jeung, N., and Arnold, J. S. (1966). Development of spontaneous mammary tumors over the life span of the female Charles River (Sprague-Dawley) rat: The influence of ovariectomy, thyroidectomy, and adrenalectomy-ovariectomy. *Cancer Res.* **26,** 400–411.

112. Emmelot, P. (1971). Some aspects of the mechanism of liver carcinogenesis. *Liver Cancer, Proc. Work. Conf., 1969* IARC Sci. Pub. No. 1, pp. 94–109.

113. Engle, E. T. (1946). Tubular adenomas and testis-like tubules of the ovaries of aged rats. *Cancer Res.* **6,** 578–582.

114. Epstein, S. S. (1976). Aldrin and dieldrin suspension based on experimental evidence and evaluation and societal needs. *Ann. N. Y. Acad. Sci.* **271,** 187–195.

115. Ertürk, E., Cohen, S. M., Price, J. M., and Bryan, G. T. (1969). Pathogenesis, histology, and transplantation of urinary bladder carcinomas induced in albino rats by oral administration of n[4(5-nitro-2-furyl)2-thiazolyl] formamide. *Cancer Res.* **29,** 2219–2228.

116. Ertürk, E., Price, J. M., Morris, J. E., Cohen, S., Leith, R. S., von Esch, A. M., and Crovetti, A. J. (1967). The production of carcinoma of the urinary bladder in rats by feeding *N*[4-(5-nitro-2-furyl)-2-thiazolyl) formamide. *Cancer Res.* **27,** 1998–2002.

117. Ertürk, E., Cohen, S. M., Price, J. M., von Esch, A. M., Crovetti, A. J., and Bryan, G. T. (1969). The production of hemangioendothelial-sarcoma in rats by feeding 5-acetamido-3-(5-nitro-2-furyl)-6*H*-1,2,4,-oxadizine. *Cancer Res.* **29,** 2212–2218.

118. Evans, A. S. (1976). Causation and disease: The Henle-Koch postulates revisited. *Yale J. Biol. Med.* **49,** 175–195.

119. Evans, I. A., and Mason, J. (1965). Carcinogenic activity of bracken. *Nature (London)* **208,** 913–914.

120. Fairchild, E. J. (1976). Guidelines for a NIOSH policy on occupational carcinogenesis. *Ann. N. Y. Acad. Sci.* **271,** 200–207.

121. Farber, E. (1976). Introductory remarks to the symposium on early lesions and the development of epithelial cancer. *Cancer Res.* **36,** 2481.

122. Farber, E. (1976). Discussion. *Ann. N. Y. Acad. Sci.* **271,** 466–467.

123. Franks, L. M. (1967). Normal and pathologic anatomy of the genital tract of rats and mice. *In* "Pathology of Laboratory Rats and Mice" (E. Cotchin and F. J. C. Roe, eds.), pp. 469–499. Davis, Philadelphia, Pennsylvania.

123a. Franks, L. M., and Maldague, P. (1976). Tumours of the accessory male sex glands. *In* "Pathology of Tumours in Laboratory Animals," (V. S. Turusov, ed.), IARC Sci. Publ. No. 6, Vol. I, Part 2, pp. 151–160. Int. Agency Res. Cancer, Lyon, France.

124. Fraumeni, J. F. (1974). Carcinogenic effects of drugs in man. *In* "Carcinogenesis Testing of Chemicals" (L. Golberg, ed.), pp. 51–54. CRC Press, Cleveland, Ohio.

125. Friedman, L. (1974). Dose selection and administration. *In* "Carcinogenesis Testing of Chemicals" (L. Golberg, Ed.), pp. 21–22. CRC Press, Cleveland, Ohio.

126. Fukushima, S., Hirose, M., Tsuda, H., Sharai, T., Hirao, K., Arai, M., and Ito, N. (1976). Histological classification of urinary bladder cancers in rats induced by *N*-butyl-*N*-(4-hydroxybutyl)nitrosamine. *Gann* **67,** 81–90.

127. Furth, J., and Clifton, K. H. (1958). Screening techniques and problems of hormone-responsive tumors. *Ann. N. Y. Acad. Sci.* **76,** 681–695.

127a. Furth, J., Natane, P. K., and Pasteels, J. L. (1976). Tumours of the pituitary gland. *In* "Pathology of Tumours in Laboratory Animals" (V. S. Turusov, ed.), IARC Sci. Pub. No. 6, Vol. I, Part 2, pp. 201–238. Int. Agency Res. Cancer, Lyon, France.

128. Gardner, W. U. (1953). Hormonal aspects of experimental tumorigenesis. *Adv. Cancer Res.* **I,** 173–232.

129. Gillman, J., Gilbert, C., and Spence, I. (1953). Phaeochromocytoma in the rat. Pathogenesis and collateral reactions and its relation to comparable tumors in man. *Cancer* **6,** 494–511.

130. Gillman, T., Kinns, M., and Cross, R. M. (1969). Hodgkin's disease: A possible experimental model in rats. *Lancet* **2,** 1421–1422.

131. Glucksmann, A., and Cherry, C. P. (1973). Tumours of the salivary glands. *In* "Pathology of Tumours in Laboratory Animals" (V. S. Turusov, ed.), IARC Sci. Pub., Vol. I, Part 1, No. 5, pp. 75–86. Int. Agency Res. Cancer, Lyon, France.

132. Golberg, L. (1974). Recommendations of the conference as stated by the conference chairman. *In* "Carcinogenesis Testing of Chemicals" (L. Golberg, ed.), pp. 123–124. CRC Press, Cleveland, Ohio.

133. Goto, K., Kurokawa, Y., Hayashi, J., and Sato, H. (1975). Transplantable adenocarcinomas from colo-rectal tumors induced by infusion of *N*-methyl-*N*^{1}-nitro-*N*-nitrosoguanidine in ACI/N rats. *Gann* **66,** 89–93.

134. Greenberg, S. D., Hurst, G. A., Matlage, W. T., Miller, J. M., Hurst, I. J., and Mabry, L. C. (1976). Tyler asbestos workers program. *Ann. N. Y. Acad. Sci.* **271,** 353–364.

135. Griswold, D. P. (1972). Consideration of the subcutaneously implanted B16 melanoma as a screening model for potential anticancer agents. *Cancer Chemother. Rep., Part 2* **3,** 315–324.

136. Griswold, D. P., and Corbett, T. H. (1975). A colon tumor model for anticancer agent evaluation. *Cancer* **36,** 2441–2444.

137. Griswold, D. P., and Corbett, T. H. (1976). Breast tumor modeling for prognosis and treatment. *In* "Breast Cancer: A Multidisciplinary Approach" (G. St. Arneault, P. Band, and L. Israël, eds.), pp. 42–58. Springer-Verlag, Berlin and New York.

138. Griswold, D. P., Skipper, H. E., Laster, W. R., Wilcox, W. S., and Schabel, F. M. (1966). Induced mammary carcinoma in the female rat as a drug evaluation system. *Cancer Res.* **26,** 2169–2180.

139. Gross, L. (1970). "Oncogenic Viruses." Pergamon, Oxford.

139a. Hard, G. C. (1976). Tumours of the kidney, renal pelvis and ureter. *In* "Pathology of Tumours in Laboratory Animals" (V. S. Turusov, ed.), IARC Sci. Publ. No. 6, Vol. I, Part 2, pp. 73–102. Int. Agency Res. Cancer, Lyon, France.

140. Harris, R. J. C. (1976). "Cancer," 3rd ed. Penguin Books Inc., Baltimore, Maryland.

141. Hartwell, J. L. (1951). "Survey of Compounds Which Have Been

Tested for Carcinogenic Activity,'' Public Health Serv. Publ. No. 149. US Govt. Printing Office, Washington, D.C.

142. Hayashi, Y., Furukawa, H., and Hasegawa, T. (1972). Pancreatic tumors in rats induced by 4-nitroquinoline 1-oxide. *In* ''Topics in Chemical Carcinogenesis'' (W. Nakahara, S. Takayama, T. Sugimura, and S. Odashima, eds.), pp. 53-72. Univ. Park Press, Baltimore, Maryland.

143. Hayashi, Y., and Hasegawa, T. (1971). Experimental pancreatic tumor in rats after intravenous injection of 4-hydroxyaminoquinoline 1-oxide. *Gann* **62**, 329-330.

144. Hecker, E. (1976). Definitions and terminology in cancer (tumor) etiology—An analysis aiming at proposals for a current internationally standardized terminology. *Int. J. Cancer* **18**, 122-129.

144a. Hicks, R. M., Wakefield, J. St.J., Vlasov, N. N., and Pliss, G. B. (1976). Tumours of the urinary bladder. *In* ''Pathology of Tumours in Laboratory Animals'' (V. S. Turusov, ed.), IARC Sci. Pub. No. 6, Vol. I, Part 2. pp. 103-134. Int. Agency Res. Cancer, Lyon, France.

145. Hieger, I. (1961). ''Carcinogenesis.'' Academic Press, New York.

146. Higginson, J. (1972). The role of environmental biology in chemical carcinogenesis. *In* ''Topics in Chemical Carcinogenesis'' (W. Nakahara, S. Takayama, T. Sugimura, and S. Odashima, eds.), pp. 511-528. Univ. Park Press, Baltimore, Maryland.

147. Higginson, J. (1967). Discussion of papers by Professor Shabad and Dr. van Esch. *In* ''Potential Carcinogenic Hazards from Drugs'' (R. Truhaut, ed.), pp. 202. Springer-Verlag, Berlin and New York.

148. Hilfrich, J. (1973). A new model for inducing malignant ovarian tumors in rats. *Br. J. Cancer* **28**, 46-54.

149. Hill, B. F., ed. (1975). Inbred rat strains: Their development, history, and use. *Charles River Dig.* **14**, (3).

150. Hoel, D. G. (1976). Statistical extrapolation methods for estimating risks from animal data. *Ann. N. Y. Acad. Sci.* **271**, 418-420.

151. Hoffmann, D., and Wynder, E. L. (1974). Positive controls in environmental respiratory carcinogenesis. *In* ''Carcinogenesis Testing of Chemicals'' (L. Golberg, ed.), pp. 35-39. CRC Press, Cleveland, Ohio.

151a. Hollander, C. F., and Snell, K. C. (1976). Tumours of the adrenal gland. *In* ''Pathology of Tumours in Laboratory Animals'' (V. S. Turusov, ed.), IARC Sci. Pub. No. 6, Vol. I, Part 2, pp. 273-294. Int. Agency Res. Cancer, Lyon, France.

152. Hosokawa, M., Gotohda, E., and Kobayashi, H. (1971). Leukemia and mammary tumor in rats administered N-nitrosobutylurea. *Gann* **62**, 557-559.

153. Hueper, W. C. (1967). Carcinogenic hazard from arsenic and metal containing drugs. *In* ''Potential Carcinogenic Hazards from Drugs'' (R. Truhaut, ed.), pp. 79-104. Springer-Verlag, Berlin and New York.

154. Hueper, W. C., and Conway, W. D. (1964). ''Chemical Carcinogenesis and Cancers.'' Thomas, Springfield, Illinois.

155. Huggins, C. (1967). Pulse-doses of carcinogens. *In* ''Carcinogenesis: A Broad Critique,'' pp. 725-730. Williams & Wilkins Company, Baltimore, Maryland.

156. Huggins, C., and Oka, H. (1972). Regression of stem-cell erythroblastic leukemia after hypophysectomy. *Cancer Res.* **32**, 239-242.

157. Ida, N., Ikawa, Y., Ogawa, K., Takada, M., and Sugano, H. (1974). Cell culture from a rat brain tumor induced by intracerebral inoculation with murine sarcoma virus. *J. Natl. Cancer Inst.* **53**, 431-447.

158. Iglesias, R., Sternberg, W. H., and Segaloff, A. (1950). A transplantable functional ovarian tumor occurring spontaneously in a rat. *Cancer Res.* **10**, 668-673.

159. Isler, H., LeBlond, C. P., and Axelrad, A. A. (1959). Influence of age and of iodine intake on the production of thyroid tumors in the rat. *J. Natl. Cancer Inst.* **21**, 1065-1081.

160. Ito, A., Moy, P., Kaunitz, H., Kortwright, K., Clarke, S., Furth, J., and Meites, J. (1972). Incidence and character of the spontaneous pituitary tumors in strain CR and W/FU male rats. *J. Natl. Cancer Inst.* **49**, 701-711.

161. Ito, N., Hiasa, Y., Toyoshima, K., Okajima, E., Kamamoto, Y., Makiura, S., Yokota, Y., Sugihara, S., and Matayoshi, K. (1972). Rat bladder tumors induced by N-butyl-N-(4-hydroxybutyl)nitrosamine. *In* ''Topics in Chemical Carcinogenesis'' (W. Nakahara, S. Takayama, T. Sugimura, and S. Odashima, eds.), pp. 175-197. Univ. Park Press, Baltimore, Maryland.

162. Ito, N., Kamaoto, Y., Hiasa, Y., Makiura, S., Marugani, M., Yokota, Y., Sugihara, S., and Hirao, K. (1971). Histopathological and ultrastructural studies on esophageal tumors in rats treated with N-nitrosopiperidine. *Gann* **62**, 445-451.

163. Ivankovic, S. (1972). Prenatal carcinogenesis. *In* ''Topics in Chemical Carcinogenesis'' (W. Nakahara, S. Takayama, T. Sugimura, and S. Odashima, eds.), pp. 463-475. Univ. Park Press, Baltimore, Maryland.

163a. Ivankovic, S. (1976). Tumours of the heart. *In* ''Pathology of Tumours in Laboratory Animals'' (V. S. Turusov, ed.), IARC Sci. Pub. No. 6, Vol. I, Part 2, pp. 313-320. Int. Agency Res. Cancer, Lyon, France.

164. Ivankovic, S., Wohlenberg, H., Mennel, H. D., and Preussmann, R. (1972). Induction of heart tumors in BD-rats by continuous oral administration of the carcinogen 1-pyridyl-3,3-diethyltriazene *Z. Krebsforsch.* **77**, 217-225. (Engl. summ.).

165. Ivankovic, S., and Zeller, W. J. (1974). Leukemia L5222 of the rat strain BDIX. *Blut* **28**, 288-292. (Engl. summ.).

166. Jacobs, B. B., and Huseby, R. A. (1967). Neoplasms occurring in aged Fischer rat, with special reference to testicular, uterine, and thyroid tumors. *J. Natl. Cancer Inst.* **39**, 303-307.

167. Jones, R., McKensie, D., Stevens, M. L., Dunning, W. F., and Curtis, M. R. (1958). Usefulness of Dunning leukemia IRC 741 for quantitative pharmacological studies of cancer chemotherapeutic agents. *Ann. N. Y. Acad. Sci.* **76**, 659-672.

168. Juhasz, J. (1967). On potential carcinogenicity of some hydrazine derivatives used as drugs. *In* ''Potential Carcinogenic Hazards from Drugs'' (R. Truhaut, ed.), pp. 180-187. Springer-Verlag, Berlin and New York.

169. Kimbrough, R. D., Squire, R. A., Linder, R. E., Strandberg, J. D., Montali, R. J., and Buise, V. W. (1975). Induction of liver tumors in Sherman strain female rats by polychlorinated biphenyl Aroclor 1260. *J. Natl. Cancer Inst.* **55**, 1453-1459.

170. Kirsten, W. H., Anderson, D. G., Platz, C. E., and Crowell, E. B. (1962). Observations on the morphology and frequency of polyoma tumors in rats. *Cancer Res.* **22**, 484-491.

171. Koestner, A., Swenberg, J. A., and Wechsler, W. (1971). Transplacental production with ethylnitrosourea of neoplasms of the nervous system in Sprague-Dawley rats. *Am. J. Pathol.* **63**, 37-56.

172. Koestner, A., Swenberg, J. A., and Wechsler, W. (1972). Experimental tumors of the nervous system induced by resorptive N-nitrosourea compounds. *Prog. Exp. Tumor Res.* **17**, 9-30.

173. Kotin, P. (1976). Dose-response relationship and threshold concepts. *Ann. N. Y. Acad. Sci.* **271**, 22-28.

174. Kuschner, M., and Laskin, S. (1971). Experimental models in environmental carcinogenesis. *Am. J. Pathol.* **64**, 183-196.

175. Kuschner, M., Laskin, S., Drew, R. T., Cappiello, V., and Nelson, N. (1975). The inhalation carcinogenicity of alpha halo-ethers. 3. Lifetime and limited period inhalation studies with bis(chloromethyl) ether at 0.1 ppm. *Arch. Environ. Health* **30**, 73-77.

176. Labrie, F., Kelly, P. A., Asselin, J., and Raynaud, J.-P. (1976). Potent inhibitory activity of a new antiestrogen, RU 16-117, on the development and growth of DMBA-induced rat mammary adenocarcinoma. *In* ''Breast Cancer: A Multidisciplinary Approach (G. St.-

Arneault, P. Band, and L. Israël, eds.),'' pp. 109–120. Springer-Verlag, Berlin and New York.

177. Laskin, S., Drew, R. T., Cappiello, V., Kuschner, M., and Nelson, N. (1975). The inhalation carcinogenicity of alpha halo-ethers. 2. Chronic inhalation studies with chloromethyl ether. *Arch. Environ. Health* **30**, 70–72.

178. Laskin, S., Kuschner, M., and Drew, R. (1970). Studies in pulmonary carcinogenesis. *AEC Symp. Ser.* **18**, 321–351.

179. Laskin, S., Kuschner, M., Drew, R. T., Cappiello, V. P., and Nelson, N. (1971). Tumors of the respiratory tract induced by bis(chloromethyl) ether. *Arch. Environ. Health* **23**, 135–136.

180. Lassiter, D. V. (1976). Vinyl chloride—Best available technology. *Ann. N. Y. Acad. Sci.* **271**, 176–178.

181. Lassiter, D. V. (1976). Prevention of occupational cancer—Toward an integrated program of governmental action. *Ann. N. Y. Acad. Sci.* **271**, 214–215.

182. Lemen, R. A., Johnson, W. M., Wagoner, J. K., Archer, V. E., and Saccomanno, G. (1976). Cytologic observations and cancer incidence following exposure to BCME. *Ann. N. Y. Acad. Sci.* **271**, 71–80.

183. Lemon, P. G. (1967). Hepatic neoplasms of rats and mice. *In* "Pathology of Laboratory Rats and Mice" (E. Cotchin and F. J. C. Roe, eds.), pp. 25–56. Davis, Philadelphia, Pennsylvania.

184. Leuchtenberger, R., Leuchtenberger, C., Stewart, S. E., and Eddy, B. E. (1961). Difference in host cell virus relationship between tubular epithelium and stroma in kidneys of mice infected with SE polyoma virus. *Cancer* **14**, 567–576.

185. Liebelt, A. G., and Liebelt, R. A. (1967). Transplantation of tumors. *Methods Cancer Res.* **1**, 143–226.

186. Lindsay, S., Nichols, C. W., and Chaikoff, I. L. (1968). Natural occurring thyroid carcinoma in the rat: Similarities to human medullary carcinoma. *Arch. Pathol.* **86**, 353–364.

187. Lindsay, S., Potter, G. D., and Chaikoff, I. L. (1957). Thyroid neoplasms in the rat: A comparison of naturally occurring and I¹³¹-induced tumors. *Cancer Res.* **17**, 183–189.

188. Lindsay, S., Sheline, G. E., Potter, G. D., and Chaikoff, I. L. (1961). Induction of neoplasms in the thyroid gland of the rat by X-irradiation of the gland. *Cancer Res.* **21**, 9–16.

189. Litvinov, N. N., and Soloviev, Ju. N. (1973). Tumours of the bone. *In* "Pathology of Tumours in Laboratory Animals" (V. S. Turusov, ed.), IARC Sci. Pub. No. 5, Vol. I, Part 1, pp. 169–184. Int. Agency Res. Cancer, Lyon, France.

190. Long, E. L., and Jenner, P. M. (1963). Esophageal tumors produced in rats by the feeding of dihydrosafrole. *Fed. Proc., Fed. Am. Soc. Exp. Biol.* **22**, 275.

191. Longnecker, D. S., and Crawford, B. G. (1974). Hyperplastic nodules and adenomas of exocrine pancreas in azaserine-treated rats. *J. Natl. Cancer Inst.* **53**, 573–577.

192. Loustalot, P., Desaulles, P. A., and Meier, R. (1958). Characterization of the specificity of action of tumor-inhibiting compounds. *Ann. N. Y. Acad. Sci.* **76**, 838–849.

193. Lum, G. S., and Schreiner, A. W. (1963). Study of a virus isolated from a chloroleukemic Wistar rat. *Cancer Res.* **23**, 1742–1747.

194. Magee, P. N. (1971). Liver carcinogens in the human environment. *Liver Cancer, Proc. Work. Conf., 1969* IARC Sci. Publ. No. 1, pp. 110–120.

195. Maltoni, C. (1976). Predictive value of carcinogenesis bioassays. *Ann. N. Y. Acad. Sci.* **271**, 431–443.

196. Maltoni, C. (1976). Precursor lesions in exposed populations as indicators of occupational cancer risk. *Ann. N. Y. Acad. Sci.* **271**, 444–447.

197. Maltry, E., ed. (1971). "Benign and Malignant Tumors of the Urinary Bladder." Medical Examination Publishing Company, Flushing, New York.

198. Marsh, W. S., and Cullen, M. R. (1958). Observations on chemotherapy of human tumors HS1 and HEp3 in the heterologous host. *Ann. N. Y. Acad. Sci.* **76**, 752–763.

199. Martin, F., Martin, M. S., Bordes, M., and Knobel, S. (1975). Antigens associated with chemically induced intestinal carcinomas of rats. *Int. J. Cancer* **15**, 144–151.

200. Martin, F., Knobel, S., Martin, M., and Bordes, M. (1975). A carcinofetal antigen located on the membrane of cells from rat intestinal carcinoma in culture. *Cancer Res.* **35**, 333–336.

201. Mauer, S. M., Lee, C. S., Najarian, J. S., and Brown, D. M. (1974). Induction of malignant kidney tumors in rats with streptozotocin. *Cancer Res.* **34**, 158–160.

202. Mauer, S. M., Sutherland, D. E. R., Steffes, M. W., Lee, C. S., Najarian, J. S., and Brown, D. M. (1974). Effects of kidney and pancreas transplantation on streptozotocin-induced malignant kidney tumors in rats. *Cancer Res.* **34**, 1643–1645.

203. Maugh, T. H. (1974). Vitamin A: Potential protection from carcinogens. *Science* **186**, 1198.

204. McCall, D. C., and Cole, J. W. (1974). Transplantation of chemically induced adenocarcinomas of the colon in an inbred strain of rats. *Cancer* **33**, 1021–1026.

205. McCann, J. (1976). Discussion. *Ann. N. Y. Acad. Sci.* **271**, 467–468.

206. McCann, J., and Ames, B. N. (1976). A simple method for detecting environmental carcinogens as mutagens. *Ann. N. Y. Acad. Sci.* **271**, 5–13.

207. McEwan, J. C. (1976). Cytological monitoring of nickel sinter plant workers. *Ann. N. Y. Acad. Sci.* **271**, 365–369.

208. Mendelsohn, W. (1934). The malignant cells of two Crocker *Cysticercus* sarcomata. *Am. J. Cancer* **21**, 571–580.

208a. Mennel, H. D. and Zülch, E. (1976). Tumours of the central and peripheral nervous systems. *In* "Pathology of Tumours in Laboratory Animals (V. S. Turusov, ed.), IARC Sci. Publ. No. 6, Vol. I, part 2, pp. 295–312. Int. Agency Res. Cancer, Lyon, France.

209. Miller, J. A., and Miller, E. C. (1953). The carcinogenic aminoazo dyes. *Adv. Cancer Res.* **I**, 340–396.

210. Miller, J. A., and Miller, E. C. (1967). Metabolism of drugs in relation to carcinogenicity. *In* "Potential Carcinogenic Hazards from Drugs" (R. Truhaut, ed.), pp. 209–223. Springer-Verlag, Berlin and New York.

211. Miwa, M., Takenaka, S., Ito, K., Fujiwara, K., Kogure, K., Tokunaga, A., Hozumi, M., Fujimura, S., and Sugimura, T., (1976). Spontaneous colon tumors in rats. *J. Natl. Cancer Inst.* **56**, 615–621.

212. Moloney, W. C., Boschetti, A. E., and King, V. P. (1970). Spontaneous leukemia in Fischer rats. *Cancer Res.* **30**, 41–43.

213. Moloney, W. C. (1974). Primary granulocytic leukemia in the rat. *Cancer Res.* **34**, 3049–3057.

214. Money, W. L., and Rawson, R. W. (1947). The experimental production of thyroid tumors in the rat exposed to prolonged treatment with thiouracil. *Cancer* **3**, 321–335.

215. Moon, H. D., Simpson, M. E., Li, C. H., and Evans, H. M. (1950). Neoplasms in rats treated with pituitary growth hormone. III. Reproductive organs. *Cancer Res.* **10**, 549–556.

216. Moon, H. D., Simpson, M. E., Li, C. H., and Evans, H. M. (1950). Neoplasms in rats treated with pituitary growth hormone. II. Adrenal glands. *Cancer Res.* **10**, 364–370.

217. Mori, K. (1964). Induction of pulmonary and uterine tumors in rats by subcutaneous injections of 4-nitroquinoline 1-oxide. *Gann* **55**, 277–282.

218. Mori, M., and Sasaki, M. (1974). Chromosome studies on rat leukemias and lymphomas, with special attention to fluorescent karyotype analysis. *J. Natl. Cancer Inst.* **52**, 153–160.

219. Morris, H. P. (1955). The experimental development and metabolism

of thyroid gland tumors. *Adv. Cancer Res.* **3,** 51–115.

220. Morris, H. P., Dubnik, C. S., and Johnson, J. M. (1950). Studies of the carcinogenic action in the rat of 2-nitro-, 2-amino-, 2-acetylamino- and 2-diacetylaminofluorene after ingestion and after painting. *J. Natl. Cancer Inst.* **10,** 1201–1213.

221. Morris, H. P., Dyer, H. M., Wagner, B. P., Miyaji, H., and Rechcigl, M. (1964). Some aspects of the development, biology and biochemistry of rat hepatomas of different growth rate. *Adv. Enzyme Regul.* **2,** 321–333.

222. Morris, H. P., Lombard, L. S., Wagner, B. P., and Weisburger, J. H. (1957). Pituitary tumors in rats ingesting diets containing *p*-fluoroacetanilide, *o*-hydroxyacetanilide and 2,4-dimethylanilide. *Proc. Am. Assoc. Cancer Res.* **2,** 234.

223. Morris, H. P., Wagner, B. P., Ray, F. E., Stewart, H. L., and Snell, K. C. (1962). Comparative carcinogenic effects of N,N^1-2,7-fluorenylenebisacetamide by intraperitoneal and oral routes of administration to rats, with particular reference to gastric carcinoma. *J. Natl. Cancer Inst.* **29,** 997–1011.

223a. Mostofi, F. K., and Bresler, V. M. (1976). Tumours of the testis. *In* "Pathology of Tumours in Laboratory Animals" (V. S. Turusov, ed.), IARC Sci. Publ. No. 6, Vol. I, Part 2, pp. 135–150. Int. Agency Res. Cancer, Lyon, France.

224. Mulay, A. S., and Eyestone, W. H. (1955). Transplantable adrenocortical adenocarcinomas in Osborne-Mendel rats fed a carcinogenic diet. *J. Natl. Cancer Inst.* **16,** 723–739.

225. Murphy, G. P., Johnston, G. S., and Melby, E. C. (1967). Comparative aspects of experimentally induced and spontaneously observed renal tumors. *J. Urol.* **97,** 965–972.

226. Nagayo, T. (1973). Tumours of the stomach. *In* "Pathology of Tumours in Laboratory Animals" (V. S. Turusov, ed.), IARC Sci. Publ. No. 5, Vol. I, Part 1, pp. 101–118. Int. Agency Res. Cancer, Lyon, France.

227. Napalkov, N. P. (1967). On blastomogenic effect of antithyroid drugs. *In* "Potential Carcinogenic Hazards from Drugs" (R. Truhaut, ed.), pp. 172–179. Springer-Verlag, Berlin and New York.

227a. Napalkov, N. P. (1976). Tumours of the thyroid gland. *In* "Pathology of Tumours in Laboratory Animals (V. S. Turusov, ed.), IARC Sci. Publ. No. 6, Vol. I, Part 2, pp. 239–272. Int. Agency Res. Cancer, Lyon, France.

228. Narisawa, T., Sato, T., Hayakawa, M., Sakuma, A., and Nakano, H. (1971). Carcinoma of the colon and rectum by rectal infusion of N-methyl-N^1-nitro-N-nitrosoguanidine. *Gann* **62,** 231–234.

229. Navarette, A., and Spjut, H. J. (1967). Effect of colostomy on experimentally produced neoplasms of the colon of the rat. *Cancer* **20,** 1466–1472.

230. Nebert, D. (1974). Metabolic activation as a factor in carcinogenic response. *In* "Carcinogenesis Testing of Chemicals" (L. Golberg, ed.), pp. 75–78. CRC Press, Cleveland, Ohio.

231. Negroni, G. (1967). The cause of human leukemia-viruses or mycoplasmas? *In* "Carcinogenesis: A Broad Critique," pp. 91–102. Williams & Wilkins, Baltimore, Maryland.

232. Nelson, A. A., Fitzhugh, O. G., Morris, H. J., and Calvery, H. O. (1942). Neurofibromas of rat ears produced by prolonged feeding of crude ergot. *Cancer Res.* **2,** 11–15.

233. Nelson, N. (1976). The chloroethers-occupational carcinogens: A summary of laboratory and epidemiological studies. *Ann. N. Y. Acad. Sci.* **271,** 81–90.

234. Newberne, P. M., and Butler, W. H. (1969). Acute and chronic effects of aflatoxin on the liver of domestic and laboratory animals, a review. *Cancer Res.* **29,** 236–250.

235. Newberne, P. M., and Rogers, A. E. (1973). Animal model of human disease: Adenocarcinoma of the rat. *Am. J. Pathol.* **72,** 541–544.

236. Newberne, P. M., and Rogers, A. E. (1973). Animal model of human disease: Hepatocellular carcinoma. *Am. J. Pathol.* **72,** 137–140.

237. Newberne, P. M. (1974). Diets. *In* "Carcinogenesis Testing of Chemicals" (L. Golberg, ed.), pp. 17–20. CRC Press, Cleveland, Ohio.

238. Newill, V. A. (1976). Methodologies of risk assessment. *Ann. N. Y. Acad. Sci.* **271,** 413–417.

239. Nicholson, W. J. (1976). Asbestos—the TLV approach. *Ann. N. Y. Acad. Sci.* **271,** 152–169.

240. Noble, R. L., Beer, C. T., and Cutts, J. H. (1958). Role of chance observations in chemotherapy: *Vinca Rosea, Ann. N. Y. Acad. Sci.* **76,** 882–894.

241. Odashima, S. (1972). Leukemogenic effect of N-nitroso-N-butylurea in rats. *In* "Topics in Chemical Carcinogenesis" (W. Nakahara, S. Takayama, T. Sugimura, and S. Odashima, eds.), pp. 477–491. Univ. Park Press, Baltimore, Maryland.

242. Olson, H. M., and Capen, C. C. (1977). Animal models of human disease—osteosarcoma. *Comp. Pathol. Bull.* **9,** 3–4.

243. Oppenheimer, B. S., Oppenheimer, E. T., Stout, A. P., Danishefsky, I., and Willhite, M. (1959). Studies of the mechanism of carcinogenesis by plastic films. *Acta Unio Int. Cancrum* **15,** 659–663.

244. Oppenheimer, B. S., Oppenheimer, E. T., Danishefsky, I., Stout, A. P., and Eirich, F. R. (1955). Further studies of polymers as carcinogenic agents in animals. *Cancer Res.* **15,** 333–340.

245. Oppenheimer, B. S., Oppenheimer, E. T., Danishefsky, I., and Stout, A. P. (1956). Carcinogenic effects of metals in rodents. *Cancer Res.* **16,** 439–441.

246. Oppenheimer, E. T., Willhite, M., Stout, A. P., Danishefsky, I., and Fishman, M. M. (1964). A comparative study of the effects of imbedding cellophane and polystyrene films in rats. *Cancer Res.* **24,** 379–382.

247. Palm, J. E., Teller, M. N., Merker, P. C., and Woolley, G. W. (1958). Host conditioning in experimental chemotherapy. *Ann. N. Y. Acad. Sci.* **76,** 812–820.

248. Peck, H. M. (1974). Design of experiments to detect carcinogenic effect of drugs. *In* "Carcinogenesis Testing of Chemicals" (L. Golberg, ed.), pp. 1–13. CRC Press, Cleveland, Ohio.

249. Peckham, B. M., and Greene, R. R. (1952). Experimentally produced granulosa-cell tumors in rats. *Cancer Res.* **12,** 25–29.

250. Perk, K., Shachat, D. A., and Moloney, J. B. (1968). Pathogenesis of a rhabdomyosarcoma (undifferentiated type) in rats induced by a murine sarcoma virus (Moloney). *Cancer Res.* **28,** 1197–1206.

251. Peters, J. A., ed. (1969). "Survey of Compounds Which Have Been Tested for Carcinogenic Activity" Public Health Serv. Publ. No. 149, Suppl. 1. US Gov. Printing Office, Washington, D.C.

252. Pfeiffer, C. A. (1949). Development of leiomyomas in female rats with an endocrine imbalance. *Cancer Res.* **9,** 277–281.

253. Pitot, H. C. (1974). Criteria of neoplastic transformation. *In* "Carcinogenesis Testing of Chemicals" (L. Golberg, ed.), pp. 113–114. CRC Press, Cleveland, Ohio.

254. Pliss, G. B. (1973). Tumours of the auditory sebaceous glands. *In* "Pathology of Tumours in Laboratory Animals" (V. S. Turusov, ed.), IARC Sci. Publ. No. 5, Vol. I, Part 1, pp. 23–30. Int. Agency Res. Cancer, Lyon, France.

255. Poel, W. E. (1967). Progesterone and mammary carcinogenesis. *In* "Potential Carcinogenic Hazards from Drugs" (R. Truhaut, ed.), pp. 162–170. Springler-Verlag, Berlin and New York.

256. Pollard, M. (1973). Spontaneous prostate adenocarcinoma in aged germfree Wistar rats. *J. Natl. Cancer Inst.* **51,** 1235–1241.

257. Pollard, M., and Luckert, P. H. (1975). Transplantable metastasizing prostate adenocarcinomas in rats. *J. Natl. Cancer Inst.* **54,** 643–649.

258. Porta, G. D. (1967). Some aspects of medical drug testing for carcinogenic activity. *In* "Potential Carcinogenic Hazards from Drugs"

(R. Truhaut, ed.), pp. 33–47. Springer-Verlag, Berlin and New York.

258a. Pour, P., Stanton, M. F., Kuschner, M., Laskin, S., and Shabad, L. M. (1976). Tumours of the respiratory tract. *In* "Pathology of Tumours in Laboratory Animals (V. S. Turusov, ed.), IARC Sci. Publ. No. 6, Vol I, Part 2, pp. 1–40. Int. Agency Res. Cancer, Lyon, France.

259. Pozharisski, K. M. (1973). Tumours of the intestines. *In* "Pathology of Tumours in Laboratory Animals" (V. S. Turusov, ed.), IARC Sci. Publ. No. 5, Vol. I, Part 1, pp. 119–140. Int. Agency Res. Cancer, Lyon, France.

260. Pozharisski, K. M. (1973). Tumours of the oesophagus. *In* "Pathology of Tumours in Laboratory Animals" (V. S. Turusov, ed.), IARC Sci. Publ. No. 5, Vol. I, Part 1, pp. 87–100. Int. Agency Res. Cancer, Lyon, France.

261. Pozharisski, K. M. (1975). The significance of nonspecific injury for colon carcinogenesis in rats. *Cancer Res.* **35,** 3824–3830.

262. Pradham, S. N., Chung, E. B., Ghosh, B., Paul, B. D., and Kapadia, G. J. (1974). Potential carcinogens. I. Carcinogenicity of some plant extracts and their tannin-containing fractions in rats. *J. Natl. Cancer Inst.* **52,** 1579–1582.

263. Prehn, R. T. (1975). Neoplasia. *In* "Principles of Pathobiology" (M. F. La Via and R. B. Hill, eds.), pp. 203–254. Oxford Univ. Press, London and New York.

264. Prejean, J. D., Peckham, J. C., Casey, A. E., Griswold, D. P., Weisburger, E. K., and Weisburger, J. H. (1973). Spontaneous tumors in Sprague-Dawley rats and Swiss mice. *Cancer Res.* **33,** 2768–2773.

265. Price, J. M. (1971). Etiology of bladder cancer. *In* "Benign and Malignant Tumors of the Urinary Bladder" (E. Maltry, ed.), pp. 189–261. Medical Examination Publishing Company, Inc., Flushing, New York.

266. Price, J. M., and Pamukcu, A. M. (1968). The induction of neoplasms of the urinary bladder of the cow and the small intestine of the rat by feeding bracken fern (*Pteris aquilina*). *Cancer Res.* **28,** 2247–2251.

267. Purves, H. D., Griesbach, W. E., and Kennedy, T. H. (1951). Studies in experimental goitre: Malignant changes in a transplantable rat thyroid tumor. *Br. J. Cancer* **5,** 301–310.

268. Rall, D. P. (1974). Summary and concluding remarks. *In* "Carcinogenesis Testing of Chemicals" (L. Golberg, ed.), pp. 119–121. CRC Press, Cleveland, Ohio.

269. Rawson, R. W. (1976). Discussion of papers. *Ann. N. Y. Acad. Sci.* **271,** 391–392.

270. Reddy, J., Svoboda, D., Azarnoff, D., and Dawar, R. (1973). Cadmium-induced Leydig cell tumors of rat testis: Morphologic and cytochemical study. *J. Natl. Cancer Inst.* **51,** 891–903.

271. Reuber, M. (1974). Criteria for tumor diagnosis and classification of malignancy. *In* "Carcinogenesis Testing of Chemicals" (L. Golberg, ed.), pp. 71–73. CRC Press, Cleveland, Ohio.

272. Ribelin, W. E., and McCoy, J. R., eds. (1965). "The Pathology of Laboratory Animals." Thomas, Springfield, Illinois.

273. Richter, C. B., Estes, P. C., and Tennant, R. W. (1972). Spontaneous stem cell leukemia in young Sprague-Dawley rats. *Lab. Invest.* **26,** 419–428.

274. Roberts, D. C. (1964). "Research Using Transplanted Tumours of Laboratory Animals: A Cross Referenced Bibliography," Vol. I. Imp. Cancer Res. Fund, London.

275. Roberts, D. C. (1965). "Research Using Transplanted Tumours of Laboratory Animals: A Cross Referenced Bibliography," Vol. II. Imp. Cancer Res. Fund, London.

276. Roberts, D. C. (1966). "Research Using Transplanted Tumours of Laboratory Animals: A Cross Referenced Bibliography," Vol. III. Imp. Cancer Res. Fund, London.

277. Roberts, D. C. (1967). "Research Using Transplanted Tumours of Laboratory Animals: A Cross Referenced Bibliography," Vol. IV. Imp. Cancer Res. Fund, London.

278. Roberts, D. C. (1968). "Research Using Transplanted Tumours of Laboratory Animals: A Cross Referenced Bibliography," Vol. V. Imp. Cancer Res. Fund, London.

279. Roberts, D. C. (1969). "Transplanted Tumours of Rats and Mice: An Index of Tumours and Host Strains," Lab. Ani. Handb. No. 3. Lab. Anim. Ltd., London.

280. Roberts, D. C. (1969). "Research Using Transplanted Tumours of Laboratory Animals: A Cross Referenced Bibliography," Vol. VI. Imp. Cancer Res. Fund, London.

281. Roberts, D. C. (1970). "Research Using Transplanted Tumours of Laboratory Animals: A Cross Referenced Bibliography," Vol. VII. Imp. Cancer Res. Fund, London.

282. Roberts, D. C. (1971). "Research Using Transplanted Tumours of Laboratory Animals: A Cross-Referenced Bibliography," Vol. VIII. Imp. Cancer Res. Fund, London.

283. Roberts, D. C. (1973). "Research Using Transplanted Tumours of Laboratory Animals: A Cross-Referenced Bibliography," Vol. X. Imp. Cancer Res. Fund, London.

284. Roberts, D. C., and Barton, M. (1972). "Research Using Transplanted Tumours of Laboratory Animals: A Cross-Referenced Bibliography," Vol. IX. Imp. Cancer Res. Fund, London.

285. Roberts, D. C., and Drobycz, B. (1974). "Research Using Transplanted Tumours of Laboratory Animals: A Cross-Referenced Bibliography," Vol. XI. Imp. Cancer Res. Fund, London.

286. Roberts, D. C., and Drobycz, B. (1975). "Research Using Transplanted Tumours of laboratory Animals: A Cross-Referenced Bibliography," Vol. XII. Imp. Cancer Res. Fund, London.

287. Roberts, N. J. (1976). Discussion of papers. *Ann. N. Y. Acad. Sci.* **271,** 388–390.

288. Roe, F. J. C. (1965). Spontaneous tumours in rats and mice. *Food Cosmet. Toxicol.* **3,** 707–720.

289. Roe, F. J. C. (1967). On potential carcinogenicity of the iron macromolecular complexes. *In* "Potential Carcinogenic Hazards from Drugs (R. Truhaut, ed.), pp. 105–118. Springer-Verlag, Berlin and New York.

290. Roe, F. J. C., and Roberts, J. D. B. (1973). Tumours of the pancreas. *In* "Pathology of Tumours in Laboratory Animals" (V. S. Turusov, ed.), IARC Sci. Publ. No. 5, Vol. I, Part 1, pp. 141–150. Int. Agency Res. Cancer, Lyon, France.

291. Ross, M. H., and Bras, G. (1973). Influence of protein under- and overnutrition on spontaneous tumor prevalence in the rat. *J. Nutr.* **103,** 944–963.

292. Rowe, N. H., Grammer, F. C., Watson, F. R., and Nickerson, N. H. (1970). A study of environmental influence upon salivary gland neoplasia in rats. *Cancer* **26,** 436–444.

293. Rowe, V. K. (1976). Discussion. *Ann. N. Y. Acad. Sci.* **271,** 479.

294. Rowlatt, U. F. (1967). Spontaneous epithelial tumours of the pancreas of mammals. *Br. J. Cancer* **21,** 82–107.

295. Rowlatt, U. F. (1967). Pancreatic neoplasms of rats and mice. *In* "Pathology of Laboratory Rats and Mice" (E. Cotchin and F. J. C. Roe, eds.), pp. 85–103. Davis, Philadelphia, Pennsylvania.

296. Rowlatt, U. F. (1967). Neoplasms of the alimentary canal of rats and mice. *In* "Pathology of Laboratory Rats and Mice" (E. Cotchin and F. J. C. Roe, eds.), pp. 57–84. Davis, Philadelphia, Pennsylvania.

297. Russfield, A. B. (1966). "Tumors of Endocrine Glands and Secondary Sex Organs," Public Health Serv. Publ. No. 1332. US Govt. Printing Office, Washington, D.C.

298. Russfield, A. B. (1967). Pathology of the endocrine glands, ovary and testis of rats and mice. *In* "Pathology of Laboratory Rats and Mice" (E. Cotchin and F. J. C. Roe, eds.), pp. 391–467. Davis, Philadelphia, Pennsylvania.

299. Sabin, A. B. (1967). Search for viral etiology of human leukemia and

lymphomas: Past efforts and future perspectives. *In* "Carcinogenesis: A Broad Critique," pp. 181–200. Williams & Wilkins, Baltimore, Maryland.

300. Saccomanno, G. (1976). Discussion. *Ann. N. Y. Acad. Sci.* **271,** 509.

301. Saccomanno, G., Archer, V. E., Saunders, R. P., Auerbach, O., and Klein, M. G. (1976). Early indices of cancer risk among uranium miners with reference to modifying factors. *Ann. N. Y. Acad. Sci.* **271,** 377–383.

302. Saffiotti, U. (1967). Discussion of papers by Professor Shabad and Dr. van Esch. *In* "Potential Carcinogenic Hazards from Drugs" (R. Truhaut, ed.), p. 202. Springer-Verlag, Berlin and New York.

303. Saffiotti, U. (1967). Discussion of some aspects of medical drug testing for carcinogenic activity. *In* "Potential Carcinogenic Hazards from Drugs" (R. Truhaut, ed.), p. 46. Springer-Verlag, Berlin and New York.

304. Saffiotti, U. (1970). Experimental respiratory tract carcinogenesis and its relation to inhalation exposures. *AEC Symp. Ser.* **18,** 27–54.

305. Saffiotti, U. (1976). Prevention of occupational cancer—toward an integrated program of governmental action: Role of the National Cancer Institute. *Ann. N. Y. Acad. Sci.* **271,** 208–213.

306. Saffiotti, U. (1976). Methodologies for risk assessment in occupational carcinogenesis. *Ann. N. Y. Acad. Sci.* **271,** 393–395.

307. Saffiotti, U. (1976). Discussion. *Ann. N. Y. Acad. Sci.* **271,** 468–470.

308. Saffiotti, U. (1976). Discussion. *Ann. N. Y. Acad. Sci.* **271,** 492.

309. Saffiotti, U. (1976). Discussion. *Ann. N. Y. Acad. Sci.* **271,** 509–510.

310. Saffiotti, U. (1976). Discussion. *Ann. N. Y. Acad. Sci.* **271,** 515–516.

311. Saffiotti, U. (1976). Introductory remarks to the symposium on early lesions and the development of epithelial cancer. *Cancer Res.* **36,** 2479–2480.

312. Samuels, S. W. (1976). Determination of cancer risk in a democracy. *Ann. N. Y. Acad. Sci.* **271,** 421–430.

313. Sass, B., Rabstein, L. S., Madison, R., Nims, R. M., Peters, R. L., and Kelloff, G. J. (1975). Incidence of spontaneous neoplasms in Fischer-344 rats throughout the natural life-span. *J. Natl. Cancer Inst.* **54,** 1449–1456.

314. Saunders, L. Z. (1967). Ophthalmic pathology in rats and mice. *In* "Pathology of Laboratory Rats and Mice" (E. Cotchin and F. J. C. Roe, eds.), pp. 349–371. Davis, Philadelphia, Pennsylvania.

315. Schacham, P., Philp, R. B., and Gowday, C. W. (1970). Antihematopoietic and carcinogenic effects of bracken fern (*Pteridium aquilinum*) in rats. *Am. J. Vet. Res.* **31,** 191–197.

316. Schardein, J. L., Fitzgerald, J. E., and Kaump, D. H. (1968). Spontaneous tumors in Holtzman source rats of various ages. *Pathol. Vet.* **5,** 238–252.

316a. Schauer, A., and Kunze, E. (1976). Tumours of the liver. *In* "Pathology of Tumours in Laboratory Animals" (V. S. Turusov, ed.), IARC Sci. Publ. No. 6, Vol I, Part 2, pp. 41–72. Int. Agency Res. Cancer, Lyon, France.

317. Schiffer, D., Fabiani, A., Grossi-Paoletti, E., and Paoletti, P. (1971). Nitrosourea-induced brain tumors in the rat. Comments on the pathogenesis of mixed and polymorphic gliomas. *Tumori* **57,** 333–342. (Engl. summ.).

318. Schilling, B. von, Frohberg, H., and Oettel, H. (1967). On the incidence of naturally occurring thyroid carcinoma in the Sprague-Dawley rat. *Ind. Med. Surg.* **36,** 678–684.

319. Schilling, B. von, Frohberg, H., and Oettel, H. (1967). Spontaneous tumors of the thyroid gland of the laboratory rat. *Z. Krebsforsch.* **70,** 161–164. (Engl. summ.).

320. Schmutz, J. A., and Chaudhry, A. P. (1969). Incidence of induced tumors in the rat submandibular gland with different doses of 7,12-dimethylbenz-α-anthracene. *J. Dent. Res.* **48,** 1316.

321. Schoental, R. (1967). Pyrrolizidine (*Senecio*) alkaloids and other natural drugs as potential carcinogens. *In* "Potential Carcinogenic Hazards from Drugs" (R. Truhaut, ed.), pp. 152–161. Springer-Verlag, Berlin and New York.

322. Schoental, R., Hard, G. C., and Gibbard, S. (1971). Histopathology of renal lipomatous tumors in rats treated with the "natural" products, pyrrolizidine alkaloids and alpha, beta-unsaturated aldehydes. *J. Natl. Cancer Inst.* **47,** 1037–1044.

323. Scholler, J. (1958). Practical and theoretical considerations in the use of induced and spontaneous mammary tumors in cancer chemotherapy. *Ann. N. Y. Acad. Sci.* **76,** 855–860.

324. Schreiber, H., Nettesheim, P., and Martin, D. H. (1972). Rapid development of bronchiolo-alveolar squamous cell tumors in rats after intratracheal injection of 3-methylcholanthrene. *J. Natl. Cancer Inst.* **49,** 541–554.

325. Shabad, L. M. (1967). Discussion of Tumeurs des Cellules de Leydig du testicule du rat presented by A. Lacassagne. *In* "Potential Carcinogenic Hazards from Drugs" (R. Truhaut, ed.), p. 242. Springer-Verlag, Berlin and New York.

326. Shabad, L. M. (1967). Plastic carcinogenesis, some experimental data and its possible importance for clinic and prophylaxis of cancer. *In* "Potential Carcinogenic Hazards from Drugs" (R. Truhart, ed.), pp. 188–195. Springer-Verlag, Berlin and New York.

327. Sher, S. P. (1974). Tumors in control mice: Literature tabulation. *Toxicol. Appl. Pharmacol.* **30,** 337–359.

328. Shimkin, M. B. (1973). "Science and Cancer," DHEW Publ. No. (NIH) 76-568. Public Health Serv. Natl. Cancer Inst., Bethesda, Maryland.

329. Shimkin, M. B. (1974). Species and strain selection. *In* "Carcinogenesis Testing of Chemicals" (L. Golberg, ed.), p. 15. CRC Press, Cleveland, Ohio.

330. Shubik, P. (1967). A program for the investigation of the possible chronic hazards of drugs. *In* "Potential Carcinogenic Hazards from Drugs" (R. Truhaut, ed.), pp. 48–52. Springer-Verlag, Berlin and New York.

331. Shubik, P. (1967). Biological mechanisms in carcinogenesis. *In* "Carcinogenesis: A Broad Critique," pp. 731–738. Williams & Wilkins, Baltimore, Maryland.

332. Shubik, P. (1974). Interpretation of test results in terms of significance to man. *In* "Carcinogenesis Testing of Chemicals" (L. Golberg, ed.), pp. 45–50. CRC Press, Cleveland, Ohio.

333. Shubik, P. (1975). Skin carcinogenesis. *In* "Animal Models in Dermatology" (H. Maibach, ed.), pp. 146–155. Churchill-Livingstone, London.

334. Shubik, P., and Hartwell, J. L. (1957). "Survey of Compounds Which Have Been Tested for Carcinogenic Activity," Public Health Serv. Publ. No. 149. US Govt. Printing Office, Washington, D.C.

335. Shubik, P., and Sice, J. (1956). Chemical carcinogenesis as a chronic toxicity test. *Cancer Res.* **16,** 728–742.

336. Sikov, M. R., Mahlum, D. D., and Clarke, W. J. (1972). Effect of age on the carcinogenicity of ^{131}I in the rat-interim report. *AEC Symp. Ser.* **29,** 25–32.

337. Simms, H. S. (1967). Longevity studies in rats. I. Relation between life span and age of onset of specific lesions. *In* "Pathology of Laboratory Rats and Mice" (E. Cotchin and F. J. C. Roe, eds.), pp. 733–747. Davis, Philadelphia, Pennsylvania.

338. Snell, K. C. (1965). Spontaneous lesions of the rat. *In* "Pathology of Laboratory Animals (W. E. Ribelin and C. T. McCoy, eds.), pp. 241–302. Thomas, Springfield, Illinois.

339. Snell, K. C. (1967). Renal disease of the rat. *In* "Pathology of Laboratory Rats and Mice" (E. Cotchin and F. J. C. Roe, eds.), pp. 105–147. Davis, Philadelphia, Pennsylvania.

340. Snell, K. C. (1969). Hematopoietic neoplasms of rats and Mastomys.

Natl. Cancer Inst., Monogr. **32**, 59–63.

341. Snell, K. C., and Stewart, H. L. (1959). Variations in histologic pattern and functional effects of a transplantable adrenal cortical carcinoma in intact hypophysectomized, and new-born rats, *J. Natl. Cancer Inst.* **22**, 1119–1155.

342. Sontag, J. M., Page, N. P., and Saffiotti, U. (1976). "Guidelines for Carcinogen Bioassay in Small Rodents," Natl. Cancer Inst. Carcinog. Tech. Rep. Ser. No. 1, U. S. Dept. of Health, Education, and Welfare Publ. No. (NIH) 76-801. Natl. Cancer Inst., Bethesda, Maryland.

343. Spjut, H. J., and Noall, M. W. (1971). Experimental induction of tumors of the large bowel of rats. A review of the experience with 3,2′-dimethyl-4-aminobiphenyl. *Cancer* **28**, 29–37.

344. Spjut, H. J., and Spratt, J. S. (1965). Endemic and morphologic similarities existing between spontaneous colon neoplasms in man and 3,2′-dimethyl-4-aminobiphenyl-induced colonic neoplasms in rats. *Ann. Surg.* **161**, 309–324.

345. Sporn, M. B. (1976). Approaches to prevention of epithelial cancer during the preneoplastic period. *Cancer Res.* **36**, 2699–2702.

346. Squire, R. A. (1969). Comparative aspects of hematopoietic neoplasms of animals and man—Summary. *Natl. Cancer Inst., Monogr.* **32**, 363–365.

347. Squire, R. A., and Levitt, M. H. (1975). Report of a workshop on classification of specific hepatocellular lesions in rats. *Cancer Res.* **35**, 3214–3223.

348. Stavrou, D. (1971). Comparative pathology of tumours of the nervous system. *Zentralbl. Veterinaermed., Reihe A* **18**, 585–645 (Engl. summ.).

349. Steele, G., and Sjögren, H. O. (1974). Embryonic antigens associated with chemically induced colon carcinomas in rats. *Int. J. Cancer* **14**, 435–444.

350. Steele, G., and Sjögren, H. O. (1974). Cross-reacting tumor-associated anitgen(s) among chemically induced rat colon carcinomas. *Cancer Res.* **34**, 1801–1807.

351. Steele, G., Sjögren, H. O., and Price, M. R. (1975). Tumor-associated and embryonic antigens in soluble fractions of a chemically-induced rat colon carcinoma. *Int. J. Cancer* **16**, 33–51.

352. Stewart, H. L., Snell, K. C., Dunham, L. J., and Schlyen, S. M. (1959). Transplantable and transmissible tumors of animals. *In* "Atlas of Tumor pathology," Sect. XII, Fasc. 40. Armed Forces Inst. Pathol., Washington, D.C.

353. Swaen, G. J. V., and Van Heerde, P. (1973). Tumours of the haematopoietic system. *In* "Pathology of Tumours in Laboratory Animals" (V. S. Turusov, ed.), IARC Sci. Pub. No. 5, Vol. I, Part 1, pp. 185–214. Int. Agency Res. Cancer, Lyon, France.

354. Symeonidis, A., and Mori-Chanez, P. (1952). A transplantable ovarian papillary adenocarcinoma of the rat with ascites implants in the ovary. *J. Natl. Cancer Inst.* **13**, 409–429.

355. Takizawa, S., and Nishihara, H. (1971). Induction of tumors in the brain, kidney, and other extra-mammary gland organs by a continuous oral administration of *N*-nitrosobutylurea in Wistar/Furth rats. *Gann* **62**, 495–503.

356. Tannenbaum, A., Vesselinovitch, S. D., Malontini, C., and Mitchell, D. S. (1962). Multipotential carcinogenicity of urethan in the Sprague-Dawley Rat. *Cancer Res.* **22**, 1362–1370.

357. Teller, M. N., Merker, P. C., Palm, J. E., and Woolley, G. W. (1958). The human tumor in cancer chemotherapy in the conditioned rat. *Ann. N. Y. Acad. Sci.* **76**, 742–751.

358. Thompson, S. W., and Hunt, R. D. (1963). Spontaneous tumors in the Sprague-Dawley rat: Incidence rates of some types of neoplasms as determined by serial section versus single section technics. *Ann. N. Y. Acad. Sci.* **108**, 832–845.

359. Thompson, S. W., Huseby, R. A., Fox, M. A., Davis, C. L., and Hunt, R. D. (1961). Spontaneous tumors in the Sprague-Dawley rat. *J. Natl. Cancer Inst.* **27**, 1037–1057.

360. Tiltman, A. J., and Friedell, G. H. (1971). The histogenesis of experimental bladder cancer. *Invest. Urol.* **9**, 218–226.

361. Tomashefsky, P., Homsy, Y. L., Lattimer, J. K., and Tannenbaum, M. (1976). A murine Wilm's tumor as model for chemotherapy and radiotherapy. *J. Natl. Cancer Inst.* **56**, 137–142.

362. Tomashefsky, P., Lattimer, J. K., Priestley, J., Furth, J., Vakili, B., and Tannenbaum, M. (1973). An experimental Wilm's tumor suitable for therapeutic and biologic studies. II. The inhibition of renal compensatory hypertrophy by a transplantable tumor. *Invest. Urol.* **11**, 141–144.

363. Tomatis, L. (1974). Inception and duration of tests. *In* "Carcinogenesis Testing of Chemicals" (L. Golberg, ed.), pp. 23–27. CRC Press, Cleveland, Ohio.

364. Tomatis, L. (1976). The IARC program on the evaluation of the carcinogenic risk of chemicals to man. *Ann. N. Y. Acad. Sci.* **271**, 396–409.

365. Tomatis, L., Turusov, V. S., and Guibbert, D. (1972). Prenatal exposure to chemical carcinogens. *In* "Topics in Chemical Carcinogenesis" (W. Nakahara, S. Takayama, T. Sugimura, and S. Odashima, eds.), pp. 445–461. Univ. Park Press, Baltimore, Maryland.

366. Toolan, H. W. (1958). The transplantable human tumor. *Ann. N. Y. Acad. Sci.* **76**, 733–741.

367. Truhaut, R., ed. (1967). "Potential Carcinogenic Hazards from Drugs," UICC Monogr. Ser., Vol. 7. Springer-Verlag, Berlin and New York.

368. Twombly, G. H., Meisel, D., and Stout, A. P. (1949). Leydig-cell tumors induced experimentally in the rat. *Cancer* **2**, 884–892.

369. Ueda, G., Cali, A., and Kunii, A. (1971). Biological features of a transplantable Leydig cell tumor in rat. *Gann* **62**, 41–48.

370. Upton, A. C. (1967). Comparative observations on radiation carcinogenesis in man and animals. *In* "Carcinogenesis: A Broad Critique," pp. 631–675. Williams & Wilkins, Baltimore, Maryland.

371. Uriel, J. (1971). Transitory liver antigens and primary hepatoma in man and rat. *Liver Cancer, Proc. Work. Conf., 1969* IARC Sci. Publ. No. 1, pp. 58–68.

372. Van Esch, G. J. (1967). Plastic carcinogenesis; suggestions for the use of plastics in surgery, orthopedics, etc. *In* "Potential Carcinogenic Hazards from Drugs" (R. Truhaut, ed.), pp. 196–203. Springer-Verlag, Berlin and New York.

373. Veenema, R. J., Fingerhut, B., and Lattimer, J. K. (1965). Experimental studies on the biological potential of bladder tumors. *J. Urol.* **93**, 202–211.

374. Vesselinovitch, S. D., and Mihailovich, N. (1968). The induction of benign and malignant liver tumors by urethane in newborn rats. *Cancer Res.* **28**, 881–887.

375. Vesselinovitch, S. D., and Mihailovich, N. (1968). The development of neurogenic neoplasms, embryonal kidney tumors, Harderian-gland adenomas, Anitschkow-cell sarcomas of the heart, and other neoplasms in urethan-treated newborn rats. *Cancer Res.* **28**, 888–897.

376. Wagner, J. C. (1962). Experimental production of mesothelial tumours of the pleura by implantation of dusts in laboratory animals. *Nature (London)* **196**, 180–181.

377. Wagner, J. C., Berry, G., and Timbrell, V. (1973). Mesotheliomata in rats after inoculation with asbestos and other materials. *Br. J. Cancer* **28**, 173–185.

378. Wagoner, J. K. (1976). Occupational carcinogenesis: The two hundred years since Percivall Pott. *Ann. N. Y. Acad. Sci.* **271**, 1–4.

379. Warwick, G. P. (1971). Metabolism of liver carcinogens and other factors influencing liver cancer induction. *Liver Cancer, Proc. Work. Conf. 1969* IARC Sci. Publ. No. 1, pp. 121–157.

380. Waxweiler, R. J., Stringer, W., Wagoner, J. K., and Jones, J. (1976). Neoplastic risk among workers exposed to vinyl chloride. *Ann. N. Y. Acad. Sci.* **271,** 40–48.

381. Weber, G. (1971). Biochemistry of liver cancer. *Liver Cancer, Proc. Work. Conf., 1969* IARC Sci. Pub. No. 1, pp. 69–93.

382. Weinstein, I. B. (1976). Discussion. *Ann. N. Y. Acad. Sci.* **271,** 461–466.

383. Weisberger, J. H., and Weisburger, E. K. (1967). Tests for chemical carcinogens. *Methods Cancer Res.* **1,** 307–398.

384. Weisburger, J. H. (1974). Inclusion of positive control compounds. *In* "Carcinogenesis Testing of Chemicals" (L. Golberg, ed.), pp. 29–34. CRC Press, Cleveland, Ohio.

385. Wilfong, R. F., Bigner, D. D., Self, D. J., and Wechsler, W. (1973). Brain tumor types induced by the Schmidt-Ruppin strain of Rous sarcoma virus in inbred Fischer rats. *Acta Neuropathol.* **25,** 196–206.

386. Willis, R. A. (1967). "Pathology of Tumours." Butterworth, London.

387. Wogan, G. N., and Newberne, P. M. (1967). Dose-response characteristics aflatoxin B$_1$ carcinogenesis in the rat. *Cancer Res.* **27,** 2370–2376.

388. Woolley, G. W. (1958). Discussion. *Ann. N. Y. Acad. Sci.* **76,** 821–825.

389. Yamada, S., Ito, M., and Nagayo, T. (1971). Histological and autoradiographical studies on intestinal tumors of rat induced by oral administration of *N,N*^{1}-2,7-fluorenylanebisacetamide. *Gann* **62,** 471–478.

390. Yamada, S., Masuko, K., Ito, M., and Nagayo, T. (1973). Spontaneous thymoma in Buffalo rats. *Gann* **64,** 287–291.

391. Yoshida, T. (1958). Screening with ascites hepatoma. *Ann. N. Y. Acad. Sci.* **76,** 610–618.

392. Young, S., and Cowan, D. M. (1963). Spontaneous regression of induced mammary tumors in rats. *Br. J. Cancer* **17,** 85–89.

393. Young, S., and Hallowes, R. C. (1973). Tumours of the mammary gland. *In* "Pathology of Tumours in Laboratory Animals" (V. S. Turusov, ed.), IARC Sci. Publ. No. 5, Vol. I, Part 1, pp. 31–73. Int. Agency Res. Cancer, Lyon, France.

394. Zackheim, H. S. (1973). Tumours of the skin. *In* "Pathology of Tumours in Laboratory Animals" (V. S. Turusov, ed.), IARC Sci. Publ. No. 5, Vol. I, Part 1, pp. 1–21. Int. Agency Res. Cancer, Lyon, France.

Chapter 7

Experimental Gerontology

Joe D. Burek and Carel F. Hollander

I. INTRODUCTION

The ultimate goals of aging research are to understand the mechanisms of aging and to modify them. Practically, the goal is not necessarily to prolong human life indefinitely, but to find the means to alleviate many disabling and painful conditions that often occur with advancing age, thereby enhancing the quality of life (45).

Detailed discussion of a proposed hypotheses of aging is not within the scope of this chapter. The theories and hypotheses involve a wide range of complex topics, including immunology (21,83,89), genetic defects (20,31), and cellular and subcellular alterations (34,35,40,57,67). Numerous books and reviews have been published on the subject, but only a few selected references are given here (25–28, 33, 34, 37, 41–43, 56, 61, 69–71, 74, 79). Research efforts in gerontology have been directed toward descriptive observations in the past, and only recently, have hypotheses or possible mechanisms of aging been advanced and serious attempts made to test them experimentally (42,71).

THE LABORATORY RAT, VOL. II

Animal models with specific age-related diseases and aging processes are needed to fulfill the broad research objectives of investigations in aging research and to study mechanisms of aging in intact animals. The animals that are needed for aging research will differ depending on the specific goals and research interests of individual investigators. If, however, the goal of a research effort is to study aging in animals in order to gain an insight into the aging of humans, then mammals with relatively short life spans are needed. They should be large enough for adequate volumes of blood or other body fluids to be obtained, but small enough that maintenance costs will be sufficiently low to permit production and maintenance of large numbers. Finally, information should be available on the expected age-associated disease patterns.

The laboratory rat is one animal that can be used for such studies as it fulfills most of the requirements listed above. It also can be useful for other long-term and life span studies, such as carcinogenesis testing, late effects of irradiation, and chronic toxicity studies. In this chapter, we emphasize problems and difficulties that may be encountered in life span studies, rather than discuss only the uses of the rat. This was done for two reasons. First, there is often little agreement on how to define an old or aged rat. For example, one study considered Sprague-Dawley rats of 14 months of age as old (49); another study defined 24-month-old (720 days) Wistar rats as old (13). As long ago as 1928, Inukai (50) defined his Wistar rats as old when over 30 months (1,017 days) of age. Second, factors that influence survival are important to investigators conducting life span studies with rats. This is especially true if the factors can explain different experimental results obtained in different laboratories. In addition to describing variables that can influence experimental results and longevity, the potential advantages of rats for life span studies

are discussed. The recent literature is reviewed, and some previously unpublished information is presented to emphasize special problems we have encountered in using rats for life span studies.

II. SOME FACTORS THAT INFLUENCE LONGEVITY AND THE INTERPRETATION OF RESULTS FROM LIFE SPAN STUDIES USING RATS

The definition of an old or senescent rat or population of rats requires detailed knowledge of the strain or stock that is studied. Longevity data are influenced by many factors, such as the strain or stock, sex, husbandry conditions, breeding history, and nutrition. All of these factors can influence the usefulness of the rat in life span studies and must be considered before a definition of an old rat can be given. They must also be considered when evaluating conflicting experimental results from different laboratories (34,38).

A. Survival and Mortality Data

To obtain accurate and dependable longevity data on aging rats requires complete survival and mortality data. Large numbers of rats must be permitted to live out their natural life span under well-controlled laboratory conditions. Only in this way can the young, adult and aged population be defined for a specific strain or stock. Figures 1 and 2 are survival curves for BN/Bi and WAG/Rij rats from the Institute for Experimental Gerontology TNO. The curves were made from cohorts of rats

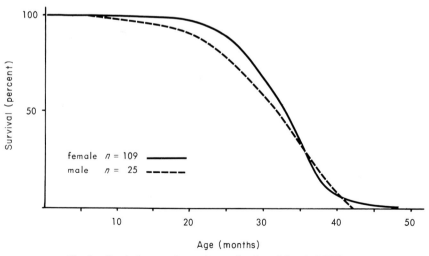

Fig. 1.   Survival curves from cohorts of male and female BN/Bi rats.

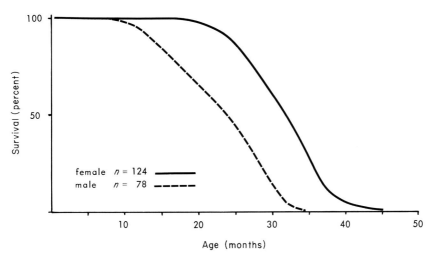

Fig. 2. Survival curves from cohorts of male and female WAG/Rij rats.

that had been subjected to identical husbandry conditions. For both strains, a postweanling plateau is present with few or no deaths. This plateau corresponds to the adult population. A bend which is followed by a period of rapid decline in the number of survivors is seen in the curves. This period of rapid decline may be defined as the onset of senescence for the population studied (25). In this manner, the 90, 50, 10%, and maximal survival ages for the different strains and different sexes can be compared. The importance of these data in the interpretation of aging experiments has been emphasized (34,38).

Investigators using rodents often select the 50% survival age (average longevity time) and older as the aged population (54,84). Such a definition is arbitrary. It serves only as a functional definition that does permit at least a superficial comparison of longevity data among different strains or stocks of rats and among different laboratories. To compare data, more information is needed to determine what has influenced the slope and shape of the survival curves. Preweanling mortality, early postweanling mortality, and maximum survival ages will affect the 50% survival age. These data should be available to compare longevity data and before one can define the period when deaths are due to age-associated disease and senescence. Such data are not always given in publications and are often not available. Additional studies are needed that provide survival and mortality data for various strains or stocks of rats from birth until death.

Survival curves are only one of several possible methods of expressing survival or mortality data. Other methods of plotting data such as Gompertz curves are useful (25,56,70). No matter how the data are expressed, it should be possible to compare results among different laboratories. We consider the minimum information needed for such comparisons to be the 50% survival and the maximal survival ages for the rats stud-

ied. Unfortunately, many publications in the field of aging fail to provide even this minimum information.

B. Infectious Diseases

Survival or mortality data alone is not sufficient to determine the age of senescence for a population of rats because death is not an absolute criterion of old age. Deaths in a strain or stock of rats can occur from acute and chronic infectious diseases. The result is different survival data in the rat colony with infectious diseases compared to the same rats free of the diseases. Specific examples of the influence of the environment and infectious diseases on longevity can be found in several examples quoted by Cohen (22). There are many examples in current publications where infectious diseases were ignored in aging and long-term studies. As illustrated by Paget and Lemon (62), "dirty" rats do not live as long as their specific-pathogen-free counterparts, even though the shape of the survival curves is nearly identical in both groups.

It is important to know what indigenous infectious diseases are present in aging rats, because any disease that causes premature deaths will change survival data. The most important infectious disease that can complicate life span studies is murine respiratory mycoplasmosis (60), dealt with in detail elsewhere in this treatise (see Volume I, Chapter 10).

Other infectious diseases that may complicate research using rats have been described by Squire (76), and elsewhere in this treatise (see Volume I, Chapters 9–12). Cohen and Anver (23), in a recent review, stressed the need for "good husbandry" for long-term studies in order to control infectious diseases. It is clear from these few examples that colonies of rats in which deaths occur from infectious diseases are not the best sources of animals for life span studies. The survival data obtained from such colonies can give a false impression of the potential longevity for the rat population that is studied.

C. Age-Associated Pathology

In an aging rat colong free of life-shortening infectious diseases, the age-associated neoplastic and nonneoplastic lesions are the most common factors associated with death. It is imperative to know what these lesions are, their incidences at various ages, and how they influence or shape the survival curves and possible experimental results. The need to determine the normal baseline pathology data and for continued monitoring of each strain and stock of rats has been emphasized in several recent publications (46,78).

Much is known about the common lesions in rats. Several investigators have described nonneoplastic lesions (1–4, 10, 14, 19, 23, 24, 29, 64, 72, 73, 75) (see also Anver and Cohen, Volume I, Chapter 14). Tumors in rats have been the subject of many recent publications and reviews (8, 10, 17, 19, 24, 30, 36, 51, 65, 80–82) (see also, Altman and Goodman, Volume I, Chapter 13). It is apparent that there is considerable information about the pathology of rats. However, little is known about how these lesions influence mortality. Few studies have described both the neoplastic and nonneoplastic lesions in strains or stocks of rats and correlated lesions with survival. Examples of such studies are found in investigations by Simms (73), Simms and Berg (72), Berg (4), Berg and Simms (1–3) on Sprague-Dawley rats; Coleman *et al.* (24) on male Fischer 344 rats; and Burek (19) on BN/Bi, WAG/Rij, and F₁ (WAG/ Rij × BN/Bi) rats.

Much of the published data on tumor incidences in various rat strains and stocks have been derived from rats less than 2 or 2.5 years of age, and many of the studies included rats that were killed before the end of their natural life span. We did not find published studies that compared differences in the incidence of lesions in groups of rats that died spontaneously with those of the same age that were killed. Rats that die with age-associated lesions, may represent different aging populations. As suggested by Simms (72), some rats acquire a lesion or lesions and then are eliminated from the surviving population by death. In a study by Burek and Hollander (17), some BN/Bi rats died without tumors, regardless of their age. Approximately 50% of the rats died with multiple tumors, even in younger rats of 13 to 18 months of age. These findings illustrate the variability of lesions in aging rats. Similar individual variability was found in aging WAG/Rij rats (10,19).

It is clear that even in highly inbred strains of rats, such as BN/Bi and WAG/Rij rats, there is considerable individual variability in the age-associated lesions found, and the degree of variability among individual rats increases as the population ages. This individual variability can influence experimental

results and must be considered when evaluating survival data and possible causes of death, and when using these rats for other aging studies.

Collection of baseline pathology data may have ramifications in areas other than aging. At the Institute for Experimental Gerontology TNO, several specific cancers have been recognized in aging rats. Examples include the following: a high incidence of spontaneous ureter and bladder carcinomas in BN/Bi rats (11); a high incidence of spontaneous cervical and vaginal tumors in female BN/Bi (16); a high incidence of medullary thyroid carcinomas in WAG/Rig (9,12) and F₁ (WAG/Rij × BN/Bi) rats (19); and the occurrence of granular cell tumors (granular cell myoblastoma) in the brain of BN/Bi and F₁ (WAG/Rij × BN/Bi) rats (48). These specific tumors are now being developed as animal models to compare them with their counterparts in man. Each of the tumor models was the direct result of determining normal background of neoplastic and nonneoplastic lesions in the aging rats.

D. Nutrition

Defining the true longevity for a population of rats is also complicated by nutritional factors. Information about the composition of the diet fed throughout life is needed before conflicting experimental results among different laboratories can be compared. The importance of the quality and quantity of the diet has been stressed in several publications (1–3,66). Dietary changes can result in differences in neoplastic and nonneoplastic lesions (1–3,66), longevity (3), and biochemical values (34). These have been reviewed in detail recently by Cohen and Anver (23) and elsewhere in this volume (Volume I, Chapter 14). As suggested by these studies and reviews, major differences can occur in the longevity, pathology, and biochemical parameters simply by changing the diet. Thus, the compositions of diets fed to rats is yet another source of variables that must be controlled in life span studies.

E. Males versus Females

Relatively few studies have been published that compare the longevity and pathology data of male and female rats of the same strain under well-controlled laboratory conditions. It has been demonstrated that female Sprague-Dawley rats live longer than males (86). In other studies there was little difference in mortality between SPF male and female rats (62).

Longevity data on male and female BN/Bi, WAG/Rij and F₁ (WAG/Rij × BN/Bi) rats from the Institute for Experimental Gerontology TNO are presented in Table I. These rats were maintained in the same rooms, fed the same diet, and cared for by the same personnel. They were maintained under "clean" conventional conditions, and were free of mycoplasma pneu-

Table I

Summary of the 90%, 50%, 10%, and Maximum Survival Ages from Complete Cohorts of
Virgin BN/Bi, WAG/Rij, and F_1 (WAG/Rij × BN/Bi) Rats

Strain	Sex	Number	Age (months)			
			90% Survival	50% Survival	10% Survival	Maximum survival
BN/Bi	Female	109	25	33	38	48
	Male	25	21	32	39	42
WAG/Rij	Female	124	24	32	37	45
	Male	78	13	24	31	34
F_1	Female	47	22	29	37	42
	Male	46	19	30	41	43

monia. Additional information concerning these rats is found in recent publications (19,47).

The ages of most groups of rats were similar. It is interesting that under identical conditions, male and female BN/Bi, female WAG/Rij, and male and female F_1 rats all had similar longevity data with the 50% survival occurring at approximately 30 months. However, male WAG/Rij had a significantly shorter life span as reflected with the 50% survival at 24 months. To explain such a difference requires a knowledge of the baseline pathology data from all groups of rats. Such information will be presented in more detail elsewhere (19).

Preliminary findings indicate that WAG/Rij males have earlier onset of several types of lesions compared to female WAG/Rij, BN/Bi, and F_1 rats. Among these are an earlier onset of pituitary tumors and medullary thyroid carcinomas, severe atrophy of the thymus, and earlier onset and increased frequency of gastric ulcers and prostatic abscesses. These observations help to explain the cause of death, but there also may be an underlying mechanism responsible for the "premature" occurrence of these lesions. Comparisons between sexes of the same strain may offer new or different information than would be gained by examining only one sex. Although sex differences in longevity data are observed in some rat strains, such differences are probably influenced by many factors, especially strain or stock.

It is clear that generalizations on the longevity of male and female rats cannot be made at this time. More specific survival data, compared with pathology data, are still needed from other rat strains or stocks that have been maintained under well controlled laboratory conditions. Only with these types of data can broad conclusions be made.

F. Virgins and Retired Breeders

Few laboratories maintain large numbers of rats that are permitted to live out their natural life span. The only sources of

old rats for most investigators are commercial breeders, and most of the animals obtained are retired breeders. This point has been stressed by Anver and Cohen elsewhere in this treatise (Volume I, Chapter 14). There is usually little or no history given about the number of litters or how many young were produced and weaned. The fact that a rat has been retired from breeding does not imply that it is an old rat. Most females can still become pregnant, and most rats are less than 15 months of age when retired from breeding colonies. Therefore, a retired breeder is not old and may have one-half to three-fourths of its potential life expectancy remaining.

There is little published information on how longevity is affected by single or multiple breedings and number of young produced. If these factors influence longevity, then such information would be of extreme importance to individuals using rats for life span studies. There is a need for such data if one is to compare young rats (usually virgins) with old rats (often retired breeders). Breeders, especially females, are exposed to different environmental and physiological stimuli because of pregnancy, birth, nursing, and weaning. Some breeders produce more litters or wean more young, and the nutritional demands of breeding rats are different from those of nonbreeding and nonlactating rats. It is not known if these factors significantly alter longevity data, tumor incidence, or incidence of other age-associated lesions in retired breeders as compared to virgins of the same strain or stock.

Wexler and Greenberg (86) found the 50% survival age for retired breeders was younger than for virgins. In addition, retired breeders had more vascular lesions, especially periarteritis nodosa, than virgins. Similarly, Paget and Lemon reported that their specific-pathogen-free (SPF) breeding rats had earlier mortality than nonbreeding rats from the same stock (62). Longevity studies on female WAG/Rij and BN/Bi rats did not demonstrate any significant differences in mortality between virgins and retired breeders (Fig. 3) (19). These differences emphasize the need for investigators to know more about the rats they are using in aging studies. Clearly, the

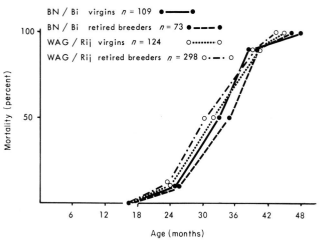

Fig. 3. Mortality curves from cohorts of female BN/Bi and WAG/Rij virgin and retired breeder rats.

It is not known if certain diseases or tumors predispose a rat to more rapid autolysis or cannibalization. For example, a rat dying with acute septicemia may have a fever and undergo more rapid autolysis than rats dying from other causes. By eliminating autolyzed rats, one may be selecting out a significant number of animals that die from acute infections. Similar speculation can be made for cannibalism. Rats dying with specific metabolic disorders may be more readily eaten by cage mates than those dying from other causes.

The elimination of these rats from aging studies may alter the percentages of specific tumors or diseases. Many of the losses occur at night, on weekends, and holidays, and these are always difficult times for personnel to provide close observations of the colony. The value of old rats may justify the cost of the additional personnel that would be needed to provide closer observations to reduce this waste of costly animals.

generalization that retired breeders do not live as long as virgin rats is not applicable to all strains and stocks of rats.

G. Autolysis and Cannibalism

Some rats will be lost from any life span study because of autolysis or cannibalism. Autolyzed rats cannot be used for most pathology or biochemical studies. Rats that have been partially eaten, even if only recently dead, represent animals that can only be partially examined. Such deaths can be used in mortality calculations but not for the determination of complete pathology data and represent a loss of valuable data.

The BN/Bi and WAG/Rij rat strains and their F_1 hybrid (WAG/Rij × BN/Bi) were maintained at the Institute for Experimental Gerontology TNO for aging studies. All rats were maintained in the same rooms and were subjected to identical husbandry conditions. All dead or sick animals were processed as defined in a standard protocol that applied to all rats. Strain differences were found in the number of rats lost because of autolysis and cannibalism. Losses of WAG/Rij and F_1 rats were about 15%, compared with 30% of the BN/Bi rats.

Such differences are difficult to explain, but cannot be ignored. The WAG/Rij and F_1 rats that died had a high percentage of pituitary tumors (approximately 80–95%), and the male F_1 rats had a high incidence of paresis and paralysis of the hindlimbs (15). Rats with these conditions are sick weeks before death and are observed more closely. The BN/Bi rats, however, seem to die more acutely. They are seldom observed to have terminal, chronic, wasting disease and are usually found dead unexpectedly. Differences in diseases and tumor patterns help to explain strain differences in the percentage of rats that were autolyzed or cannibalized.

III. HOW OLD IS AN OLD RAT?

To define the age of senescence for a rat, or a population of rats, requires background information for each strain or stock as discussed above. Such information is not usually available or provided in publications. It is difficult, therefore, to compare longevity data for different strains or stocks of rats. However, a few generalizations will be made.

Rats, and a few other rodent species, are the shortest-lived mammals readily available for aging research (25). Detailed longevity studies are found in several references (19, 24, 39, 47, 62, 68, 73). From these investigations, it is clear that rats are not necessarily old at 18 to 24 months as commonly believed. Data from recent investigations on several rat strains and stocks are now available and include female (retired breeders and virgins together) WAG/Rij strain (10), male and female BN/Bi strain (virgins) (17), female (retired breeders) BN/Bi (19), female Sprague-Dawley stock (retired breeders) (B. J. Cohen and M. R. Anver, personal communication), male and female Sprague-Dawley stock (4), APF male and female Alderly Park strain 1 rats (62), female Wistar-derived rats of Gsell (39,68), and male Fischer 344 rats (24). All have mean survival values of about 27 to 30 months. This means, of course, that one-half of the rats survived longer than 27 to 30 months, and many survived longer than 3 years. As early as 1924, there was a reference to a rat living 4 years and 8 months (25). Therefore, if 50% survival age is used as an indication of "old," then for most strains and stocks, one needs rats that are ± 30 months of age for aging studies. In Fig. 1 and 2, the female WAG/Rij and male and female BN/Bi rats of 18 to 24 months of age are at approximately the 90% survival age. That is, 90% of the rats are still alive and only 10% have died. Rats of 18 or even 24 months of age may have more than half their life expectancy remaining. As presented in Fig. 3, the same is true for retired breeder female WAG/Rij and BN/Bi rats.

Male WAG/Rij rats (Fig. 2) had a 50% survival age of 22 months. This was similar to the data presented by Gsell (39,68), where Wistar-derived male rats had a 50% survival age of 23 months in contrast to 27 months for the females. Using the definition we presented earlier, the WAG/Rij males must be considered "aged" 6 to 8 months earlier than other strains. Additional investigations are needed to determine if the earlier mortality represents an earlier onset of senescence. A discussion of this problem is not within the scope of this chapter, but has been discussed by Burek (19). The shorter survival of male WAG/Rij, clearly indicates the need to establish normal baseline longevity and pathology data for each sex and each rat strain or stock used in aging research.

IV. COST OF OLD RATS

The highest quality rats are needed for any research, but this is especially true for life span studies. However, there must be a point where the costs are too great to justify potential benefits. This point must be determined by each individual investigator based on specific research goals and monetary resources available. Weisbroth (85) recently reevaluated barrier isolation methods for laboratory rats, and the major problem was the cost involved as compared to expected results. To give some indication of the costs to produce old rats, data were recently calculated at the Institute for Experimental Gerontology TNO (Table II). The cost for a 30-month-old female BN/Bi rat was about $116, while a 35-month-old was over $400. These are costs calculated for feed, housing, and general care. Considered in the calculations was the loss of rats involved to produce sufficient numbers of rats that would survive until 30 to 35

Table II

Production Costs (in Dollars) per Rat[a,b] Calculated for BN/Bi and WAG/Rij Rats of Various Ages at the Institute for Experimental Gerontology TNO

Age (months)	WAG/Rij		BN/Bi	
	Male	Female	Male	Female
10	20	20	21	20
15	32	32	32	31
20	62	44	49	43
25	250	59	75	61
30	—	120	150	116
35	—	389	467	412

[a] 1976 data.

[b] Calculations corrected for mortality at a given age for each sex and strain.

months of age. Inasmuch as the 50% survival is about 30 months, the costs of rats that die prior to 30 months of age must be divided among the survivors. Therefore, the older the rat, the proportionately more costly it becomes. Obviously, the costs are high to obtain sufficient numbers of old rats to use in aging studies. Such rats should be used with great care in multidisciplinary, coordinated efforts by several investigators to assure the best possible use of each aged rat.

V. MODELS OF AGING USING THE RAT

Selection of suitable models of organ aging requires knowledge of the age-associated pathology of the rat as a whole and of the specific organ to be studied. Furthermore, the pathology data needs to be correlated with information on mortality and morbidity in the strain selected. Such information is important to prevent selection of a seemingly normal organ whose function might be hampered by pathological lesions elsewhere in the body. It is beyond the scope of this chapter to discuss this problem extensively or to mention every organ of the rat as a model for aging, but features of organ aging in the rat will be illustrated using a few selected examples.

Functional and morphological aspects of aging of the rat liver have been studied at the Institute for Experimental Gerontology TNO for several years. De Leeuw-Israël (59) has shown that the rat liver is a suitable model for studying reduction of functional reserve capacity with age. As the study progressed, the need to have information on the age-associated pathology of rats was recognized. The outcome of the pathology studies proved the BN/Bi rat to be the strain of choice for these studies. In two other strains studied [WAG/Rij and the F_1 (WAG/Rij × BN/Bi) hybrid], there was a high incidence of lesions of the parenchymal cells that consisted of foci or areas of cellular change, neoplastic nodules, or hepatocellular carcinomas (77). In BN/Bi rats, such lesions were infrequent where the most common lesion was cysts.

Since an organ consists of different cell types, a particular aging phenomenon observed in an intact organ, or measured with a homogenate of that organ, may be an expression of all of the cells constituting that organ. Alternatively, different cell types may show different aging patterns (54). Furthermore, organ function also may be influenced by changes in circulation with age. Liver cells can be divided into two large groups: parenchymal cells (hepatocytes) and nonparenchymal cells (endothelial and Kupffer cells) (53). Parenchymal cells represent about 60 to 65% of the total number of liver cells and more than 90% of the liver mass. Procedures have been developed to isolate viable parenchymal and nonparenchymal cells from young and old rats (7). Recently, techniques have been developed to separate nonparenchymal endothelial cells and

Kupffer cells (55). The availability of populations of viable cells of the two liver cell classes is essential to study cell metabolism and subcellular changes with aging. Isolated cell populations are needed to define the contribution of each cell class to the phenomenon of liver aging. Recently, Knook and Sleijster (55) compared lysosomal enzyme activities of homogenates from whole livers with separated parenchymal and nonparenchymal cells. They found major differences in lysosomal enzyme activity in homogenates of whole livers as compared to populations of separated parenchymal and nonparenchymal cells. This suggests a need for careful interpretation of data, since conflicting patterns of changes can be observed for enzyme activities in parenchymal compared to nonparenchymal cells.

Information is needed, not only on the pathology of the target organ to be studied but also on other organs. This can be illustrated by the attempts to study functional aspects of aging in the kidney using a previously developed method of orthotopic renal transplantation within the same inbred strain of rats (44). It seemed feasible to transplant kidneys from old rats into young recipients to study the maximum life span of the transplanted kidney, as well as functional aspects of the transplanted organ. The transplanted "old" kidney should be free of nephrosclerotic lesions if the impact on the life span of the recipient is to be determined. Furthermore, renal function tests should be available to be used in a longitudinal study of rats receiving a renal transplant. On the basis of limited existing information, the BN/Bi rat was chosen for this study because nephrosclerotic lesions were absent or only minimal in these old rats. Detailed results of this study have been published (6,44). The major pitfall in designing the above study was the absence at that time of detailed information on the age-associated pathology in the BN/Bi rat as a whole. It was not known that this strain had a high incidence of spontaneous urinary bladder and ureter tumors (11). The occurrence of the tumors might have led to progressive hydronephrosis in a number of rats receiving a renal transplant, and this might explain the high frequency of unexplained dilated bladders with concomitant hydronephrosis observed in some transplanted rats. In this study, the WAG/Rij rat would have been a better strain for kidney transplantation studies because the incidence of ureter and bladder tumors is very low. The outcome of this experiment once more stresses the necessity to have available baseline pathology and life span data before starting experiments on different aspects of organ aging.

An awareness of the age-associated pathology permits the selection of suitable models for studying the relationship between aging and disease. For example, the observed high incidence of bladder tumors in the aging BN/Bi rats triggered additional studies into the relationships between the immune system and carcinogenesis. Knowledge of the incidence of pituitary tumors, which is high in the WAG/Rij (>90%) and relatively low in the BN/Bi rat (<25%), is also of importance in choosing a rat strain for studying endocrinological aspects of aging. Furthermore, the incidence of pituitary tumors is so high in the WAG/Rij rats that they cannot be used for behavioral or aging studies involving the central nervous system. The tumors are often large and produce secondary hydrocephalus and compression of the brain parenchyma. Both of these lesions can influence life span studies on the rat brain. The recent observations by Burek and Meihuizen (18) of cortical atrophy of the thymus of old female BN/Bi rats with concomittant hyperplastic changes of the thymic epithelium suggests that female BN/Bi rats would be a suitable model for studying the influence of age-associated changes in the thymus on the immune system.

There is little specific information concerning tumor transplantation in aging rats. In a study by Kerkvliet and Kimeldorf (52), older rats (17 months) had reduced acceptance of Walker 256 carcinosarcoma than did younger rats. In a pilot study by J. D. Burek and A. L. Nooteboom (unpublished observations) five young (6-week-old) and five old (28-month-old) rats received subcutaneous tumor transplants of an adrenal cortical carcinoma. The transplants grew rapidly in all five of the young recipients, but did not grow in any of the older rats after 6 months. These findings were surprising in light of the high incidence of naturally occurring tumors in old rats.

Enzyme activities are considered to correlate with specific physiological states of an organ throughout life. Therefore, numerous studies have been devoted to compare changes in specific enzymes with age in different organs. Among the rat strains used in these studies, the Wistar, McCollum, and Sprague-Dawley are the most common. The organs most frequently studied are liver, kidney, and brain. In a recent review article, Finch (34) listed all enzymes studied, as well as the factors which may influence such studies.

The relationship of endocrine system to aging have been studied in the rat. The role of the hypothalamic–pituitary complex in regulation of aging and particularly in relation to environmental factors, such as nutrition, temperature, and stress, has been reviewed in a number of chapters in the monograph "Hypothalamus, Pituitary and Aging" (32). References are made to such studies as the role of the pituitary gland, collagen aging, metabolism, and renal function, reproduction and aging, and aging and adrenocortical functions.

Berlin and Wallace (5) have reviewed the literature on aging and the central nervous system. Among the rat strains mentioned are the Long-Evans, Wistar, Sprague-Dawley, and an albino rat. In some of these studies Wistar rats as old as 40 months have been used, and the authors conclude that "there is a rather considerable lack of consistency in the data both within the frame of reference of a single species, and with regard to intraspecies comparisons." Furthermore, they alluded to how strain, age, or both might produce different results and, in addition, that different histological techniques to process the brain tissue may influence the outcome of experimental studies.

Many observations have been reported regarding learning and memory in relation to age in rats. Wolthuis *et al.* (88) showed acquisition deficits in 30-month-old female WAG/Rij rats. Landfield *et al.* (58) studied Fischer rats and found a sizable loss of pyramidal cells and remarkably hypertrophic astrocytes in the hippocampus of 24 to 25-month-old rats. They specualted that pathological lesions in the hippocampus were related to impaired memory, and furthermore that the astrocytic changes observed may be an analogue of senile plaques in man. Additional studies in several different rat strains are needed to further elucidate the observed morphological changes and their correlation with function.

The usefulness of the rat in the study of pigments associated with aging, particularly lipofuscin and lipofuscin-like pigments, has been stressed by Wolman (87) and Porta and Hartroft (63). They discussed the locations and the types of pigments that have been observed. These pigments occur in many organs in rats, but especially in the liver and central nervous system. Similarly, Burek (19) found lipofuscin-like pigments in most organs of aging BN/Bi, WAG/Rij, and F_1 (WAG/Rij × BN/Bi) rats, but accumulations were most obvious in liver, brain, kidney, and accessory male genitalia.

There are many more examples where the laboratory rat has been used in aging research. The few examples given in this section serve only to illustrate that most facets of aging research have utilized the rat and that the rat clearly has a useful place in the broad areas of study in experimental gerontology.

VI. SOURCES OF OLD RATS

There are few commercial sources of old rats in the United States. So called "retired breeders" often become available for purchase at 8–12 months of age. However, the 50% survival of many rat stocks approaches 30 months so it is clear that a "retired breeder" cannot be considered old in relation to the normal life expectancy. Commercial rodent breeders are unlikely to devote space that otherwise could serve for ordinary rat production to rearing aged rats unless a market for the rats is assured and the risks are covered in maintaining rats to an age where large numbers of spontaneous deaths become likely. In recognition of this problem, several investigators have contracted with commercial breeders to provide the service of rearing rats to the age of 2 years or older. Similarly, the U.S. National Institute of Aging (NIA) has contracted with several commercial breeders to rear modest numbers of aged inbred Fischer 344 (F344) rats and outbred rats of Sprague-Dawley (SD) origin. These rats are being made available for pilot

studies in aging to investigators who apply for them through the National Institute of Aging (D.C. Gibson, personal communication).

A number of rat stocks and strains are being maintained to old age in university, federal, and other laboratories. Investigators can locate possible noncommercial sources of aged rats by appropriate reference to the scientific literature on aging. The high cost of aged rats (at least $75 per 2-year-old rat) is a strong incentive to collaborative efforts in research, so as to make the most productive possible use of the limited numbers of old animals available.

VII. SUMMARY

The laboratory rat has many potential uses in aging research. It can be used to study aging in a population or in an intact individual rat, to evaluate aging in organ systems, and to study cellular and subcellular alterations with aging. The greatest disadvantage is the high cost of obtaining old rats, but the costs are not too high to prohibit their usefulness.

In this chapter, we have stressed the need for investigators to know the quality and background of their rats. Baseline longevity and pathology data are also essential. Such data are needed from each strain or stock of rats and for each sex that is studied. The diet fed to rats can influence biochemical parameters, neoplastic and nonneoplastic lesions, and longevity. Infectious diseases, breeding history, and sex can also influence longevity. Longevity data are required before the young, adult, and aged population of rats can be defined. Pathology data are important because they provide a means of selecting a sex or strain of rat that is best suited for a specific investigation. Also, some strains can be eliminated from certain studies because of the pathology findings. Baseline longevity and pathology data together provide the possibility to correlate the onset of "age-associated" lesions, risk of death with a specific lesion, and possible causes of death in rats.

Experimental gerontology is a relatively young science with many theories and hypotheses that must be tested. The rat is only one of many possible animal models to use in experimentally testing new ideas. It clearly has a place in these experiments, and the results obtained will likely be linked to the degree of integrated, multidisciplinary studies on animals such as the rat.

ACKNOWLEDGMENTS

The authors wish to thank Dr. C. Zurcher and Dr. B. J. Cohen for their advice and assistance and Ms. D. Euser-Gaanderse for her help in preparing this chapter.

REFERENCES

1. Berg, B. N., and Simms, H. S. (1960). Nutrition and longevity in the rat. II. Longevity and onset of disease with different levels of food intake. *J. Nutr.* **71**, 255–263.
2. Berg, B. N., and Simms, H. S. (1961). Nutrition and longevity in the rat. III. Food restriction beyond 800 days. *J. Nutr.* **74**, 23–32.
3. Berg, B. N., and Simms, H. S. (1965). Nutrition, onset of disease, and longevity in the rat. *Can. Med. Assoc. J.* **93**, 911–913.
4. Berg, B. N. (1967). Longevity studies in rats. II. Pathology of aging rats. *In* "Pathology of Laboratory Rats and Mice" (E. Cotchin and F. J. C. Roe, eds.), Chapter 23, pp. 749–786. Blackwell, Oxford.
5. Berlin, M., and Wallace, R. B. (1976). Aging and the central nervous system. *Exp. Aging Res.* **2**, 125–164.
6. Bezooijen, C. F. A. van, de Leeuw-Israël, F. R., and Hollander, C. F. (1974). Long-term functional aspects of syngeneic, orthotopic rat kidney grafts of different ages. *J. Gerontol.* **29**, 11–19.
7. Bezooijen, C. F. A. van, van Noord, M. J., and Knook, D. L. (1974). The viability of parenchymal liver cells isolated from young and old rats. *Mech. Ageing Dev.* **3**, 107–119.
8. Boorman, G. A., and Hollander, C. F. (1972). Occurrence of spontaneous cancer with aging in an inbred strain of rats. *TNO Nieuws* **27**, 692–695.
9. Boorman, G. A., van Noord, M. J., and Hollander, C. F. (1972). Naturally occurring medullary thyroid carcinoma in the rat. *Arch. Pathol.* **94**, 35–41.
10. Boorman, G. A., and Hollander, C. F. (1973). Spontaneous lesions in the female WAG/Rij (Wistar) rat. *J. Gerontol.* **28**, 152–159.
11. Boorman, G. A., and Hollander, C. F. (1974). High incidence of spontaneous urinary bladder and ureter tumors in the brown Norway rat. *J. Natl. Cancer Inst.* **52**, 1005–1008.
12. Boorman, G. A., and Hollander, C. F. (1976). Medullary carcinoma of the thyroid in the rat. Animal model of human disease. *Am. J. Pathol.* **83**, 237–240.
13. Brownson, R. H. (1955). Perineuronal satellite cells in the motor cortex of aging brains. *J. Neuropathol. Exp. Neurol.* **14**, 424–432.
14. Bullock, B. C., Banks, K. L., and Manning, P. J. (1968). Common lesions in the aged rat. *In* "The Laboratory Animal in Gerontological Research" (T. W. Harris, ed.), Publ. No. 1591, pp. 62–82. Natl. Acad. Sci., Washington, D. C.
15. Burek, J. D., van der Kogel, A. J. and Hollander, C. F. (1976). Degenerative myelopathy in three strains of aging rats. *Vet. Pathol.* **13**, 321–331.
16. Burek, J. D., Zurcher, C., and Hollander, C. F. (1976). High incidence of spontaneous cervical and vaginal tumors in an inbred strain of rats. *J. Natl. Cancer Inst.* **56**, 549–554.
17. Burek, J. D., and Hollander, C. F. (1977). Incidence patterns of spontaneous tumors in BN/Bi rats. *J. Natl. Cancer Inst.* **58**, 99–105.
18. Burek, J. D. and Meihuizen, S. P. (1977). Age-Associated Changes in the thymus of Aging BN/Bi Rats. *In* "Proceedings of the 5th European Symposium on Basic Research in Gerontology" (Schmidt, U. J., Brüschke, G., Lang, E. Viidik, A., Platt, D., Frolkins, V. V., and Schulz, F. H., eds.), pp. 167–170. Verlag, Dr. med D. Straube, Erlangen, Weimar, East Germany.
19. Burek, J. D. (1978). "Pathology of Aging Rats: A Morphological and Experimental Study of the Age-Associated Lesions in Aging BN/Bi, WAG/Rij and (WAG × BN)F₁ Rats." CRC Press, Inc., West Palm Beach, Florida.
20. Burnet, F. M. (1973). A genetic interpretation of aging. *Lancet* **2**, 480–483.
21. Cheney, K. E., and Walford, R. L. (1975). Minireview: Immune function and disfunction in relation to aging. *Life Sci.* **14**, 2075–2083.
22. Cohen, B. J. (1968). Effects of environment on longevity in rats and mice. *In* "The Laboratory Animal in Gerontological Research" (T. W. Harris, ed.), Publ. No. 1591, pp. 21–29. Natl. Acad. Sci., Washington, D.C.
23. Cohen, B. J., and Anver, M. R. (1976). Pathological changes during aging in the rat. *In* "Special Review of Experimental Aging Research: Progress in Biology" (Elias, M. F., Eleftheriou, B. E., and Elias, P. K. eds), pp. 379–403. Experimental Aging Research, Bar Harbor, Maine.
24. Coleman, G. L., Barthold, S. W., Osbaldiston, G. W., Foster, S. J., and Jonas, A. M. (1977). Pathological changes during aging in barrier-reared Fischer 344 male rats. *J. Gerontol* **32**, 258–278.
25. Comfort, A. (1964). "Aging, The Biology of Senescence." Routledge & Kegan Paul, London.
26. Comfort, A. (1968). Feasibility in age research. *Nature (London)* **217**, 320–322.
27. Comfort, A. (1970). Basic Research in Gerontology. *Gerontologia* **16**, 48–64.
28. Comfort, A. (1974). The position of aging studies. *Mech. Ageing Dev.* **3**, 1–31.
29. Cotchin, E., and Roe, F. J. C., eds. (1967). "Pathology of Laboratory Rats and Mice." Blackwell, Oxford.
30. Crain, R. C. (1958). Spontaneous Tumors in the Rochester strain of Wistar rat. *Am. J. Pathol.* **34**, 311–323.
31. Curtis, H. J. (1971). Genetic factors in aging. *Adv. Genet.* **16**, 305–324.
32. Everitt, A. V., and Burgess, J. A. (1976). "Hypothalamus, Pituitary and Aging." Thomas, Springfield, Illinois.
33. Finch, C. E. (1971). Comparative biology of senescence-evolutionary and developmental considerations. *In* "Animal Models for Biomedical Research", Vol. IV, pp. 47–67. NAS, Washington, D.C.
34. Finch, C. E. (1972). Enzyme activities, gene function and aging in mammals (review). *Exp. Gerontol.* **7**, 53–67.
35. Franks, L. M. (1973). Structural changes in aging cells. *Z. Alternsforsch.* **27**, 237–245.
36. Gilbert, C., and Gillman, J. (1958). Spontaneous neoplasms in the albino rat. *S. Afr. J. Med. Sci.* **23**, 257–272.
37. Goldstein, S. (1971). The biology of aging. *N Engl. J. Med.* **285**, 1120–1129.
38. Gore, I. Y. (1973). Methodology in gerontological research. *Gerontol. Clin.* **15**, 133–140.
39. Gsell, D. (1964). Absterbekurven und Wachstumscharakteristika einer "Alterszucht" von Wistar-Ratten. Int. *Z. Vitaminforsch.* **9**, 114–125.
40. Hahn, H. P. von (1970). Structural and functional changes in nucleoprotein during the aging of the cell. Critical review. *Gerontologia* **16**, 116–128.
41. Hayflick, L. (1970). Aging under glass. *Exp. Gerontol.* **5**, 291–303.
42. Hayflick, L. (1973). Invited review. The biology of human aging. *Am. J. Med. Sci.* **265**, 432–445.
43. Henderson, E. (1972). The aging process. *J. Am. Geriatr. Soc.* **20**, 565–571.
44. Hollander, C. F. (1971). Age limit for the use of syngeneic donor kidneys in the rat. *Transplant. Proc.* **3**, 594–597.
45. Hollander, C. F. (1973). Experimental gerontological research in the Netherlands. *Proc. Aust. Assoc. Gerontol.* **2**, 5–8.
46. Hollander, C. F. (1973). Guest editorial: Animal models for aging and cancer research. *J. Natl. Cancer Inst.* **51**, 3–5.
47. Hollander, C. F. (1976). Current experience using the laboratory rat in aging studies. *Lab Anim. Sci.* **26**, 320–328.
48. Hollander, C. F., Burek, J. D., Boorman, G. A., Snell, K. C., and Laqueur, G. L. (1976). Granular cell tumors of the central nervous system of the rat. *Arch. Pathol.* **100**, 445–447.
49. Hruza, A., and Zbuzkova, V. (1975). Cholesterol turnover in plasma, aorta, muscles and erythrocytes in young and old rats. *Mech. Ageing Dev.* **4**, 169–179.

50. Inukai, T. (1928). On the loss of Purkinje cells, with advanced age from the cerebellar cortex of the albino rat. *J. Comp. Neurol.* **45,** 1–31.

51. Jacobs, B. B., and Huseby, R. A. (1967). Neoplasms occurring in aged Fischer rats, with special reference to testicular, uterine and thyroid tumors. *J. Natl. Cancer Inst.* **39,** 303–309.

52. Kerkvliet, N. L., and Kimeldorf, D. J. (1973). The effect of host age on the transplantability of the Walker 256 carcinosarcoma. *J. Gerontol.* **28,** 276–280.

53. Knook, D. L., Sleijster, E. C., and van Noord, M. J. (1975). Changes in lysosomes during aging of parenchymal and nonparenchymal liver cells. *Adv. Exp. Med. Biol.* **53,** 155–169.

54. Knook, D. L., and Hollander, C. F. (1978). Embryology and aging of the rat liver. *In* "Rat Hepatic Neoplasia," (P. M. Newberne and W. H. Butler, eds.), pp. 8–40. MIT Press, Cambridge, Massachusetts.

55. Knook, D. L., and Sleijster, E. C. (1976). Separation of Kupffer and endothelial cells of rat liver by centrifugal elutriation. *Exp. Cell Res.* **99,** 444–449.

56. Kohn, R. R. (1971). "Principles of Mammalian Aging." Found. Dev. Biol. Ser. Prentice-Hall, Englewood Cliffs, New Jersey.

57. Kormendy, C. G., and Bender, A. D. (1971). Experimental modification of the chemistry and biology of the aging process. *J. Pharm. Sci* **60,** 167–180.

58. Landfield, P. W., Rose, G., Sandles, L., Wohlstadter, T. C., and Lynch, G. (1977). Patterns of astroglial hypertrophy and neuronal degeneration in the hippocampus of aged, memory-deficient rats. *J. Gerontol.* **32,** 3–12.

59. Leeuw-Israël, F. R. de (1971). Aging changes in the rat liver. Ph. D. Thesis, University of Leiden.

60. Lindsey, J. R., Baker, H. J., Overcash, R. G., Cassell, G. H., and Hunt, C. E. (1971). Murine chronic respiratory disease. Significance as a research complication and experimental production with *mycoplasma pulmonis. Am. J. Pathol.* **64,** 675–716.

61. Marx, J. (1974). Aging research. II. Pacemakers for aging? *Science* **186,** 1196–1197.

62. Paget, G. E., and Lemon, P. G. (1965). The interpretation of pathology data. *In* "Pathology of Laboratory Animals" (W. E. Ribelin and J. R. McCoy, eds.), Chapter 14, pp. 382–405. Thomas, Springfield, Illinois.

63. Porta, E. A., and Hartroft, W. S. (1969). Lipid pigments in relation to aging and dietary factors (Lipofuscins). *In* "Pigments in Pathology" (M. Wolman, ed.), pp. 191–235. Academic Press, New York.

64. Ribelin, W. E., and McCoy, J. R., eds. (1965). "Pathology of Laboratory Animals." Thomas, Springfield, Illinois.

65. Roe, F. J. C. (1965). Spontaneous tumors in rats and mice. *Food Cosmet. Toxicol.* **3,** 707–720.

66. Ross, M. H., and Bras, G. (1965). Tumor incidence patterns and nutrition in the rat. *J. Nutr.* **87,** 245–260.

67. Sacher, G. A. (1968). Molecular versus systemic theories on the genesis of aging. *Exp. Gerontol.* **3,** 265–271.

68. Schlettwein-Gsell, D. (1970). Survival curves of an old age rat colony. *Gerontologia* **16,** 111–115.

69. Shock, N. W. (1961). Physiological aspects of aging in man. *Annu. Rev. Physiol.* **23,** 97–122.

70. Shock, N. W. (1966). "Perspectives in Experimental Gerontology." Thomas, Springfield, Illinois.

71. Shock, N. W. (1968). Current trends in biological research on aging. *Jpn. J. Gerontol.* **5,** 21–27.

72. Simms, H. S., and Berg, B. N. (1957). Longevity and the onset of lesions in male rats. *J. Gerontol.* **12,** 244–252.

73. Simms, H. S. (1967). Longevity studies in rats. I. Relation between life span and age of onset of specific lesions. *In* "Pathology of Laboratory Rats and Mice" (E. Cotchin and F. J. C. Roe, eds.), Chapter 22, pp. 733–748. Blackwell, Oxford.

74. Smith, J. M. (1963). Review lectures on senescence I. The causes of aging. *Proc. R. Soc. London, Ser. B* **157,** 115–127.

75. Snell, K. C. (1965). Spontaneous lesions of the rat. *In* "Pathology of Laboratory Animals" (W. E. Ribelin and J. R. McCoy, eds.), Chapter 10, pp. 241–302. Thomas, Springfield, Illinois.

76. Squire, R. A. (1971). "A Guide to Infectious Diseases of Mice and Rats," Comm. Lab. Ani. Dis., Inst. Lab. Ani. Resour. Natl. Acad. Sci., Washington, D.C.

77. Squire, R. A., and Levitt, M. H. (1975). Report of a workshop on classification of specific hepatocellular lesions in rats. *Cancer Res.* **35,** 3214–3223.

78. Stewart, H. L. (1975). Comparative aspects of certain cancers. *In* "Cancer" (F. F. Becker, ed.), Vol. 4, Chapter 10, pp. 303–374. Plenum, New York.

79. Strehler, B. L. (1975). Implications of aging research for society. *Fed. Proc., Fed. Am. Soc. Exp. Biol.* **34,** 5–8.

80. Thompson, S. W., Huseby, R. A., Fox, M. A., Davis, C. L., and Hunt, R. D. (1961). Spontaneous tumors in the Sprague-Dawley rat. *J. Natl. Cancer Inst.* **27,** 1037–1057.

81. Turusov, V. S., ed. (1973). "Pathology of Tumors in Laboratory Animals," IARC Sci. Publ. No. 5, Vol. I, Part 1. Int. Agency Res. Cancer, Lyons, France.

82. Turusov, V. S., ed. (1976). "Pathology of Tumors in Laboratory Animals," IARC Sci. Publ. No. 6, Vol. I, Part 2. Int. Agency Res. Cancer, Lyons, France.

83. Walford, R. L. (1969). "The Immunologic Theory of Aging." Williams & Wilkins, Baltimore, Maryland.

84. Walford, R. L. (1976). When is a mouse "old"? Letters to the Editor. *J. Immunol.* **117,** 352–353.

85. Weisbroth, S. H. (1972). Pathogen-free substrates for Gerontological research: Review, sources, and comparison of barrier-sustained vs. conventional rats. *Exp. Gerontol.* **7,** 417–426.

86. Wexler, E. J., and Greenberg, L. J. (1970). Co-existent arteriosclerosic, PAN, and premature aging. *J. Gerontol.* **25,** 373–380.

87. Wolman, M. (1975). The pigment of aging. *I. J. Med. Sci.* **11,** Suppl., 147–162.

88. Wolthuis, O. L., Knook, D. L., and Nickolson, V. J. (1976). Age-related acquisition deficits and activity in rats. *Neurosci. Lett.* **2,** 343–348.

89. Yunis, E. J., and Greenberg, L. J. (1974). Immunopathology of aging. *Fed. Proc., Fed. Am. Soc. Exp. Biol.* **33,** 2017–2019.

Chapter 8

Cardiovascular Research

Sanford P. Bishop

I. INTRODUCTION

The rat has received general acceptance and is widely used for many types of studies involving the cardiovascular system. In many of these studies specific advantages are provided by the rat, such as small size of the heart in comparison to other commonly used species. However, economic factors are also important in selecting rats, particularly when large numbers of animals are required. The purpose of this chapter is to indicate those areas of cardiovascular research in which the rat has been particularly useful, to describe those features of the rat cardiovascular system which may influence experimental studies, and to discuss the advantages and disadvantages of the rat for specific research objectives.

II. MACROSCOPIC ANATOMY

A. Heart

The heart of the rat is a four-chambered organ, as in other mammals, without normal communications between the left and right chambers. The heart is enclosed within a thin transparent pericardium which is attached to the major arteries and veins at the base of the heart. The topographic features of the heart are similar to those of other mammals, with a clearly defined atrio-ventricular groove separating the atria from the ventricles.

Anterior and posterior papillary muscles in the left ventricle have relatively long slender apical portions projecting into the ventricular lumen which anchor the chordae tendineae of the mitral valve. The anterior papillary muscle of the rat is located more lateral than in the dog or man. Papillary muscles in the right ventricle are slender elongated structures varying from 2 to 5 or more in number.

The cardiac valves are similar to those in other species. The aortic and pulmonic valves have three leaflets; the mitral valve, an anterior and posterior leaflet; and the tricuspid, a posterior septal leaflet and two lateral leaflets on the free wall portion of the right ventricular wall. There are multiple thin chordae tendineae connecting the mitral and tricuspid leaflets to the papillary muscles.

B. Heart Shape during Growth

Left ventricular morphology has been studied in detail by Grimm *et al.* (44) in rats from 85 to 445 gm body weight to evaluate the mechanical capabilities of the growing myocardium. During normal growth, increasing radii of the left ventricular lumen were accompanied by a proportional increase in wall thickness and valve to apex measurement, so that during growth the shape of the heart remained unchanged.

C. Heart Weight

The weight of the rat heart at various stages during growth is of interest in many studies which are designed to alter normal heart weight. Since the rat continues to gain weight through the first year of life, the heart also is continually increasing in weight. Therefore, studies designed to increase the heart weight must take into consideration not only the experimentally induced increase in heart weight, but also the increase due to normal growth. If the experimental design also alters normal body weight gain, this must also be considered in evaluation of the results.

Several investigators have reported total heart or ventricular weights for rats of body weights from 50 to over 600 gm, but there are few reports of heart weights in young animals under 50 gm body weight. Rakusan *et al.* (114) have studied heart weights in rats from birth through old age. They found a total ventricular weight to body weight ratio of 5.20 in neonates which decreased to 4.48 by 35 days of age (50 gm body weight). After about 40 days of age, the heart grows less rapidly than the rest of the body, resulting in a more rapid decrease in heart weight to body weight ratio. These investigators and others (5, 24, 43, 45, 65, 104, 124) report heart weight to body weight values for various size rats within the following ranges: 100 gm body weight, 3.3–3.8 mg/gm; 200 gm, 2.9–3.5 mg/gm; 300 gm, 2.4–2.9 mg/gm; 400 gm, 2.4–2.8 mg/gm; 500 gm, 2.25–2.6 mg/gm. Regression equations calculated from data published by these several authors give a Y intercept (heart weight in mg) of 165 to 235 and a slope of 1.9 to 2.6X (body weight in grams). Beznak (11) reported a regression equation of $Y = 186.2 + 2.59X$.

Weights of the right and left ventricles and the combined atria have been determined for 289 male Sprague-Dawley rats weighing 23–600 gm. These data with the computer-derived regression lines are shown in Fig. 1A–D. These data illustrate

the decreasing rate of growth relative to increasing body weight for the various cardiac chambers.

Differences in values reported by various authors are due to different strains, environmental conditions during growth, and methods of dissection of the heart. Each investigator must establish a normal growth curve for each strain under specific environmental conditions.

D. Major Arteries and Veins Leaving and Entering the Heart

The anatomy of the rat circulatory system has been described by Greene and Halpern (41, 49, 51). The aorta and pulmonary artery originate from the left and right ventricles at the base of the heart. The pulmonary artery divides into left and right branches in the usual manner. The left and right coronary artery ostia are located at the root of the ascending aorta. There

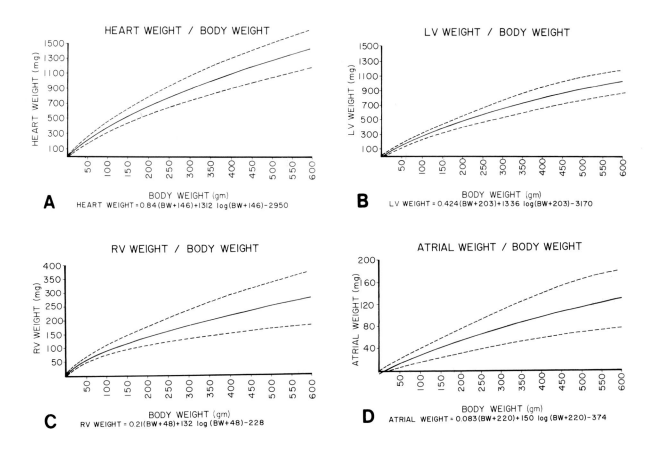

Fig. 1. Heart weight to body weight ratios determined on 289 male Sprague-Dawley rats. The solid line is the computer derived mean for the group with the regression equation given under each graph. The dashed lines were drawn to include 95% of all observations. LV, left ventricle; RV, right ventricle. (Data contributed by David G. Penney, Department of Physiology, Wayne State University, Detroit, Michigan, and obtained from a colony maintained at the University of Illinois, Chicago Circle.)

are three major branches arising from the arch of the aorta in the rat, the innominate (brachiocephalic), the left common carotid, and the left subclavian arteries. The innominate artery gives rise to the right common carotid, right subclavian, and right internal mammary arteries. The left subclavian artery gives rise to the left internal mammary artery.

There are three major veins entering the right atrium of the rat heart (49). The inferior vena cava enters the right atrium at the posterior right border, and the right superior vena cava enters cranially. In addition, the rat heart has a left superior vena cava which enters the coronary sinus region with the inferior vena cava after crossing the posterior surface of the heart from the left. The persistence of two superior venae cavae is normal in the rat. The azygous vein is on the left and drains into the left superior vena cava, quite the opposite to the situation in higher mammals without a left superior vena cava. A hemiazygous vein may be present, and drains into the azygous vein.

E. Blood Supply of the Heart

Both coronary and extracoronary sources supply blood to the rat heart (51). The right coronary artery arises from the aortic root and courses diagonally across the lateral wall of the right ventricle. A large septal artery commonly is given off near the origin of the right coronary artery and is the major blood supply to the septum. Small branches supply the right atrium adjacent to the atrioventricular sulcus. The right coronary artery also sends a branch to the interatrial septum and the region of the AV node.

The left coronary artery descends over the anterior lateral wall of the left ventricle parallel to the anterior longitudinal sulcus. It may give rise to a major septal artery near its origin. Branches of the left coronary extend to the pulmonary conus region and ramify laterally over the left ventricular free wall, including the posterior wall of the left ventricle. There is no clearly defined left circumflex coronary artery branch as in higher mammals, but a small branch often courses through the left ventricle parallel to the atrioventricular sulcus. Small branches reach the left atrium. There may be two major arteries arising from one or both aortic sinuses. The tendency for variant patterns is more common in some strains, apparently as an inherited trait (18).

The left and right internal mammary arteries are the origin of the left and right cardiacomediastinal arteries, which are important extracardiac sources of blood to the heart. The right cardiacomediastinal artery courses along the right superior vena cava to the right atrium where it supplies the sinus node, the bulk of the right atrium, and portions of the left atrium. The left cardiacomediastinal artery follows the left superior vena cava and supplies the remaining portion of the left atrium.

Thus, the atria including the sinus node receive their major blood supply from a noncoronary artery source. This fact must be recognized when comparing the atria and ventricles of the rat heart and may be useful for certain types of vascular studies.

The retention of a dual blood supply in the rat heart is similar to the situation found in lower animals including reptiles and amphibians. Although this dual blood supply is better developed in the rat than in higher mammals, it is by no means nonexistent in higher forms. Petelenz (105) studied 100 human specimens and demonstrated branches of the bronchial artery reaching the sinus node region in 44%, and in 26% the extent of vascularization was judged to be of functional significance.

III. MICROSCOPIC ANATOMY

A. Ventricular Myocardium

Cardiac muscle cells are arranged in parallel bundles joined to each other at the intercalated disc regions at the ends and along the sides of the cells. In general, orientation of muscle fibers within the left ventricle is vertical, running from base to apex in the subendocardium, with parallel fibers running longitudinally within the papillary muscles and trabecular muscles of both ventricles. Successively deeper layers in the left ventricle from the endocardium are oriented more horizontally, with fibers in the midwall region approximatly 70°–90° to those of the subendocardium. At the subepicardium fibers have turned approximately 120° from those in the subendocardium. Thus, a full wall thickness section cut horizontally through the midportion of the left ventricle will have fibers oriented in cross section in subendocardium and longitudinal section in the midwall region. The longitudinal orientation of subendocardial fibers in both ventricles generally follows the lines of blood flow.

The cell type, volume, and numerical composition of the myocardium are of importance in many studies utilizing heart tissue. Morphometric studies of rat myocardium have shown the volume composition to be 83% myocytes, with the remainder consisting of interstitial connective tissue, fibrocytes, capillary endothelial cells, and various other cell types (3). The capillary to myofiber ratio is approximately 1:1 in the mature rat (1, 112, 113) as in other species (125, 139). However, on a strictly numerical basis, heart muscle tissue in the 200-gm rat consists of only 20–25% heart muscle cells, the bulk of cells consisting of the much smaller endothelial cells and other interstitial connective tissue cells (46, 87). This fact is particularly important for studies dependent on cell number rather than volume, such as DNA synthesis or other nuclear-dependent phenomena.

The cross-sectional diameter of heart muscle cells has been determined by several investigators. Several reports in which formalin-fixed, paraffin-embedded tissues were used indicated mean cross-sectional diameter of mature ventricular myocardial cells to be 11–14 μm (1, 24, 26, 32, 55, 112, 113). However, more recent studies utilizing either fresh preparations of myocytes or epoxy-embedded tissues in which shrinkage artifact is considerably reduced, report mean cross-sectional diameters of 17–18.5 μm (3, 65). The length of heart muscle cells is difficult to obtain in histologic sections due to orientation of the plane of section. Isolated heart muscle cells obtained from 240-gm (84 days of age) rats in our laboratory had a mean length (at 1.6 μm sarcomere length) of 94 ± 18 (SD) μm, a mean cell width of 17.8 ± 4 μm, and a mean length–width ratio of 5.6 (16). Isolated myocytes from rats weighing 320 gm (150 days of age) were 102 ± 20 μm long and 19.7 ± 5 μm wide. During the early neonatal period cardiac myocytes change from an ovoid shape to a more elongated shape as the myocardium converts from the hyperplastic growth characteristic of the fetal and early neonatal period to growth by increase in cell size, or hypertrophy. Calculated myocyte volume remains constant at approximately 800 μm^3 during the first week of postnatal life (hyperplastic growth); mean cell volume increases to 42,000 μm^3 by 250 days of age (hypertrophic growth). Myocytes from the inner half of the left ventricle are slightly larger in both width and length than those from the outer half of the left ventricular wall. Right ventricular myocytes are significantly smaller than left ventricular myocytes (16a).

The fractional organelle volume of ventricular myocytes determined by morphometric analysis of electron micrographs is 35–40% mitochondria, 45–56% myofibrils, 1.0–2.2% T tubular system, and 3.0–3.5% sarcoplasmic reticulum (3, 32, 65, 101, 102). A well-documented, but seldom appreciated, fact is the presence of two nuclei in the great majority of myocytes from normal rat heart (89, 120). Studies in our laboratory using isolated rat heart myocytes have demonstrated that at the time of birth, nearly all myocytes contain a single nucleus, but during postnatal growth there is a rapid increase in percentage of myocytes with two nuclei. By 15 days of age, 85% of isolated myocytes contain two nuclei, and in an individual animal, there is a direct relationship between mean cytoplasmic volume and nuclear number.

B. Atrial Myocardium

Myocytes in the atrium are smaller and more loosely arranged than ventricular myocytes (19, 60, 110). Although some reports indicate an absence of T tubules in atrial myocardium (59), studies with horseradish peroxidase tracer have identified two populations of atrial cells: one without T tubules and the other possessing T tubules (36). Atrial myocytes are reported to have only a single nucleus with large mitochondrial collections at the nuclear poles (60). However, unpublished studies in our laboratory with isolated myocytes show that atrial myocytes have two nuclei as in ventricular myocytes. Jamieson and Palade (60) described the ultrastructure of the rat atrium and specifically drew attention to the electron-dense, membrane-bound atrial-specific granules present not only in rat atrium but in all mammalian atrial myocytes. In the rat heart, atrial granules are larger and more numerous than in larger mammals. The function of atrial specific granules is not known, but they apparently do not contain catecholamines (59). First studied in detail in the rat heart by Bompiani et al. (19), atrial granules have been studied extensively by Bencosme and co-workers (7, 8, 25).

C. Conduction System

The sinus node of the rat heart is described by Halpern (50) and King (67) as a relatively large horseshoe-shaped structure located at the junction of the right superior vena cava and the right atrium. A significant feature of the sinus node in the rat is the extracoronary origin of the sinus node artery described above. Ultrastructurally, sinus node cells are reported to always contain a single nucleus, are smaller in size, and have fewer contractile elements than atrial myocytes (84). Sinus node cells contain large numbers of atrial-specific granules. The AV node cells also contain a single nucleus (84) and have no distinctive features different from AV node cells of other species. T tubules are not present in either sinus node, AV node, or Purkinje fibers (127). AV node cells of the rat have abundant amounts of glycogen as do the peripheral Purkinje fibers; sinus node cells are relatively devoid of glycogen (84).

IV. METHODS FOR *in Vivo* EVALUATION OF THE CARDIOVASCULAR SYSTEM

A. Electrocardiogram

Electrocardiograms are obtained using standard electrocardiographic equipment. Needle electrodes are the most satisfactory means to obtain skin contact if the animal can be anesthetized. If the animal is not anesthetized, a restraining device is required to hold the animal during the recording (100). Heart rate may be determined from the electrocardiogram.

B. Blood Pressure

Blood pressure is obtained from the tail using a specially designed cuff equipped with either an ultrasound microphone or a pressure transducer attached to the cuff and a recording system (23, 80, 140, 148). Systems are also available for continuous recording of blood pressure in unrestrained animals (73). In anesthetized animals, blood pressure may be recorded by cannulation of a peripheral artery, such as the femoral or carotid.

V. METHODS FOR *in Vitro* STUDY OF MYOCARDIUM

A variety of *in vitro* techniques have been utilized to study normal structure, function and metabolism of rat myocardium. The rat has been the most widely used species for most of these *in vitro* techniques, since an adequate amount of tissue is available, function is easily maintained, and a standardized animal is readily available at relatively low cost.

A. Tissue Culture

Establishment of beating heart muscle cells in culture has been accomplished by trypsinization of fetal or neonatal rat hearts. In early studies by Harary (53, 54), heart muscle cells continued to beat for up to 9 weeks, and more recently, many laboratories have maintained beating cultures of fetal myocardial cells for longer periods of time (81). During the fetal and early neonatal period, heart muscle cells from the rat are metabolically adapted to a highly anaerobic type of environment in contrast to the oxidative metabolism required by adult myocardial cells. In addition, fetal and neonatal cells retain the capability of growth by cellular division, or hyperplasia, which is lost by a few weeks after birth (155). While these characteristics of the neonatal myocyte provide the possibility of maintaining beating heart muscle cells in culture which are useful for a variety of studies, it must be kept in mind that such myocytes are structurally, functionally, and metabolically different from adult cardiac myocytes. It is well known that cells maintained in culture tend to revert toward a more immature type of cell (72). Furthermore, mutations may occur resulting in very different cells in the culture after a few passages compared to the originally derived cells. Contamination of the culture with other cell types such as fibroblasts which may overgrow the culture is also a common occurrence.

Heart muscle cells grown in tissue culture usually tend to form single layers of synchronously beating cells. Some cells will beat independently of others in the culture, while others will not beat unless in actual contact with beating cells (53). Harary and Farley (52) described a tendency toward fiberlike development in cultures of embryonic heart muscle cells, and Halbert *et al.* (47) described organoid development *in vitro*.

These latter authors found that cells grown on polystyrene surfaces tended to form three-dimensional structures after several days of growth. These heart cell masses, or "minihearts," could be very useful for many types of studies.

B. Isolated Myocytes from Adult Myocardium

A variety of techniques have been reported for preparation of isolated beating myocytes from adult rat myocardium. Since these cells are highly dependent on oxidative metabolism and have lost their capability of cellular division, long-term cultures of adult cardiac cells have not been achieved.

Early attempts to isolate adult heart muscle cells used trypsinization of minced myocardium. Low yields of beating cardiac muscle cells were obtained (17, 136). In later studies, the yield of cells was improved by using collagenase and hyaluronidase in the cell isolation media with repeated washing and incubation of minced tissue (38). Purification of the heart muscle cells from the contaminating endothelial and connective tissue cells was achieved by density gradient centrifugation (111). Cell yield and time required for isolation was greatly improved by perfusion of the isolated heart with calcium-free medium containing collagenase and hyaluronidase which caused disassociation of the cells (9). Further refinement of enzyme concentration, ionic strength, pH, osmolality, and temperature of perfusion media have resulted in improvement in the number and quality of adult cardiac myocytes which may be obtained in an intact state (16, 28a, 88, 109). Low-speed centrifugation through 3% Ficoll removes the vast majority of noncontracting, damaged myocytes and other cell types.

Isolated heart muscle cells from postneonatal rats will beat rhythmically for up to several hours in oxygenated media at 37°C, or may be maintained in a noncontracting state at 4°C. Morphologically intact cells exclude trypan blue and have the characteristic structural features of myocytes *in situ* (Fig. 2).

C. Organ Culture

Organ culture of fetal hearts from the mouse (146, 147) has been used in several laboratories, and the fetal rat heart should be adaptable to this system. Fetal mouse hearts are suspended on a mesh at the air–culture media interface and will continue to beat for many days.

D. Isolated Organelles

The rat myocardium has been used by a number of investigators for preparation of isolated cellular organelles, such as

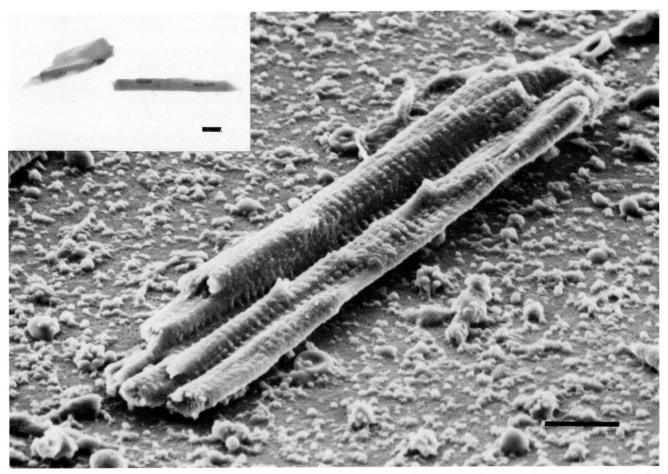

Fig. 2.   Scanning electron micrograph of a cardiac myocyte isolated from a 60-day-old Fischer rat. Cell isolated by perfusion of heart with Ca^{2+}-free, collagenase containing media, fixed in 2% glutaraldehyde, mounted on a Nuclepore filter, and critical point dried. Transverse ridges correspond to underlying sarcomeres. Intercalated discs are seen at the ends of the cell and at several lateral locations. Scale, 10 μm. Inset is light microscopic appearance of isolated cells stained with hematoxylin and eosin. Note presence of two nuclei in each cell. Scale, 10 μm.

nuclei, mitochondria, ribosomes, or sarcoplasmic reticulum. There appears to be no particular advantage or disadvantage in using the rat myocardium for this type of study except that due to the small size of the heart, relatively small amounts of organelle preparation are obtained.

A factor which must be considered when studying functions of the myocardium related to cellular organelles is that heart muscle is composed of several cell types. Although the bulk of myocardial mass is cardiac muscle tissue, on a numerical basis, myocytes are a definite minority. In the mature rat myocardium, only 20 to 30% of the total cell number is composed of cardiac myocytes (46, 87), the largest number of cells being the much smaller endothelial cells and connective tissue cells such as fibrocytes. Therefore, studies measuring nuclear function, for example, would principally detect endothelial cell or fibroblast nuclear function rather than myocyte nuclear function.

E. Isolated Heart Preparation

The rat has been the principal species used for studies utilizing isolated heart preparations. The size of the heart is ideal for incorporation studies since only small amounts of material are required, yet sufficient tissue is available for analysis. Under appropriate conditions, the isolated rat heart will continue beating for several hours with a reasonably steady state of function.

Two types of preparations are commonly used: (a) the nonworking heart or Langendorff preparation, and (b) the working heart preparation. In either case, the heart is suspended in a chamber from a cannula inserted into the aortic arch. In the Langendorff preparation, oxygenated perfusion media is pumped at a controlled rate retrogradely into the aorta and through the coronary arteries (74, 86, 141, 149). The effusate from the coronary veins drips off the end of the heart and may either be collected for analysis of metabolic products, residual substrates, etc., or recirculated. Although the heart continues to beat during the procedure, there is no inflow to the left ventricle, lumenal pressure does not exceed aortic perfusion pressure, and no fluid is ejected from the left ventricle. Thus the preparation is nonworking.

In the working heart preparation, an additional cannula is secured in the left atrium, and the oxygenated perfusion media enters the left atrium under a controlled pressure. The left ventricle receives fluid from the left atrium during diastole and ejects it into the aorta during systole. The left ventricle performs work by pumping fluid against any desired hydrostatic pressure or other resistance in the line.

Isolated heart preparations have the advantage over *in vivo* studies in that all neurogenic and hormonal influences are removed from control of the heart. Perfusion fluid volume and content are easily controlled in the isolated heart preparation, and metabolic substrates are conserved since only a small amount of tissue is being perfused compared to whole animal preparations. Function of the heart is easily controlled and monitored. Light microscopic histology is unaltered except for mild interstitial edema after 1–3 hr at 37°C (21), and myocardial ultrastructure is maintained for at least 90 min at 32°C if glucose and insulin are present in the perfusion media (141). Disadvantages of the isolated heart preparation are that hemodynamic and other functional conditions are altered from the *in vivo* situation, and after several hours, functional and structural alterations occur leading to myocardial failure. Although function and structure appear stable for several hours, in fact, degenerative changes are probably occurring throughout the perfusion period.

F. Heart–Lung Preparation

The heart–lung preparation has received wide use in larger species of animals, but has only rarely been used in the rat due to the small size of vessels which must be cannulated. A technique for the heart–lung preparation in the rat has been described, however (85), and has been used for studies in the myocardium (63, 70). The technique has the advantage over isolated heart preparations in that *in vivo* conditions are more closely simulated; nevertheless, marked alterations in function are produced and the method is technically difficult.

G. Morphologic and Biochemical Studies

The rat is an ideal animal because of size for many types of studies involving morphologic or biochemical determinations. A frequent difficulty encountered with larger species of animals for morphologic studies involving light or electron microscopy is that many sections are required to obtain adequate

sampling. In the case of the rat, a relatively large proportion of the myocardium may be sampled with a small number of sections. The same is true for many biochemical studies, and the additional expense of larger species is thus avoided.

VI. SPONTANEOUS MODELS OF HUMAN CARDIOVASCULAR DISEASE

A. Spontaneously Hypertensive Rats (SHR)

1. Historical Development

A strain of rats with persistently elevated blood pressure was developed by Okamoto and Aoki in 1963 (96) and has been maintained since that time by selective sibling matings. By 1969 the strain had reached the twenty-first generation. This strain, designated as SHR, is maintained by the Veterinary Resources Branch of the National Institutes of Health and numerous commercial breeders. The SHR strain has become one of the more widely used inbred strains of rats in the United States. General information relative to background, sources of animals, description of the animal, and colony care has been published (135), and two symposia on SHR have been published (37, 95).

2. Hypertension and Complicating Lesions

Hypertension is present in 100% of SHR. Increased blood pressure (greater than 140 mm Hg) is present by 5 weeks of age in many animals and by 10 weeks of age is maintained at over 180 mm Hg, frequently exceeding 200 mm Hg. In a review of the model, Okamoto (93) described complicating lesions associated with the hypertension in several organs. Cerebral softening due to infarction, subarachnoid hemorrhage, and microscopic cerebral hemorrhages occurred in 8%. Myocardial necrosis and fibrosis, mainly detected by microscopy, was found in 19%, nephrosclerosis in 17%, and vascular lesions including fibrinoid necrosis and hyaline degeneration in renal, pancreatic, hepatic and, less commonly, coronary arteries were present in 39%. In a strain with severe hypertension derived from the parent SHR strain, the incidence of these complicating lesions was greatly increased (98).

The metabolic basis for the development of the hypertension remains unclear. Okamoto has reported that noradrenaline content of the adrenal medulla in SHR is about twice that of normal rats and is present in a readily releasable form (93, 132). It has been shown that immunosympathectomy with nerve growth factor antiserum and peripheral catecholamine depletion prevent the development of hypertension in SHR (28, 31). Destruction of monoaminergic structures in the CNS by administration of 6-hydroxydopamine into the lateral ventricles of neonatal rats can prevent the development of hypertension in the SHR (34a, 46a). Therefore, destruction of either central or peripheral monoaminergic pathways by chemical means prevents the hypertensive syndrome in the SHR. Hypertension may also be prevented by antihypertensive drugs (123).

3. Genetic Basis of Hypertension

There is clearly a genetic basis for the hypertensive trait in SHR, the strain having been originated from animals selected for higher than normal blood pressure. There has been a tendency for succeeding generations to have higher blood pressure (95). Mating SHR with normotensive rats of the parent strain (WKY) produces offspring with moderately elevated blood pressure, suggesting a polygenic mode of inheritance for the hypertensive trait (153). Strain-specific esterase isozymes of nonspecific α-naphthyl acetate esterase serve as a marker for the strain, are single gene-controlled, and codominantly transmitted (152).

4. Cardiac Hypertrophy

Cardiac hypertrophy is a characteristic feature of SHR and is expressed as an increased ventricular weight to body weight ratio. Although Sen et al. (123) reported an increased heart weight to body weight ratio as early as 3–4 weeks of age prior to the development of increased blood pressure, Kawamura et al. (65) did not find increased heart weight until 13 weeks of age. Current studies have found increased left ventricular weight to body weight ratios compared to controls in SHR at all ages studied including newborn (31a).

Using epoxy-embedded tissue, Kawamura et al. (65) found the shortest cross-sectional diameter of myocytes at the nucleus in the posterior papillary muscle increased compared to control animals at 5 weeks of age, and in the left ventricular lateral wall as well by 11 weeks of age. By 20 weeks of age, myofibers measured 23 μm in diameter in SHR compared to 17 μm in Wistar-Kyoto control rats.

By electron microscopy, Kawamura et al. (65) found alterations of the Z bands and intercalated discs similar to those found in other mammals with cardiac hypertrophy (15, 35, 56). These changes consisted of expansions of the Z band material beneath the sarcolemma or within myofibrils up to the full length of the sarcomere, and increased folding and tortuosity of the intercalated disc (Figs. 3–5). SHR with these changes did not show signs of congestive failure. The Z band and intercalated disc changes were interpreted as evidence for active contractile protein synthesis.

Cells from very large hearts often contained bundles of myofibrils oriented in different directions from each other, a condition also reported to occur in human hearts with idiopathic hypertrophic subaortic stenosis (35).

Kawamura also reported finding capillaries incorporated into "tunnels" within enlarged cardiac muscle cells, an apparent attempt of the enlarged muscle fiber to reduce diffusion distance from the capillary to the contractile tissue.

5. Pathogenesis of Hypertrophy

The pathogenesis of the cardiac hypertrophy is unknown, but appears to be unrelated to the hypertension. Sen et al. (123) prevented the hypertension in SHR by treatment with hydralazine or α-methyldopa added to the drinking water. Heart weight was not different in the hydralazine-treated animals from untreated SHR. Heart weight to body weight ratio of α-methyldopa-treated SHR was reduced from untreated SHR; however, α-methyldopa-treated normotensive control rats also had a decreased heart weight–body weight ratio. These find-

Fig. 3. Electron micrograph of cardiac myocyte with normal structure from an 11-week-old SHR. The intercalated disc (ICD) and other cell organelles have normal structure in SHR less than 11 weeks of age. Note uniform alignment of sarcomeres and Z lines. C, capillary; M, mitochondria; Z, Z line. Scale, 1 μm. [From Kawamura et al. (65), by permission.]

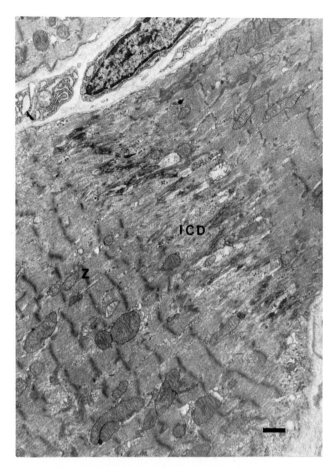

Fig. 4. Electron micrograph of heart muscle fiber from a 15-week-old SHR with marked increased folding of the intercalated disc (ICD). There is distortion and misalignment of the sarcomeres adjacent to the intercalated disc. Z, Z-line. Scale, 1 μm. [From Kawamura *et al.* (65), by permission.]

ings suggest α-methyldopa may have an effect on growth independent of its effect on blood pressure. Treatment of SHR with nerve growth factor antiserum produced peripheral sympathectomy and prevented the development of hypertension, but not cardiac hypertrophy (31). It has been proposed that myocardial hypertrophy in SHR may develop as a result of a genetic cardiovascular abnormality that does not require increased systemic pressure for its expression (31).

6. Renin

The role of renin in the development of increased blood pressure and cardiac hypertrophy in SHR is unclear. Renin activity is reportedly increased in the young SHR coinciding with increased systemic pressure, then becomes suppressed once the hypertension has stabilized (122). Hydralazine-treated rats had increased plasma renin activity, normal blood pressure, and persistence of cardiac hypertrophy. α-Methyldopa

treatment, however, reduced plasma renin activity as well as blood pressure (123). In nerve growth factor antiserum-treated rats in which hypertension but not hypertrophy was prevented, plasma renin activity was the same as in controls (31). Okamoto has stated that his laboratory has found no evidence for a role of renin in the pathogenesis of increased blood pressure in SHR (93). Further study is needed to fully resolve this issue.

7. Substrains of SHR

Two important substrains of SHR have recently been developed: the stroke-prone SHR (94, 97) and the obese SHR (68). Stroke-prone SHR were developed by breeding animals from litters with a high incidence of cerebral hemorrhages. In this substrain developed by Okamoto (97), 80% of male rats over 100 days of age and 60% of females over 150 days of age develop cerebral hemorrhages. They develop severe hypertension of about 240 mm Hg. Cerebral hemorrhages are most common in the left occipital area of the cortex and subcortex. Hemorrhages also occur in frontal and medial cortical areas and the basal ganglia.

The obese SHR substrain was developed from litters in which some animals were obese (68). By selective breeding of heterozygous parents, approximately 25% of the offspring were obese. The obese trait appears to be inherited as an autosomal recessive gene, expressed only in the homozygous recessive.

Obese SHR have a marked increase in food consumption and a more rapid body weight gain than nonobese SHR. Fat is deposited in subcutaneous tissues, retroperitoneally, and in the mesentery, while the face and paws are spared. The rats take on a rounded body shape and may weigh up to 1000 gm. They have a high incidence of lipid-containing atherosclerotic plaques in the aorta, coronary arteries, and other vessels. There is marked hyperlipemia, detected as early as 6 weeks of age. There are marked increases in both serum cholesterol and, especially, triglycerides compared to control nonobese SHR.

B. New Zealand Strain of Hypertensive Rats

Another strain of rats derived from the Wistar line has been developed which have blood pressure significantly greater than control animals (106, 107). Blood pressure is reported to be about 170 mm Hg by 60 days of age, somewhat lower than found in SHR. By 45 days of age, cardiac hypertrophy is present. Immunosympathectomy with nerve growth factor antiserum reduces the degree of hypertension in this strain as in SHR (28). There is evidence in this strain of rats of a polygenic mode of inheritance for hypertension (108) as has also been shown for SHR (77).

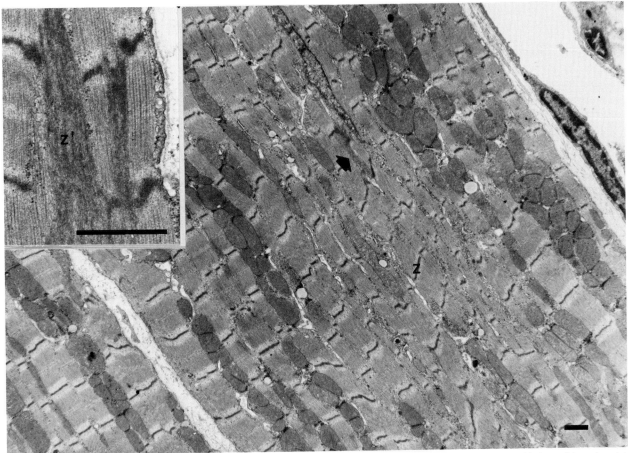

Fig. 5. Longitudinal section of cardiac myocytes from a 15-week-old SHR. There is distortion of the Z bands (Z) with loss of lateral register of sarcomeres. There are diffuse expansions of dense Z band material (arrow). Scale, 1 μm. Inset shows higher magnification of expanded Z band material (Z') present in SHR with cardiac hypertrophy. Scale, 1 μm. [From Kawamura *et al.* (65), by permission.]

C.   Hypertension in Rats with Hereditary Hypothalamic Diabetes Insipidus

The Brattleboro strain of rats with hereditary hypothalamic diabetes insipidus has been reported to develop blood pressure in excess of 150 mm Hg (48). Hypertension occurred in unilaterally nephrectomized rats given either distilled water or 0.6% NaCl as the sole drinking fluid, but did not occur in unilaterally nephrectomized normal rats (Long-Evans strain) given saline or distilled water to drink. Blood pressure returned to normal values when the rats were treated with vasopressin. Heart rate was increased in the diabetes insipidus rats compared to normal rats, and heart weight to body weight ratio was increased.

D.   Salt Sensitive (S) and Salt Resistant (R) Rats of Dahl

Dahl and co-workers (31b, 31c) have developed two lines of rats by selective inbreeding which are either resistant (R) or sensitive (S) to the effects of a high (8%) salt diet in regard to

development of hypertension. Neither of these lines has hypertension if salt is not included in the diet, but the S rats develop hypertension in excess of 180 mm Hg when fed the high salt diet. Although these lines have been maintained since 1962, the lines do not breed true and must be maintained by continued selection for or against the development of high blood pressure when fed the high salt diet. Selective breeding studies have demonstrated a genetic basis for salt sensitivity in the Dahl rats (67a), and the model is a useful one for studies involving both genetic and environmental factors. Different genetic loci apparently are involved in the Dahl rats than in the SH rats of Okamoto. The strain-specific aryl esterase isoenzymes found in the SHR (152) were not present in either the S or R lines of Dahl rats (115).

E.   Spontaneous Arteriosclerosis in Repeatedly Bred Rats

Wexler has described a type of arteriosclerotic lesion in the aorta and its major branches in several strains of repeatedly bred rats (142–144). Females are affected more severely and frequently than males. The severity of the lesion is directly correlated with the number and frequency of breedings and is more severe in rats which have suckled large litters with abrupt weaning at 23 days age (62). Rats with smaller litters, prolonged lactation with natural weaning, or longer rest periods between litters had less severe arteriosclerosis.

Lesions are identified grossly in about 40% of female rats with four or more litters, and microscopic lesions are found in 80–90% of aortas. Gross lesions are not found in virgin rats of the same age. Male breeder rats seldom exhibit macroscopic lesions, but microscopic lesions are present in 70–80% of aortas. The lesions occur first in the abdominal aorta followed by the aortic arch and the thoracic aorta. In the more severe cases, lesions are found in the carotid and peripheral arteries. Focal ectasia and saccular aneurysms occur in the aortic arch.

Histologically, the lesions first appear as swelling of the media with acid mucopolysaccharide material. There is cartilaginous metaplasia, occasionally a few foam cells, but seldom any lipid material. Mineralization of the lesion occurs with bone formation in the aortic arch and innominate and carotid arteries. There is associated endothelial cell proliferation and fibrosis of the intima, particularly in the peripheral arteries. The internal elastic membrane is often disrupted. Ulceration and thrombosis are rare.

Although this lesion in the rat represents an arteriosclerotic change, there is little resemblance to human arteriosclerosis. The lack of lipid, presence of cartilagenous and osseous metaplasia, and higher incidence in females than in males are all distinctly dissimilar to the human lesion.

F. Polyarteritis Nodosa of Mesenteric Arteries

Sclerosis of the mesenteric arteries has been described in August red-hooded rats, occurring with greater incidence in this strain than in Wistar or Long-Evans rats (99). The lesion is found in approximately 45% of rats over 400 days of age, occurring with increasing frequency in older animals of the August strain. Similar lesions were rare in animals of similar ages of the other two strains, suggesting a genetic influence in the development of the lesion.

The mesenteric vessels are characterized by replacement of the media with fibrous connective tissue, dilation of the vessel, tortuosity, and, occasionally, aneurysmal dilation. Thrombosis may occur.

A similar lesion has been described by Skold (126) as panarteritis in approximately 10% of Wistar and Sprague-Dawley rats over 500 days old. The mesenteric arteries are principally affected, but testicular arteries are also involved. The lesion is characterized by adventitial proliferation and fibrosis of the media. There is a dense fibrinoid material beneath the endothelium which is replaced by collagen in later stages.

G. Miscellaneous Spontaneous Disease of the Cardiovascular System in the Rat

There are a number of naturally occurring conditions affecting the cardiovascular system of the rat in addition to those described in the preceding sections. However, most occur sporadically and thus are of limited value for use in cardiovascular research. Focal left ventricular endocardial fibroplasia of unknown origin has been described in the rat (20).

Trypanosoma cruzi myocarditis was encountered in 50% of a colony of rats studied in Panama (34); this could serve as a useful model of infectious myocarditis.

VII. EXPERIMENTALLY INDUCED MODELS FOR CARDIOVASCULAR RESEARCH

A. Systemic Hypertension

Systemic hypertension is produced in rats by a variety of techniques, the most common of which are related to stimulation of the renin–angiotensin system by manipulation of blood flow to the kidneys. The most commonly used systems employ a modification of the classic model originally described by Goldblatt (39). In the two-kidney Goldblatt model, partial occlusion of one renal artery is produced and the other kidney left intact. In the one-kidney Goldblatt model, the renal artery to one kidney is stenosed and the second kidney is removed. Other variations of this model include removal of one kidney with cellophane wrapping of the other, and complete occlusion of the aorta between the renal arteries producing ischemic necrosis of one kidney and increased blood flow to the other (75, 150). Acute hypertension has been produced by direct infusion of angiotensin (145). Systemic arterial pressure usually exceeds 150 mm Hg with these models, and in some, exceeds 200 mm Hg.

A frequently used model for induction of systemic hypertension is the uninephrectomized rat treated with desoxycorticosterone acetate (DOCA) and saline for drinking water. One kidney is removed at 4 to 6 weeks of age and weekly subcutaneous injections of 10 mg DOCA are administered (47a). Blood pressure gradually increases and by 3 weeks stabilizes at approximately 200 mm Hg. Heart weight to body weight ratios are increased compared to nontreated controls.

Banding of the abdominal aorta below the diaphragm has

also been used to produce hypertension proximal to the partial obstruction (10, 11, 91a).

B. Pulmonary Hypertension

Pulmonary hypertension with resulting right ventricular hypertrophy has been produced in rats by administration of *Crotalaria spectabilis* seeds or extracted monocrotaline pyrrole and by chronic or intermittent exposure to hypobaric pressure.

1. Crotalaria

Severe and gradually increasing pulmonary hypertension is produced by feeding ground *Crotalaria spectabilis* seeds (66, 134) or by a single intravenous injection of the extracted monocrotaline pyrrole (27). The pulmonary vessels become thickened by medial hyperplasia and intimal fibrosis, leading to increased pulmonary vascular resistance. Injection of 4 mg/kg monocrotaline pyrrole intravenously produced peak systolic right ventricular pressure of 91 mm Hg, no change in aortic pressure, increased hemoglobin and hematocrit, and two- to threefold increase in right ventricular free wall weight compared to control rats. Animals died with congestive heart failure.

2. High Altitude

Chronic exposure to high altitude (3000–4000 m or higher) or simulated high altitude in a hypobaric chamber results in pulmonary hypertension mediated by pulmonary vasoconstriction and the development of right ventricular hypertrophy. In one study, rats maintained for 5 weeks at a simulated altitude of 5500 m nearly doubled right ventricular weight from control weight of 216 ± 10 to 426 ± 22 mg (154). In many reported studies, however, severe weight loss during chronic exposure to hypobaric conditions has caused difficulty in interpretation of data. This technique has been used by many investigators to study various aspects of the cardiopulmonary system subjected to pulmonary hypertension (14, 79, 83).

C. Left Ventricular Hypertrophy

A variety of stresses, generally resulting in an increased workload on the myocardium, have been used to produce an increase in mass of the left ventricle. The major methods used in the rat are summarized below.

1. Swimming

One of the most common methods of producing an increased work load on the heart is to subject the rats to swimming exercise several hours per day for a period of several weeks (76, 130, 131). If the animals are forced to keep swimming and not rest on the sides or bottom of the tank, an approximate 10–20% increase in heart weight is produced after several weeks of exercise. Rats have also been forced to exercise by running on a treadmill for prolonged periods (133, 138). Both of these methods, however, require a great deal of time and produce only a mild degree of hypertrophy. In addition, stress factors associated with forced swimming or running may produce effects on the animal other than those due to exercise alone.

2. Banding of the Aorta

Pressure overload, or increased afterload on the left ventricle has been produced by banding of the aorta. Beznak *et al.* (10, 11) introduced this method by placing a suture on the aorta below the diaphragm. The degree of obstruction was controlled by placing a wire of known gauge against the aorta while making the tie. The wire was then removed, leaving a lumen the size of the wire. While the method is easily performed without the necessity of thoracotomy, only mild cardiac hypertrophy is produced, often less than 10% increase in mass, and there is increased systemic pressure to the cranial portion of the animal.

Banding of the ascending aorta has been developed by Nair *et al.* (91) to produce increased pressure load on the heart without increased pressure to the rest of the body. A silver clip is placed on the ascending aorta with specially designed forceps which allows constriction of the aorta to the maximum degree tolerated. The procedure is rapidly completed through a small thoracotomy. There is a modest decrease in body weight gain for several days followed by normal weight gain. By 4–7 days after the procedure, there is a 10–40% increase in heart weight. This procedure has achieved wide acceptance in many laboratories.

3. Volume Overload—Anemia

Volume overload of the heart occurs when the heart is required to pump a larger than normal amount of blood in order to meet the requirements of the body. Surgical preparations creating AV shunts have been used in the rat, but are technically difficult and not easily quantified. A more satisfactory procedure has been to produce iron deficiency anemia in weanling rats (69, 71). After the weaning, rats are fed a milk and sugar diet which is deficient in iron and copper. Hemoglobin concentration drops to 3.5–5.0 gm/dl blood while control rats with iron and copper added to the milk and sugar diet maintain hemoglobin of 13–16 gm/dl blood. Heart weights are increased by more than 100% by 11 to 19 weeks; both left and right ventricles are hypertrophied.

4. Cardiac Hypertrophy due to Carbon Monoxide Exposure

Exposure of neonatal rats to 500 ppm carbon monoxide has been shown to result in approximately a 70% increase in heart weight after 35 days of exposure (103). In this model approximately 40% of the hemoglobin is in the form of carboxyhemoglobin, and hemoglobin values are increased to 17 gm/dl compared to 13 gm/dl in nonexposed controls. While the mechanism for production of cardiac hypertrophy in carbon monoxide toxicity is not clear, it is probably related to increased blood viscosity with secondarily increased vascular resistance and to the increased volume of blood which must be pumped by the heart due to its decreased oxygen carrying capacity. This appears to be an excellent model for induction of cardiac hypertrophy in young animals.

5. Thyroxine

Cardiac hypertrophy of approximately 38% increase in total heart weight may be produced by daily administration of thyroxine (T_4) subcutaneously over a 10-week period (12, 13, 118). There is complete regression of the thyroxine-induced cardiac hypertrophy by 2 weeks after cessation of thyroxine administration (13, 137).

6. Isoproterenol

Cardiac hypertrophy has also been produced in the rat by administration of low, subnecrotizing doses of isoproterenol to the rat (117, 129, 151). Administration of large doses of isoproterenol, 80–100 mg/kg, results in myocardial necrosis, but repeated administration of 5 mg/kg body weight results in stimulation of RNA synthesis in the myocardium and development of cardiac hypertrophy (40).

D. Atherosclerosis

The normotensive rat is very resistant to development of severe atheromatous lesions when placed on high lipid-containing diets. Arterial lesions containing extensive lipid deposits are difficult to produce, and complicated lesions with hemorrhage, thrombosis, and resulting myocardial infarction are very rare. Therefore, the rat is seldom used for studies of atherosclerosis.

A notable exception to the resistance of the rat to development of atherosclerosis is found in the spontaneously hypertensive rat. Okamoto and co-workers (96, 97) have demonstrated that certain substrains of the SHR are particularly susceptible to development of lipid-containing lesions in the arteries and show a pronounced susceptability to cholesterol-containing di-

ets. These workers have concluded that certain substrains have genetic differences in addition to the hypertension which account for the susceptibility to development of atheromatous lesions.

E. Ischemic Heart Disease

The rat has been used by a number of investigators to study ischemic heart disease induced by ligation of a major coronary artery (4, 22, 64, 78). The left coronary artery is accessible through a left thoracotomy for ligation (61, 121). As discussed above (Section II, E), the left coronary artery does not usually have a large circumflex branch. Ligation of the proximal left coronary will result in 20–60% necrosis of the left ventricle. The procedure is usually reported to have a 40% or greater mortality, although Bajusz (4) reported no mortality following occlusion of the left coronary artery 1 mm below its origin.

F. Endocarditis

Bacterial endocarditis may be produced in the rat experimentally by introduction of bacteria into the animal either as the sole agent, or in combination with some other stress, such as cold exposure or tumbling. Under these conditions, bacteria will localize on the mitral and aortic valves, producing a fibrinothrombotic lesion resembling spontaneous bacterial endocarditis in man (29).

Nonbacterial thrombotic endocarditis has been extensively studied in the rat by Angrist and co-workers (2, 92). The condition is induced by stress situations, such as cold, trauma, or corticosteroid administration, without concomitant introduction of pathogenic bacteria.

VIII. USE OF THE RAT FOR MISCELLANEOUS STUDIES

Embryologic and fetal development of the vascular system has been extensively studied in the rat, including a variety of morphologic, biochemical, and functional studies. The fetal rat is an ideal species for studies of teratogenic effects of drugs.

A major application of the rat has been its use in determining cardiotoxicity of a variety of toxic agents. Notable examples include such agents as rapeseed oil and other brominated oils (6, 90), cobalt, the toxic agent in beer drinkers cardiomyopathy (42, 58), isoproterenol (30, 119), and various combinations of electrolytes and steroids (57, 82).

IX. SPECIAL CONSIDERATIONS FOR USE OF THE RAT IN CARDIOVASCULAR RESEARCH

A. Anatomic Differences

There are several anatomic features of the rat heart which differ from the human heart or larger species of animals commonly used for cardiovascular research. These are described in more detail in Section II and only mentioned here. The rat normally has a left superior vena cava in addition to the usual right superior vena cava found in most larger mammals (the rabbit also normally has both left and right superior venae cavae). Although because of its small size, the rat is seldom used for hemodynamic studies or measurements of blood flow within the heart, it should be kept in mind that coronary sinus blood is composed of blood draining from the left superior vena cava as well as from the coronary vein.

The rat heart has a dual blood supply, the atria including the sinus node region, receiving its blood supply from the cardiacomediastinal arteries which arise from the internal mammary arteries. This fact would have special significance in certain types of preparations; for example, the isolated heart perfused retrogradely through the aorta presumably would not have the sinus node region directly perfused.

Coronary artery distribution is different in the rat compared to higher mammals. The main branches of the right and left coronary arteries descend over the free wall portions of the respective ventricles rather than follow the atrioventricular grooves or the anterior interventricular sulcus as in higher mammals.

The male rat continues to increase in body weight throughout most of its normal life span, and the heart weight increases at a somewhat lesser rate than body weight, resulting in a constantly decreasing heart weight to body weight ratio. This is particularly important when conducting studies related to changes in heart weight, since heart weight to body weight ratios from animals of different body weight cannot be compared without reference to a set of normal values for animals of various weights and from the same colony.

B. Electrocardiography and Electrophysiology

Evaluation of the electrocardiogram of the rat is difficult due to the very rapid heart rate (350–450 beats/min), and poorly defined wave form of the body surface ECG. The S–T segment is markedly slurred in the rat ECG with poorly defined S and T waves, making interpretation of changes in these waves difficult.

The action potential recorded from rat myocardial cells is well known to have different characteristics from that of other mammals. The membrane action potential has essentially no plateau as in other mammals. Further, it has been demonstrated that stretch-induced excitation is different from other mammals, behaving in a graded and asynchronous manner (128). Stretch appears to induce a partial conduction block in isolated rat papillary muscle producing delayed activation, decremental conduction, subthreshold electrotonic potentials, and other electrical and mechanical abnormalities not found in papillary muscle preparations from other species.

C. Pharmacological Differences

It is well known that the rat is very resistant to digitalis glycosides, being many times more resistant to its effects than the dog or man. Detweiler (33) has reviewed the comparative pharmacology of the cardiac glycosides and notes a variety of differences between the rat and other species in the handling of digitalis glycosides. The rat, for example, has a very high level of Na^+, K^+ ATPase compared to other species, which has been postulated as a major reason for the species difference in glycoside action (116).

X. CONCLUSIONS

The rat has been extensively used for a large variety of cardiovascular studies in spite of many peculiarities of this species compared to other mammals. In selecting an animal species for use in particular research problem, the investigator must carefully weigh the advantages and disadvantages of the species for the particular type of research. The economic advantages or availability of background data for the species must be weighed against the possible disadvantages of differences in anatomy, function, or metabolism compared to other species.

The rat has been particularly useful for studies of myocardial structure and metabolism. The response of the myocardial cell to a variety of physiologic and pathologic stimuli has been studied in both the *in vivo* and *in vitro* situation, and thus provided basic information on the response of the myocardial cell to such stimuli. Another particularly promising area of research using the rat is the study of hypertension, an area which has been greatly facilitated by the development of the spontaneously hypertensive rat and the various substrains of SHR. Major areas in which the rat has been less useful include electrophysiology of the myocardium and atherosclerosis. Selection of the rat as a species for cardiovascular research, therefore, must be based on consideration of the peculiarities of this species and how they will affect the realization of the objectives of the particular study.

ACKNOWLEDGMENT

Grateful recognition is given to Keishiro Kawamura, Kyoto University Faculty of Medicine, Kyoto, Japan, for generous assistance in preparation of the material on spontaneously hypertensive rats and the illustrations used in Figs. 3–5.

REFERENCES

1. Angelakos, E. T., Bernardini, P., and Barrett, W. C., Jr. (1964). Myocardial fiber size and capillary–fiber ratio in the right and left ventricles of the rat. *Anat. Rec.* **149**, 671–676.

2. Angrist, A. A., Oka, M., Nakao, K., and Marquiss, J. (1960). Studies in experimental endocarditis. I. Production of valvular lesions by mechanisms not involving infection or sensitivity factors. *Am. J. Pathol.* **36**, 181–200.

3. Anversa, P., Loud, A. V., and Vitali-Mazza, L. (1976). Morphometry and autoradiography of early hypertrophic changes in the ventricular myocardium of adult rat. An electron microscopic study. *Lab. Invest.* **35**, 475–483.

4. Bajusz, E., and Jasmin, G. (1964). Histochemical studies on the myocardium following experimental interference with coronary circulation in the rat. I. Occlusion of coronary artery. *Acta Histochem.* **18**, 222–237.

5. Bartosova, D., Chvapil, M., Korecky, B., Poupa, O., Rakusan, R., Turek, Z., and Visek, M. (1969). The growth of the muscular and collagenous parts of the rat heart in various forms of cardiomegaly. *J. Physiol. (London)* **200**, 285–295.

6. Beare-Rogers, J. L., and Nera, E. A. (1972). Cardiac fatty acids and histopathology of rats, pigs, monkeys, and gerbils fed rapeseed oil. *Comp. Biochem. Physiol. B* **41**, 793–800.

7. Bencosme, S. A., and Berger, J. M. (1971). Specific granules in mammalian and non-mammalian vertebrate cardiocytes. *Methods Achie. Exp. Pathol.* **5**, 173–213.

8. Berger, J. M., and Bencosme, S. A. (1971). Fine structural cytochemistry of granules in atrial cardiocytes. *J. Mol. Cell. Cardiol.* **3**, 111–120.

9. Berry, M. N., Friend, D. S., and Scheuer, J. (1970). Morphology and metabolism of intact muscle cells isolated from adult rat heart. *Circ. Res.* **26**, 679–687.

10. Beznak, M. (1952). The effect of the pituitary and growth hormone on the blood pressure and on the ability of the heart to hypertrophy. *J. Physiol. (London)* **116**, 74–83.

11. Beznak, M. (1955). The effect of different degrees of subdiaphragmatic aortic constriction on heart weight and blood pressure of normal and hypophysectomized rats. *Can. J. Biochem. Physiol.* **33**, 985–994.

12. Beznak, M. (1962). Cardiovascular effects of thyroxine treatment in normal rats. *Can. J. Biochem. Physiol.* **40**, 1647–1654.

13. Beznak, M., Korecky, B., and Thomas, G. (1969). Regression of cardiac hypertrophies of various origin. *Can. J. Physiol. Pharmacol.* **47**, 579–586.

14. Bischoff, M. B., Dean, W. D., Bucci, T. J., and Frics, L. A. (1969). Ultrastructural changes in myocardium of animals after five months residence at 14,100 feet. *Fed. Proc., Fed. Am. Soc. Exp. Biol.* **28**, 1268–1673.

15. Bishop, S. P., and Cole, C. R. (1969). Ultrastructural changes in canine myocardium with right ventricular hypertrophy and congestive heart failure. *Lab. Invest.* **20**, 219–229.

16. Bishop, S. P., and Drummond, J. L. (1979). Surface morphology and cell size measurement of isolated rat cardiac myocytes. *J. Mol. Cell. Cardiol.* **11**, 423–433.

16a. Bishop, S. P., Oparil, S., Reynolds, R. H., and Drummond, J. L. (1979). Regional myocyte size in normotensive and spontaneously hypertensive rats. *Hypertension* **1**, 378–383.

17. Bloom, S. (1970). Spontaneous rhythmic contraction of separated heart muscle cells. *Science* **167**, 1727–1729.

18. Bloor, C. M., Leon, A. S., and Pitt, B. (1967). The inheritance of coronary artery anatomic patterns in rats. *Circulation* **36**, 771–776.

19. Bompiani, G. D., Rouiller, C., and Hatt, P. Y. (1959). Le tissu de conduction de coeur chez le rat. Etude au microscope électronique. I. Le tranc commun du faisceau de His et les cellules claires de l'oreillette droite. *Arch. Mal. Coeur Vaiss.* **52**, 1257–1274.

20. Boorman, G. A., Zurcher, C., Hollander, C. F., and Feron, V. J. (1973). Naturally occurring endocardial disease in the rat. *Arch. Pathol.* **96**, 39–45.

21. Brown, J. W., Cristian, D., and Paradise, R. R. (1968). Histological effects of procedural and environmental factors on isolated rat heart preparations. *Proc. Soc. Exp. Biol. Med.* **129**, 455–462.

22. Bryant, R. E., Thomas, W. A., and O'Neal, R. M. (1958). An electron microscopic study of myocardial ischemia in the rat. *Circ. Res.* **6**, 699–709.

23. Bunag, R. D. (1973). Validation in awake rats of a tail-cuff method for measuring systolic pressure. *J. Appl. Physiol.* **34**, 279–282.

24. Carney, J. A., and Brown, A. L. (1964). Myofilament diameter in the normal and hypertrophic rat myocardium. *Am. J. Pathol.* **44**, 521–530.

25. Chang, W. W. L., and Bencosme, S. A. (1969). Quantitative electron microscopic analysis of the specific granule population of rat atrium. *Can. J. Physiol. Pharmacol.* **47**, 483–485.

26. Chanutin, A., and Barksdale, E. E. (1933). Experimental renal insufficiency produced by partial nephrectomy. *Arch. Intern. Med.* **52**, 739–751.

27. Chesney, C. F., Allen, J. R., and Hsu, I. C. (1974). Right ventricular hypertrophy in monocrotaline pyrrole treated rats. *Exp. Mol. Pathol.* **20**, 257–268.

28. Clark, D. W. J. (1971). Effects of immunosympathectomy on development of high blood pressure in genetically hypertensive rats. *Circ. Res.* **28**, 330–336.

28a. Clark, M. G., Gannon, B. J., Bodkin, N., Patten, G. S., and Berry, M. N. (1978). An improved procedure for the high-yield preparation of intact beating heart cells from the adult rat. Biochemical and morphologic study. *J. Mol. Cell. Cardiol.* **10**, 1101–1121.

29. Clawson, B. J. (1950). Experimental endocarditis with fibrinoid degeneration in the heart valves of rats. *Arch. Pathol.* **50**, 68–74.

30. Csapo, Z., Dusek, J., and Rona, G. (1972). Early alterations of the cardiac muscle cells in isoproterenol-induced necrosis. *Arch. Pathol.* **93**, 356–365.

31. Cutilletta, A. F., Erinoff, L., Heller, A., Low, J., and Oparil, S. (1977). Development of left ventricular hypertrophy in young spontaneously hypertensive rats after peripheral sympathectomy. *Circ. Res.* **40**, 428–434.

31a. Cutilletta, A. F., Benjamin, M., Culpepper, W. S., and Oparil, S. (1978). Myocardial hypertrophy and ventricular performance in the absence of hypertension in spontaneously hypertensive rats. *J. Mol. Cell. Cardiol.* **10**, 689–703.

31b. Dahl, L. K., Heine, M., and Tassinari, L. (1962). Role of genetic factors in susceptibility to experimental hypertension due to chronic excess salt ingestion. *Nature (London)* **194**, 480–482.

31c. Dahl, L. K., Heine, M., and Tassinari, L. (1962). Effects of chronic excess salt ingestion. Evidence that genetic factors play an important role in susceptibility to experimental hypertension. *J. Exp. Med.* **115**, 1173–1190.

32. Datta, B. N., and Silver, M. D. (1975). Cardiomegaly in chronic anemia in rats. An experimental study including ultrastructural histometric, and stereologic observations. *Lab. Invest.* **32**, 503–514.

33. Detweiler, D. K. (1967). Comparative pharmacology of cardiac glycosides. *Fed. Proc., Fed. Am. Soc. Exp. Biol.* **26**, 1119–1124.

34. Edgcomb, J. H., Walker, D. H., and Johnson, C. M. (1973). Pathological features of *Trypanosoma cruzi* infection of Rattus rattus. *Arch. Pathol.* **96**, 36–38.

34a. Erinoff, L., Heller, A., and Oparil, S. (1975). Prevention of hypertension in the SH rat: Effects of differential central catecholamine depletion. *Proc. Soc. Exp. Biol. Med.* **150**, 748–754.

35. Ferrans, V. F., Morrow, A. G., and Roberts, W. C. (1972). Myocardial ultrastructure in idiopathic hypertrophic subaortic stenosis. A study of operatively excised left ventricular outflow tract muscle in 14 patients. *Circulation* **45**, 769–792.

36. Forssmann, W. G., and Girardier, L. (1970). A study of the T system in rat heart. *J. Cell Biol.* **44**, 1–19.

37. Geller, R. G., ed. (1976). "Spontaneous Hypertension: Its Pathogenesis and Complications." US Govt. Printing Office, Washington, D.C.

38. Glick, M. R., Burns, A. H., and Reddy, W. J. (1974). Dispersion and isolation of beating cells from adult rat heart. *Anal. Biochem.* **61**, 32–42.

39. Goldblatt, H., Lynch, J., Hanzal, R. F., and Summerville, W. W. (1934). Studies on experimental hypertension. I. The production of persistent elevation of systolic blood pressure by means of renal ischemia. *J. Exp. Med.* **59**, 347–379.

40. Gordon, A. L., Inchiosa, M. A., Jr., and Lehr, D. (1969). Myocardial actomyosin and total protein in isoproterenol-induced hypertrophy. *Bull. N.Y. Acad. Med.* [2] **45**, 98.

41. Greene, E. C. (1959). "Anatomy of the Rat." Hafner, New York.

42. Grice, H. C., Goodman, T., Munro, I. C., Wiberg, G. S., and Morrison, A. B. (1969). Myocardial toxicity of cobalt in the rat. *Ann. N.Y. Acad. Sci.* **156**, 189–194.

43. Grimm, A. F., de la Torre, L., and La Porta, M., Jr. (1970). Ventricular nuclei-DNA relationships with myocardial growth and hypertrophy in the rat. *Circ. Res.* **26**, 45–52.

44. Grimm, A. F., Katele, K. V., Klein, S. A., and Lin, H. (1973). Growth of the rat heart: Left ventricular morphology and sarcomere lengths. *Growth* **37**, 189–208.

45. Grimm, A. F., Kubota, R., and Whitehorn, W. V. (1963). Properties of myocardium in cardiomegaly. *Circ. Res.* **12**, 118–124.

46. Grove, D., Nair, K. G., and Zak, R. (1969). Biochemical correlates of cardiac hypertrophy: III. Changes in DNA content; the relative contributions of polyploidy and mitotic activity. *Circ. Res.* **25**, 463–471.

46a. Haeusler, G., Finch, L., and Thoenen, H. (1972). Central adrenergic neurones and the initiation and development of experimental hypertension. *Experientia* **28**, 1200–1203.

47. Halbert, S. P., Bruderer, R., and Lin, T. M. (1971). *In vitro* organization of dissociated rat cardiac cells into beating three-dimensional structures. *J. Exp. Med.* **133**, 677–695.

47a. Hall, C. E., and Hall, O. (1965). Hypertension and hypersalimentation. II. Desoxycorticosterone hypertension. *Lab. Invest.* **14**, 1727–1735.

48. Hall, C. E., Ayachi, S., and Hall, O. (1973). Spontaneous hypertension in rats with hereditary hypothalamic diabetes insipidus (Brattleboro strain). *Tex. Rep. Biol. Med.* **31**, 471–487.

49. Halpern, M. H. (1953). Extracoronary cardiac veins in the rat. *Am. J. Anat.* **92**, 307–328.

50. Halpern, M. H. (1955). The sinoatrial node of the rat heart. *Anat. Rec.* **123**, 425–436.

51. Halpern, M. H. (1957). The dual blood supply of the rat heart. *Am. J. Anat.* **101**, 1–16.

52. Harary, I., and Farley, B. (1960). *In vitro* organization of single beating rat heart cells into beating fibers. *Science* **132**, 1839–1840.

53. Harary, I., and Farley, B. (1963). *In vitro* studies on single beating rat heart cells. I. Growth and organization. *Exp. Cell Res.* **29**, 451–465.

54. Harary, I., and Sato, E. (1964). Studies *in vitro* on single beating heart cells. V. Changes in adenosine-triphosphate-induced contractions of extracted models. *Biochim. Biophys. Acta* **82**, 614–616.

55. Harrison, T. R., Ashman, R., and Larson, R. M. (1932). Congestive heart failure. XII. The relation between the thickness of the cardiac muscle fiber and the optimum rate of the heart. *Arch. Intern. Med.* **49**, 151–164.

56. Hatt, P. Y., Berjal, G., Moravec, J., and Swynghedauw, B. (1970). Heart failure: An electron microscopic study of the left ventricular papillary muscle in aortic insufficiency in the rabbit. *J. Mol. Cell. Cardiol.* **1**, 235–247.

57. Heggtveit, H. A., Herman, L., and Mishra, R. K. (1964). Cardiac necrosis and calcification in experimental magnesium deficiency. A light and electron microscope study. *Am. J. Pathol.* **45**, 757–782.

58. Heggtveit, H. A., Grice, H. C., and Wiberg, G. S. (1970). Cobalt cardiomyopathy: Experimental basis for the human lesion. *Pathol. Microbiol.* **35**, 110–113.

59. Hibbs, R. G., and Ferrans, V. J. (1969). An ultrastructural and histochemical study of rat myocardium. *Am. J. Anat.* **124**, 251–280.

60. Jamieson, J. D., and Palade, G. E. (1964). Specific granules in atrial muscle cells. *J. Cell Biol.* **23**, 151–172.

61. Johns, T. N. P., and Olson, B. J. (1954). Experimental myocardial infarction: I. A method of coronary occlusion in small animals. *Ann. Surg.* **140**, 675–682.

62. Judd, J. T., and Wexler, B. C. (1969). The role of lactation and weaning in the pathogenesis of arteriosclerosis in female breeder rats. *J. Atheroscler. Res.* **10**, 153–172.

63. Kako, K., and Minelli, R. (1969). Regulation of leucine incorporation into cardiac protein by work loads. *Experientia* **25**, 34–36.

64. Kaufman, N., Gavan, T. L., and Hill, R. W. (1959). Experimental myocardial infarction in the rat. *AMA Arch. Pathol.* **67**, 482–488.

65. Kawamura, K., Kashii, C., and Imamura, K. (1976). Ultrastructural changes in hypertrophied myocardium of spontaneously hypertensive rats. *Jpn. Circ. J.* **40**, 1119–1145.

66. Kay, J. M., Smith, P., and Heath, D. (1969). Electron microscopy of *crotalaria* pulmonary hypertension. *Thorax* **24**, 511–526.

67. King, T. S. (1954). Sinuatrial node in rat. *Am. Heart J.* **48**, 785–786.

67a. Knudsen, K., Dahl, L. K., Thompson, K., Iwai, J., Heine, M., and Leitl, G. (1970). Effects of chronic salt ingestion. Inheritance of hypertension in the rat. *J. Exp. Med.* **132**, 967–1000.

68. Koletsky, S. (1973). Obese spontaneously hypertensive rats—A model for study of atherosclerosis. *Exp. Mol. Pathol.* **19**, 53–60.

69. Korecky, B., and French, I. W. (1967). Nucleic acid synthesis in enlarged hearts of rats with nutritional anemia. *Circ. Res.* **21**, 635–640.

70. Korecky, B., Beznak, M., and Korecka, M. (1966). Cardiac performance in heart-lung preparations of rats with experimental cardiac hypertrophy. *Can. J. Physiol. Pharmacol.* **44**, 21–27.

71. Korecky, B., Rakusan, K., and Paupa, O. (1964). The effect of anaemia due to iron deficinecy during early postnatal development of the rat on growth and body composition later in life. *Physiol. Bohemoslov.* **13**, 72–77.

72. Kuramitsu, H., and Harary, I. (1964). Studies *in vitro* on single beating rat-heart cells. III. Enzyme changes and loss of specific function in

culture. *Biochim. Biophys. Acta* **86**, 65–73.

73. Laffan, R. J., Peterson, A., Hitch, S. W., and Jeunelot, C. (1972). A technique for prolonged, continuous recording of blood pressure of unrestrained rats. *Cardiovasc. Res.* **6**, 319–324.

74. Langendorff, O. (1895). Untersuchungen am uberlebender Saugetierherzen. *Pfluegers Arch. Gesamte Physiol. Menschen Tiere* **61**, 291–332.

75. Latta, H., White, F. N., Osvaldo, L., and Johnston, W. H. (1975). Unilateral renovascular hypertension in rats: Measurements of medullary granules, juxtaglomerular granularity and cellularity, and areas of adrenal zones. *Lab. Invest.* **33**, 379–390.

76. Leon, A. S., and Bloor, C. M. (1968). Effects of exercise and its cessation on the heart and its blood supply. *J. Appl. Physiol.* **24**, 485–490.

77. Louis, W. J., Tabei, R., Sjoerdsma, A., and Spector, S. (1969). Inheritance of high blood pressure in the spontaneously hypertensive rat. *Lancet* **1**, 1035–1036.

78. MacLean, D., Fishbein, M. C., Maroko, P. R., and Braunwald, E. (1975). Hyaluronidase-induced reductions in myocardial infarct size. *Science* **194**, 199–200.

79. Mager, M., Blatt, W. F., Natale, P. J., and Blattei, C. M. (1968). Effects of high altitude on lactic dehydrogenase isozymes of neonatal and adult rats. *Am. J. Physiol.* **215**, 8–13.

80. Maistrello, I., and Matscher, R. (1969). Measurement of systolic blood pressure of rats: Comparison of intraarterial and cuff values. *J. Appl. Physiol.* **26**, 188–193.

81. Mark, G. E., and Strasser, F. F. (1966). Pacemaker activity and mitosis in culture of newborn rat heart ventricle cells. *Exp. Cell Res.* **44**, 217–233.

82. Maurat, J. P., Mercier, J. N., Ledoux, C., and Hatt, P. Y. (1965). Le myocarde dans les dépletions expérimentales en potassium chez le rat. Etude au microscope électronique. *Arch. Mal. Coeur Vaiss.* **7**, 1004–1021.

83. McGrath, J. J., and Bullard, R. W. (1968). Altered myocardial performance in response to anoxia after high altitude exposure. *J. Appl. Physiol.* **25**, 761–764.

84. Melax, H., and Leeson, T. S. (1970). Fine structure of the impulse-conducting system in rat heart. *Can. J. Zool.* **48**, 837–839.

85. Minelli, R., and Casella, C. (1967). Influence of load and work on the high energy phosphates content in the myocardium (rat heart–lung preparation). *Pfluegers Arch. Gestamte Physiol. Menschen Tiere* **295**, 119–126.

86. Morgan, H. E., Henderson, M. J., Regen, D. M., and Park, C. R. (1961). Regulation of glucose uptake in muscle. I. The effects of insulin and anoxia on glucose transport and phosphorylation in the isolated, perfused heart of normal rats. *J. Biol. Chem.* **236**, 253–261.

87. Morkin, E., and Ashford, T. P. (1968). Myocardial DNA synthesis in experimental cardiac hypertrophy. *Am. J. Physiol.* **215**, 1409–1413.

88. Moustafa, E., Skomedal, T., Osnes, J., and Oye, I. (1976). Cyclic AMP formation and morphology of myocardial cells isolated from adult heart: Effect of Ca^{2+} and Mg^{2+}. *Biochim. Biophys. Acta* **421**, 411–415.

89. Muir, A. R. (1965). Further observations on the cellular structure of cardiac muscle. *J. Anat.* **99**, 27–46.

90. Munro, I. C., Hasnain, S., Salem, F. A., Goodman, T., Grice, H. C., and Heggtveit, H. A. (1972). Cardiotoxicity of brominated vegetable oils. *Recent Adv. Stud. Cardiac Struct. Metab.* **1**, 588–595.

91. Nair, K. G., Cutilletta, A. F., Zak, R., Koide, T., and Rabinowitz, M. (1968). Biochemical correlates of cardiac hypertrophy. I. Experimental model; changes in heart weight, RNA content, and nuclear RNA polymerase activity. *Circ. Res.* **23**, 451–462.

91a. Nolla-Panades, J. (1963). Hypertension and increased hindlimb vascular reactivity in experimental coarctation of the aorta. *Circ. Res.* **12**, 3–9.

92. Oka, M., Girerd, R. J., Brodie, S. S., and Angrist, A. (1966). Cardiac valve and aortic lesions in beta-aminopropionitrile fed rats with and without high salt. *Am. J. Pathol.* **48**, 45–64.

93. Okamoto, K. (1969). Spontaneous hypertension in rats. *Int. Rev. Exp. Pathol.* **7**, 227–270.

94. Okamoto, K. (1972). Establishment of the stroke-prone spontaneously hypertensive rat (SHR). *Circ. Res., Suppl.* **1**, I-143–I-153.

95. Okamoto, K. (1972). "Spontaneous Hypertension. Its Pathogenesis and Complications." Igaku Shoin Ltd., Tokyo.

96. Okamoto, K., and Aoki, K. (1963). Development of a strain of spontaneously hypertensive rats. *Jpn. Circ. J.* **27**, 282–293.

97. Okamoto, K., Yamori, Y., Ooshima, A., and Tanaka, T. (1972). Development of substrains in spontaneously hypertensive rats: Geneology, isozymes and effect of hypercholesterolemic diet. *Jpn. Circ. J.* **36**, 461–470.

98. Ooshima, A., Yamori, Y., and Okamoto, K. (1972). Cardiovascular lesions in the selectively-bred group of spontaneously hypertensive rats with severe hypertension. *Jpn. Circ. J.* **36**, 797–812.

99. Opie, E. L., Lynch, C. J., and Tershakovec, M. (1970). Sclerosis of the mesenteric arteries of rats. Its relation to longevity and inheritance. *Arch. Pathol.* **89**, 306–313.

100. Osborne, B. E. (1973). A restriaining device facilitating electrocardiogram recording in rats. *Lab. Anim.* **7**, 185–188.

101. Page, E., and McCallister, L. P. (1973). Quantitative electron microscopic description of heart muscle cells. Application to normal, hypertrophied and thyroxine-stimulated hearts. *Am. J. Cardiol.* **31**, 172–181.

102. Page, E., Earley, J., and Power, B. (1974). Normal growth of ultrastructures in rat left ventricular myocardial cells. *Circ. Res.* **34** and **35**, II-12–16.

103. Penney, D., Dunham, E., and Benjamin, M. (1974). Chronic carbon monoxide exposure: Time course of hemoglobin, heart weight and lactate dehydrogenase isozyme changes. *Toxicol. Appl. Pharmacol.* **28**, 493–497.

104. Penney, D., Sakai, J., and Cook, K. (1974). Heart growth: Interacting effects of carbon monoxide and age. *Growth* **38**, 321–328.

105. Petelenz, T. (1965). Extracoronary blood supply of the sinu-atrial (Keith-Flack's) node. *Cardiologia* **47**, 57–67.

106. Phelan, E. L. (1966). Cardiovascular reactivity in rats with spontaneous inherited hypertension and constricted renal artery hypertension. *Am. Heart J.* **71**, 50–57.

107. Phelan, E. L. (1968). The New Zealand strain of rats with genetic hypertension. *N. Z. Med. J.* **67**, 334–344.

108. Phelan, E. L. (1970). Genetic and autonomic factors in inherited hypertension. *Circ. Res.* **26** and **27**, Suppl. II, II-65–II-74.

109. Powell, T., and Twist, V. W. (1976). A rapid technique for the isolation and purification of adult cardiac muscle cells having respiratory control and a tolerance to calcium. *Biochem. Biophys. Res. Commun.* **72**, 327–333.

110. Prakash, R. (1954). The heart of the rat with special reference to the conducting system. *Am. Heart J.* **47**, 241–251.

111. Pretlow, T. G., II, Glick, M. R., and Reddy, W. J. (1972). Separation of beating cardiac myocytes from suspensions of heart cells. *Am. J. Pathol.* **67**, 215–226.

112. Rakusan, K. (1971). Quantitative morphology of capillaries of the heart. *Methods Achiev. Exp. Pathol.* **5**, 272–286.

113. Rakusan, K., and Poupa, O. (1963). Changes in the diffusion distance in the rat heart muscle during development. *Physiol. Bohemoslov.* **12**, 220–227.

114. Rakusan, K., Korecky, B., Roth, Z., and Poupa, O. (1963). Development of the ventricular weight of the rat heart with special reference to the early phases of postnatal ontogenesis. *Physiol. Bohemoslov.* **12**, 518–525.

115. Rapp, J. P., and Dahl, L. K. (1972). Arylesterase isoenzymes in rats bred for susceptibility and resistance to the hypertensive effect of salt. *Proc. Soc. Exp. Biol. Med.* **139**, 349–352.

116. Repke, K., and Portius, H. J. (1963). Uber die Identitat der Ionenpumpen-ATPase in der Zellmembran des Herzmuskels mit einen Digitalis-Receptorenzym. *Experientia* **19**, 452–458.

117. Rona, G., Chappel, C. I., Balazs, T., and Gaudry, R. (1959). An infarctlike myocardial lesion and other toxic manifestations produced by isoproterenol in the rat. *AMA Arch. Pathol.* **67**, 443–455.

118. Sandler, C., and Wilson, G. M. (1959). The production of cardiac hypertrophy by thyroxine in the rat. *Q. J. Exp. Physiol. Cogn. Med. Sci.* **44**, 282–289.

119. Saroff, J., and Wexler, B. C. (1970). Isoproterenol-induced myocardial infarction in rats: Distribution of corticosterone. *Circ. Res.* **27**, 1101–1109.

120. Sasaki, R., Morishita, T., Ichikawa, S., and Yamagata, S. (1970). Autoradiographic studies and mitosis of heart muscle cells in experimental cardiac hypertrophy. *Tohoku J. Exp. Med.* **102**, 159–167.

121. Selye, H., Bajusz, E., Grasso, S., and Mendell, P. (1960). Simple techniques for the surgical occlusion of coronary vessels in the rat. *Angiology* **11**, 398–407.

122. Sen, S., Smeby, R. R., and Bumpus, F. M. (1972). Renin in rats with spontaneous hypertension. *Circ. Res.* **31**, 876–880.

123. Sen, S., Tarazi, R. C., Khairallah, P. A., and Bumpus, F. M. (1974). Cardiac hypertrophy in spontaneously hypertensive rats. *Circ. Res.* **35**, 775–781.

124. Setnikar, I., and Magistretti, M. J. (1965). Relationships between organ weight and body weight in the male rat. *Arzneim.-Forsch.* **15**, 1042–1048.

125. Shipley, R. A., Shipley, L. J., and Wearn, J. T. (1937). The capillary supply in normal and hypertrophied hearts of rabbits. *J. Exp. Med.* **65**, 29–42.

126. Skold, B. H. (1961). Chronic arteritis in the laboratory rat. *J. Am. Vet. Med. Assoc.* **138**, 204–207.

127. Sommer, J. R., and Johnson, E. A. (1970). Comparative ultrastructure of cardiac cell membrane specializations. A review. *Am. J. Cardiol.* **25**, 184–194.

128. Spear, J. F., and Moore, E. N. (1972). Stretch-induced excitation and conduction disturbances in the isolated rat myocardium. *J. Electrocardiol.* **5**, 15–24.

129. Stanton, H. C., Brenner, G., and Mayfield, E. D., Jr. (1969). Studies on isoproterenol-induced cardiomegaly in rats. *Am. Heart J.* **77**, 72–80.

130. Stevenson, J. A. F. (1967). Exercise, food intake and health in experimental animals. *Can. Med. Assoc. J.* **96**, 862–867.

131. Stevenson, J. A. F., Box, B. M., Feleki, V., and Beaton, J. R. (1966). Bouts of exercise and food intake in the rat. *J. Appl. Physiol.* **21**, 118–122.

132. Tabei, R., Maruyama, T., Kumada, M., and Okamoto, K. (1972). *In* "Spontaneous Hypertension—Its Pathogenesis and Complications" (K. Okamoto, ed.), pp. 185–193. Igaku Shoin Ltd., Tokyo.

133. Thomas, B. M., and Miller, A. T., Jr. (1958). Adaptation to forced exercise in the rat. *Am. J. Physiol.* **193**, 350–354.

134. Turner, J. H., and Lalich, J. J. (1965). Experimental cor pulmonale in the rat. *Arch. Pathol.* **79**, 409–418.

135. Undenfriend, S., Bumpus, F. M., Foster, H. L., Freis, E. D., Hansen, C. T., Lovenberg, W. M., and Yamori, Y. (1976). Spontaneously hypertensive (SHR) rats: Guidelines for breeding, care and use. *Inst. Lab. Anim. Resour. News* **19**, G1–G20.

136. Vahouny, G. V., Wei, R., Starkweather, R., and Davis, C. (1970). Preparation of beating heart cells from adult rats. *Science* **167**, 1616–1618.

137. Van Liere, E. J., and Sizemore, D. A. (1971). Regression of cardiac hypertrophy following experimental hyperthyroidism in rats. *Proc. Soc. Exp. Biol. Med.* **136**, 645–648.

138. Van Liere, E. J., Krames, B. B., and Northup, D. W. (1965). Differences in cardiac hypertrophy in exercise and hypoxia. *Circ. Res.* **16**, 244–248.

139. Wearn, J. T. (1928). The extent of the capillary bed of the heart. *J. Exp. Med.* **47**, 273–292.

140. Weeks, J. R., and Jones, J. A. (1960). Routine measurement of arterial blood pressure in unanesthetized rats. *Proc. Soc. Exp. Biol. Med.* **104**, 646–648.

141. Weissler, A. M., Kruger, F. A., Baba, N., Scarpelli, D. G., Leighton, R. F., and Gallimore, J. K. (1968). Role of anaerobic metabolism in the preservation of functional capacity and structure of anoxic myocardium. *J. Clin. Invest.* **47**, 403–416.

142. Wexler, B. C. (1964). Spontaneous arteriosclerosis in repeatedly bred male and female rats. *J. Atheroscler. Res.* **4**, 57–80.

143. Wexler, B. C. (1964). Spontaneous arteriosclerosis of the mesenteric, renal and peripheral arteries of repeatedly bred rats. *Circ. Res.* **15**, 485–496.

144. Wexler, B. C. (1964). Spontaneous coronary arteriosclerosis in repeatedly bred male and female rats. *Circ. Res.* **14**, 32–43.

145. Wiener, J., and Giacomelli, F. (1973). The cellular pathology of experimental hypertension. VII. Structure and permeability of the mesenteric vasculature in angiotensin-induced hypertension. *Am. J. Pathol.* **72**, 221–232.

146. Wildenthal, K. (1970). Factors promoting the survival and beating of intact foetal mouse hearts in organ culture. *J. Mol. Cell. Cardiol.* **1**, 101–104.

147. Wildenthal, K. (1971). Long-term maintenance of spontaneously beating mouse hearts in organ culture. *J. Appl. Physiol.* **30**, 153–157.

148. Williams, J. R., Harrison, T. R., and Grollman, A. (1939). A simple method for determining the systolic blood pressure of the unanesthetized rat. *J. Clin. Invest.* **18**, 373–376.

149. Williamson, J. R., and Krebs, H. A. (1961). Acetoacetate as fuel of respiration in the perfused rat heart. *Biochem. J.* **80**, 540–547.

150. Wilson, C., Ledingham, J. M., and Floyer, M. A. (1971). *In* "The Kidney" (C. L. Rouiller and A. F. Muller, eds.), pp. 156–248. Academic Press, New York.

151. Wood, W. G., Lindenmayer, G. E., and Schwartz, A. (1971). Myocardial synthesis of ribonucleic acid. I. Stimulation by isoproterenol. *J. Mol. Cell. Carciol.* **3**, 127–138.

152. Yamori, Y., and Okamoto, K. (1970). Zymogram analyses of various organs from spontaneously hypertensive rats. A genetico-biochemical study. *Lab. Invest.* **22**, 206–211.

153. Yamori, Y., Ooshima, A., and Okamoto, K. (1972). Genetic factors involved in spontaneous hypertension in rats. An analysis of F_{-2} segregate generation. *Jpn. Circ. J.* **36**, 561–568.

154. York, J. W., Penney, D. G., Weeks, T. A., and Stagno, P. A. (1976). Lactate dehydrogenase changes following several cardiac hypertrophic stresses. *J. Appl. Physiol.* **40**, 923–926.

155. Zak, R. (1974). Development and proliferative capacity of cardiac muscle cells. *Circ. Res.* **35** and **35**, Suppl. II. II-17–II-26.

Chapter 9

Immunology

Philip B. Carter and Hervé Bazin

I. INTRODUCTION

The immune response of the rat toward pathogenic agents and tumors requires a complex interaction between antibodies, lymphocytes, and macrophages. While this chapter is not intended to be a comprehensive review of work on the rat immune response, an attempt will be made to describe the major contributions to the understanding of the rat's immune mechanisms and to indicate the present state of knowledge and existing problems.

II. HUMORAL ASPECTS

A. Immunoglobulin Classes

1. Nomenclature of Rat Immunoglobulins

The nomenclature of the various immunoglobulin classes or subclasses underlines the difficulties met by the authors of this chapter. Confusion occurred mainly between IgA and IgG1, in the rat as well as in the mouse. The diversity of names which are still in use, proves that an agreement between scientists is difficult to obtain, even in such a restricted field. Table I gives the different nomenclatures employed for rat immunoglobulin (sub)classes. According to the World Health Organization (WHO) recommendations (300), a nomenclature using "gamma" or "γ" for the whole immunoglobulin molecule is to be rejected, and the "Ig" sign should be adopted instead. Likewise, all the letters have to be written on the line, thus IgG1 and not IgG$_1$. The division of the IgG classes on the basis of electrophoretic mobility (a division established by its use in most of the animal species) can be maintained at the present stage of knowledge, at least for historical reasons. The new IgR appellation, given to the IgG1 (sub)class by Binaghi and Boussac-Aron (27) seems questionable. They claim that rat IgG1 (sub)class molecules have no antigenic determinants in common with human IgG. However, Bazin et al. (19) and Orlans (202) have published cross-reactions they obtained by using human anti-IgG sera and LOU IgG1 monoclonal immunoglobulin or normal rat IgG1. Orlans (202) wrote: "in mouse, and also rat, the IgG1 is more closely related to human IgG than the electrophoretically slower subclasses." Last, class specific antimouse IgG1 serum strongly cross-reacted with purified rat monoclonal IgG1 (H. Bazin, unpublished results). Therefore, in the absence of more advanced knowledge on the chemical structure of these proteins, it seems difficult to change the name given by Jones (126) to this Ig class: $7S\gamma1$, i.e., IgG1 according to the present recommendations of the WHO. Rat IgM, IgA, and IgE seem to have properties quite analogous to their human homologues. The rat equivalent to human IgD is still under study, but data suggest that there is a rat homologue of human IgD.

At the present stage of knowledge, it is impossible to divide the various IgG species (IgG1, IgG2a, IgG2b, IgG2c) into classes or subclasses, or to relate any precise similarities in the (sub)classes between rat and IgG classes of any other animal species, especially of the human or mouse.

2. Historical

The first data concerning rat serum proteins, obtained by immunoelectrophoretic analysis, were published by Grabar and Courcon (84) and Escribano and Grabar (57). These workers distinguished two precipitation lines in the γ-globulin area. Some years later, Arnason and colleagues (5, 6) identified three proteins with antibody activities in rat serum. The authors referred to them as IgM, IgX (and later as IgA) and IgG.

a. IgM Immunoglobulin. Arnason et al. (5, 6) characterized this Ig class by its susceptibility to cysteine and later by its molecular weight.

b. IgG Immunoglobulin. The nomenclature of rat IgG used to be confused. At present, it is known that the IgA (or IgX) and IgG immunoglobulin classes distinguished by Arnason et al. (5, 6), are both IgG (sub)classes.

c. Fast IgG Immunoglobulin. This Ig class was identified by Arnason et al. (5, 6) and named IgX and later IgA to draw attention to the possible correspondence with human IgA. The determination was essentially based on antibodies present in saliva, and had a minor homologue in the serum. Nash et al. (191) reported at the same time the possibility of identifying a component of rat milk with a rabbit anti-mouse IgA serum. Antiserum raised against this rat immunoglobulin, using the cross-reaction between rat IgA and a rabbit anti-mouse IgA serum, allowed the demonstration of a large number of IgA-

Table I

Nomenclature of the Rat Immunoglobulin (Sub)Classes

Reference	IgM[a] (γM, γ_1M, β_2M)	IgA (β_2A, γ_1A, γA)	IgD (γD)	IgE (γE)	IgG1	IgG2a ($7S\gamma$, γ_{SS}, or γG)	IgG2b	IgG2c
5	IgM				IgX	IgG		
6	IgM				IgA	IgG		
8	IgM				IgA	IgGa	IgGb	
10	γM	U1		U2	γ1	γ2		
32, 33, 191		IgA						
126	γM	γA			$7S\gamma1$	$7S\gamma2$		
256				IgE				
16	IgM	IgA		IgE	IgG1	IgG2a	IgG2b	IgG2c
27	IgM	IgA		IgE	IgR	IgGa	IgGb	IgGc
237			IgD					

[a] Synonyms (old usage) are given in parentheses.

containing cells in rat *lamina propria*. The latter authors called this Ig class, IgA, and demonstrated that its anodal mobility was much faster than the previously termed ''IgA'', which was, in fact, the IgG1 class. Orlans and Feinstein (203) obtained a cross-reaction between human and rat IgA.

d. IgE Immunoglobulins. Reaginic antibodies have been shown to exist in the rat (26, 188) and to have properties similar to those of man (36, 129). Stechschulte *et al.* (255) and Jones and Edwards (127) have published what appears to be the rat equivalent for human IgE. Their work has been fully confirmed by Bazin *et al.* (23), who used LOU (see next section) monoclonal rat proteins.

e. IgD Immunoglobulins. Using a chicken antiserum to human IgD, Ruddick and Leslie (237) have demonstrated an antigenically cross-reactive homologue to human IgD on rat lymphocyte membranes. Moreover, Bazin *et al.* (18) have recently found four LOU rat IgD monoclonal immunoglobulins.

f. T Cell Recognition Structures. The antigen receptor of T lymphocytes is still largely unknown. However, Binz and Wigzell (30) have shown that, in rats, T and B cells specific for the same antigen express similar idiotypes. Moreover, in the same species, idiotypic T cell receptors, with specificity for major histocompatibility antigens, are coded for by genes (31). Therefore, it seems that the antigen receptor of T lymphocytes has a strong relationship to presently known immunoglobulin classes, although its precise nature has not yet been determined.

Historical discovery and identification of the various rat immunoglobulin (sub)classes is given in Table II.

3. LOU Rat Malignant Immunocytomas

The accumulation of present knowledge concerning mouse immunoglobulins was greatly facilitated by the discovery of spontaneous C3H mouse plasmacytomas (54), and above all, the plasmacytomas induced in BALB/c mice (223). Similarly,

in rats, the discovery of LOU ileocecal malignant immunocytomas contributed to the development of knowledge on immunoglobulins of this species.

a. Origin of the LOU Strain. In 1955, Maisin and his colleagues, working at the Cancer Institute of the Faculty of Medicine, University of Louvain, had their attention drawn to a special type of lymphoid tumor appearing in the ileocecal area in rats (159, 160). Deckers (51) and Maisin *et al.* (158) reported that some of these ileocecal tumors were secreting immunoglobulins. In 1970, Bazin and Beckers started breeding rats from various nuclei from the University of Louvain, and gave them the name LOU, in honor of their native University. The rats were graciously given to them by Drs. J. and H. Maisin, C. Deckers, and R. De Meyer. They bred 28 different and distinct lines, in which they observed the fertility level and the tumor incidence in the ileocecal lymph node. The line presenting the maximum malignant immunocytoma incidence was chosen as histocompatibility reference and called LOU/C. The line presenting the best reproduction rate and histocompatibility identical to that of the LOU/C rats was called LOU/M. Both these lines, LOU/C and LOU/M are perfectly histocompatible as all the LOU/M males are selected according to their endless acceptance of skin grafts from LOU/C. Wsl refers to the proper subline designation for the breeding colonies of Dr. H. Bazin and A. Beckers (59).

b. Ileocecal Malignant Immunocytomas. Malignant immunocytomas or immunoglobulin-secreting tumors appear spontaneously in the ileocecal lymph node of 8-month-old LOU/C rats or older. Their incidence is about 34 and 17% in males and females, respectively. Tumors apparently very similar have been described sporadically in various breeding colonies of rats all over the world, but their incidence was never above 2 or 3% (21). Since no study of serum proteins has been performed on these ileocecal tumor-bearing rats, it is impossible to be sure of their identity with LOU malignant immunocytomas. However, their lymphosarcoma or reticulum cell sarcoma nature suggests that these ileocecal tumors are relatively frequent in rats and that LOU rats are not the only

Table II

Historical Designation of the Various Rat Immunoglobulin (Sub)Classes

Reference	IgM	IgA	IgD	IgE	IgG1	IgG2a	IgG2b	IgG2c
5	IgM				IgX	——IgG——		
29						IgGa	IgGb	
126					7Sγ1	—— 7Sγ2 ——		
33, 191		IgA						
256				IgE				
16								IgG2c
237			IgD					

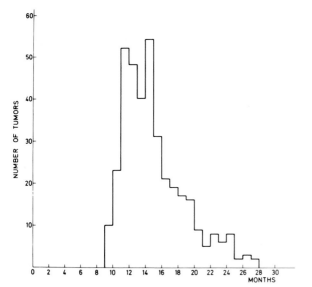

Fig. 1. Age distribution of 374 spontaneous ileocecal immunocytomas in the LOU/C/Wsl rats.

more nodules in the ileocecal region. When the tumor has progressed for a long time, all abdominal viscera can be infiltrated with metastases. The tumors are always highly vascularized and often necrotic in places. Metastases can be found in mediastinal lymph nodes or in the pleural cavity. The histological aspect of these tumors, as they appeared in LOU rat ancestors, has been described by Maldague *et al.* (160). At that time, as now, in most cases proliferating cells could not often be described as plasma cells; they are poorly differentiated. The secreting tumor cells exhibit a marked uniformity in size, a granular nucleus with relatively small nucleoli, and a rim of deeply basophilic and pyroninophilic cytoplasm (16, 21). The etiology of the LOU rat malignant immunocytomas is still unknown (184). A viral origin and a favorable genetic background seem probable. Table III shows that in both histocompatible rat strains, the LOU/C/Wsl and the LOU/M/Wsl, the tumor incidence is very different. Crosses between LOU/C/Wsl female rats and AUG/Wsl or AxC/Wsl males produce F_1 hybrids with a high tumor incidence. In contrast, the same cross with OKA/Wsl male rats gives hybrids in which no tumor develops. One or more dominant loci of resistance to the tumor incidence could exist in the OKA strain (15).

ones to show an incidence of these ileocecal malignant immunocytomas.

Figure 1 shows the age distribution of LOU immunocytomas. The incidence is maximum at 12–15 months. Some tumors appear in 27-month-old rats, maximum age for these animals (16, 21). The sites of origin of LOU malignant immunocytomas are, in most of the cases, the ileocecal lymph node (183). The tumors appear in the peritoneal cavity as solid masses. In live rats, they are mobile under the fingers and can easily be detected by palpation of the abdomen. They are fast growing, and the tumor bearers generally die within the month following appearance of their immunocytomas. Ascitic fluid is sometimes present and can be of help in detecting the primitive tumor. At autopsy, the tumor appears nearly always as one or

c. Biosynthesis of Monoclonal Proteins. Table IV gives the distribution of serum monoclonal immunoglobulins and Bence-Jones proteins synthesized by 250 LOU immunocytomas, which appeared consecutively in the Wsl breeding colony. The class distribution of LOU monoclonal immunoglobulins is given in Table V. The distribution is quite similar in different LOU substrains and LOU/C F1 hybrids (15, 17).

d. Production of Monoclonal Proteins. About 90% of the primitive LOU immunocytomas can be transplanted in LOU/ C/Wsl, LOU/M/Wsl, or their F_1 hybrids. The tumors are re-

Table III

Incidence of Ileocecal Immunocytomas in LOU Rats and Their Crosses with the AUG/Wsl, AxC/Wsl, and OKA/Wsl Rats

	Number of immunocytomas/number of rats under observation and percentage			
	Male		Female	
LOU/C/Wsl	145/422	(34.4%)	72/424	(17%)
LOU/M/Wsl	1/138	(0.7%)	7/328	(2.1%)
(LOU/C/Wsl ♀ × AUG/Wsl ♂)F₁[a]	12/72	(16.7%)	6/77	(7.8%)
(LOU/C/Wsl ♀ × AxC/Wsl ♂)F₁[a]	10/65	(15.4%)	9/78	(11.6%)
(LOU/C/Wsl ♀ × OKA/Wsl ♂)F₁[a]	0/61	(0.0%)	0/57	(0.0%)

[a] Animals still under observation. The percentages given are a minimum evaluation of the incidence.

Table IV

Distribution of the Monoclonal Immunoglobulins Synthesized by 250
Consecutive LOU/C/Ws1 Ileocecal Immunocytomas[a]

	Secreting tumors	Monoclonal Ig with or without Bence-Jones proteins	Bence-Jones proteins only
Number of tumors	185	159	26
Absolute percentage	74	64	10
Percentage of secreting tumors	100	86	14

[a] See also ref. 12.

moved under sterile conditions and teased apart in a petri dish.
The tumor tissue is minced until it can be aspirated through a
19-gauge needle into a syringe. Into one side of the chest or
into the peritoneal cavity 0.1 to 0.2 ml is injected subcutaneously. In the latter case, an ascitic tumor can be obtained after
few passages which can then be transplanted intraperitoneally
by 0.2 to 0.5 ml of ascitic fluid. The immunoglobulin-
secreting properties of the LOU immunocytomas are quite variable, but most of the tumors maintain their ability to synthesize monoclonal proteins through a large number of *in vivo*
passages. IgG1 and IgG2a secreting tumors have been transplanted for 3 years without any change in the production of
monoclonal immunoglobulins. Sometimes, however, the secreting properties are lost more or less rapidly. In most cases,
IgE immunocytoma production is considerably reduced after
one or two passages, and disappears completely after a few
additional transplantations. Monoclonal IgE serum levels from
10 to 20 mg/ml can normally be obtained. The LOU rat immunocytomas may be preserved in liquid nitrogen according to

the technique described by Bazin *et al.* (21). Another possibility of producing monoclonal Ig is the culture of *in vitro* cells
(40). These continuous *in vitro* cell lines can be transferred
into LOU rats and produce immunocytomas which are in no
way different from the original tumors, at least after short
periods of culture.

4. Production of Antisera to Rat Immunoglobulin (Sub)Classes

Methods are identical, in their basic principles, to those used
for other animal species. The use of monoclonal immunoglobulins, originated by LOU immunocytomas, greatly facilitates the work.

a. Purification of Rat Immunoglobulins. i. IgM. Differ-
ent techniques have been described by Oriol *et al.* (201),
Van Breda Vriesman, and Feldman (271), Cremer *et al.* (48),
Bazin *et al.* (19) and McGhee *et al.* (164). They refer to
euglobulin properties and to the molecular weight of this Ig.
Figure 2 gives a method of purification (19). (If using mono-

Rat serum containing, or not containing, monoclonal rat IgM.

> Serum diluted twice with saline and precipitated by ammonium sulfate at 40% final concen-
> tration. The precipitate is washed twice with 45% ammonium sulfate.

Ammonium sulfate precipitate

> Dissolved in Tris-HCl buffer (0.005 M, pH 8.0) and dialyzed overnight against the same
> buffer at 4°C.

Euglobulin precipitate

> Resuspended in 2% (w/v) NaCl buffered with 0.002 M Tris-HCl, pH 8.0, and containing
> 0.1% (w/v) NaN$_3$. Filtration on Sephadex G200 (Pharmacia-Sweden) and collection of the
> first peak.

IgM

Fig. 2. Technique of purification of rat IgM.

Table V

Class Distribution of Complete Monoclonal Immunoglobulins Synthesized by Immunocytomas from Various Crosses of LOU Rats with Different Inbred Strains

	Number of Ig-secreting immunocytomas	Number of immunocytomas	IgM	IgA	IgD	IgE	IgG1	IgG2a	IgG2b	IgG2c	ND[a]
LOU/C/Ws1	642	408	11	13	4	171	159	24	3	20	3
LOU/M/Ws1	8	4	1	—	—	2	1	—	—	—	—
(LOU/C/Ws1 ♀ × AUG/Ws1 ♂) F$_1$	18	12	—	—	—	10	2	—	—	—	—
(LOU/C/Ws1 ♀ × AxC/Ws1 ♂) F$_1$	23	16	1	—	—	7	5	3	—	—	—
(LOU/C/Ws1 ♀ × OKA/Ws1 ♂) F$_1$	0	—	—	—	—	—	—	—	—	—	—
Total	691	440	13	13	4	190	167	27	3	20	3
Percentage	—	100	2.95	2.95	0.91	43.18	37.9	6.13	0.68	4.54	0.68

[a] Not determined.

clonal IgM, it is necessary to check whether this peculiar protein reacts as an euglobulin or a pseudoglobulin, as some LOU rat monoclonal IgM proteins are pseudoglobulins.) Another technique, which is based on the common antigens carried by human and rat IgM molecules, has been described by Sapin and Druet (239).

ii. IgA. Isolation of rat monoclonal IgA or normal IgA is quite difficult as the molecular weight of this Ig class is heterogenous. A technique derived from that described by Bazin *et al.* (20) is represented in Fig. 3.

iii. IgE. A technique of purification of normal (poly-clonal) rat IgE has been described by Isersky *et al.* (110). Bazin *et al.* (23) and Bazin and Beckers (15) published a technique which is summarized in Fig. 4, for LOU monoclonal IgE.

iv. IgG (sub)classes. Normal rat IgG2a can be obtained by DEAE–cellulose chromatography (33, 164), using a phosphate buffer at pH 7.5, 0.005 M. The same buffer, but at 0.01 M gives an IgG fraction containing both IgG2a and IgG2b. Monoclonal Ig of the IgG1, IgG2a classes can be purified by ammonium sulfate precipitation, DEAE–cellulose chromatography, and gel filtration on AcA34 (LKB-Sweden) or G200 Sephadex (Pharmacia-Sweden). A better degree of purity can be obtained by electrophoresis in pevikon, agarose, or poly-acrylamide gel, as a last step. Different examples are given in Bazin *et al.* (19).

Normal and most of the monoclonal rat IgG2c possess strong euglobulin properties and can be obtained easily by diluting 1 volume of the serum in 19 volumes of distilled water adjusted to pH 6.0. Precipitation is performed three times. Then, the precipitated proteins are dissolved in an acetate buffer at pH 5.5 and filtrated on a G200 Sephadex or AcA34 column, equilibrated with the same buffer (19; G. A. Medgyesi, per-

Rat serum containing a monoclonal IgA protein

> Diluted twice with saline and precipitated by ammonium sulfate at 40% saturation. The pre-cipitate is washed with 50% saturated ammonium sulfate.

Ammonium sulfate precipitate

> Precipitate resuspended in saline, dialyzed against 0.02 Tris-HCl buffer, pH 8.0, and applied on a DEAE-cellulose column equilibrated with the same buffer and eluted with a similar buffer at 0.1 Tris-HCl, 0.08 HaCl.

Crude fraction of rat IgA

> After concentration by ultracentrifugation and a short dialysis against Tris-HCl buffer, pH 8.0, 0.1 M, the fraction is applied on an AcA34 or AcA22 (LKB, Sweden) column according to the molecular weight of the IgA to be purified. The fractions containing the IgA are identified by Ouchterlony test with an antiserum against rat IgA. The last gel filtration could be repeated to increase the purity of the IgA.

Rat monoclonal IgA

Fig. 3. Technique of purification of rat monoclonal IgA.

Rat serum containing monoclonal IgE

> Diluted twice and precipitated by ammonium sulfate at 50% final concentration. The precipitate is washed twice with 50% saturated ammonium sulfate.

Ammonium sulfate precipitate

> Precipitate resuspended in saline, dialyzed against 0.05 M Tris-HCl buffer, pH 8.0, and applied to a DEAE-cellulose column and eluted with a convex gradient. The starting buffer is 0.05 M Tris-HCl, pH 8.0, and the limit buffer 0.10 M Tris-HCl, 0.40 M NaCl, pH 8.0.

Crude fraction of rat serum containing the IgE monoclonal protein

> After concentration and dialysis against a 0.10 M Tris-HCl, 0.15 M NaCl, 1% NaN$_3$, pH 8.0 buffer, the fraction is applied on a AcA34 (LKB-Sweden) column. The first peak is the monoclonal IgE.

Rat monoclonal IgE

Fig. 4. Rat serum containing monoclonal IgE.

sonal communication). Another technique is to precipitate the serum from which IgG2c has to be taken with 40% ammonium sulfate. The precipitate is washed with sodium acetate buffer, 0.1 M, pH 5.45, and put on a carboxymethyl cellulose (CM-cellulose) column equilibrated with the same buffer. Chromatography is performed with a sodium acetate gradient of 0.1 to 0.5 M, pH 5.45. The IgG2c can be run a second time on AcA34 in the same acidic buffer (R. Rousseaux, H. Bazin, and J. Biserte, personal communication).

b. Immunization of Animals. Antisera containing the anti-rat Ig classes can be obtained by inoculation of the whole molecules or even better the heavy chains only. The technique by Metzger and Houdayer (174) has proved to be very useful in obtaining sera with the rat IgG anti-(sub)classes. Briefly, the polypeptidic heavy chains were prepared from mildly S-sulfonated monoclonal Ig classes by gel filtration on G100 Sephadex (Pharmacia) in 0.01 M formic acid with 5 M urea. The chains were transferred into a 0.2% ammonium carbonate solution (pH 8.5) by gel filtration on a G25 Sephadex column and lyophilized. The solubilization of partially S-sulfonated chains was performed as described by Metzger *et al.* (175), in 6.0 M guanidine hydrochloride dissolved in 0.5 M ammonium chloride buffer, pH 8.6. This step was followed by a gel filtra-tion over a G25 Sephadex column in 0.05 M sodium carbonate buffer, pH 8.6.

Antisera are prepared by injecting 0.1 to 1.0 mg of antigen in 0.4 ml of saline with 0.4 ml of complete Freund's adjuvant (Difco) into footpads of rabbits or guinea pigs, at days 1 and 15. Goats are injected intramuscularly with the same mixture, on the same schedule. Animals are bled 10 to 15 days after the last injection. Figure 5 shows the immunoelectrophoretic pat-tern obtained against normal rat serum with rabbit anti-rat im-munoglobulin.

5. Physicochemical Properties

Monoclonal proteins synthesized by LOU rat im-munocytomas provided an opportunity for exhaustive studies

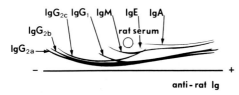

Fig. 5. Immunoelectrophoretic pattern of normal rat serum developed against rabbit anti-rat immunoglobulin [taken from Bazin *et al.* (19), with permission]. (Present conventions of nomenclature dictate that the name be written on one line, e.g., IgG2a.)

in recent years. The major data obtained with the different rat immunoglobulin (sub)classes are given in Table VI. Chemical characterization of both κ and λ rat light chain types has been firmly established by Quérinjean *et al.* (226, 228). N-terminal sequence analysis of a series of LOU rat κ chains showed a high degree of variability in this region (287, 288). Similar analysis of monoclonal heavy chains confirmed the presence of human $V_H III$ subgroup homologue in the rat species. Comparison between monoclonal and normal pooled heavy chains were also discussed by Quérinjean *et al.* (227). The complete sequence of a LOU rat κ Bence-Jones protein was established by Starace and Quérinjean (254). Comparing this sequence with corresponding sequences of mouse, rabbit, and human, different phylogenetic relationships between *V*- and *C*-region genes were delineated: the rat *V* region being more homologous to the human $V_\kappa I$ subgroup, rat and mouse being the closest when the *C* region was considered. An unusual frag-

mentation in the vicinity of the *V-C* bridge of four monoclonal heavy chains was reported by Quérinjean (225), but its origin (biosynthetic or catabolic) is not yet clarified.

6. Biological Properties

The main biological properties presently known concerning the rat immunoglobulin (sub)classes are given in Table VII. These data make it clear that the various experimental models are all peculiar. Comparison is impossible without considering the differences which can exist between them. Rat IgG1 and IgG2a antibodies bind complement; mouse IgG1 antibodies do not, although its IgG2a antibody does (171, 209).

Rat lymphocyte membrane immunoglobulin has been studied by Ladoulis *et al.* (141) and Smith *et al.* (250). Secretion of Bence-Jones protein has been studied by Nelson *et al.* (192) as a model for the study of cell population growth by direct measurement of urine excretion of this tumor characteristic protein.

7. Allotypes of Rat Immunoglobulins

Molecules of an Ig class or subclass do not always bear identical antigenic determinants in all individuals from the same species. Sometimes, they can present two or more antigenic forms, each of them being called "allotypes" accord-

Table VI

Physicochemical Properties of Rat Immunoglobulin (Sub)Classes

Designation	Reference	IgM	IgA	IgD	IgE	IgG1	IgG2a	IgG2b	IgG2c
Heavy chains	—	μ	α	δ	ϵ	$\gamma 1$	$\gamma 2a$	$\gamma 2b$	$\gamma 2c$
Light chains	—	κ or λ	κ or λ	—	κ or λ	κ or λ	κ or λ	κ or λ	κ or λ
Molecular formula	19, 201	$(2L+2H)_5$	$(2L+2H)_n$	—	$(2L+2H)$	$(2L+2H)$	$(2L+2H)$	$(2L+2H)$	$(2L+2H)$
Presence of J chain	132	Yes	Yes (in polymeric molecule)	—	—	—	—	—	—
Presence of secretory component	270	—	Yes (in secretion)	—	—	—	—	—	—
Sedimentation coefficient	17, 19, 33, 48, 255	18.2 S	7–19 S	—	7.6 S	6.7 S	6.4 S	6.5 S	6.7 S
Molecular weight		—	163,000[a]	—	183,000 to 193,000	156,000	156,000	—	156,000

[a] Monomer.

Table VII

Biological Properties of Rat Immunoglobulin (Sub)Classes

	Reference	IgM	IgA	IgD	IgE	IgG1	IgG2a	IgG2b	IgG2c
Concentration range in normal serum (mg/ml)	19	0.56 to	0.13	—	0.02	5.85	8.00	—	2.60
	48	0.15–0.90	—	—	—	—	—	—	—
	43, 44, 112, 116, 118, 216	—	—	—	0.0005 to 0.02[a]	—	—	—	—
	164, 171	0.95 ± 0.07[b]	0.18 ± 0.03	—	—	—	6.91 ± 0.21	0.89 ± 0.04	—
	27	—	—	—	—	0.5 to 7.0	—	—	—
Valency for antigen binding	19, 201	5+5	2	—	2	2	2	2	2
Half-life (in days)	17, 48, 263	2.6	—	—	0.5	—	5.0	—	—
Antibody activity	29, 197, 201	Yes	Yes	—	Yes	Yes	Yes	Yes	Yes
Complement fixation									
Total complement	171[c]	Yes	Yes	—	Yes	Yes	Yes	—	No
C1		Yes	No	—	No	Yes	Yes	—	No
Cross placenta	28, 164, 176	No	No	—	No	Yes	Yes	—	—
Presence in colostrum and milk	27, 28, 164, 176	No	Yes	—	Yes	Yes	Yes	—	—
Absorption by the intestinal tract (in suckling rats)	27, 28, 164, 176	—	No	—	Yes	Yes	Yes	—	—
Fix to homologous mast cells	15, 23, 95, 256	—	—	—	Yes	No	No[d]	No	No
Half-life in homologous skin (days)	263	—	—	—	7.40 ± 0.89	—	2.40 ± 0.25	—	—
Passive cutaneous anaphylaxis in rat	23, 27, 127, 187	—	—	—	Yes	No	Yes	No	—
Thermolability (at 56°C)	23	No	No	—	Yes	No	No	No	No

[a] In *Nippostrongylus brasiliensis*-infected rats, the IgE serum levels can rise up to 0.4 mg/ml.

[b] Mean ± standard error.

[c] Based on monoclonal proteins: IgM, IR202, IR319, IR473; IgA, IR22; IgE, IR162; IgG1, IR31, IR243, IR314; IgG2a, IR230, IR241; IgG2c, IR64, IR221.

[d] Very weak fixation could not be excluded.

ing to the Oudin terminology (208). In this case, subjects from a unique species could be divided into two or more allotypic groups. Allotypic specificities are supported by structural differences determined by genes, the inheritance of which is Mendelian. All the known genes are codominant, i.e., expressed phenotypically in the heterozygote. However, each Ig-producing cell expresses only one of the allelic genes for pair; they seem to be functionally haploid.

The first allotypic difference of rat Ig was reported by Barabas and Kelus (11), who immunized "black and white hooded" rats with Wistar Ig. Later, Wistar (299) demonstrated the presence of an allotypic marker on rat light chains. These data were confirmed by various authors (3, 90, 91, 107). Rokhlin *et al.* (234), suggested that this allotypic difference was taking place on the κ light chains. This result was formally confirmed by Beckers *et al.* (25), who used Bence-Jones proteins from κ type originating from LOU/C/Wsl immu-

nocytomas, and by Nezlin *et al.* (194), who localized the allotypic markers of κ light chains of rat immunoglobulins in the constant part of the chains. Table VIII shows the equivalence between the various nomenclatures of the rat κ allotype.

Rat κ chains have been sequenced. The complete sequence of the S211 LOU κ light chain has been published by Starace and Quérinjean (254). Gutman *et al.* (89) have found one sequence gap and many amino acid substitutions between the κ chains of the DA(RI-1a) and the LEW(RI-1b) rat strains. They described the rat κ allotype as a complex one and suggested that these genes are not true alleles. Moreover, there are two amino acid differences between the LOU S211 and the LEW κ chains (89, 254), both being IK(1a). This suggests an isotypic variation.

Two other allotypes have been described. Bazin *et al.* (20) found an allotype located on the rat α chain and determined by a gene they called *Iα(1)*. The antigenic determinant(s) corre-

Table VIII

Nomenclature of the Rat K-chain Allotypic Alleles.[a]

Beckers et al. (25)	Barabas and Kelus (11)	Wistar (299)	Gutman and Weisman (90)	Rokhlin et al. (234)	Humphrey and Santos (107)	Armerding (3)
Ik(1a)	Ra-1	non-RI-1	RI-1b	RL2	—	W-1
IK(1b)	non-Ra-1	RI-1	RI-1a	RL1	+	SD-1

[a] The correspondence between the appellations of Beckers *et al.* (25), Gutman and Weisman (90), Rokhlin *et al.* (234) and Amerding (3) have been confirmed, the others are probable.

sponding to this allotype was found on 100% of the alpha chains from homozygote rats for the *Iα(1)* locus. Another allotypic difference has been described by Beckers and Bazin (24) for the γ2b heavy chain, with two allotypes: *Iγ2b(1a)* and *Iγb(1b)* determined by a gene they called *Iγ2b(1)*. Table IX gives the distribution of the allotypes disclosed up to now in the various strains of rats. Table X gives the number of different strains of rats grouped according to their respective allotypic marker. Only two strains possess the *Iγ2b(1b)* allotype. It has to be noticed that both these inbred strains (nonhistocompatible as we checked them) are hypertensive and issued from the same laboratory where they were selected.

The *Iα(1)* and *Iγ2b(1)* loci are linked to each other (24), but they are not linked at the *IK(1)* locus (25). Moreover, they were shown to segregate independently from genes concerned with sex, coat pattern and color, and eye color (20). The locus *IK(1)* is not linked to the *RT-1(Ag-B* or *H-1)* locus, three-coat color genes (90), or a gene controlling the immunological responsiveness to the lactate dehydrogenase antigen (3). Different strains of rat, congenic to inbred strains, have already been created. S. V. Hunt and J. L. Gowans (personal communication) have derived congenic lines of the PVG/c inbred rats (RI-1b) by backcrossing the *RI-1a* locus of the DA rats onto the PVG/c background. Homozygous PVG.1a lines were established after the fifth, tenth, and fifteenth backcross, giving PVG.1a (N5), PVG.1a (N10), and PVG.1a (N15), respectively. PVG/c and PVG.1a (N10 or 15) rats are perfectly histocompatible. Bazin and Beckers (15) selected four congenic lines to LOU/C rats by inserting into their genomes the loci of light chain of OKA/Wsl rats or the heavy chains of OKA/Wsl, AUG/Wsl, and AxC/Wsl rats, respectively. These strains are called LOU/C.IK(OKA)/Wsl (N10), LOU/C.IH(OKA)/Wsl (N7), LOU/C.IH(AUG)/Wsl (N7), and LOU/C.IH(A×C)/Wsl (N7), where I represents Ig and K and H, respectively, the κ and heavy chains. All these congenic lines are histocompatible with the LOU/C/Wsl rats. The allotypic differences can serve as markers for the Ig. In the rats, the κ chain allotypes make it possible to follow immunoglobulins from any class and has already been used by Hunt and Williams (109),

Williams and Gowans (293), and Nezlin and Rokhlin (193). Rat α chain allotype has also been used as a marker for this immunoglobulin (13) or for the Ig heavy chains loci of synthesis (31).

B. Humoral Immunological Responses

Humoral immunological responses of rats have been studied for a long time. However, the physicochemical characteristics of the antigen, its dose, the adjuvant used or not used, the route of inoculation, the genetic constitution of the inoculated animal, and the class of the studied antibodies are all factors which make it difficult to describe the humoral immunological responses in rats as well as in other species.

1. Immunoglobulin and Antibody Synthesizing Cells

Different techniques have been used in order to identify these cells in normal or immunized rats. Examples of such studies can be found in Nast *et al.* (191), Kraft and Wistar (137), Kuhlmann and Avrameas (138), Williams and Gowans (293), Pierce and Gowans (222), Miller *et al.* (1978), Kuhlman and Avrameas (139), Antoine and Avrameas (2), Mayrhofer *et al.* (163), and Ishizaka *et al.* (112).

2. Immunoglobulin and Antibody Serum Levels after Antigenic Stimulation

a. Conventional Immunological Responses. Individuals from the rat species seem to be as good as those from other laboratory rodent species in producing IgM and IgG (sub)classes of antibodies. Bovine γ-globulin, bovine serum albumin, and dinitrophenylated bovine γ-globulin have been used at 100 μg to 2.5 mg doses with complete or incomplete Freund's adjuvant (6, 10, 35, 201, 221, 264). Depending on the antigen and the physiological status of the rat, the dose of antigen to be inoculated in order to obtain precipitating antibodies is variable, but doses between 100 μg and 1 mg with complete Freund's adjuvant are generally convenient. Many other antigens have been used including sheep red blood cells, 0.35 ×

Table IX

Distribution of the Allotypes Currently Determined in Strains and Stocks of Rats[a]

Strains and Stocks	Reference	IK(1)	I α(1)	I γ2b(1)
ACI	25, 90, 107, 228	b	—	a
ACJ[b]	24, 25	b	—	a
ACP	24, 25	a	—	a
AGUS	20, 24, 25	a	—	a
ALB	24, 25	b	a	a
AO	109	a	ND	ND
AS	3, 20, 24, 25	a	a	a
AS2	3, 20, 24, 25	a	a	a
Atrichis	20, 24, 25	a	a	a
AUG	3, 20, 24, 25, 107, 225	a	—	a
AVN	3, 24, 25, 234	a	—	a
AxC[b]	3, 20, 24, 25	a	—	a
BDI	234	b	ND	ND
BD V	24, 25	b	—	a
BD V	3, 234	a	ND	ND
BD IX	2, 24	a	—	a
BD X	24, 25	a	—	a
BDE	3	a	ND	ND
BH	107	a	ND	ND
BICR (Marchall)[b]	5	a	ND	ND
BIRMA	234	a	ND	ND
BIRMB	234	a	ND	ND
Black and White Hooded	11	b	ND	ND
BN	3, 20, 24, 25, 90, 107, 234	a	a	a
BP	234	a	ND	ND
BS	3, 24, 25	a	a	a
BUF	3, 24, 25, 90, 107, 234	a	a	a
CAP	3, 24, 25	a	a	a
COP	2, 24, 25	b	—	a
DA	3, 20, 24, 25, 90, 107, 109, 234	b	—	a
E-3	3	a	ND	ND
F-45[b]	234	a	ND	ND
F-344	3, 20, 24, 25, 90, 107, 234	b	a	a
Gunn	20, 24, 25	b	—	a
Gowans (Albino)	11	a	ND	ND
LIS	(H. Bazin and A. Beckers, unpublished)	a	a	a
HS	3	a	ND	ND
Hypodactyle	24, 25	a	a	a
HW[b]	3	a	ND	ND
KGH	(A. Beckers and H. Bazin, unpublished)	a	a	a
LE	20, 24, 25	a	a	a
L/E[b]	234	a	ND	ND
LEP	3, 25	a	—	ND
LEW	3, 20, 24, 25, 90, 107, 299	a	a	a
LOU/C	20, 24, 25, 234	a	a	a
LOU/M	24, 25	a	a	a
LOU/C.IH(AUGUST)	24	a	—	a
LOU/C.IH(AxC)	(A. Beckers and H. Bazin, unpublished)	a	—	a
LOU/C.IH(OKAMOTO)	24, 25	a	—	b

(*continued*)

Table IX (Continued)

Strains and Stocks	Reference	IK(1)	Iα(1)	Iγ2b(1)
LOU/C.IK(OK)	24	b	a	a
M520	24, 25, 234	a	a	a
MAXX	107	a	ND	ND
MSU[b]	24, 25, 234	b	a	a
NBR	25	ND	—	ND
OFA[b]	20, 24, 25	b	—	a
OKA	20, 24, 25	b	—	b
OM	25	ND	—	ND
PB[b]	3	a	ND	ND
PVG/c	20, 24, 25	a	a	a
R	234	a	ND	ND
RA[b]	234	a	ND	ND
S5B	25	ND	—	ND
SEL	24, 25	a	a	a
SH[b]	24, 25	b	—	b
Sherman	20, 24, 25	a	a	a
Sprague-Dawley	3, 24, 25, 234	b	—	a
WF	3, 24, 25, 90, 107	a	a	a
WFH[b]	107	a	ND	ND
Wistar	3, 11	a	ND	ND
Wistar outbred (ASH/W)[b]	234	a	ND	ND
Wistar Af	3	a	ND	ND
WAG	3, 20, 24, 25, 234	a	a	a
Wistar BB[b]	3	a	ND	ND
R	20, 24, 25	a	a	a
W.P.[b]	3	a	ND	ND
Yoshida	24, 25	b	a	a
Z61	24, 25	a	—	a

[a] ND, not determined; a, allele *a*; b, allele *b*; —, absence of allele *a*.
[b] Strain designations not reconciled to international nomenclature. See Chapter 2, Table I.

10^8 to 14×10^8 by injection (249); *Salmonella typhimurium* (249); lactic dehydrogenase isozyme LKH-A₄ (302); synthetic polypeptide antigens (70, 71, 88); encephalitogenic protein (69, 294); and group A streptococcal vaccine (148). Circulating antibodies raised by allografts have been studied by Wright *et al.* (301). In all the cases, the immunological response seems to be under the influence of genetic factors (4, 70, 71, 87, 140, 142, 238, 253, 261).

b. IgE and Reaginic Serum Levels after Antigenic Stimulation. Rats were considered poor producers of reaginic antibodies after stimulation with conventional antigens. Their reaginic responses were accepted as short lived and relatively impossible to boost (26, 189). Jarrett and Steward (121) demonstrated that, at least in some combinations of strains, antigens, and adjuvant and immunization schedule, high and long-lived primary reaginic responses and even more, true secondary responses can be obtained. They used the LIS strain of rat which seems especially able to mount reaginic immunological responses. To a lesser extent, the LOU/M strain seems to be able to produce high titers of reaginic antibodies too (22). But in each case, the use of adjuvant

(*Bordetella pertussis*, aluminum hydroxide, complete Freund's adjuvant, bacterial lipopolysaccharide, etc.) is necessary to induce the primary but not the secondary response. Tertiary responses are difficult to obtain (119, 121). The route of inoculation most often used is certainly the intraperitoneal route, although others (the intradermal or the oral routes) can be employed with great efficiency (22, 121). The discovery made by Levine and Vaz (149) that minute amounts of antigen inoculated into mice are more effective in inducing a high and persistent reaginic response is also true for rats (121). Doses of 10

Table X

Number of Different Inbred Strains of Rats Grouped According to Their Respective Allotypic Markers

I γ2b Allotype	IK(1a)		IK(1b)	
	I α(1a)	I α(1.)	I α(1a)	I α(1.)
I γ2b(1a)	14[a]	7	3	6
I γ2b(1b)	0	0	0	2

[a] Number of rat strains found with this allotype combination.

μg to 1 mg are generally used in this species. Higher doses have an inhibitory effect (121, 262). Examples of reaginic antibody responses in rats to conventional antigens can be found in Murphey *et al.* (190), Jarret and Steward (121), Jarrett *et al.* (119), Bazin and Platteau (22), and Bazin (14), and to alloantigens in Broom and Alexander (39). Another technique of inducing reaginic antibodies in the rat is experimental or natural infections with live worm parasites. The most common is the nematode *Nippostrongylus brasiliensis* (128). After inoculation of infective larvae into the rat, a parasite specific reaginic response occurs (198, 199); there is an increase of the IgE responses to unrelated antigens (205), and finally a massive elevation of total serum IgE (117). These phenomena also have been described by Jarrett and Steward (120), Jarrett (115), Orr *et al.* (206), Carson *et al.* (44), Ishizaka *et al.* (111), and Jarrett *et al.* (118).

III. CELLULAR ASPECTS

A. Lymphocytes

1. Kinetics

The rate of formation and the life span of rat lymphocytes has been the object of vigorous study since 1945 when Reinhardt developed a method for collecting lymphocytes free of contamination by other blood cells (231). The early work of Reinhardt and Li (232) was extended by Caffrey *et al.* (41), among others, who employed tritiated thymidine labeling techniques to demonstrate the presence of two populations of small lymphocytes in thoracic duct lymph, different in terms of life span and rate of formation. These workers showed that the absolute number of newly formed small lymphocytes was the same in animals from which lymph had been drained for 30 hr as in control animals without prior lymph drainage. The proportion of newly formed lymphocytes in the animals drained 30 hr was increased, however (10 versus 1% in control animals). These data indicated that the rate of formation was constant and that the mobilizable lymphocyte pool was proportional to body weight and calculated to be 7.8×10^6 lymphocytes per gram of body weight.

Numerous observations indicated that the rate of formation of new lymphocytes could not account for all the lymphocytes present in rat lymph at any point in time and that prolonged thoracic duct drainage caused a drop in the number of cells in the lymph (79, 161). Also, reinfusion of thoracic duct lymphocytes, but not lymph fluid, into the blood resulted in maintenance of normal levels of thoracic duct lymphocytes (79). These observations led Gowans (80) to postulate a recirculation of lymphocytes from blood to lymph.

2. Recirculation of Lymphocytes

In 1959, Gowans demonstrated that small lymphocytes in the rat recirculate from blood to lymph (80). This discovery was later confirmed by Shorter and Bollman (244) and Caffrey *et al.* (41) and extended by Gowans and various co-workers.

The technique used by Gowans and others to obtain pure populations of lymphocytes was that developed by Bollman *et al.* (38) for cannulation of the abdominal thoracic duct [for cannulation of the superior thoracic duct see Reinhardt and Li (232)]. The technique allows lymph to be drained for more than 5 days and is illustrated in Fig. 6 [see also Ford and Hunt (64)]. The major difficulty in the collection of thoracic duct lymphocytes is not the cannulation itself, but rather ensuring a constant flow of lymph over 1 day or longer. Figures 7 and 8 show a method for housing cannulated rats which allows a good output of lymph.

Since the number of lymphocytes issuing from a lymph node was many times the number draining from the tissues into the node (afferent lymph), a mode by which lymphocytes from the blood might enter the lymph nodes was sought. In 1964, Gowans and Knight published a bench mark paper (82) in which they observed that transfused small lymphocytes homed rapidly to the lymph nodes, splenic white pulp, and the Peyer's patches. Radiolabeled lymphocytes were observed to traverse postcapillary venules located in the paracortex of lymph nodes. Since the cells forming the endothelium in these venules were exceptionally deep, they were referred to as "high" endothelium. Employing electron microscopy, Marchesi and Gowans (162) were able to observe lymphocytes actually traversing the endothelium of postcapillary venules. Schoefl (241) later presented evidence to indicate that lymphocytes traverse the endothelium by squeezing between the endothelial cells rather than passing through them. A high incidence of lymphocyte clusters suggested that particularly weak spots in the cohesion of the endothelial cell sheet existed which favored the passage of lymphocytes. It was then postulated that the high endothelium was an adaptation in lymphoid tissue, excluding the thymus, which allowed sustained cell traffic without excessive fluid loss.

As indicated, recirculating lymphocytes enter the lymph in lymph nodes via postcapillary venules in the paracortex. The venules possessing high endothelium are in an analogous location in Peyer's patches (75). In the spleen, recirculating lymphocytes enter the white pulp via postcapillary venules in the marginal zone and situate themselves in the periarteriolar sheath (75). The cells in the periarteriolar sheath appear then to arise primarily from blood-borne lymphocytes rather than from splenic follicles (61). Within 6 hr after entry, lymphocytes leave the periarteriolar sheath and move to the red pulp (62). Data suggest that the lymphoid cells which localize in the periarteriolar sheath are primarily thymus-dependent lympho-

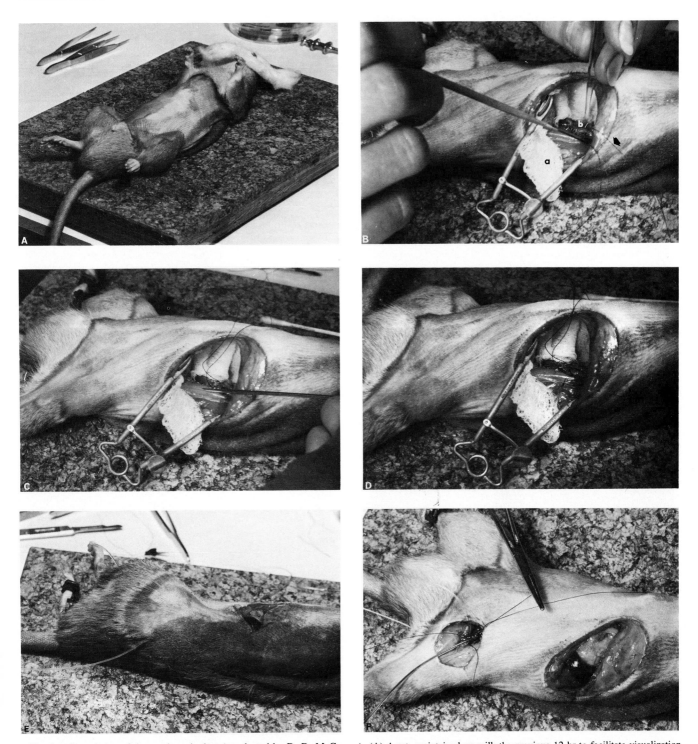

Fig. 6. Cannulation of the rat thoracic duct (as adapted by D. D. McGregor). (A) A rat, maintained on milk the previous 12 hr to facilitate visualization of the thoracic duct, is anesthetized with ether and its abdomen shaved. (Drapes are omitted in this series of demonstrations using a cadaver.) (B) An incision is made on the left ventral side of the rat below the diaphragm and the internal organs held out of the way by saline-soaked gauze packing (a). The thoracic duct (b) is then dissected free of the aorta and surrounding connective tissue with care. 5-O suture silk (arrow) is then slipped under and around the thoracic duct. (C) A 16-gauge trochar needle is used to penetrate the left rear dorsum posteriorly to the renal blood vessels and the cannula (not shown) passed through. The cannula (with the end bent to form a J so the orifice faces the flow of lymph) is then inserted into the thoracic duct approximately 2 cm below the diaphragm. (D) The cannula is held in place by sutures passed through surrounding muscle. (E) View of dorsum showing cannula exiting from the rat. (F) A cannula is introduced into the femoral vein to replace fluid loss in rats from which lymph is collected for more than a few hours. [See also Chapter 23 in Weir (289)].

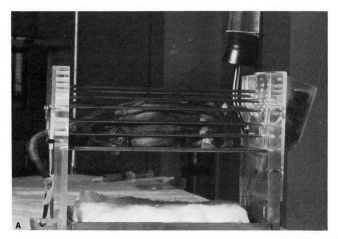

Fig. 7. Restraining cage for cannulated rats is constructed of plexiglass for the base and sides. Holes are drilled into the sides to allow stainless steel rods to be passed through to form the flooring and the sides. Five holes are drilled in every direction from the center of the restrainer to allow accommodation of rats of various sizes. A food hopper and a water bottle are arranged to allow easy access by the animal whose movements are controlled somewhat by two sutures in the back which are laced around the top restraining bar and by taping the tail which is passed through the hole for that purpose. [After Bollman (37).] (B) Restraining cage with cover in place. The cover contributes to maintenance of body heat and keeps the animal calm.

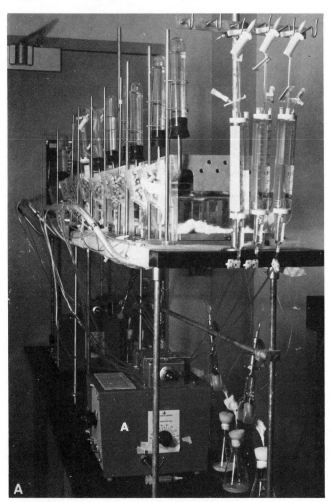

Fig. 8. The front (A) and rear (B) of the platform on which the restraining cages are placed to allow collection of lymph in flasks (arrow). Intravenous injection of saline is facilitated by peristaltic pumps (A).

cytes (75), while thymus-independent lymphocytes localize in the follicle from where they may move into the red pulp (63).

3. Lymphocyte Localization or "Homing"

The localization of recirculating lymphocytes in different tissue compartments and the factors affecting this localization or "homing" as it has been called, is presently the subject of a good deal of study. Although a considerable number of questions remain unanswered and the subject of debate, some basic findings in the rat have withstood the test of time and gained acceptance.

Studies involving lymphocyte migration and localization have most commonly employed *in vitro* labeling of lymphocytes taken from different tissue compartments with either [³H]thymidine, to selectively label blast cells, or [⁵¹Cr]- or [³H]uridine or other RNA precursor, to label all of the lymphocytes. The labeled cells are then injected into recipient rats which are necropsied at various times and the tissue localization of the radioactivity determined either by radiometry or autoradiography. Using such techniques, one of the earliest findings, since confirmed in a number of laboratories, is the predilection of large lymphocytes from the thoracic duct of normal rats to localize in the lamina propria of the intestinal tract (82). The large blast cells have been shown to localize strictly in the lamina propria of the small intestine rather than in the Peyer's patches (86). This was interesting in view of the fact that over 95% of the cells in thoracic duct lymph come from nodes draining the intestinal tract (161), suggesting a route by which cells stimulated by intestinal antigens could be distributed throughout the gut mucosa. Later, however, Halstead and Hall (96) demonstrated that the large lymphoid blast cells from the thoracic ducts of donor rats preferentially localized in the intestinal wall even when injected into unsuckled, neonatal rats presumably lacking food and gut flora antigens. Hall *et al.* (93) found that by 24 hr, after an early detention in the lungs, labeled large lymphocytes localized predominantly in the small intestine with some apparent labeling in the Peyer's patches. Careful examination using autoradiography, however, showed that the labeled cells were present only in the mucosal tissue surrounding the Peyer's patches. It appears, then that lymphoid cells in the gut can be stimulated by intestinal antigens, probably in the Peyer's patches which then enter the lymph and localize in the lamina propria (222). The cells moving in this way are primarily IgA-containing cells (293). A small percentage of cells from peripheral lymph nodes localize in the lamina propria of the gut (94), but for the most part, lymphocytes from peripheral nodes localize in lymph nodes and spleen when infused into recipients (143).

4. B and T Cells

The discovery of the existence of two functionally distinct populations of lymphocytes spanned species barriers and depended upon studies in the chicken (72), mouse (179), rat (7, 114, 286), and rabbit (78).

The early studies of Jankovic *et al.* (114), Arnason *et al.* (7), and Waksman *et al.* (286), and later by Rieke (233), demonstrated that two distinct populations of lymphocytes were present in rats. These workers, using rats thymectomized or splenectomized at various times after birth, observed that antibody formation, immediate and delayed hypersensitivity, cellular development of lymphoid tissues, and the ability of transferred lymphocytes to cause runting were differentially affected by such surgery. In the intervening years since these basic observations, work has been directed toward the characterization of the different lymphocyte populations, those bone marrow-derived lymphocytes that depend upon the thymus to function fully, T cells, and those which do not, the B cells.

a. B Cell Markers. One of the characteristics longest used to delineate between T cells and B cells is the differential incorporation of uridine *in vitro*. Rieke (233) first suggested such a difference when he reported that lymphocytes from thymectomized rats incorporated less uridine and cytidine *in vitro* than lymphocytes from normal rats. This observation has since been made repeatedly in a number of laboratories (8, 65, 101, 104, 106, 108, 243) and appears to be a true and consistent characterisitc. Howard *et al.* (104) used thoracic duct lymphocytes (TDL) from rats which had been thymectomized, irradiated with 1000 rads 2–4 weeks later, and then reconstituted with 2×10^8 dissociated bone marrow cells, or so-called "B" rats (102), to demonstrate a greatly depressed *in vitro* incorporation of [³H]uridine compared to the lymphocytes from normal rats. Cells from these rats appeared very lightly labeled in autoradiographs and localized in B areas of lymphoid tissue (Section III,A,6), whereas TDL from normal rats contained lymphocytes of two types, one labeling heavily with [³H]uridine and localizing in thymus-dependent areas of lymphoid tissue and the other having the characteristics of lymphocytes from "B" rats. All RNA precursors do not produce this differential incorporation; Gowans and Knight (82) did not observe a striking difference using [³H]adenosine, and Hollingsworth and Carr (101) observed equivalent labeling of lymphocytes in peritoneal exudates using [³H]cytidine.

The discovery that T and B lymphocytes metabolized uridine differentially *in vitro* allowed the two populations to be followed *in vivo* and characterized further. A summary of characteristics possessed by rat T and B lymphocytes is given in Table XI. Using uridine-labeled lymphocytes from "B" rats, Howard (102) found that B lymphocytes recirculate from blood

Table XI

Differentiating Characteristics of Rat Lymphocytes

Characteristics	B cell	T cell	Reference
Recirculate from blood to lymph	+	+	65, 103
Life span (days)	5	5	58, 103
In vitro utilization of uridine and thymidine	Low	High	8, 65, 101, 104, 233
Recirculation time (hr) in splenectomized rats	24–28	14–18	65
Thymus-independent; marrow origin	+	−	103
Thymus-specific alloantigen	−	+	154, 44a
Fc and C3 receptors	+	−	211
Surface immunoglobulin	+	±	74, 109, 211

to lymph and that a large proportion (90%) are long-lived, being present in lymph for more than 7 days, confirming earlier estimates of Everett *et al.* (58). The frequency of short-lived B cells did not differ from that in normal lymph. It is postulated that these short-lived B cells are the immediate precursors of antibody-forming cells (102). The recirculation times of B lymphocytes in splenectomized rats is estimated to be between 24 and 28 hr (65).

Populations of lymphocytes have little morphological heterogeneity, so Parish and Hayward (210, 211) used cell surface markers to differentiate subpopulations of lymphocytes. Lymphocytes from normal and "B" rats were divided into subpopulations of cells bearing either Fc receptors, C3 receptors, or surface immunoglobulin by forming rosettes with sheep red blood cells sensitized to each receptor or surface immunoglobulin which could then be isolated on Isopaque-

Table XII

Characterization of Cells in Rat Thoracic Duct Lymph Collected within 24 hr of Cannulation

Type	Mean (%)	Reference
T cells	87	76
	75	101
B cells	11	76
	42	211
Fc receptor cells	29.5	211
C3 receptor cells	24.1	211
Surface immunoglobulin	31.5	211
Large lymphocytes	9.1	293
Contain internal IgA	5.7	293
Contain surface IgA	8–15	293
Contain surface Fab	38–40	293
	25–30	109
	40–50	109, 122

Ficoll columns (210) (see Section III, A, 7). It was found that 29.5% of TDL collected within 24 hr of cannulation demonstrated Fc receptor (the receptor for antigen–antibody complexes), 24.1% possessed C3 receptor, and 31.5% contained surface immunoglobulin (211) (see Table XII). Cells possessing B cell antigens but lacking Fc, C3, or surface immunoglobulin constituted less than 5% of normal lymph but amounted to 32–42% of TDL collected from "B" rats (211). Most lymphocytes possessing Fc and C3 receptors also had surface immunoglobulin, but only 75–88% of cells with surface immunoglobulin also possessed C3 receptor. Estimates of the proportion of B cells in normal TDL varies considerably depending on technique and strain of rat (Table XII). Goldschneider and McGregor (76) found that 11% (range 8–14%) of the cells in normal TDL collected within 24 hr from (LEW × DA)F$_1$ rats reacted with anti-lymphocyte serum made specific for B cells. Using a similar approach, Parish and Hayward (211) found that 42.4% of undepleted, 24-hr TDL from HO rats were killed by anti-B serum. This proportion increased with the length of thoracic duct drainage. If the presence of surface Fab fragments is used to define B cells, the range of B cells in normal lymph extends from 25% in DA rats (109) to approximately 50% in conventional PVG/c rats (122). Hunt and Williams (109) used the sensitive and quantitative method of reacting TDL from DA and PVG/c rats with ^{125}I-labeled anti-Fab and ^{125}I-labeled anti-light chain allotype sera to determine the proportion of B cells in normal lymph. Only cells demonstrating between 10,000 and 100,000 molecules of antibody per cell were considered B cells. Their data indicate definite strain differences in the proportion of B cells in normal lymph since 25–30% of TDL from DA rats reacted with anti-Fab serum, whereas 40–50% of TDL from PVG/c rats of comparable age and sex reacted (109, 122). This strain difference is apparently unrelated to microbial status of the animals, since Williams and Gowans (293) found no difference in the percentage of TDL reacting with anti-Fab antibody in specific pathogen-free (SPF) and conventional PVG/c rats. Thus, the difference in estimates of the proportion of B cells in normal TDL between Goldschneider and McGregor (76), 11%, and Hunt and Williams (109), 40–45%, may be a combination of strain differences and the relative sensitivity of the techniques used. The proportion of TDL demonstrating IgA immunoglobulin was also observed to be less in SPF rats than in conventional, "dirty" rats, this being approximately one-half of all large thoracic duct lymphocytes (293). The immunoglobulin found in large TDL was exclusively of the IgA type, with neither IgG2a + 2b nor IgM being found (123).

An additional marker which is present on T cells as well as B cells has been termed LC antigen (leukocyte-common) (58b, 260a). It has recently been discovered that this antigen is identical to rat Ly-1 described by Fabre and Morris (58a) and is very similar to the A.R.T. antigen (44a) (see page 198).

b. T Cell Markers. As indicated in the discussion of B lymphocytes, it has been consistently observed that T cells actively incorporate uridine *in vitro*. T cells become heavily labeled when exposed to [³H]uridine *in vitro* making it possible to study their life history, proportion, and distribution in lymphoid tissues (Table XIII).

T lymphocytes are known to be long-lived (58) and to recirculate from blood to lymph with a recirculation time in splenectomized rats of 14–18 hr (65). The determination of cell surface markers on T cells is not as clear as for B cells. Parish and Hayward (211) found that 5–10% of the cells in normal TDL which carry Fc receptors do not demonstrate surface immunoglobulin. However, in the absence of a dependable peripheral T cell marker in the rat, it is not clear whether these Fc-containing cells are T cells or immature B cells. It also is a moot question whether T cells carry surface immunoglobulin. Goldschneider and Cogen (74) have reported finding immunoglobulin molecules of undetermined origin on the membranes of activated T cells. Using the sensitive, quantitative technique of reacting TDL with ¹²⁵I-labeled anti-Fab, Hunt and Williams (109) detected two populations of lymphocytes delineated on the basis of the degree of labeling. One population absorbed 10,000–100,000 molecules of antibody per cell (B cells) and the other, which they considered T cells, 100–2000 molecules per cell. Through the use of an anti-light chain allotype, these workers were able to show that the immunoglobulin on the lightly labeled cells was passively acquired and

primarily of the IgM type. Jensenius and Williams (123), however, found a surprising amount of immunoglobulin in detergent-solubilized thymocytes. The immunoglobulin detected was primarily of the IgA and IgG2a + 2b classes.

Since the report of an alloantigen (Thy-1) in mice (230) [originally called θ-AKR and θ-C3H, found on T cells but absent in B cells (229)], a considerable effort has been directed toward finding a similar T cell antigen in the rat. The occurrence of thymus-specific antigens in the rat was reported early from a number of laboratories (9, 47, 113, 224). Fuji *et al.* (66) first suggested the presence of Thy-1.1 (θ-AKR) antigen on rat thymocytes when they reported that lymphoid cells from C3H mice immunized with AKR thymus produced plaques on rat thymus cell cultures from six different strains. The finding of Thy-1 in rats was later supported by studies from other laboratories (46, 52, 177, 219). Douglas (53) showed that rat brain and thymocytes would absorb anti-θ-AKR (Thy-1.1) serum but not anti-θ-C3H (Thy-1.2). He observed, by determining the absorptive capacity of lymphoid tissues for anti-Thy-1 serum, that the distribution of Thy-1 antigen in rats closely paralleled that found in mice. An exception to the similarity in distribution was that peripheral lymphoid tissues and cells of the rat exhibited very little Thy-1 antigen. It also was observed that rat bone marrow contained more Thy-1 positive cells than mouse (177). Acton *et al.* (1) confirmed earlier work on Thy-1 and also showed that lymphocytic leukemia cells as well as thymocytes possessed Thy-1.1; in addition, it

Table XIII

Percentage of T and B Cells in Lymphoid Compartments of the Rat

Tissue	Mean (%) of lymphocytes		Technique	Reference
	T	B		
Thymus	99	0.1	ALS	76
	93	—	Anti-Thy-1	292
	90	—	Anti-Thy-1.1	1
Spleen	50	50	ALS	76
	65	—	Uridine incorporation	101
	12	—	Anti-Thy-1.1	1
Peripheral blood	70	—	Uridine incorporation	101
	61	40	ALS	76
Lymph nodes	81	18	ALS	76
	8	—	Anti-Thy-1	292
	2	—	Anti-Thy-1.1	1
Peyer's patches	—	—	ND[b]	
Thoracic duct	87	11	ALS	76
	75	—	Uridine incorporation	101
	8	—	Anti-Thy-1	292
	2	—	Anti-Thy-1.1	1
Thoracic duct (TxB rats)	14[a]	84	ALS	76
Bone marrow	13	38	ALS	76
	45	—	Anti-Thy-1	292

[a] Two percent of normal.
[b] ND, no data.

was shown that the rat Thy-1.1 antigen was so similar to mouse Thy-1.1 that the two could not be distinguished by tissue absorption. It has since been observed that rat Thy-1 is composed of several xenoantigenic determinants as well as Thy-1.1 (185, 186), and appears to be a glycoprotein of approximately 25,000 MW (266).

As has been mentioned, rats differ from mice in the distribution of lymphocytes carrying Thy-1 in that 30–45% of rat bone marrow cells react with anti-Thy-1 antibody (292), while 12% of spleen and only 2% of lymph node or thoracic duct lymphocytes react (1). Williams (292) has postulated that the Thy-1-positive lymphocytes in the spleen are peripheral T cell precursors issuing from bone marrow rather than thymus. This would be the converse of what is observed in mice, wherein prothymocytes lack Thy-1 while peripheral T cells express it (133). Using parameters mentioned above by which T cells in the rat are defined, it is clear that the majority of mature rat T cells lack Thy-1, and conversely it is unlikely that the bone marrow cells possessing Thy-1 are mature T cells (292). This is further supported by the fact that less than 2.5% of rat bone marrow lymphocytes are immunocompetent (105). It is quite clear, therefore, that the Thy-1 antigenic marker is not useful in the detection of mature T cells in the peripheral lymphoid tissue of the rat as it is in the mouse. Attempts have, therefore, been made to find an antigenic determinant in the rat possessed by mature T cells but absent from B lymphocytes. Lubaroff (154) and Howard and Scott (106) have reported two possible alloantigenic systems. Howard and Scott (106) developed cytotoxic antisera which detect a polymorphic alloantigen system which may contain part of the Lubaroff system. The antigens detected are present on rat peripheral thymus-derived lymphocytes but reduced or absent on thymocytes.

More recently, Lubaroff (155) has characterized an alloantigenic marker found on rat thymus and thymus-derived lymphocytes which he terms A.R.T. (antigen of rat thymus). The absence of this antigen from rat and mouse brain, as well as the augmentation of serum titers against this antigen by substituting rabbit serum for guinea pig serum as a complement source, suggests that A.R.T. is more analogous to mouse Ly antigens than to Thy-1. Weak A.R.T. antiserum can be produced by using several Ag-B compatible strains, but NBR anti-LEW antiserum (both compatible at the *RT-1^l* locus) demonstrated a cytotoxic index which was a magnitude greater than that of antiserum produced in two other A.R.T.-negative strains. Whereas A.R.T. may be more similar to an Ly antigen than to Thy-1, Thompson and Morris (265) have reported the ability of rabbits to detect two xenoantigenic determinants of Thy-1 (but not Thy 1.1) when injected with rat thymus. Two other antigens were recognized by rabbits injected with thymoma, one of which was found on thymus and peripheral lymphocytes but only on 1% of bone marrow cells. These systems await further definition.

Table XIV
Turnover of Monocytes in Normal and *Salmonella*-Infected LBN Rats

Parameter examined	Normal	*Salmonella*-infected	Reference
Total monocyte count (cells/mm^3)	1200	4100	276
Total blood monocyte pool	6.3×10^7	21.6×10^7	273
Half-time loss of monocytes from the blood (hours) based on regression analysis of percent labeled monocytes:			
Highly labeled cohorts^a	35	12	276
All labeled monocytes	61	25	276
Average life span (hr)	88	35	276

^{a}The more highly labeled monocytes are believed to represent a more homogeneous population and hence yield a more accurate estimate.

An approximation of the proportion of T and B cells in the various lymphoid compartments of the rat as determined by reaction with anti-lymphocyte sera made specific for T or B cells (76), differential uridine incorporation (101), or reactivity with anti-Thy-l (292) and anti-Thy-1.1 (1) antisera are shown in Table XIII. Not listed is the percentage of T cells in peritoneal exudates which Hollingsworth and Carr (101) have found to vary considerably (3–38%), depending on the inducing agent.

5. B and T Cell Cooperation

Cooperation between B and T cells has been studied primarily in regard to initiation and enhancement of the humoral response. Only recently has B and T cell interaction in cell-mediated immune responses received attention (see Section III,B,2). Johnston and Wilson (124) first reported the need for thymus-derived helper cells, as well as bone marrow-derived antibody-forming precursor cells in the initiation of an optimal response to sheep red blood cells in the rat. This confirmed the existence in the rat of a B and T cell interaction similar to that reported in the mouse (183).

The presence of functional T and B cells in bone marrow (105) and in thoracic duct lymph (242) has been demonstrated in the rat. Although the proportion of T cells in rat bone marrow is estimated at less than 2.5%, this is still sufficient to restore the hemolytic response to sheep red blood cells in thymectomized, X-irradiated recipients (105). This restorative effect can be abrogated by chronic thoracic duct drainage of the donor, indicating that the T "helper" cell recirculates from blood to lymph and is long-lived (105, 258). The immediate precursor to the IgG or IgM antibody-forming cell is a large B lymphocyte which is the cell capable of adoptively transferring both types of hemolytic plaque-forming responses (103). This has been further defined by Gowans (81) who showed that removal of lymphocytes possessing Fc receptor had no effect on the ability of donor lymph to transfer a plaque-forming cell response to irradiated recipients. Removal of lymphocytes with surface immunoglobulin or C3 receptors abrogated the 7 S response but not the 19 S response in recipients. It was postulated that precursor B cells for 19 S antibodies are resistant to 850 rads irradiation and thus mounted a response with T cells from the donor (81). Unprimed or virgin B cells do not recirculate (258) and are short-lived, with a turnover rate which appears dependent on the antigen (258, 260).

In addition to lymphocytes which transfer a primary response, populations exist which are capable of transferring a memory or secondary response to an antigen (83, 170, 258) after priming. Both B and T memory cells exist in the rat which are long-lived and able to recirculate, although the rate of the two is different (259, 260).

The reader is referred to reviews by Miller and Mitchell (181) and Miller *et al.* (180) for a more general discussion of B and T cell cooperation.

6. B and T Regions of Lymphoid Tissues

The first indication that lymphoid tissues in the rat possessed rather discrete areas which were populated predominantly by B cells or T cells came when Waksman *et al.* (286) observed depletion of lymphocyte numbers in certain areas of lymphoid tissues from thymectomized rats. Studies of B and T cells in the rat lagged, somewhat, behind those of the mouse until Howard *et al.* (104) developed the necessary reagents and technology to define B and T areas in rat lymphoid tissue as Parrott *et al.* (212) had done in the mouse. Howard and his colleagues noted that two populations of small lymphocytes could be delineated in thoracic duct lymph on the basis of sedimentation rate and differential *in vitro* incorporation of uridine (104). Thoracic duct cells which labeled lightly *in vitro* with [³H]uridine and which sedimented slowly in albumin gradients were greatly enriched in numbers when taken from thymectomized, irradiated, and bone marrow reconstituted rats, therefore corresponding to B cells. These lightly labeled cells, when transfused into normal recipients were observed to localize predominantly in the follicular areas of the lymph nodes and spleen and in the corona of lymphocytes around the germinal center in Peyer's patches. Small thoracic duct lymphocytes, corresponding to T cells, which labeled heavily *in vitro* with [³H]uridine and which sedimented quickly in albumin gradients, localized predominantly in areas surrounding splenic arterioles, the deep cortex of lymph nodes, and the interfollicular areas of Peyer's patches.

Exhaustively absorbed anti-lymphocyte sera (ALS) directed against B cell surface antigens were used by Goldschneider and McGregor (76) to show that the cells which corresponded to B

and T cells in the study of Howard *et al.* (104) did indeed possess the antigenic characteristics of B and T cells. These workers observed that B lymphocyte-specific ALS reacted specifically with cells in the follicular cortex of lymph nodes and cells surrounding germinal centers in the spleen. On the other hand, ALS directed against T cells reacted with lymphocytes in the deep cortex (paracortex), medullary cords, and lymph sinuses of lymph nodes, and the periarteriolar lymphocyte sheath in the splenic white pulp. Peyer's patches were not looked at in this study. Neither Goldschneider and McGregor (76) nor Howard *et al.* (104) observed cells with characteristics of either B or T cells in germinal centers of lymph nodes, as Gutman and Weissman (90) did in finding T cells in the germinal centers of mice. Osogoe *et al.* (207) did find, however, that germinal center cells of spleen, Peyer's patches, and lymph nodes are similar to thymus cells in that these cells incorporate [³H]deoxycytidine *in vivo* much better than [³H]thymidine, which is readily utilized by the other lymphoid cells. A schematic drawing of rat lymphoid tissues incorporating these findings is shown in Fig. 9.

7. Separation of B and T Cells

Both physical and immunochemical methods have been employed in attempts to separate and isolate subclasses of lymphocytes to allow their further characterization. Fractionation of rat lymphocytes on glass bead columns was used by Shortman (245), while Mel *et al.* (173) employed the STAFLOW system to fractionate bone marrow cells generally. Later workers employed differences in cell density to separate lymphocyte classes physically. In 1968, Shortman (246) reported successfully separating classes of thymus and thoracic duct cells from rats on the basis of their buoyant density. Discrete bands or peaks were obtained by the centrifugation of cells in isotonic albumin gradients. Centrifugation of the gradients at 3500–4000 *g* for 45 min yielded optimum resolution of lymphocyte peaks with a minimum of cell damage. Cell viability remained high, and the average cell recovery from the gradient was 89% in the case of rat thymus and thoracic duct cells. Furthermore, no alteration in the response of the cells to antigen (98) or their performance in the hemolytic plaque assay could be attributed to the separation procedure (99).

Howard *et al.* (104) used the velocity sedimentation methods described by Miller and Phillips (182) to delineate two subpopulations of lymphocytes in rat thoracic duct lymph. Lymphocytes were allowed to sediment in a shallow bovine serum albumin gradient at 1 *g* for 8–9 hr at 4°C; partial separation of two distinct populations of lymphocytes was achieved in this way. One fraction, corresponding to B lymphocytes, had a low sedimentation velocity, low uridine uptake, and migrated to thymus-independent areas of lymph nodes, and the other, with a relatively high sedimentation rate, a high degree of uridine

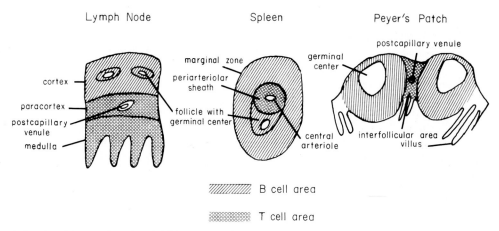

Fig. 9.   A schematic drawing of the areas in rat lymphoid tissues which are populated primarily by B or T lymphocytes. [Partially adapted from Goldschneider and McGregor (76).]

incorporation, which localized in thymus-dependent areas of lymph nodes, corresponded to T cells. Hunt (108) discovered that TDL passed through siliconized bead columns would selectively remove B cells. The use of uridine incorporation and alloantigenic markers indicated that B cells would stick to the column, whereas T cells would pass through quickly.

More recently, immunochemical methods have been used to separate and characterize subclasses of lymphocytes. In 1973, Schlossman and Hudson (240) used rabbit anti-mouse Fab antibody coupled to Sephadex G-200 to separate populations of mouse spleen cells. Later, Crum and McGregor (49) adapted this technique for use in the rat. Antibody directed against rat F(ab')₂ was produced in rabbits and coupled to Sephadex G-200. Rat thoracic duct cells passed over the column in Hank's balanced salt solution (Fig. 10) separated into two fractions. The nonadherent fraction, consisting of lymphocytes without detectable surface immunoglobulin (T cells and "null" cells), was recovered in the effluent, while lymphocytes with readily demonstrable surface immunoglobulin (B cells) were retained on the column. These latter lymphocytes could be recovered by flooding the column with rat immunoglobulin. Immunoglobulin-bearing cells in the nonadherent fraction were reduced to 2% or less following one passage through the 50-mm column and could be removed completely with a second passage. Over 94% of the cells retained on the column had readily demonstrable surface immunoglobulin. Cells recovered from the column were over 99% viable and were unaffected immunologically as determined by transferring resistance to *Listeria monocytogenes* infection, graft versus host reactivity, the plaque-forming response to sheep erythrocytes (SRBC), and the transfer of a memory response to phage φX174.

Parish and Hayward (210, 211) used a combination of physical and immunochemical methods to characterize lymphocytes

further. Rat lymphocytes from spleen, lymph nodes or the thoracic duct were first induced to form rosettes with SRBC which had been sensitized to detect either Fc receptor, C3 receptor, or surface immunoglobulin. The lymphocytes forming rosettes and those not forming rosettes were separated from each other by layering the cells on a mixture of 5 parts of 32.8% Isopaque (Nyegaard; Oslo, Norway) and 12 parts of 14% Ficoll (Pharmacia; Upssala, Sweden) and centrifuging at 1200 *g* for 30 min at 20°C. The cells forming rosettes sedimented to the bottom of the tube, while nonrosetting lymphocytes remained suspended. In the case of TDL, the rosette-forming lymphocytes were 93–98% pure, i.e., less than 7% unrosetted lymphocytes. Total recovery of cells using this method was greater than 90% with a viability exceeding 95%. Use of this method allowed B cell depletion of lymphocyte populations, since all three receptors occurred predominantly on B cells (211).

Recently, the fluorescence-activated cell sorter (FACS) has proven to be very useful in the separation of lymphocyte populations (100a). The discovery of monoclonal antibodies specific for subpopulations of rat lymphocytes (290a) has added greatly to the potential of FACS analysis.

8. Delayed Hypersensitivity Responses

Although delayed-type hypersensitivity is almost defined in terms of the guinea pig response to PPD following immunization with mycobacteria, many workers have used the rat in studies of delayed hypersensitivity. Initially, it was considered that the rat was incapable of developing a delayed skin hypersensitivity (73, 100, 204), but in 1941, Wessels (290) described the successful production of tuberculin sensitivity in the rat. Since that time footpad and flank skin (60) thickness have been used to demonstrate delayed-type hypersensitivity in means of calipers and areas of erythema by eye, some workers have estimated footpad swelling by fluid displacement.

Lefford (145) has attempted to reduce the subjectivity involved in measuring delayed-type hypersensitivity reactions in the rat by using radiolabeling techniques. The knowledge that monocytes entering sites of delayed hypersensitivity reactions were newly formed in the bone marrow (279) was exploited by pulse labeling the test animal with an injection of [^{3}H]thymidine 24 hr before administering antigen. The test antigen is injected into the pinna of one ear and diluent injected into the other. After 24 hr (i.e., 48 hr after pulse labeling), the injection sites are biopsied by means of a Keyes punch. The radioactivity in the biopsied tissues is assessed using a liquid scintillation counter, and the ratio of counts between the antigen injected ear and the control ear, receiving diluent alone, is determined.

An *in vitro* assay for delayed hypersensitivity (50) has also been adapted for the rat which correlates with measurements of skin sensitivity (136). In this assay, peritoneal exudate cells are removed from test animals 3 days after induction with sodium caseinate. The cells from individual animals are packed in capillary tubes and placed in chambers containing culture medium with or without the test antigen, and migration, or inhibition of migration, of the cells is measured following incubation. This assay proved more sensitive than other methods of assessing delayed hypersensitivity if high concentrations of antigens were used.

B. Monocytes and Mononuclear Phagocytes

1. Kinetics

It still is a moot question whether blood-borne monocytes and mononuclear phagocytes,* primarily Kupffer cells of the liver and alveolar macrophages of the lung, in the rat are or are not part of the same cell line. The work of Volkman and Gowans (279) and Spector *et al.* (251) has indicated that blood-borne monocytes are the source of immigrant macrophages in inflammatory foci in the rat. The circulating monocytes are known to be produced primarily in the bone marrow and to a lesser extent in the spleen (252, 280). They circulate in the blood of the rat with a half-life in the range of 13–75 hr and blood-borne life span in the range of 19–108 hr with the more representative figures being 42 and 61 hr, re-

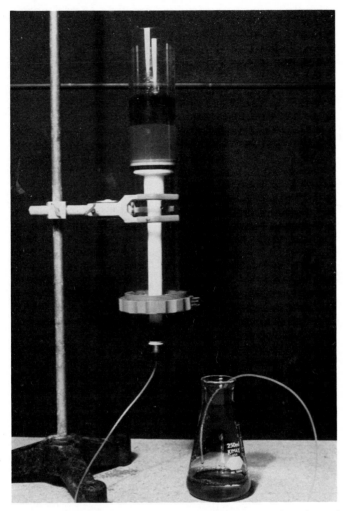

Fig. 10. Rat thoracic duct lymphocytes being passed through a column of rabbit anti-rat F (ab′)$_2$ antibody coupled to Sephadex G-200. Cells without surface immunoglobulin are retained on the column. The latter are eluted with homologous serum.

the rat. Measurements of delayed skin hypersensitivity are extremely subjective and vary considerably between observers; this is especially true when calipers are used to measure skin thickness. The thick rat dermis leads to variation in measurements between observers due to the variation in compressibility of edema fluid in the skin. In spite of this, careful workers can obtain reproducible results with little variation using calipers, since it demands an unequivocal inflammatory reaction. The calipers employed should spring open rather than spring close, since the latter type will exert increased tension as the induration increases thus introducing a source of error. If edema fluid is considered as much a part of the reaction as the hyperemia underlying the erythema and the cellular infiltrate responsible for the induration, the reaction site should not be compressed (275). In addition to measuring induration by

---

*A variety of terms have been used in reference to the mononuclear phagocytes found primarily in the lungs, liver, tissue spaces, and peritoneal cavity. These terms have included fixed, sessile, littoral, tissue, and resident macrophage, none of which accurately describes these macrophages. Thus, in this chapter, mononuclear phagocytic cells which are localized in organs or tissue spaces will be referred to simply as mononuclear phagocytes; this term excludes blood-borne monocytes which are not phagocytic while in the circulation.

spectively (272, 273). These times are distinctive for monocytes; the granulocytes have a much shorter life span in the blood, and the lymphocytes generally have a much longer life span. Bone marrow precursors of rat monocytes possess a cell cycle time of between 18 (273) and 21 hr (291) and emerge from the bone marrow a minimum of 6–8 hr after the last precursor cell division (273, 291). Neither the output (cells per kilogram per hour) of monocytes nor the total precursor pool has been determined for the rat. The turnover and distribution of monocytes in the rat can be affected by an infectious disease (276) as is illustrated in Table XIV.

The early work of Ebert and Florey (55) in the rabbit presented solid evidence that the circulating lymphocyte was not the precursor of the tissue macrophage. The work cited above indicates that the same is true in the rat; labeled lymphocytes injected into histocompatible recipients never yielded labeled macrophages on skin windows (279). Any apparent lymphocytic origin of macrophages in the rat can be accounted for by the small percentage of macrophage precursors present in rat thoracic duct lymph (236). A controversy still exists, however, as to whether the blood-borne monocyte which enters areas of inflammation is also the antecedent of tissue and alveolar macrophages and Kupffer cells. From studies in mice, van Furth (67) contends that the monocyte is the precursor of the alveolar macrophage, peritoneal macrophage, histiocyte, and Kupffer cells. Van Furth et al. (68) feel this applies to mammalian systems generally. Volkman (274) maintains that data from the rat do not support such a view at the present time. It is argued that present information neither supports nor wholly disproves a single cell origin of macrophages and that further study may reveal several subclasses of macrophage precursors analogous to what has already been found among the lymphocytes. In this connection, Wisse (296) gives an excellent review of the work published on the differentiation of rat endothelial and Kupffer cells, which some consider to be different stages of development of the same cell type. He finds that differences between rat monocytes and Kupffer cells, particularly in peroxidase activity, does not support a transition of monocytes to Kupffer cells. In ultrastructural studies, Wisse (297) indicates "that Kupffer cells constitute a homogeneous population of macrophages, recognizable by characteristic morphological features."

Studies performed in the mouse by North (195) have shown that Kupffer cells are capable of division, in this case during a response to a systemic *Listeria monocytogenes* infection; thus all Kupffer cells cannot be considered direct progeny of a blood-borne monocyte precursor. Kupffer cell division is known to occur normally in the rat (274). Moreover, Wisse and Daems (298) have shown that Kupffer cell division occurs after partial hepatectomy as is also the case with hepatocytes. Resolution of the controversy awaits further characterization of mononuclear phagocytes and blood-borne monocyte(s) in the

rat as well as embryological studies. The latter may well indicate a hematogenous origin for mononuclear phagocytes during fetal development which later becomes a self-renewing system occasionally fortified in instances of excessive cell loss.

2. Functional Aspects

In addition to granulocytes, the blood-borne monocytes and the tissue macrophages are the phagocytic cells in the rat which act directly to combat infection and play an important role in anti-tumor immunity. The macrophages in the rat are primarily the alveolar macrophages of the lung, Kupffer cells of the liver, and histiocytes of tissue and the peritoneal cavity. The dendritic cells which are found in the spleen, lymph nodes, and Peyer's patches of the mouse have been observed in the rat as well (257), but these cells most probably do not function as phagocytes.

It appears that the macrophages of the rat exist in a more activated state than those of the mouse in terms of normal phagocytic and bactericidal activity (45, 282, 283) which may account for the rat's comparative basic resistance to many infectious agents. Studies of immune activation of macrophages over the past two decades, done particularly in the rat and the mouse, have led to the development of a proposed mechanism of cellular immunity (156, 157). A comprehensive summary of the work in various animal species which has contributed to the understanding of cellular immune mechanisms has recently been published by Campbell (42). As mentioned, immune activation of macrophages appears to develop faster in the rat than in the mouse as indicated by studies with tubercle bacilli (146). This activation is mediated by specifically sensitized lymphocytes as shown in work by McGregor and Koster (166) who transferred TDL from specifically immunized rats to protect cyclophosphamide-treated rats against a *Listeria* challenge. It was shown that macrophages of the recipient rat were activated by the transferred lymphocytes to produce accelerated destruction of the challenge dose. In the mouse, the lymphocyte which produces immune activation of macrophages is thymus derived (34, 144, 196). Although not specifically demonstrated in the rat, due to the lack of an acceptable T cell marker, the lymphocyte responsible for immune activation of macrophages is almost certainly a T cell.* The lymphocytes which function in rat cell-mediated immune responses share so many of the biological characteristics of similarly functioning mouse lymphocytes, shown to be T lymphocytes, that it is unlikely these are other than T cells in the rat. Using thoracic duct cells separated by means of anti-immunoglobulin columns (see Section III,A,5), Crum and McGregor (49) demonstrated

---

*Markers specific for peripheral T lymphocytes in the rat have recently been described (290a), and McGregor et al. (164a) have succeeded in transferring anti-*Listeria* immunity with T lymphocytes in rats.

impressive protection against a *Listeria* challenge with an infusion of the nonadherent, T cell-enriched fraction as compared to the adherent fraction. Slight protection was transferred with cells from the adherent population which can be accounted for by the sticking of activated T cells, since these cells can possess immunoglobulin on their surface (74). Another possibility which has received little attention in the rat, is the ability, in guinea pigs, of a population of B cells to activate macrophages (284, 295, 303). These findings in the guinea pig depend upon the presence of complement receptors on lymphocyte membranes to differentiate B cells from T cells since, as in the rat, the guinea pig presently lacks an acceptable peripheral T cell antigenic marker. If it is true that guinea pig B lymphocytes alone, when presented with certain antigens, can produce lymphokines which cause activation of macrophages, a similar system could possibly be found to exist in the rat. Two pathways leading to immune activation of macrophages, one via T cell mediators and the other via B cell mediators, could account for the ability of cells which adhere to anti-immunoglobulin columns (49) to transfer cellular immunity. Such a possibility may also help to explain the observation of McGregor *et al.* (165) that rats, depleted of T cells by thymectomy and irradiation, rapidly regained the ability to defend themselves against a primary infection with *Listeria*. In this work, the defense mechanism was shown to be immunologically based by protecting cyclophosphamide-treated recipients against listeriosis with cells from the thymectomized, irradiated donors. It was shown that TDL from normal animals improved the response of the thymectomized rats, whereas bone marrow cells did not, suggesting the need for T cells.

It has been shown for a number of species (220), as well as the rat (136), that products from sensitized lymphocytes (lymphokines) are the direct mediators of macrophage mobilization and activation as part of the cellular immune response. In guinea pigs, lymphokine-mediated activation is characterized *in vitro* by an increased uptake of glucosamine by macrophages (97). Although presently not known to exist in rats, Wahl *et al.* (285) have shown *in vitro* that guinea pig macrophages are required by T lymphocytes to produce monocyte chemotactic factor and macrophage activating factor and to proliferate in response to certain antigens. If this proves to be true for rats as well, immune activation of macrophages may occur via the pathways described in Fig. 11 for the cellular response to an endotoxin-containing organism using *Salmonella enteritidis* as an example.

The initial inflammatory response to infection results in the early appearance of rapidly dividing short-lived lymphocytes which are incapable of recirculating from blood to lymph (167) but which possess the capacity to accumulate in sites of specific or nonspecific inflammation (134). The advent of these cells in a lesion results in an influx of blood-borne monocytes. A secondary effect is the production of long-lived small lymphocytes which are capable of recirculating from blood to lymph and carry immunological memory (83, 170, 258). A second exposure to antigen yields an accelerated influx of cells at the antigenic focus consisting first of sensitized lymphocytes followed by macrophages (169). The sensitized lymphocytes which localize in sites of secondary infection release products which amplify inflammation causing an outpouring of monocyte-derived macrophages (135) into the lesion and inducing a release of monocytes into the circulation as was observed in rats undergoing a rechallenge to *Salmonella* (276).

The size of the memory cell population which is produced in response to a primary infection appears to be related to the

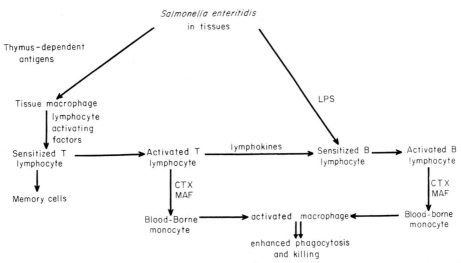

Fig. 11. Possible pathways by which a cellular immune response may be initiated toward live *Salmonella enteritidis* in the rat. CTX, chemotactic factors; MAF, macrophage activating factors.

chronicity of the primary infection. Rats infected with BCG, an organism able to cause a chronic infection, produces a cell-mediated response of great longevity compared to that produced by acute *Listeria* or subacute *Francisella* infection (168). The microbicidal activity of rat activated macrophages upon reinfection with a specific organism is nonspecific in nature, demonstrating enhanced phagocytosis and killing of organisms unrelated to that which induced the secondary response (136).

IV. AUTOIMMUNE DISEASE

Spontaneously occurring diseases of an immunopathological or autoimmune nature are relatively infrequent in rats. However, the rat has proved to be an excellent animal for studies of experimental autoimmune diseases as models of human immunopathological conditions. Conditions which have received particular emphasis in the rat include experimental autoimmune thyroiditis (EAT), experimental allergic encephalitis (EAE), and glomerulonephritis.

A. Thyroiditis

Thyroiditis in the rat has been observed to occur spontaneously in BUF rats (92, 247). The spontaneous disease is similar in all respects to autoimmune thyroiditis, being characterized by a typical mononuclear cell infiltrate of the thyroid stroma of lymphocytes, macrophages, and plasma cells. The first report of a consistent, reproducible thyroiditis in rats was made by Jones and Roitt (125) who found evidence of EAT within 2–4 weeks after injecting rats with homologous or heterologous thyroglobulin or thyroid extracts which had been emulsified in Freund's adjuvant. The pathological response ultimately regressed completely. More recently, progressive disease without evidence of regression, thus more closely resembling the human condition, can be achieved by incorporating pertussis vaccine (167). Also, studies have since demonstrated the production of EAT without the need for Freund's adjuvant (248). As is typical of the experimental model for autoimmune disease in the rat, the pathological effects can be transferred by lymphoid cells (268) and inhibited by pretreatment or concomitant treatment with anti-lymphocyte serum (131). The appearance of EAT in susceptible rat strains is enhanced by T cell depletion which has long been interpreted as being the result of loss of suppressor T cells allowing large amounts of antithyroid antibody to be produced (217). Such an interpretation is supported by the observation that cyclophosphamide treatment causes a permanent remission of active thyroiditis in rats (214).

A definite genetic relationship has been reported for development of both spontaneous thyroiditis (247) and the experimental thyroiditis (218, 235). The strain demonstrating spontaneous thyroiditis (BUF) is not the best for EAT (LEW) (247).

B. Autoimmune Encephalitis

Experimental autoimmune encephalitis is the oldest and most extensively studied of experimental autoimmune diseases, and the rat has been important as a model for these studies (213). The disease can be produced in rats by the injection of encephalitogenic protein without the need for Freund's adjuvant (152). It is characterized by a mononuclear infiltrate (213), although increased vascular permeability with accumulation of immunoglobulin may precede these cells (200). The disease can be produced by the passive transfer of sensitized lymphoid cells (153, 213). The ability of the transferred lymphoid cells to induce disease can be affected by irradiation of the recipient (215), yielding enhanced disease.

Unlike EAT, depletion of T lymphocytes inhibits the development of EAE and the production of antibody against myelin basic protein (77). A similar effect can be achieved by using the T cell suppressive drug, tilorone (172), although later, more severe manifestations of EAE may develop (150). As is the case for thyroiditis, however, different rat strains do exhibit a differential susceptibility to EAE (151) which is closely linked to the major histocompatibility locus, probably an immune response gene affecting T cell function (294).

C. Other Immunopathological Diseases

Additional diseases of an autoimmune or immunopathological nature for which the rat represents a valuable animal model include experimental glomerulonephritis (269) and experimental autoimmune myasthenia gravis (147). Both are mediated by antibody produced against host antigens, kidney tissue in the former and acetylcholine receptor in the latter. The occurrence of *Salmonella*-associated arthritis in rats (277, 278) appears dependent on the host's immune response and may represent a rat form of rheumatoid arthritis.

V. INFECTION AND TUMOR IMMUNITY

It is not within the scope of this chapter to describe the rat's immune response to specific antigens, pathogens, or tumor model systems. The reader is directed to the chapters dealing with bacteria (Volume I, Chapter 9), viral (Volume I, Chapter

11), rickettsial and mycoplasmal (Volume I, Chapter 10), parasitic (Volume I, Chapter 12) and neopolastic (Volume I, Chapter 13) diseases for a more detailed account of particular host–parasite relationships.

VI. CONCLUSION

This chapter has concentrated on a discussion of the structure and function of the immune system in the rat. Taken together with chapter 10, this volume, on immunogenetics, the authors hope the reader will receive the basic parameters by which mechanisms of the rat immune response can be studied. For a more general discussion of immune mechanisms and immunological techniques transcending species barriers, the reader is directed to such texts as those by Eisen (56), Kabat and Mayer (130), and Weir (289).

ACKNOWLEDGMENTS

The authors wish to express their thanks to their colleagues Drs. A. Beckers, F. Collins, M. Lefford, J. McGhee, D. McGregor, R. North, P. Quérinjean, and A. Volkman for their important help in the development of this chapter. They are indebted to Mrs. Jenny De Mets, Miss Bernadette Platteau, Ms. Patricia Scheefer, and Mr. J. P. Kints for their excellent technical assistance and Mrs. Daniele Amthor for participation in writing the text. This work was supported by the "Fonds Cancérologique de la Caisse Générale d'Epargne et Retraite" (Belgium), by the "Fonds de la Recherche Scientifique et Médicale" (Belgium) Contract No. 3.4518.76, by EURATOM, Contract No. 250-77-1Bio B, and by the National Institutes of Health (United States), Contract No. RO1-AI-12840.

REFERENCES

1. Acton, R. T., Morris, R. J., and Williams, A. F. (1974). Estimation of the amount and tissue distribution of rat Thy-1.1 antigen. *Eur. J. Immunol.* **4**, 598–602.
2. Antoine, J. C., and Avrameas, S. (1976). Correlations between immunoglobulin and antibody-synthesizing cells during primary and secondary immune responses of rats immunized with peroxidase. *Immunology* **30**, 537–547.
3. Armerding, D. (1971). Two allotypic specificities of rat immunoglobulins. *Eur. J. Immunol.* **1**, 39–45.
4. Armerding, D., Katz, D. H., and Benacerraf, B. (1974). Immune response genes in inbred rats. I. Analysis of responder status to synthetic polypeptides and low doses of bovine serum albumin. *Immunogenetics* **1**, 329–339.
5. Arnason, B. G., deVaux St. Cyr, C., and Grabar, P. (1963). Immunoglobulin abnormalities of the thymectomized rats. *Nature (London)* **199**, 1199–1200.
6. Arnason, B. G., deVaux St. Cyr, C., and Relvveld, E. H. (1964). Role of the thymus in immune reactions in rats. IV. Immunoglobulins and

antibody formation. *Int. Arch. Allergy Appl. Immunol.* **25**, 206–224.
7. Arnason, B. G., Jankovic, B. D., Waksman, B. H., and Wennersten, C. (1962). Role of the thymus in immune reactions in rats. II. Suppressive effect of thymectomy at birth on reactions of delayed (cellular) hypersensitivity and the circulating small lymphocytes. *J. Exp. Med.* **116**, 177–186.
8. Austin, C. M. (1968). Patterns of migration of lymphoid cells. *Aust. J. Exp. Biol. Med. Sci.* **46**, 581–593.
9. Bachvaroff, R., Galdiero, F., and Grabar, P. (1969). Anti-thymus cytotoxicity: Identification and isolation of rat thymus-specific membrane antigens and purification of the corresponding antibodies. *J. Immunol.* **103**, 953–961.
10. Banovitz, J., and Ishizaka, K. (1967). Detection of five components having antibody activity in rat antisera. *Proc. Soc. Exp. Biol. Med.* **125**, 78–82.
11. Barabas, A. Z., and Kelus, A. S. (1967). Allotypic specificity of serum protein in inbred strains of rats. *Nature (London)* **215**, 155–156.
12. Bazin, H. (1974). Tumeurs secrétant des immunoglobulines chez le rat LOU/Wsl: Etude portant sur 200 immunocytomes. *Ann. Immunol. (Paris)* **125c**, 277–279.
13. Bazin, H. (1976). The secretory antibody system. *In* "Immunological Aspects of the Gastro-intestinal Tract and Liver" (A. Ferguson and R. N. M. McSween, eds.), pp. 32–82. Med. Tech. Publ. Co. Ltd., Lancaster, England.
14. Bazin, H. (1977). Synthesis and distribution of IgE in some experimental models. *Proc. Int. Congr. Allergol., 9th, 1976.*
15. Bazin, H., and Beckers, A. (1976). IgE-myelomas in rats. *Mol. Biol. Aspects Acute Allergic Reaction, Proc. Nobel Symp., 33rd, 1976* pp. 125–152.
16. Bazin, H., Beckers, A., Deckers, C., and Moriame, M. (1973). Transplantable immunoglobulin-secreting tumours in rats. V. Monoclonal immunoglobulins secreted by 250 ileocecal immunocytomas of the LOU/Wsl rats. *J. Natl. Cancer Inst.* **51**, 1359–1361.
17. Bazin, H., Beckers, A., Moriame, M., Platteau, B., Naze-De Mets, J., and Kints, J. P. (1974). LOU/C/Wsl rat monoclonal immunoglobulins. *FEBS Meet.* **86**, 117–121.
18. Bazin, H., Beckers, A., Platteau, B., Urbain, G., Urbain, J., and Pauwels, R. (1977). Transplantable immunoglobulin-secreting tumors in rats. VII. Five IgE secreting malignant immunocytoma tumors. (In preparation.)
19. Bazin, H., Beckers, A., and Quérinjean, P. (1974). Three classes and four (sub)classes of rat immunoglobulins: IgM, IgA, IgE and IgG1, IgG2a, IgG2b, IgG2c. *Eur. J. Immunol.* **4**, 44–48.
20. Bazin, H., Beckers, A., Vaerman, J. P., and Hereman, J. F. (1974). Allotypes of rat immunoglobulins. I. An allotype at the alpha-chain locus. *J. Immunol.* **112**, 1035–1041.
21. Bazin, H., Deckers, C., Beckers, A., and Heremans, J. F. (1972). Transplantable immunoglobulin-secreting tumours in rats. I. General features of LOU/Wsl strain rat immunocytomas and their monoclonal proteins. *Int. J. Cancer* **10**, 568–580.
22. Bazin, H., and Platteau, B. (1976). Production of circulating reaginic (IgE) antibodies by oral administration of ovalbumin to rats. *Immunology* **30**, 679–684.
23. Bazin, H., Quérinjean, P., Beckers, A., Heremans, J. F., and Dessy, F. (1974). Transplantable immunoglobulin-secreting tumours in rats. IV. Sixty-three IgE-secreting immunocytoma tumours. *Immunology* **26**, 713–723.
24. Beckers, A., and Bazin, H. (1975). Allotypes of rat immunoglobulins. III. An allotype of the gamma 2b-chain locus. *Immunochemistry* **12**, 671–675.
25. Beckers, A., Bazin, H., and Quérinjean, P. (1974). Allotypes of rat immunoglobulins. II. Distribution of the allotypes kappa and alpha chain loci in different inbred strains of rats. *Immunochemistry* **11**,

605-609.

26. Binaghi, R. A., and Benacerraf, B. (1964). Production of anaphylactic antibody in the rat. *J. Immunol.* **92**, 920-926.

27. Binaghi, R. A., and Boussac-Aron, Y. (1975). Isolation and properties of 17 S rat immunoglobulin different from IgG. *Eur. J. Immunol.* **5**, 194-197.

28. Binaghi, R. A., Oettgen, H. F., and Benacerraf, B. (1966). Anaphylactic antibody in the young rat. *Int. Arch. Allergy Appl. Immunol.* **29**, 105-111.

29. Binaghi, R. A., and Sarando de Merlo, E. (1966). Characterization of rat IgA and its non-identity with the anaphylactic antibody. *Int. Arch. Allergy Appl. Immunol.* **30**, 589-596.

30. Binz, H., and Wigzell, H. (1975). Shared idiotypic determinants on B and T lymphocytes reactive against the same antigenic determinants. I. Demonstration of similar or identical idiotypes on IgG molecules and T-cell receptors with specificity for the same alloantigens. *J. Exp. Med.* **142**, 197-211.

31. Binz, H., Wigzell, H., and Bazin, H. (1976). T-cell idiotypes are linked to immunoglobulin heavy chain genes. *Nature (London)* **264**, 639-642.

32. Bistany, T. S., and Tomasi, T. B. (1969). Characterization of rat immunoglobulin A. *Fed. Proc., Fed. Am. Soc. Exp. Biol.* **28**, 280.

33. Bistany, T. S., and Tomasi, T. B. (1970). Serum and secretory immunoglobulins of the rat. *Immunochemistry* **7**, 453-460.

34. Blanden, R. V. (1970). Mechanisms of recovery from a generalized viral infection: Mousepox. I. The effects of anti-thymocyte serum. *J. Exp. Med.* **132**, 1035-1054.

35. Bloch, K. J., Morse, H. C., and Austen, K. K. (1968). Biologic properties of rat antibodies. I. Antigen binding by four classes of anti-DNP antibodies. *J. Immunol.* **101**, 650-657.

36. Bloch, K. J., and Wilson, R. J. M. (1968). Homocytotropic antibody response in the rat infected with the nematode *Nippostrongylus brasiliensis*. III. Characteristics of the antibody. *J. Immunol.* **100**, 629-636.

37. Bollman, J. L. (1948). A cage which limits the activity of rats. *J. Lab. Clin. Med.* **33**, 1348.

38. Bollman, J. L., Cain, J. C., and Grindlay, J. H. (1948). Techniques for the collection of lymph from the liver, small intestine, or thoracic duct of the rat. *J. Lab. Clin. Med.* **33**, 1349-1352.

39. Broom, B. C., and Alexander, P. (1979). Rat tumour allografts evoke anaphylactic antibody responses. *Immunology* (in press).

40. Burtonboy, G., Bazin, H., Deckers, C., Beckers, A., Lamy, M., and Heremans, J. F. (1973). Transplantable immunoglobulin-secreting tumours in rats. III. Establishment of immunoglobulin-secreting cell lines from LOU/Wsl strain rats. *Eur. J. Cancer* **9**, 259-262.

41. Caffrey, R. W., Rieke, W. O., and Everett, N. B. (1962). Radioautographic studies of small lymphocytes in the thoracic duct of the rat. *Acta* Haematol. **28**, 145-154.

42. Campbell, P. A. (1976). Immunocompetent cells in resistance to bacterial infections. *Bacteriol. Rev.* **40**, 284-313.

43. Capron, A., Dessaint, J. P., Capron, M., and Bazin, H. (1975). Specific IgE antibodies in immune adherence of normal macrophages to *Schistosoma mansoni* schistosomules. *Nature (London)* **253**, 474-475.

44. Carson, D., Metzger, H., and Bloch, K. J. (1975). Serum IgE levels during the potentiated reagin response to egg albumin in rats infected with *Nippostrongylus brasiliensis*. *J. Immunol.* **114**, 521-523.

44a. Carter, P. B., and Sunderland, C. A. (1979). Rat alloantisera A.R.T. and Ly-1 detect a polymorphism of a leukocyte-common antigen. *Transplant. Proc.* **11**, 1646-1647.

45. Cassell, G. H., Lindsey, J. R., Overcash, R. G., and Baker, H. J. (1973). Murine mycoplasma respiratory disease. *Ann. N.Y. Acad Sci.* **225**, 395-412.

46. Clagett, J., Peter, H.-H., Feldman, J. D., and Weigle, W. O. (1973). Rabbit antiserum to brain-associated thymus antigens of mouse and rat. II. Analysis of species-specific and cross-reacting antibodies. *J. Immunol.* **110**, 1085-1089.

47. Colly, D. G., Malakian, A., and Waksman, B. H. (1970). Cellular differentiation in the thymus. II. Thymus-specific antigens in rat thymus and peripheral lymphoid cells. *J. Immunol.* **104**, 585-592.

48. Cremer, N. E., Taylor, D. O. N., Lennette, E. H., and Hagens, S. J. (1973). IgM production in rats infected with Moloney leukemia virus. *J. Natl. Cancer Inst.* **51**, 905-915.

49. Crum, E. D., and McGregor, D. D. (1976). Functional properties of T and B cells isolated by affinity chromatography from rat thoracic duct lymph. *Cell. Immunol.* **23**, 211-222.

50. David, J. R., Al-Askari, S., Lawrence, H. S., and Thomas, L. (1964). Delayed hypersensitivity *in vitro*. I. The specificity of inhibition of cell migration by antigens. *J. Immunol.* **93**, 264-273.

51. Deckers, C. (1964). Etude électrophorétique et immunoélectrophorétique des protéines du rat atteint de leucosarcome. *Protioles Biol. Fluids, Proc. Colloq.* pp. 105-108.

52. Douglas, T. C. (1972). Occurrence of a theta-like antigen in rats. *J. Exp. Med.* **136**, 1054-1062.

53. Douglas, T. C. (1973). A rat analogue of the mouse theta antigen. *Transplant. Proc.* **5**, 79-82.

54. Dunn, T. B. (1957). Plasma cell neoplasms beginning in the ileocecal area in strain C3H mice. *J. Natl. Cancer Inst.* **19**, 371-391.

55. Ebert, R. H., and Florey, H. W. (1939). The extravascular development of the monocyte observed *in vivo*. *Br. J. Exp. Pathol.* **20**, 342.

56. Eisen, H. N. (1974). "Immunology: An Introduction to Molecular and Cellular Principles of the Immune Responses." Harper & Row, Hagerstown, Maryland.

57. Escribano, M. J., and Grabar, P. (1962). L'analyse immunoélectrophorétique du sérum de rat normal. *C.R. Hebd. Seances Acad. Sci.* **255**, 206-208.

58. Everett, N. B., Caffrey, F. W., and Rieke, W. O. (1964). Recirculation of lymphocytes. *Ann. N.Y. Acad. Sci.* **113**, 887-897.

58a. Fabre, J. W., and Morris, P. J. (1974). The definition of a lymphocyte-specific alloantigen system in the rat (Ly-1). *Tissue Antigens* **4**, 238-246.

58b. Fabre, J. W., and Williams, P. J. (1977). Quantitative serological analysis of rabbit anti-rat lymphocyte serum and preliminary biochemical characterisation of the major antigen recognised. *Transplantation* **23**, 349-359.

59. Festing, M., and Staats, J. (1973). Standardized nomenclature for inbred strains of rats. *Transplantation* **16**, 221-245.

60. Flax, M. H., and Waksman, B. H. (1962). Delayed cutaneous reactions in the rat. *J. Immunol.* **89**, 496-504.

61. Fliedner, T. M., Keese, M., Cronkite, E. P., and Robertson, J. S. (1965). Cell proliferation in germinal centers of the rat spleen. *Ann. N.Y. Acad. Sci.* **113**, 578-594.

62. Ford, W. L. (1969). The immunological and migratory properties of the lymphocytes recirculating through the rat spleen. *Br. J. Exp. Pathol.* **50**, 257-269.

63. Ford, W. L. (1975). Lymphocyte migration and immune responses. *Prog. Allergy* **19**, 1-59.

64. Ford, W. L., and Hunt, S. V. (1973). The preparation and labelling of lymphocytes. *Hand. Exp. Immunol., 2nd Ed.* **2**, 233-236.

65. Ford, W. L., and Simmonds, S. J. (1972). The tempo of lymphocyte recirculation from blood to lymph in the rat. *Cell Tissue Kinet.* **5**, 175-189.

66. Fugi, H., Zaleski, M., and Milgrom, F. (1971). Immune response to allo-antigens of thymus studies in mice with plaque assay. *J. Immunol.* **106**, 56-64.

67. Furth, R. van (1970). The origin and turnover of promonocytes, monocytes, and macrophages in normal mice. *In* "Mononuclear Phagocytes" (R. van Furth, ed.), pp. 151–165. Blackwell, Oxford.

68. Furth, R. van, Cohn, Z. A., Hirsch, J. G., Humphrey, J. H., Spector, W. G., and Langevoort, H. L. (1972). The mononuclear phagocyte system: A new classification of macrophages, monocytes, and their precursor cells. *Bull. W.H.O.* **46**, 845–852.

69. Gasser, D. L., Newlin, C. M., Palm, J., and Gonatas, N. K. (1973). Genetic control of susceptibility to experimental allergic encephalomyelitis in rats. *Science* **181**, 872–873.

70. Gill, T. J., III, and Kunz, H. W. (1970). Genetic and cellular factors in the immune response. II. Evidence for the polygenic control of the antibody response from further breeding studies and from pedigree analysis. *J. Immunol.* **106**, 980–992.

71. Gill, T. J., III, H. W. Kunz, Stechschulte, D. J., and Austen, K. F. (1970). Genetic and cellular factors in the immune response. I. Genetic control of the antibody response to poly GLU^{52}LYS^{33}TYR^{15} in the inbred strains ACI and F344. *J. Immunol.* **105**, 14–28.

72. Glick, B., Chang, T. S., and Jaap, R. G. (1956). The bursa of Fabricius and antibody production. *Poult. Sci.* **35**, 224–225.

73. Gloyne, S. R., and Page, D. S. (1923). The reaction to *B. tuberculosis* in the albino rat. *J. Pathol. Bacteriol.* **26**, 224–233.

74. Goldschneider, I., and Cogen, R. B. (1973). Immunoglobulin molecules on the surface of activated T lymphocytes in the rat. *J. Exp. Med.* **138**, 163–175.

75. Goldschneider, I., and McGregor, D. D. (1968). Migration of lymphocytes and thymocytes in the rat. I. The route of migration from blood to spleen and lymph nodes. *J. Exp. Med.* **127**, 155–168.

76. Goldschneider, I., and McGregor, D. D. (1973). Anatomical distribution of T and B lymphocytes in the rat. Development of lymphocyte-specific antisera. *J. Exp. Med.* **138**, 1443–1465.

77. Gonatas, N. K., and Howard, J. C. (1974). Inhibition of experimental allergic encephalomyelitis in rats severely depleted of T cells. *Science* **186**, 839–841.

78. Good, R. A., Dalmasso, A. P., Martinez, C., Archer, O. K., Pierce, J. C., and Papermaster, B. W. (1962). The role of the thymus in development of immunologic capacity in rabbits and mice. *J. Exp. Med.* **116**, 773–796.

79. Gowans, J. L. (1957). The effect of the continuous re-infusion of lymph and lymphocytes on the output of lymphocytes from the thoracic duct of unanaesthetized rats. *Br. J. Exp. Pathol.* **38**, 67–78.

80. Gowans, J. L. (1959). The recirculation of lymphocytes from blood to lymph in the rat. *J. Physiol (London)* **146**, 54–69.

81. Gowans, J. L. (1974). Differentiation of the cells which synthesize immunoglobulins. *Ann. Immunol. (Paris)* **125c**, 201–211.

82. Gowans, J. L., and Knight, E. J. (1964). The route of recirculation of lymphocytes in the rat. *Proc R. Soc. London, Ser. B* **159**, 257–282.

83. Gowans, J. L., and Uhr, J. W. (1966). The carriage of immunological memory by small lymphocytes in the rat. *J. Exp. Med.* **124**, 1017–1030.

84. Grabar, P., and Courcon, J. (1958). Etude des sérums de cheval, lapin, rat et souris par l'analyse immunoélectrophorétique. *Bull. Soc. Chim. Biol.* **40**, 1993–2003.

85. Gray, D. F., Noble, J. L., and O'Hara, M. (1961). Allergy in experimental rat tuberculosis. *J. Hyg.* **59**, 427–436.

86. Griscelli, C., Vassalli, P., and McClusky, R. T. (1969). The distribution of large dividing lymph node cells in syngeneic recipient rats after intravenous injection. *J. Exp. Med.* **130**, 1427–1451.

87. Günther, E., and Rüde, E. (1975). Genetic complementation of histocompatibility-linked *Ir* genes in the rat. *J. Immunol.* **115**, 1387–1393.

88. Günther, E., Rüde, E., and Stark, O. (1972). Antibody response in rats to synthetic polypeptide (T,G)-A-L genetically linked to their major histocompatibility system. *Eur. J. Immunol.* **2**, 151–155.

89. Gutman, G. A., Lah, E., and Hood, L. (1975). Structure and regulation of immunoglobulins: Kappa allotypes in the rat have multiple amino acid differences in the constant region. *Proc. Natl. Acad. Sci. U.S.A.* **72**, 5046–5050.

90. Gutman, G. A., and Weisman, I. L. (1971). Inheritance and strain distribution of a rat immunoglobulin allotype. *J. Immunol.* **107**, 1390–1393.

91. Gutman, G. A., and Weissman, I. L. (1972). Lymphoid tissue architecture. Experimental analysis of the origin and distribution of T-Cells and B-cells. *Immunology* **23**, 465–479.

92. Hajdu, A., and Rona, G. (1969). Spontaneous thyroiditis in laboratory rats. *Experientia* **25**, 1325–1327.

93. Hall, J. G., Parry, D. M., and Smith, M. E. (1972). The distribution and differentiation of lymph-borne immunoblasts after intravenous injection into syngeneic recipients. *Cell Tissue Kinet.* **5**, 269–281.

94. Hall, J. G., and Smith, M. E. (1970). Homing of lymph-borne immunoblasts to the gut. *Nature (London)* **226**, 262–263.

95. Halpern, J., and Metzger, H. (1976). The interaction of IgE with rat basophilic leukemia cells. VI. Inhibition by IgGa immune complexes. *Immunochemistry* **13**, 907–913.

96. Halstead, T. E., and Hall, J. G. (1972). The homing of lymph-borne immunoblasts to the small gut of neonatal rats. *Transplantation* **14**, 339–346.

97. Hammond, E. M., and Dvorak, H. F. (1972). Antigen-induced stimulation of glucosamine incorporation by guinea pig peritoneal macrophages in delayed hypersensitivity. *J. Exp. Med.* **136**, 1518–1532.

98. Haskill, J. S. (1969). Density distribution analysis of antigen-sensitive cells in the rat. *J. Exp. Med.* **130**, 877–893.

99. Haskill, J. S., Legge, D. G., and Shortman, K. (1969). Density distribution analysis of cells forming 19S hemolytic antibody in the rat. *J. Immunol.* **102**, 703–712.

100. Hehre, E., and Freund, J. (1939). Sensitization, antibody formation and lesions produced by tubercle bacilli in the albino rat. *Arch. Pathol.* **27**, 287–306.

100a. Herzenberg, L. A., and Herzenberg, L. A. (1978). Analysis and searation using the fluorescence activated cell sorter (FACS). *In* "Handbook of Experimental Immunology" (D. M. Weir, ed.), Blackwell, Oxford.

101. Hollingsworth, J. W., and Carr, J. (1973). ^{3}H-uridine incorporation as a T lymphocyte indicator in the rat. *Cell. Immunol.* **8**, 270–279.

102. Howard, J. C. (1972). The life-span and recirculation of marrow-derived small lymphocytes from the rat thoracic duct. *J. Exp. Med.* **135**, 185–199.

103. Howard, J. C., and Gowans, J. L. (1972). The role of lymphocytes in antibody formation. III. The origin from small lymphocytes of cells forming direct and indirect haemolytic plaques to sheep erythrocytes in the rat. *Proc. R. Soc. London, Ser. B* **182**, 193–209.

104. Howard, J. C., Hunt, S. V., and Gowans, J. L. (9172). Identification of marrow-derived and thymus-derived small lymphocytes in the lymphoid tissue and thoracic duct lymph of normal rats. *J. Exp. Med.* **135**, 200–219.

105. Howard, J. C., and Scott, D. W. (1972). The role of recirculating lymphocytes in the immunological competence of rat bone marrow cells. *Cell. Immunol.* **3**, 421–429.

106. Howard, J. C., and Scott, D. W. (1974). The identification of sera distinguishing marrow-derived and thymus-derived lymphocytes in the rat thoracic duct. *Immunology* **27**, 903–922.

107. Humphrey, R. L., and Santos, G. S. (1971). Serum protein allotype markers in certain inbred rat strains. *Fed. Proc., Fed. Am. Soc. Exp.*

Biol. **30**, 248.

108. Hunt, S. V. (1973). Separation of thymus-derived and marrow-derived rat lymphocytes on glass bead columns. *Immunology* **24**, 699–705.

109. Hunt, S. V., and Williams, A. F. (1974). The origin of cell surface immunoglobulin of marrow-derived and thymus-derived lymphocytes of the rat. *J. Exp. Med.* **139**, 479–496.

110. Isersky, C., Kulczcki, A., and Metzger, H. (1974). Isolation of IgE from reaginic rat serum. *J. Immunol.* **112**, 1909–1919.

111. Ishizaka, T., Konig, W., Kurata, M., Mauser, L., and Ishizaka, K. (1975). Immunologic properties of mast cells from rat infected with *Nippostrongylus brasiliensis*. *J. Immunol.* **115**, 1078–1083.

112. Ishizaka, T., Urban, J. F., and Ishizaka, K. V. (1976). IgE formation in the rat following infection with *Nippostrongylus brasiliensis*. I. Proliferation and differentiation of IgE-B cells. *Cell. Immunol.* **22**, 248–261.

113. Iverson, J. G. (1970). Specific antisera against recirculating and non-recirculating lymphocytes in the rat. *Clin. Exp. Immunol.* **6**, 101–108.

114. Jankovic, B. D., Waksman, B. H., and Arnason, B. G. (1962). Role of the thymus in immune reactions in rats. I. The immunologic response to bovine serum albumin (antibody formation, arthus reactivity, and delayed hypersensitivity) in rats thymectomized or splenectomized at various times after birth. *J. Exp. Med.* **116**, 159–176.

115. Jarrett, E. E. E. (1972). Potentiation of reaginic (IgE) antibody to ovalbumin in the rat following sequential trematode and nematode infections. *Immunology* **22**, 1099–1101.

116. Jarrett, E. E. E. (1978). Production of IgE and reaginic antibody in rats in relation to worm infections. *Mol. Biol. Aspects Acute Allergic Reaction, Proc. Nobel Symp., 1976* (in press).

117. Jarrett, E. E. E., and Bazin, H. (1974). Elevation of total serum IgE in rats following helminth parasite infection. *Nature (London)* **251**, 613–614.

118. Jarrett, E. E. E., Haig, D. M., and Bazin, H. (1976). Time course studies on rat IgE production in *Nippostrongylus brasiliensis* infection. *Clin. Exp. Immunol.* **24**, 346–351.

119. Jarrett, E. E. E., Haig, D. M., McDouglas, W., and McNulty, E. (1976). Rat IgE production. II. Primary and booster reaginic antibodies responses following intradermal or oral immunization. *Immunology* **30**, 671–677.

120. Jarrett, E. E. E., and Steward, D. (1972). Potentiation of rat reaginic (IgE) antibody by helminth infection. Simultaneous potentiation of separate reagins. *Immunology* **23**, 749–755.

121. Jarrett, E. E. E., and Steward, D. (1974). Rat IgE production. I. Effect of dose antigen on primary and secondary reaginic antibody response. *Immunology* **27**, 365–381.

122. Jensenius, J. C., and Williams, A. F. (1974). The binding of antiglobulin antibodies to rat thymocytes and thoracic duct lymphocytes. *Eur. J. Immunol.* **4**, 91–97.

123. Jensenius, J. C., and Williams, A. F. (1974). Total immunoglobulin of rat thymocytes and thoracic duct lymphocytes. *Eur. J. Immunol.* **4**, 98–105.

124. Johnston, J. M., and Wilson, D. B. (1970). Origin of immunoreactive lymphocytes in rats. *Cell. Immunol.* **1**, 430–444.

125. Jones, H. E. H., and Roitt, I. M. (1961). Experimental auto-immune thyroiditis in the rat. *Br. J. Exp. Pathol.* **42**, 546–557.

126. Jones, V. E. (1969). Rat 7 S immunoglobulins: Characterization of γ2- and γ1-anti-hapten antibodies. *Immunology* **16**, 589–599.

127. Jones, V. E., and Edwards, A. J. (1971). Preparation of an antiserum specific for rat reagin (ratγE?). *Immunology* **21**, 383–385.

128. Jones, V. E., Edwards, A. J., and Ogilvie, B. M. (1970). The circulating immunoglobulins involved in protective immunity to the intestinal stage of *Nippostrongylus brasiliensis* in the rat. *Immunology* **18**, 621–633.

129. Jones, V. E., and Ogilvie, B. M. (1967). Reaginic antibodies and immunity to *Nippostrongylus brasiliensis* in the rat. II. Some properties of the antibodies and antigens. *Immunology* **12**, 583–597.

130. Kabat, E. A., and Mayer, M. M. (1961). "Experimental Immunochemistry." Thomas, Springfield, Illinois.

131. Kalden, J. R., James, K., Williamson, W. G., and Irvine, W. J. (1969). The effect of anti-lymphocyte IgG on established autoallergic thyroiditis in rats. *Clin. Exp. Immunol.* **5**, 597–606.

132. Kobayashi, K., Vaerman, J. P., Bazin, H., Lebacq-Verheyden, A. M., and Heremans, J. F. (1973). Identification of J-chain in polymeric immunoglobulins from a variety of species by cross-reaction with rabbit antisera to human J-chain. *J. Immunol.* **111**, 1590–1594.

133. Komuro, K., and Boyse, E. A. (1973). *In vitro* demonstration of thymic hormone in the mouse by conversion of precursor cells into lymphocytes. *Lancet* **1**, 740–743.

134. Koster, F. T., McGregor, D. D., and Mackaness, G. B. (1971). The mediator of cellular immunity. II. Migration of immunologically committed lymphocytes into inflammatory exudates. *J. Exp. Med.* **133**, 400–409.

135. Kostiala, A. A. I., McGregor, D. D., and Lefford, M. J. (1976). The mediator of cellular immunity. XI. Origin and development of MIF producing lymphocytes. *Cell. Immunol.* **24**, 318–327.

136. Kostiala, A. A. I., McGregor, D. D., and Logie, P. S. (1975). Tularemia in the rat. I. The cellular basis of host resistance to infection. *Immunology* **28**, 855–869.

137. Kraft, N., and Wistar, R. (1971). A modified antibody forming cell assay for detecting cells producing a particular class of antibody. *Aust. J. Exp. Biol. Med. Sci.* **49**, 11–20.

138. Kuhlmann, W. D., and Avrameas, S. (1972). Cellular differentiation and antibody localization during the primary immune response in peroxidase stimulated lymph nodes of rat. *Cell. Immunol.* **4**, 425–441.

139. Kuhlmann, W. D., and Avrameas, S. (1975). Lymphocyte differentiation and antibody synthesis in the secondary immune response of peroxidase sitmulated lymph nodes of rat. *Cell Tissue Res.* **156**, 391–402.

140. Kunz, H. W., Gill, T. J., III, Hansen, C. T., and Poloskey, P. E. (1975). Genetic studies in inbred rats. III. Histocompatibility type and immune response of some mutant stocks. *J. Immunogenet.* **2**, 51–54.

141. Ladoulis, C. T., Misra, D. N., Estes, L. W., and Gill, T. J., III (1974). lymphocyte plasma membranes. I. Thymic and splenic membranes from inbred rats. *Biochim. Biophys. Acta* **356**, 27–35.

142. Ladoulis, C. T., Shonnard, J. W., Kunz, H. W., and Gill, T. J., III (1973). Genetic control of the induction of the antibody response. *Protides Biol. Fluids, Proc. Colloq.* **21**, 283–289.

143. Lance, E. M., and Taub, R. N. (1969). Segregation of lymphocyte populations through differential migration. *Nature (London)* **221**, 841–843.

144. Lane, F. C., and Unanue, E. R. (1972). Requirement for thymus (T) lymphocytes for resistance to listeriosis. *J. Exp. Med.* **135**, 1104–1112.

145. Lefford, M. J. (1974). The measurement of tuberculin hypersensitivity in rats. *Int. Arch. Allergy Appl. Immunol.* **47**, 570–585.

146. Lefford, M. J., McGregor, D. D., and Mackaness, G. B. (1973). Immune response to *Mycobacterium tuberculosos* in rats. *Infect. Immun.* **8**, 182–189.

147. Lennon, V. A., Lindstrom, J. M., and Seybold, M. E. (1975). Experimental autoimmune myasthenia: A model of myasthenia gravis in rats and guinea pigs. *J. Exp. Med.* **141**, 1365–1375.

148. Leslie, G. A., and Carwile, H. F. (1973). Immune response of rats to group A streptococcal vaccine. *Infect. Immun.* **7**, 781–785.

149. Levine, B. B., and Vaz, N. M. (1970). Effect of combinations of inbred strain, antigen and antigen dose on immune responsiveness and reagin production in the mouse. A potential mouse model for immune

aspects of human atopic allergy. *Int. Arch. Allergy Appl. Immunol.* **39**, 156–171.

150. Levine, S., and Sowinski, R. (1976). Necrotic myelopathy (myelomalacia) in rats with encephaloymelitis treated with tilorone. *Am. J. Pathol.* **82**, 381–392.

151. Levine, S., and Wenk, E. J. (1961). Studies on the mechanism of altered susceptibility to experimental allergic encephalomyelitis. *Am. J. Pathol.* **39**, 419–441.

152. Levine, S., and Wenk, E. J. (1963). Allergic encephalomyelitis: Rapid induction without the aid of adjuvants. *Science* **141**, 529–530.

153. Levine, S., Wenk, E. J., and Hoenig, E. M. (1967). Passive transfer of allergic encephalomyelitis between inbred rat strains: Correlation with transplantation antigens. *Transplantation* **5**, 534–541.

154. Lubaroff, D. M. (1973). An alloantigenic marker on rat thymus and thymus-derived cells. *Transplant. Proc.* **5**, 115–118.

155. Lubaroff, D. M. (1977). Antigenic markers on rat lymphocytes. I. Characterization of A.R.T., an alloantigenic marker on rat thymus and thymus-derived cells. *Cell. Immunol.* **29**, 147–158.

156. Mackaness, G. B. (1962). Cellular resistance to infection. *J. Exp. Med.* **116**, 381–406.

157. Mackaness, G. B. (1971). Delayed hypersensitivity and the mechanism of cellular resistance to infection. *Prog. Immunol., Int. Congr. Immunol., 1st, 1971* pp. 413–424.

158. Maisin, J., Maldague, P., Deckers, C., and Gond-Que, P. (1964). Le leucosarcome du rat. *Symp. Lymph. Tumours Afr., 1963* pp. 341–354.

159. Maisin, J., Maldague, P., Dunjic, A., and Maisin, H. (1957). Syndromes mortels et effets tardifs des irradiations totales et subtotales chez le rat. *J. Belge Radiol.* **40**, 346–398.

160. Maldague, P., Maisin, J., Dunjic, A., Pham-Hong-Que and Maisin, H. (1958). Le leucosarcome chez le rat. L'incidence de ce néoplasme après irradiation totale, subtotale et locale. *Sang* **24**, 751–763.

161. Mann, J. D., and Higgins, G. M. (1950). Lymphocytes in the thoracic duct, intestinal and hepatic lymph. *Blood* **5**, 177–190.

162. Marchesi, V. T., and Gowans, J. L. (1964). The migration of lymphocytes through the endothelium of venules in lymph nodes: An electron microscopic study. *Proc. R. Soc. London, Ser. B* **159**, 283–290.

163. Mayrhofer, G., Bazin, H., and Gowans, J. L. (1976). The nature of cells binding anti-IgE in rats immunized with *Nippostrongylus brasiliensis:* IgE synthesis in regional nodes and concentration in mucosal mast cells. *Eur. J. Immunol.* **6**, 537–545.

164. McGhee, J. R., Michalek, S. M., and Ghanta, V. K. (1975). Rat Immunoglobulins in serum and secretions: Purification of rat IgM, IgA and IgG and their quantitation in serum, colostrum, milk and saliva. *Immunochemistry* **12**, 817–823.

164a. McGregor, D. D., Crum, E. D., Jungi, T. W., and Bell, R. G. (1978). Transfer of immunity against *Listeria monocytogenes* by T cells purified by a positive selection technique. *Infect. Immun.* **22**, 209–218.

165. McGregor, D. D., Hahn, H. H., and Mackaness, G. B. (1973). The mediator of cellular immunity. V. Development of cellular resistance to infection in thymectomized irradiated rats. *Cell. Immunol.* **6**, 186–199.

166. McGregor, D. D., and Koster, F. T. (1971). The mediator of cellular immunity. IV. Cooperation between lymphocytes and mononuclear phagocytes. *Cell. Immunol.* **2**, 317–325.

167. McGregor, D. D., Koster, F. T., and Mackaness, G. B. (1971). The mediator of cellular immunity. I. The life span and circulation dynamics of the immunologically committed lymphocyte. *J. Exp. Med.* **133**, 389–399.

168. McGregor, D. D., and Kostiala, A. A. I. (1976). Role of lymphocytes in cellular resistance to infection. *Contemp. Top. Immunobiol.* **5**, 237–266.

169. McGregor, D. D., and Logie, P. S. (1975). The mediator of cellular immunity. X. Interaction of macrophages and specifically sensitized lymphocytes. *Cell. Immunol.* **18**, 454–465.

170. McGregor, D. D., and Mackaness, G. B. (1975). The properties of lymphocytes which carry immunologic memory of φX174. *J. Immunol.* **114**, 336–342.

171. Medgyesi, G. A., Füst, G., Bazin, H., Ujhelyi, E., and Gergely, J. (1974). Interaction of rat immunoglobulins with complement. *FEBS Meet., 1974* **86**, 123–128.

172. Megel, H., Raychaudhuri, A., Goldstein, S., Kinsolving, C. R., Shemano, I., and Michael, J. G. (1974). Tilorone: Its selective effects on humoral and cell-mediated immunity. *Proc. Soc. Exp. Biol. Med. Sci.* **145**, 513–518.

173. Mel, H. C., Mitchell, L. T., and Thorell, B. (1965). Continouous free flow fractionation of cellular constituents in rat bone marrow. *Blood* **25**, 63–72.

174. Metzger, J. J., and Houdayer, M. (1972). Subclasses of porcine immunoglobulins G. Production of subclass specific antisera with S-sulfonated gamma chains. *Eur. J. Immunol.* **2**, 127–130.

175. Metzger, J. J., Novotny, J., and Franek, F. (1971). Heterogeneity of pig immunoglobulin γ-chains. *FEBS Lett.* **14**, 237–240.

176. Michalek, S. M., Rahman, A. F. R., and McGhee, J. R. (1975). Rat immunoglobulins in serum and secretions: Comparison of IgM, IgA and IgG in serum, colostrum, milk and saliva of protein malnourished and normal rats. *Proc. Soc. Exp. Biol. Med.* **148**, 1114–1118.

177. Micheel, B., Pasternak, G., and Steuden, J. (1973). Demonstration of theta-AKR antigen in rat tissue by mouse alloantiserum. *Nature (London), New Biol.* **241**, 221–222.

178. Miller, H. R. P., Ternynct, T., and Avrameas, S. (1975). Synthesis of antibody and immunoglobulins without detectable antibody function in cells responding to horseradish peroxidase. *J. Immunol.* **114**, 626–629.

179. Miller, J. F. A. P. (1961). Immunological function of the thymus. *Lancet* **2**, 748–749.

180. Miller, J. F. A. P., Basten, A., Sprent, J., and Cheers, C. (1971). Interaction between lymphocytes in immune responses. *Cell. Immunol.* **2**, 469–495.

181. Miller, J. F. A. P., and Mitchell, F. G. (1969). Thymus and antigen-reactive cells. *Transplant. Rev.* **1**, 3–42.

182. Miller, R. G., and Phillips, R. A. (1969). Separation of cells by velocity sedimentation. *J. Cell. Physiol.* **73**, 191–201.

183. Mitchell, G. F., and Miller, J. F. A. P. (1968). Immunological activity of thymus and thoracic duct lymphocytes. *Proc. Natl. Acad. Sci. U.S.A.* **59**, 296–303.

184. Moriame, M., Beckers, A., and Bazin, H. (1977). Decrease in the incidence of malignant ileo-caecal immunocytoma in LOU/C rats after surgical removal of the ileo-caecal lymph nodes. *Cancer Lett.* **3**(3–4), 139–143.

185. Morris, R. J., Letarte-Muirhead, M., and Williams, A. F. (1975). Analysis in deoxycholate of three antigenic specificities associated with the rat Thy-1 molecule. *Eur. J. Immunol.* **5**, 282–285.

186. Morris, R. J., and Williams, A. F. (1975). Antigens on mouse and rat lymphocytes recognized by rabbit antiserum against rat brain: The quantitative analysis of a xenogeneic antiserum. *Eur. J. Immunol.* **5**, 274–281.

187. Morse, H. C., Bloch, K. J., and Austen, K. F. (1968). Biologic properties of rat antibodies. II. Time-course of appearance of antibodies involved in antigen-induced release of slow reacting substance of anaphylaxis (SRS-A); association of this activity with rat IgGa. *J. Immunol.* **101**, 658–663.

188. Mota, I. (1963). Mast cells and anaphylaxis. *Ann. N.Y. Acad Sci.* **103**, 264–277.

189. Mota, I. (1964). The mechanism of anaphylaxis. I. Production and biological properties of mast cell sensitizing antibody. *Immunology* **7**, 681–699.

190. Murphey, S. M., Brown, S., Miklos, N., and Fireman, P. (1974). Reagin synthesis in inbred strain of rats. *Immunology* **27**, 245-253.

191. Nash, D. R., Vaerman, J. P., Bazin, H., and Heremans, J. F. (1969). Identification of IgA in rat serum and secretions. *J. Immunol.* **103**, 145-148.

192. Nelson, W., Zinneman, H., Selden, J., Schaper, K., Halberg, F., and Bazin, H. (1974). Circadian rhythm in Bence Jones protein excretion by LOU rats bearing a transplantable immunocytoma, responsive to adriamycin treatment. *Int. J. Chronobiol.* **2**, 359-365.

193. Nezlin, R., and Rokhlin, O. (1976). Allotypes of light chains of rat immunoglobulins and their application to the study of antibody biosynthesis. *Contemp. Top. Mol. Immunol.* **5**, 161-184.

194. Nezlin, R. S., Vengerova, T. I., Rokhlin, O. V., and Machulla, H. K. G. (1974). Localization of allotypic markers of kappa light chains of rat immunoglobulins in the constant part of the chain. *Fed. Eur. Biochem. Soc. Meet. [Proc.]* **36**, 93-107.

195. North, R. J. (1969). The mitotic potential of fixed phagocytes in the liver as revealed during the development of cellular immunity. *J. Exp. Med.* **130**, 315-326.

196. North, R. J. (1973). Importance of thymus-derived lymphocytes in cell-mediated immunity to infection. *Cell. Immunol.* **7**, 166-176.

197. Nussenzweig, V., and Binaghi, R. A. (1965). Heterogeneity of rat immunoglobulins. *Int. Arch. Allergy Appl. Immunol.* **27**, 355-360.

198. Ogilvie, B. M. (1964). Reagin-like antibodies in animals immune to helminth parasites. *Nature (London)* **204**, 91-92.

199. Ogilvie, B. M. (1967). Reagin-like antibodies in rats infected with the nematode parasite *N. brasiliensis. Immunology* **12**, 113-131.

200. Oldstone, M. B. A., and Dixon, F. J. (1968). Immunohistochemical study of allergic encephalomyelitis. *Am. J. Pathol.* **52**, 251-263.

201. Oriol, R., Binaghi, R., and Coltorti, E. (1971). Valence and association constant of rat macroglobulin antibody. *J. Immunol.* **104**, 932-937.

202. Orlans, E. (1975). The antigenic interrelations of some mammalian IgG subclasses detected with cross-reacting fowl antisera to human and mouse IgG-Fc. *Immunology* **28**, 761-772.

203. Orlans, E., and Feinstein, A. (1971). Detection of alpha, kappa and lambda chains in mammalian immunoglobulins using fowl antisera to human IgA. *Nature (London)* **233**, 45-47.

204. Ornstein, G. G., and Steinbach, M. M. (1925). The resistance of the albino rat to infection with tubercle bacilli. *Am. Rev. Tuberc.* **12**, 77-86.

205. Orr, T. S. C., and Blair, A. M. J. N. (1969). Potentiated reagin response to egg-albumin and conalbumin in *N. brasiliensis* infected rats. *Life Sci.* **8**, 1073-1077.

206. Orr, T. S. C., Riley, P. A., and Doe, J. E. (1972). Potentiated reagin response to egg-albumin in *N. brasiliensis* infected rats. III. Further studies in the time course of the reagin response. *Immunology* **22**, 211-217.

207. Osogoe, B., Tyler, R. W., and Everett, N. B. (1973). The patterns of labelling of germinal center cells with tritiated cytidine. *J. Cell Biol.* **57**, 215-220.

208. Oudin, J. (1956). ''L'allotypie'' de certains antigènes protéidiques du sérum. *C. R. Hebd. Seances Acad. Sci.* **242**, 2606-2608.

209. Ovary, Z. (1966). The structures of various immunoglobulins and their biologic activities. *Ann. N.Y. Acad. Sci.* **129**, 776-786.

210. Parish, C. R., and Hayward, J. A. (1974). The lymphocyte surface. I. Relation between Fc receptors, C'3 receptors and surface immunoglobulin. *Proc. Soc. London, Ser. B* **187**, 47-63.

211. Parish, C. R., and Hayward, J. A. (1974). The lymphocyte surface. II. Separation of Fc receptor, C'3 receptor and surface immunoglobulin-bearing lymphocytes. *Proc. R. Soc. London, Ser. B* **187**, 65-81.

212. Parrott, D. M. W., de Sousa, M. A. B., and East, J. (1966). Thymus-dependent areas in the lymphoid organs of neonatally thymectomized mice. *J. Exp. Med.* **123**, 191-217.

213. Paterson, P. Y. (1966). Experimental allergic encephalomyelitis and autoimmune disease. *Adv. Immunol.* **5**, 131-208.

214. Paterson, P. Y., and Drobish, D. G. (1975). Reversal of experimental allergic thyroiditis in cyclophosphamide-treated rats. *Clin. Exp. Immunol.* **20**, 125-129.

215. Paterson, P. Y., Richardson, W. P., and Drobish, D. G. (1975). Cellular transfer of experimental allergic encephalomyelitis: Altered disease pattern in irradiated recipient Lewis rats. *Cell. Immunol.* **16**, 48-59.

216. Pauwels, R., Bazin, H., Platteau, B., and Van der Straeten, F. (1977). Total serum IgE levels in rats. *J. Immunol. Methods* **18**, 133.

217. Penhale, W. J., Farmer, A., and Irvine, W. J. (1975). Thyroiditis in T-cell depleted rats. Influence of strain, radiation dose, adjuvants, and anti-lymphocyte serum. *Clin. Exp. Immunol.* **21**, 362-375.

218. Penhale, W. J., Farmer, A., Urbaniak, S. J., and Irvine, W. J. (1975). Susceptibility of inbred rat strains to experimental thyroiditis: Quantitation of thyroglobulin-binding cells and assessment of T-cell function in susceptible and non-susceptible strains. *Clin. Exp. Immunol.* **19**, 179-191.

219. Peter, H.-H., Clagett, J., Feldman, J. D., and Weigle, W. O. (1973). Rabbit antiserum to brain-associated thymus antigens of mouse and rat. I. Demonstration of antibodies cross-reacting to T cells of both species. *J. Immunol.* **110**, 1077-1084.

220. Pick, E., and Turk, J. L. (1972). Biological activities of soluble lymphocyte products. *Clin. Exp. Immunol.* **10**, 1-23.

221. Pierce, C. W. (1967). The effects of endotoxin on the immune response in the rat. I. Antibodies formed against bovine γ-globulin in the intact rat. *Lab. Invest.* **16**, 768-781.

222. Pierce, N. F., and Gowans, J. L. (1975). Cellular kinetics of the intestinal immune response to cholera toxoid. *J. Exp. Med.* **142**, 1550-1563.

223. Potter, M. (1972). Immunoglobulin-producing tumours and myeloma proteins ofnmice. *Phys. Rev.* **52**, 631-719.

224. Potworowski, E. F., and Nairn, R. C. (1967). Origin and fate of a thymocyte-specidiv antigen. *Immunology* **13**, 597-602.

225. Querinjean, P. (1975). A possible experimental model to test the hypothesis of the integration of the V- and C-genes in the synthesis of an immunoglobulin chain. *Arch. Int. Physiol. Biochim.* **83**, 988-989.

226. Querinjean, P., Bazin, H., Beckers, A., Deckers, C., Heremans, J. F., and Milstein, C. (1972). Transplantable immunoglobulin-secreting tumours in rats. Purification and chemical characterization of four kappa-chains from LOU/Wsl rats. *Eur. J. Biochem.* **31**, 354-359.

227. Querinjean, P., Bazin, H., Kehoe, J. M., and Capra, J. D. (1975). Transplantable immunoglobulin-secreting tumours in rats. VI. N-terminal sequence variability in LOU/Wsl rat monoclonal heavy chains. *J. Immunol.* **114**, 1375-1378.

228. Querinjean, P., Bazin, H., Starace, V., Beckers, A., Deckers, C., and Heremans, J. F. (1973). Lambda light chains in rat immunoglobulins. *Immunochemistry* **10**, 653-654.

229. Raff, M. C. (1971). Surface antigenic markers for distinguishing T and B lymphocytes in mice. *Transplant. Rev.* **6**, 52-80.

230. Reif, A. E., and Allen, J. M. V. (1964). The AKR thymic antigen and its distribution in leukemias and nervous tissues. *J. Exp. Med.* **120**, 413-433.

231. Reinhardt, W. O. (1945). Rate of flow and cell count of rat thoracic duct lymph. *Proc. Soc. Exp. Biol. Med.* **58**, 123-124.

232. Reinhardt, W. O., and Li, C. H. (1945). Cell count, rate of flow, and protein content of cervical lymph in the rat. *Proc. Soc. Exp. Biol. Med.* **58**, 321-323.

233. Rieke, W. O. (1966). Lymphocytes from thymectomized rats: Im-

munologic, proliferative, and metabolic properties. *Science* **152**, 535–538.

234. Rokhlin, O. V., Vengerova, T. I., and Nezlin, R. S. (1971). RL allotypes of light chains of rat immunoglobulins. *Immunochemistry* **8**, 525–538.

235. Rose, N. R. (1975). Differing responses of inbred rat strains in experimental autoimmune thyroiditis. *Cell. Immunol.* **18**, 360–364.

236. Roser, B. J. (1976). The origin and significance of macrophages in thoracic duct lymph. *Aust. J. Exp. Biol. Med. Sci.* **54**, 541–550.

237. Ruddick, J. H., and Leslie, G. A. (1977). Structure and biological functions of human IgD. XI. Identification and ontogeny of a rat lymphocyte immunoglobulin having antigenic cross-reactivity with human IgD. *J. Immunol.* **118**, 1025–1031.

238. Ruscetti, S. K., Gill, T. J., III, and Kunz, H. W. (1976). The genetic control of the antibody response in inbred rats. *Mol. Cell. Biochem.* **7**, 145–156.

239. Sapin, C., and Druet, P. (1976). Isolation of rat and rabbit IgM from normal serum using antihuman $m\mu$ antibody-polyacrylamide beads immunoabsorbents. *J. Immunol. Methods* **12**, 355–363.

240. Schlossman, S. F., and Hudson, L. J. (1973). Specific purification of lymphocyte populations on a digestible immunoabsorbent. *J. Immunol.* **110**, 313–315.

241. Schoefl, G. I. (1972). The migration of lymphocytes across the vascular endothelium in lymphoid tissue. A reexamination. With an appendix by R. E. Miles. *J. Exp. Med.* **136**, 568–588.

242. Scott, D. W., and Howard, J. C. (1972). Collaboration between thymus-derived and marrow-derived thoracic duct lymphocytes in the hemolysin response of the rat. *Cell. Immunol.* **3**, 430–441.

243. Scott, D. W., and McKenzie, R. G. (1974). Rat B cells: Identity of immunoglobulin bearing cells with a population of lightly uridine-labeled cells. *Cell. Immunol.* **12**, 61–65.

244. Shorter, R. G., and Bollman, J. L. (1960). Experimental transfusion of lymphocytes. *Am. J. Physiol.* **198**, 1014–1018.

245. Shortman, K. (1966). The separation of different cell classes from lymphoid organs. I. The use of glass bead columns to separate small lymphocytes, remove damaged cells and fractionate cell suspensions. *Aust. J. Exp. Biol. Med. Sci.* **44**, 271–286.

246. Shortman, K. (1968). The separation of different cell classes from lymphoid organs. II. The purification and analysis of lymphocyte populations by equilibrium density gradient centrifugation. *Aust. J. Exp. Biol. Med. Sci.* **46**, 375–396.

247. Silverman, D. A., and Rose, N. R. (1975a). Spontaneous and methylcholanthrene enhanced thyroiditis in BIF rats. I. The incidence and severity of the disease, and the genetics of susceptibility. *J. Immunol.* **114**, 145–147.

248. Silverman, D. A., and Rose, N. R. (1975b). Spontaneous and methylcholanthrene-enhanced thyroiditis in BUF rats. II. Induction of experimental autoimmune thyroiditis without completed Freund's adjuvant. *J. Immunol.* **114**, 148–150.

249. Simic, M. M., Sljivic, V. S., Petrovic, M. Z., and Cirkovic, D. M. (1965). Antibody formation in irradiated rats. *Bull. Boris Kidric Inst. Nucl. Sci.* **16**, Suppl. 1, 1–151.

250. Smith, W. I., Ladoulis, C. T., Misra, D. N., Gill, T. J., III, and Bazin, H. (1975). Lymphocyte plasma membranes. III. Composition of lymphocyte plasma membranes from normal and immunized rats. *Biochim. Biophys. Acta* **382**, 506–525.

251. Spector, W. G., Walter, M. N.-I., and Willoughby, D. A. (1965). The origin of the mononuclear cells in inflammatory exudates induced by fibrinogen. *J. Pathol. Bacteriol.* **90**, 181–192.

252. Spector, W. G., and Willoughby, D. A. (1968). The origin of mononuclear cells in chronic inflammation and tuberculin reactions in the rat. *J. Pathol. Bacteriol.* **96**, 389–399.

253. Stankus, R. P., and Leslie, G. A. (1975). Genetic influences on the immune responses of rats to streptococcal A carbohydrate. *Immunogenetics* **2**, 29–38.

254. Starace, V., and Querinjean, P. (1975). The primary structure of a rat K Bence-Jones protein: Phylogenetic relationships of V- and C-region genes. *J. Immunol.* **115**, 59–62.

255. Stechschulte, D. J., and Austen, K. F. (1970). Immunoglobulins of rat colostrum. *J. Immunol.* **104**, 1052–1062.

256. Stechschulte, D. J., Orange, R. P., and Austen, K. F. (1970). Immunochemical and biologic properties of rat IgE. I. Immunochemical identification of rat IgE. *J. Immunol.* **104**, 1082–1086.

257. Steinman, R. M., and Cohn, Z. A. (1973). Identification of a novel cell type in peripheral lymphoid organs of mice. I. Morphology, quantitation, tissue distribution. *J. Exp. Med.* **137**, 1142–1162.

258. Strober, S. (1972). Initiation of antibody responses by different classes of lymphocytes. V. Fundamental changes in the physiological characteristics of virgin thymus-independent ("B") lymphocytes and "B" memory cells. *J. Exp. Med.* **136**, 851–871.

259. Strober, S., and Dilley, J. (1973). Biological characteristics of T and B memory lymphocytes in the rat. *J. Exp. Med.* **137**, 1275–1292.

260. Strober, S., and Dilley, J. (1973). Maturation of B lymphocytes in the rat. I. Migration pattern, tissue distributions and turnover rate of unprimed B lymphocytes involved in the adoptive anti-dinitrophenyl response. *J. Exp. Med.* **138**, 1331–1344.

260a. Sunderland, C. A., McMaster, W. R., and Williams, A. F. (1979). Purification with monoclonal antibody of a predominant leukocyte-common antigen and glycoprotein from rat thymocytes. *Eur. J. Immunol.* **9**, 155–159.

261. Tada, N., Itakura, K., and Aizawa, M. (1974). Genetic control of the antibody response in inbred rats. *J. Immunogenet.* **1**, 265–275.

262. Tada, T. (1975). Regulation of reaginic antibody formation in animals. *Prog. Allergy* **19**, 122–194.

263. Tada, T., Okumura, K., Platteau, B., Beckers, A., and Bazin, H. (1975). Half-lives of two types of rat homocytotropic antibodies in the circulation and in the skin. *Int. Arch. Allergy Appl. Immunol.* **48**, 116–131.

264. Tada, T., Okumura, K., and Tanigushi, M. (1973). Regulation of homocytotropic antibody formation in the rat. VII. An antigen specific T cell factor that regulates anti-hapten homocytotropic antibody response. *J. Immunol.* **111**, 952–961.

265. Thompson, A., and Morris, R. J. (1977). Rat lymphocyte differentiation antigens detected by rabbit antiserum to thymocytes and leukaemic cells. *Immunology* **32**, 419–425.

266. Trowbridge, I. S., Weissman, I. L., and Bevan, M. J. (1975). Mouse T-cell surface glycoprotein recognised by heterologous anti-thymocyte sera and its relationship to Thy-1 antigen. *Nature (London)* **256**, 652–654.

267. Twarog, F. J., and Rose, N. R. (1969). The adjuvant effect of pertussis vaccine in experimental thyroiditis in the rat. *Proc. Soc. Exp. Biol. Med.* **130**, 435–439.

268. Twarog, F. J., and Rose, N. R. (1970). Transfer of autoimmune thyroiditis of the rat with lymph node cells. *J. Immunol.* **104**, 1467–1475.

269. Unanue, E. R., and Dixon, F. J. (1967). Experimental glomerulonephritis: Immunological events and pathogenetic mechanisms. *Adv. Immunol.* **6**, 1–90.

270. Vaerman, J. P., Heremans, J. F., Bazin, H., and Beckers, A. (1975). Identification and some properties of rat secretory component. *J. Immunol.* **114**, 265–269.

271. Van Breda Vriesman, P. T. C., and Feldman, J. D. (1972). Rat γM immunoglobulin: Isolation and some biological characteristics. *Immunochemistry* **9**, 525–534.

272. Volkman, A. (1966). The origin and turnover of mononuclear cells in peritoneal exudates in rats. *J. Exp. Med.* **124**, 241–253.

273. Volkman, A. (1976). Monocyte kinetics and their changes in infection. *In* "Immunobiology of the Macrophage" (D. S. Nelson, ed.), pp. 291–322. Academic Press, New York.

274. Volkman, A. (1976). Disparity in origin of mononuclear phagocyte populations. *RES, J. Reticuloendothel. Soc.* **19**, 249–268.

275. Volkman, A., and Collins, F. M. (1968). Recovery of delayed-type hypersensitivity in mice following suppressive doses of X-radiation. *J. Immunol.* **101**, 846–859.

276. Volkman, A., and Collins, F. M. (1974). The cytokinetics of monocytosis in acute *Salmonella* infection in the rat. *J. Exp. Med.* **139**, 264–277.

277. Volkman, A., and Collins, F. M. (1975). Pathogenesis of *Salmonella*-associated arthritis in the rat. *Infect. Immun.* **11**, 222–230.

278. Volkman, A., and Collins, F. M. (1976). Role of host factors in the pathogenesis of *Salmonella*-associated arthritis in rats. *Infect. Immun.* **13**, 1154–1160.

279. Volkman, A., and Gowans, J. L. (1965). The production of macrophages in the rat. *Br. J. Exp. Pathol.* **46**, 50–61.

280. Volkman, A., and Gowans, J. L. (1965). The origin of macrophages from bone marrow in the rat. *Br. J. Exp. Pathol.* **46**, 62–70.

282. Wagner, W.-H. (1974). Activation of macrophages from different rodent species. *In* "Activation of Macrophages" (W.-H. Wagner and H. Hahn, eds.), Vol. 2, pp. 123–130. Excerpta Med. Found., Amsterdam.

283. Wagner, W.-H. (1975). Host–parasite interactions with peritoneal macrophages of mice and rats *in vitro* and *in vivo*. *Infect. Immun.* **12**, 1295–1306.

284. Wahl, S. M., and Rosenstreich, D. L. (1976). Role of B lymphocytes in cell-mediated immunity. I. Requirement for T cells or T-cell products for antigen-induced B-cell activation. *J. Exp. Med.* **144**, 1175–1187.

285. Wahl, S. M., Wilton, J. M., Rosenstreich, D. L., and Oppenheim, J. J. (1975). The role of macrophages in the production of lymphokines by T and B lymphocytes. *J. Immunol.* **114**, 1296–1301.

286. Waksman, B. H., Arnason, B. G., and Jankovic, B. D. (1962). Role of the thymus in immune reactions in rats. III. Changes in the lymphoid organs of thymectomized rats. *J. Exp. Med.* **116**, 187–206.

287. Wang, A. C., Fudenberg, H. H., and Bazin, H. (1975). The nature of "species-specific" amino acid residues. *Immunochemistry* **12**, 505–509.

288. Wang, A. C., Fudenberg, H. H., and Bazin, H. (1979). Partial amino acid sequences of kappa chains of rat immunoglobulins. Genetic and evolutionary implications. *Biochem. Genet* (in press).

289. Weir, D. M., ed. (1978). "Handbook of Experimental Immunology," 3rd ed., Vols. 1, 2, and 3. Blackwell, Oxford.

290. Wessels, C. C. (1941). Tuberculosis in the rat. I. Gross organ changes and tuberculin sensitivity in rats infected with tubercle bacilli. *Am. Rev. Tuberc.* **43**, 449–458.

290a. White, R. A. H., Mason, D. W., Williams, A. F., Galfre, G., and Milstein, C. (1978). T-lymphocyte heterogeneity in the rat: Separation of functional subpopulations using a monoclonal antibody. *J. Exp. Med.* **148**, 664–673.

291. Whitelaw, D. M., Bell, M. F., and Batho, H. F. (1968). Monocyte kinetics: Observations after pulse labeling. *J. Cell. Physiol.* **72**, 65–71.

292. Williams, A. F. (1976). Many cells in rat bone marrow have cell-surface Thy-1 antigen. *Eur. J. Immunol.* **6**, 526–528.

293. Williams, A. F., and Gowans, J. L. (1975). The presence of IgA on the surface of rat thoracic duct lymphocytes which contains internal IgA. *J. Exp. Med.* **142**, 335–345.

294. Williams. R. M., and Moore, M. J. (1973). Linkage of susceptibility to experimental allergic encephalomyelitis to major histocompatibility locus in rats. *J. Exp. Med.* **138**, 775–783.

295. Wilton, J. M., Rosenstreich, D. L., and Oppenheim, J. J. (1975). Activation of guinea pig macrophages by bacterial lipopolysaccharide requires bone marrow-derived lymphocytes. *J. Immunol.* **114**, 388–393.

296. Wisse, E. (1974). Observations on the fine structure and peroxidase cytochemistry of normal rat liver Kupffer cells. *J. Ultrastruct. Res.* **46**, 393–426.

297. Wisse, E. (1974). Kupffer cell reactions in rat liver under various conditions as observed in the electron microscope. *J. Ultrastruct. Res.* **46**, 499–520.

298. Wisse, E., and Daems, W. T. (1970). Fine structural study on the sinusoidal lining cells of rat liver. *In* "Mononuclear Phagocytes" (R. van Furth, ed.), pp. 200–210. Blackwell, Oxford.

299. Wistar, R., Jr. (1969). Immunoglobulin allotype in the rat. Localization of the specificity to the light chain. *Immunology* **17**, 23–32.

300. World Health Organization (1973). Nomenclature of human immunoglobulins. *Bull. W. H. O.* **48**, 373.

301. Wright, P. W., Hargreaves, R. E., Bernstein, I. D., and Hellström, I. (1974). Fractionation of sera from operationally tolerant rats by DEAE-cellulose chromatography; evidence that serum blocking factors are associated with IgG. *J. Immunol.* **112**, 1267–1270.

302. Wurzburg, U. (1971). Correlation between the immune response to an enzyme and histocompatibility type in rats. *Eur. J. Immunol.* **1**, 496–497.

303. Yoshida, T., Sonozaki, H., and Cohen, S. (1973). The production of migration inhibition factor by B and T cells of the guinea pig. *J. Exp. Med.* **138**, 784–797.

Chapter 10

Immunogenetics

Donald V. Cramer, Heinz W. Kunz, and Thomas J. Gill, III

The field of immunogenetics covers two broad areas of investigation. The first is the development of immunological techniques to identify biologically important molecules, especially cell surface components, and their use to study the genetic control of the synthesis and function of these molecules. The second major area is the study of the genetically controlled mechanisms that regulate immunologically important functions. In the mouse and man these studies have advanced rapidly, and they are summarized in recent books by Klein (44) and by Snell *et al*. (72). The rat is the experimental animal of choice for many studies in immunogenetics because of its relatively large size, its high reproductive capacity, and the availability of large numbers of highly inbred strains. This chapter will be concerned primarily with an overview of the immunogenetics of the rat, and it will emphasize various cell surface antigenic systems and the genetic mechanisms controlling immunological responsiveness.

I. DEFINITIONS

In discussing the various aspects of immunogenetics it is important to have an understanding of the terms commonly used.

The *gene* is a functional unit of inheritance, usually considered to represent a segment of a DNA molecule responsible for controlling the expression of a single phenotypic trait. An *al-*

lele is an alternative form of a gene at a specific location in homologous chromosomes, and a *genetic locus* is the position of a gene or one of its alleles on a chromosome. *Gene segregation* refers to the separation of genes on maternal and paternal chromosomes into different gametes during meiosis. If two or more genes are close together on a single chromosome, they are linked, and their movement together as a single unit is referred to as a *linkage group*. A group of such linked genes is referred to as a *haplotype*. Figure 1 is an illustration of the breeding protocol useful for determining the inheritance patterns and expression of alleles at the same genetic locus.

When homologous chromosomes exchange portions of DNA resulting in the expression of phenotypic traits not previously seen in either parent, a *genetic recombination* has taken place. By using genetic recombinations derived under controlled conditions, it is possible to determine the order and relative distance between the genes present on a chromosome; these data are used to construct a *linkage map*.

Histocompatibility genes are genes which determine the ability of a recipient to tolerate or reject tissue grafts from a donor. If the donor and recipient are genetically identical, they are referred to as *syngeneic* (or *isogeneic*), and grafts exchanged between them are *isografts*. Grafts exchanged between genetically different individuals or strains of the same species are called *allografts*. Under normal circumstances, the recipient tissues is recognized as being foreign by the donor and rejected. This is presumably due to the presence of cell surface antigens (*alloantigens*) on the donor tissue which are foreign to the recipient and which stimulate an immune response. Genetically different individuals from different species are *xenogeneic,* and tissue grafts exchanged between the two species are *xenografts*.

There are multiple genes involved in controlling histocompatibility differences between allogeneic individuals, usually divided into the *major histocompatibility complex* (MHC) and *minor histocompatibility loci,* which are responsible for weak transplantation reactions. These histocompatibility loci control cell surface antigens which are the predominant influence on the success or failure of transplanted tissue. In most species the MHC is composed of multiple, closely linked genetic loci, controlling not only tissue rejection but a number of other important immunological functions.

II. CELL SURFACE ALLOANTIGENS

A. Methods of Detection

The surfaces of cells contain antigens, primarily glycoproteins. The function of the various alloantigens is unknown, but their association with phenomena, such as tissue graft rejection, cell-to-cell interactions, immune responses, and maternal–fetal interactions, and their widespread occurrence in different animal species imply an important biological and evolutionary function. In the rat two antigenic systems, the classical histocompatibility antigens (RT1.A) and the blood group antigens (RT2) have received the most attention.* The histocompatibility antigens display a wider tissue distribution than blood group antigens, and they are present on erythrocytes, lymphocytes, kidney, brain, liver, spleen, eye, and skin (40).

The detection of these antigenic systems depends upon the immunization of an animal with the appropriate tissue from an allogeneic or xenogeneic donor. Antibodies directed against the foreign antigens are produced and can be used to identify the cell surface antigens. Since many different antigens may be present in the donor cells, an antiserum must be absorbed with various cells to eliminate cross-reacting antibody specificities. The result is an antiserum which detects only the desired antigen(s) and is referred to as an "operationally monospecific" typing reagent. It should be recognized that if tightly linked genes are responsible for coding different cell surface antigens, antisera produced in this fashion may not discriminate between the two antigens, hence, the term "operationally monospecific."

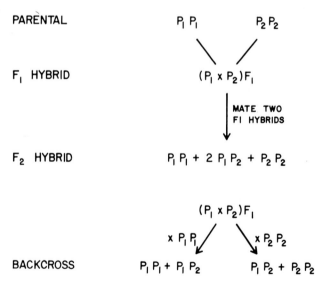

Fig. 1. Typical breeding combinations used to determine the inheritance pattern and expression of the genes that control the phenotypical traits P₁ and P₂. Using this mating protocol it is possible to determine the dominance versus recessive expression of the genes and to establish whether or not these genes segregate independently.

*The nomenclature used in this chapter is that recently adopted at the Rat Histocompatibility Workshop on September 4, 1978, in Rome, Italy. Table I includes the nomenclature conversions for the common alloantigenic systems in the rat. Detailed information about inbred rat strains may be obtained in "Inbred and Genetically Defined Strains of Laboratory Animals," Biological Handbook III. Federation of American Societies for Experimental Biology, 1979. (See also this volume, Chapter 9, and Volume I, Chapter 13.)

Our procedure for the production of typing sera for the blood group antigens (RT2) of rats is based upon the production of xenoantisera in rabbits repeatedly immunized with rat erythrocytes (56, 62). Rabbits are immunized with erythrocytes from $RT2^a$ or $RT2^b$ rats every other day for twelve injections and bled 6–7 days after the last injection. The resulting antisera must be repeatedly absorbed with the appropriate rat erythrocytes to remove antibodies directed against "species-specific" antigens and histocompatibility antigens. Typically, the rabbit antiserum is absorbed with erythrocytes from an unrelated rat strain to remove the "species-specific" antibodies followed by absorption with cells from *RT1* incompatible, *RT2* compatible strain to remove the anti-RT1 antibodies. Recently, alloantibodies against the RT2 system have been produced by multiple erythrocyte injections between different rat strains, demonstrating the feasibility of using these reagents for typing purposes (48). The observation that these alloantibodies can be seen following immunization procedures, as described below for the RT1 alloantigens, illustrates the need to carefully control donor–recipient matching for producing alloantibodies against MHC antigens.

Typing reagents for detecting RT1 histocompatibility antigens are prepared by immunizing allogeneic strains with skin grafts and splenic lymphocytes (47, 49). Split-thickness skin grafts (approximately 2 cm square) or 1 cm tail skin grafts are sutured to the lateral thorax. Three weeks later a second skin graft from the same individual or strain is placed on the opposite side. The recipients are then immunized 3 weeks thereafter with one or two injections of splenic lymphocytes. Whenever possible donor and recipient pairs are selected on the basis of the smallest number of tissue incompatibilities, including erythrocyte antigens, in order to reduce the amount of absorption necessary to produce a monospecific reagent. Once the antiserum has been obtained, it is tested against a panel of cells representing different histocompatibility types by hemagglutination or lymphocytotoxicity. The antiserum will contain antibodies directed against antigenic specificities shared by different histocompatibility groups as well as those directed at the antigenic sites that specifically characterize a single group. The former antibodies are removed by absorption until the serum is monospecific, divided into small aliquots, and stored frozen.

B. Alloantigenic Systems

Inbred strains of rats used in the laboratory are derived from a limited number of wild rats, which were initially captured for the sport of rat-baiting. They have been established in colonies for experimental work since the 1850's. The majority of inbred strains were derived from a common stock of animals, developed at the Wistar Institute. The result has been a severe limitations in the amount of genetic diversity between inbred

strains (26,39,59). Early work in describing the antigens present on erythrocytes resulted in the establishment of separate antigenic groups. In a description of the work Owen (56) referred to the antigens as groups A, C & D, E and F. Subsequently, the better established antigenic systems in these groups were reclassified by Palm and Black (61) employing the symbol "Ag" combined with a capital letter denoting the genetic locus. This produced the designations Ag-A, Ag-B, and Ag-C (formerly C-D). In addition, a fourth group, Ag-D [formerly antigen number 2 of Palm (57)] was described with the Ag-D1 allele originally present on the erythrocytes of WF rats.

The *Ag-A* locus was defined by the rare occurrence of a natural agglutinin in the sera of individual rats (15). The expression of the antigen or, in its absence, the production of the antibody "alpha," is controlled by two alleles with the gene for A antigen expression dominant. The inbred strains commonly employed in the laboratory all have the *A* allele expressed and react positively with the "alpha" antibody. This allele is apparently rare, and it has not been studied further.

The Ag-B alloantigen system was originally shown to be expressed by cells of an ascites tumor (AA). Rejection of the tumor was accompanied by the production of an antibody response against erythrocyte antigens shared by several strains (13). Resistance to the ascites tumor was associated with a lack of the specific *B* locus antigen and the production of antibody after rejection (14). These results demonstrated that in the rat, as in the mouse, a specific erythrocyte antigen group could be associated with major histocompatibility differences between strains. Subsequent reciprocal immunizations between inbred strains led to the discovery of multiple alleles at the *Ag-B* locus present as erythrocyte antigens (27,57). During the development of this work a second nomenclature for this alloantigenic system arose; Štark and his co-workers proposed the term *H-1* for the strong histocompatibility locus in parallel with the histocompatibility system of mice (74). This nomenclature and the classification of a number of inbred strains has been presented in reviews of rat immunogenetics (31,34), and direct comparisons have been made between inbred rat strains independently classified by these two nomenclature systems (50).

The description of the B blood groups as a major histocompatibility antigen led to the discovery of a number of immunological functions controlled by genes closely linked to this locus (see below). Recently genetic recombinations have been observed within the MHC of the rat, dividing this complex of genes into at least three regions (16,45,75). Because of increasing complexity in the genetic composition of the rat MHC, a uniform nomenclature replacing the *Ag-B* and *H-1* designations was adopted at the Rat Histocompatibility Workshop, September 4, 1978, in Rome, Italy. The prefix *RT* will be assigned to all allogeneic gene complexes in the rat, and individual systems will be assigned sequential numbers. The

Table I

Nomenclature Conversions for Rat Alloantigenic Systems

	RT1			RT2		RT3	
RT1	Ag-B	H-1	RT2	Ag-C	RT3	Ag-D	
a	4	a	a	1	a	1	
b	6	b	b	2	b	0	
c	5	c					
d	9	d					
f	10	f					
g	7	g					
h		h					
k	8	k					
l	1	l					
m	13	m					
n	3	n					
u	2	w					

MHC has been designated *RT1* (formerly *Ag-B* or *H-1*), and Table I compares this nomenclature with previous systems.

The *A* region of the MHC contains the genes that code for the traditional serologically defined histocompatibility (Ag-B/H-1) antigens. Tables I and II present the nomenclature conversions for *RT1* complex and the classification of the common rat strains for three alloantigenic systems. Within a single *RT1* haplotype group all of the individual rat strains share one or more antigenic specificites that are characteristic for the group. Polyspecific antisera raised by skin grafting and/or lymphocyte immunization can be absorbed to yield operationally monospecific antisera. These specific antisera have positive hemagglutination reactions with the strains of an *RT1* group and can be completely absorbed with red cells or lymphocytes from any single member of the group. This antigenic specificities are coded for by semidominant genes present within the *A* region of the MHC (*RT1.A*) and are expressed on the surfaces of many different types of cells (40). Segregation studies of these genes in F_2 and backcross combinations are compatible with a single genetic locus with multiple alleles controlling the "private" RT1.A specificities [summarized in Gill *et al.* (49,67)]. In addition, other antigenic specificities are detected by polyspecific antisera which are shared between strains of different *RT1* haplotypes. These are similar to the public antigenic specificities of the mouse and man. Although the antigenic specificities which define an *RT1* group behave genetically as a single gene with multiple alleles, it is reasonable to assume that these specificities actually represent the products of closely linked genes. Other species, such as the mouse, man, monkey, and dog, have either two or three gene-

Table II

Classification of Selected Alloantigenic Systems in Inbred Rats

Strain[a]	RT1			RT2	RT3	Strain	RT1			RT2	RT3
	Haplotype	A region	B region				Haplotype	A region	B region		
LEW	l	l	l	a	a	DA	a	a	a	b	a
F344	l	l	l	a	b	ACI	a	a	a	b	a
NBR	l	l	l	b	a	ACP	a	a	a	a	a
CAR	l	l	l	a	—[b]	COP	a	a	a	b	—
CAS	l	l	l	b	b	AUG28807	c	c	c	a	a
Z61	l	l	—	a	—	PVG	c	c	c	b	a
S5B	l	l	l	b	—	BUF	b	b	b	a	b
WF	u	u	u	b	a	ALB	b	b	b	b	b
YO	u	u	u	b	a	M520	b	b	b	a	a
LGE	u	u	u	b	a	NSD/N	b	b	b	a	a
OM/N	u	u	u	b	a	LA	b	b	—	b	—
RHA/N	u	u	—	a	—	KGH	g	g	l	b	a
RLA/N	u	u	—	b	—	WKA	k	k	k	a	a
PETH	u	u	—	b	—	OKA	k	k	k	b	a
LOU/C	u	u	u	b	—	SHR	k	k	k	b	a
WAG	u	u	u	b	—	BDV	d	d	d	b	a
BN	n	n	n	a	b	MR	o	d	a	a	a
MAXX	n	n	n	a	b	AS2	f	f	f	b	a
BI[c]	i	n	a	b	a	HW	h	h	—	a	a
RP[d]	p	u	l	—	—	MNR	m	m	c	a	a
						BDVII	e	a	c	b	—

[a] Prototypic strains for RT1 haplotype designation are underlined.

[b] Not tested(—).

[c] F_{14} generation (formerly called B3).

[d] Formerly R1.

tic loci which code for serologically detectable histocompatibility antigens. In the rat, as in these other species, the antigens coded for by genes of the *A* region are glycoprotein dimers, consisting of two noncovalently linked polypeptide chains of approximately 37,000 and 11,000 molecular weight (43).

Following the initial reports by Bogden and Aptekman (13) and Frenzl *et al.* (27) of the association between the *RT1* complex and tissue incompatibilities, several investigators have clearly linked RT1 incompatibility with strong and reproducible rejection of a variety of tissue allografts. Primary skin grafts between RT1 incompatible strains of rats are rejected in an acute fashion, usually less than 12 days (7,10,58). Some skin grafts exchanged between animals with same RT1 phenotype have long (41) to moderate (58) prolongation of skin graft survival. The subacute and acute skin rejection observed in some RT1 compatible combinations was interpreted to indicate differences at multiple minor loci capable of cumulatively producing rejection. By using inbred LEW and BN strain rats, Billingham *et al.* (12) demonstrated that these two strains differ at 14 to 16 individual histocompatibility loci, relatively few of which are associated with acute rejection.

The association of RT1 incompatibility with allograft rejection is also involved in survival times for heart and kidney. These organs apparently enjoy an immunological advantage over skin and can show prolonged survival in spite of minor *H* loci incompatibilities (7,10,11,77). Under conditions of RT1 incompatibility, kidney grafts are usually rejected within 7 days of transplantation (25,38). When RT1 compatible LEW and F344 strains are used for donor–recipient pairs, skin grafts are rejected uniformly in 10–12 days, but kidney survival is variable and prolonged with several grafts being permanently successful. Similar results have been obtained using cardiac allografts (10,11). Barker and Billingham (7) have suggested that the heart may be a more important marker than skin for studying the influence of gene dosage at minor histocompatibility loci on graft rejections. Skin is rejected maximally, even with few *H* loci differences, and allows little opportunity to determine the effect of non-MHC loci on allograft rejection.

The RT2 blood group antigens are coded by a single genetic locus with two alleles *RT2*^{a} and *RT2*^{b} (formerly *Ag-C1* and *Ag-C2*). The alleles are expressed codominantly and control the expression of antigens on the surface of erythrocytes, liver, and spleen (56,57) but not in lymphocytes or platelets (62). The *RT2* locus segregates independently of the *RT1* locus (47, 60,67) and is apparently linked to loci controlling the expression of plasma esterases in a fashion analogous to the *Ea-1* blood group locus of mice (28,80). The blood group antigens controlled by the *RT2* alleles are not associated with histocompatibility differences between RT2 incompatible strains. The BN strain of rats has been maintained in *RT2*^{a} and *RT2*^{b} sublines and reciprocal skin grafts between these two lines are permanently accepted (57). The *in vitro* mixed lymphocyte reaction, a measure of histocompatibility differences, is negative between RT2 incompatible strains (62). Similar antigenic systems have been described by Štark and his colleagues (46). These have been designated H-2 and H-3, but the relationship between RT2 and H-2 and H-3 cannot be established.

The RT3 alloantigen system (formerly *Ag-D*) controls the expression of an antigen that was first described in the WF strain (57,60). The *RT3* locus is apparently not associated with *RT1*, since the antigenic determinants controlled by this locus are not seen in animals congenic for the MHC. Antisera are produced by hyperimmunization with lymphocytes and are capable of distinguishing by hemagglutination at least 29 strains which share the *RT3*^{a} allele. An additional seven strains do not react, due to the presence of an allele *RT3*^{b} that codes for a serologically silent antigen (48,61). Breeding studies have demonstrated that the *RT3* locus is not linked to *RT1* or *RT2* or to the recently described Ag-S system (48).

Recently, a lymphocyte alloantigenic locus with several alleles, *Ag-F*, has been described in several inbred strains (22). The locus controls the expression of antigens present on the surface of thymic-derived lymphocytes which can be detected by lymphocytotoxic reactions. The *Ag-F* locus is linked to the coat color gene for albinism and may be closely related or identical to the Pta system of Howard and Scott (42), the Ly-1 system of Fabre and Morris (24a), or the A.R.T. system of Lubaroff (51a).

The *B* region of the *RT1* complex contains genes that code for a second set of alloantigens that are distinctly different from the histocompatibility antigens of the *A* region. These antigens are considered to be the rat homologues of the *I* region-associated (Ia) antigens of mice in that they (a) are glycoproteins with molecular weights of approximately 28,000 to 35,000 daltons (66), (b) show a limited tissue distribution, primarily on B lymphocytes (63,66), and (c) are coded for by genes closely linked to or identical to MHC genes controlling immune responsiveness and/or the mixed lymphocyte response (16,63). In the rat, as in other species, it has not been possible to establish clearly whether or not Ia antigens play a role in mediating immune responses, such as those to soluble or cell surface antigens, which are controlled by loci in the *B* region.

III. GENETICALLY CONTROLLED IMMUNE RESPONSES

The establishment of the blood group B antigen as the major histocompatibility locus in the rat led to a series of investigations to establish the genetic control of the MHC over the immunological functions. These included the immune re-

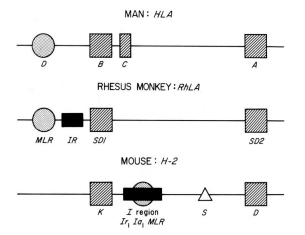

MAN: *HLA*

RHESUS MONKEY: *RhLA*

MOUSE: *H-2*

Fig. 2. A schematic representation of the MHC in three species for which a tentative positioning of the genes within the complex has been established. The similarity in the genetic structure of the MHC between these species is apparent. Each specific chromosomal region contains genes that control the expression of serologically defined histocompatibility loci (hatched blocks), immune response genes (closed blocks), and the major mixed lymphocyte response (stipled circle).

sponse to simple and complex protein antigens, cell-to-cell interactions, such as the mixed lymphocyte reactions (MLR) and graft-versus-host reactions (GVHR), and the expression of specific lymphocyte (Ia) alloantigen systems.

Studies of natural and laboratory-derived recombinant strains of rats have shown that the genes that control these responses are present in the *B* region of the MHC (16,19,75). The association of analogous series of genes controlling these immunological responses has been described in virtually all vertebrate species studied. Figure 2 is a schematic representation of the MHC in three species, man, mouse and the monkey, for which tentative positioning of these loci have been established, and Fig. 3 is the current map of the rat MHC.

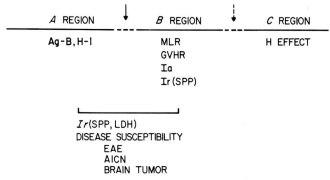

Fig. 3. The major histocompatibility complex (RT1) of the rat. The position of the A, B, and C regions relative to the centromere has not been established. SPP, synthetic polypeptide; LDH, lactic dehydrogenase; EAE, experimental allergic encephalomyelitis; AICN, autologous immune complex nephritis; H effect, histocompatibility effect.

A. Immune Response to Simple Antigens

The immune response in the rat to a number of simple polypeptide or protein antigens is under the control of genes located within the MHC. The initial experiments on the genetic control of the immune response demonstrated that simple polypeptide antigens were capable of eliciting wide variation in antibody responses in different inbred strains (71). Within an inbred strain the response to one antigen did not correlate with a second, unrelated antigen. The uniformity of the response within a strain and the large quantitative differences between strains lent additional support to the concept that immune response is under genetic control. Amerding and Rajewsky (1) immunized outbred Wistar and Sprague-Dawley rats and confirmed that similar differences in the level of antibody production occurred between two strains. In both cases the ability to respond to the antigen was transmitted as a dominant characteristic. Subsequent investigative work has more clearly defined the genetic control of specific immune responses in rats which are under control of several genes, at least one of which is part of the major histocompatibility complex (Table III).

The synthetic polypeptides, including the random linear polymer poly($Glu^{52}Lys^{33}Tyr^{15}$) and the branched polypeptides (T,G)-A—L, (H,G)-A—l, and (Phe,G) A—L have been most extensively studied (34,65). There are marked differences in the ability of inbred strains to mount an antibody response to each antigen (35,71). In the case of antibody production of poly ($Glu^{52}Lys^{33}Tyr^{15}$), strains of the *RT1* haplotypes *l, n, b, g* and *k* are low responders, producing less than 65 μg of antibody/ml serum. The members of the *$RT1^u$* group are moderate responders (200–500 μg antibody/ml), while members of the *$RT1^a$* and $RT1^c$ groups are all high responders

Table III

Genetically Controlled Immune Responses to Specific Antigens

Antigen	Reference
A. Histocompatibility linked	
Lactic dehydrogenase (LDH-A4)	81,82
Poly(Tyr, Glu)-poly(DLAla)--poly(Lys)	35
Encephalitogenic protein	29,78
Poly($Glu^{52}Lys^{33}Tyr^{15}$)	30,49
Poly(His, Glu)-poly(DLAla)--poly(Lys)	64
Poly($Glu^{50}Ala^{50}$)	2
Poly($Glu^{50}Tyr^{50}$)	2
($Tyr^{25}Glu^{25}Ala^{25}Gly^{25}$)$_n$	52
Poly(Phe, Glu)-poly(DLAla)--poly(Lys)	64
Poly(Tyr, Glu)-poly(Pro)--poly(Lys)	37
Poly(Phe, Glu)-poly(Pro)--poly(Lys)	37
Poly Pro--poly(Lys)	37
B. Not histocompatibility linked	
Sheep RBC	76
Bovine γ-globulin (BGG)	76

($> 700 \ \mu$g antibody/ml) (47,49). Similar patterns of immune responsiveness can be seen with the branched polypeptides and high, moderate, and low levels of antibody production to each antigen are associated with different histocompatibility types (64). In each case the ability to respond to a specific antigen does not serve as a means of predicting the response to a second, unrelated antigen, demonstrating the specific nature of the genetic control mechanism.

Segregation studies using high and low responder strains showed that the ability to respond to synthetic polypeptide antigens is inherited as a dominant trait. The F_1 hybrid offspring of low ($RT1^l$) and high ($RT1^a$) responder strains to poly(Glu^{53}Lys^{33}Tyr^{15}) are capable of making an antibody response to the antigen but at a level lower than the responder parent (34,67). The ability to respond by producing antibody segregates with the MHC and the quantity of antibody produced is influenced by another autosomal gene(s) segregating independently of the *RT1* complex. Figure 4 illustrates this point, demonstrating that parental DA strain rats have a signif-

icantly higher antibody response than the $RT1^a$ homozygous animals in the F_2 generation ($P < 0.01$). Similarly the $RT1^{1/a}$ heterozgotes in the two backcross combinations have significantly different antibody responses: the $RT1^{1/a}$ heterozygote from the cross into the DA background has a higher antibody response than animals with the same haplotype derived from the backcross into the low responder F344 strain. The delayed hypersensitivity response by skin testing displays a similar pattern of expression. These results are consistent with the hypothesis that the genetic control of the antibody response is polygenic: one gene responsible for antigen recognition and a second, independently segregating gene influencing the quantity of antibody produced. Certain mating combinations between low responders can restore the ability of their offspring to respond to an antigen, presumably by acquiring each of the required genetic components separately from the parents (34).

The immune response to other simple antigens is consistent with this genetic control mechanism. The antibody response to a sequential polypeptide (Tyr^{25}Glu^{25}Ala^{25} Gly$^{25})_n$ is controlled by two autosomal genes. One is a dominant, MHC-linked gene controlling the ability to respond and a second gene affects the level of antibody produced (52). The antibody response to the branched synthetic polypeptides (T,G)-A—L, (H,G)-A—L, and (Phe,G)-A—L is under dominant genetic control linked to the MHC (34,36), and on the basis of the amount of antibody produced, it is possible to distinguish among high, intermediate, and low responding strains.

The antibody response to (H,G)-A—L in F_1 hybrid offspring from the intermediate responder LEW strain ($RT1^l$) and two of four low responding strains, LEW.IN ($RT1^n$) and LEW.ID ($RT1^d$), resulted in unexpectedly high responses (32). The authors suggest that in the case of this antigen, MHC-linked genes involved in the control of the immune response to this antigen are capable of complementation. These results are similar to those observed in mice (23).

The specific MHC-associated immune response (Ir) genes in both mice and rats have been shown to depend upon intact T cell populations for their expression (9,21). It has been suggested that the Ir genetic control is the result of the expression of a structural gene that mediates or modulates the specific interaction of the T cell and antigen (8). When cells from immunized rats are stimulated *in vitro* with the antigen, there is a specific stimulation of T-dependent lymphocytes, resulting in blast transformation and proliferation. When rats are neonatally thymectomized, the proliferative response to synthetic polypeptides is almost completely abolished (21). Similarly, separation of splenic T and B cell subpopulations on nylon wool columns results in a T cell-rich population of lymphocytes capable of responding to polypeptide antigens and a B cell-enriched population that does not respond (18).

The specificity of the rat *in vitro* proliferative response to antigen is very similar to that reported for the mouse. In both

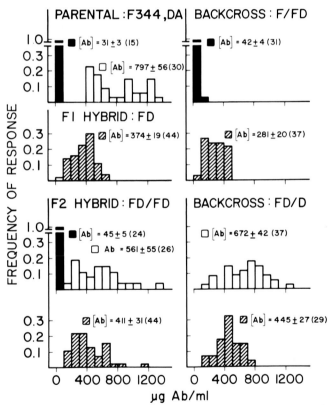

Fig. 4. A frequency histogram illustrating the genetic control of the antibody response to poly(Glu^{52}Lys^{33}Tyr^{15}) among crosses between the highly responding DA strain and the poorly responding F344 strain. The solid bars represent *RTI(l/l)*, the hatched bars *RTI(l/a)* and the open bars *RTI(a/a)* genotypes. For each genotype, the mean antibody concentration (μg Ab/ml $\pm$ S.E.) is given. The numbers in parentheses are the numbers of animals in each group. (From Shonnard *et al.,* 1976; reproduced from *J. Immunogenet.* by permission of Blackwell, Oxford.)

species the *in vitro* proliferative response to branched synthetic polypeptides appears to be more specific than *in vivo* antibody production (33,51). Immunization with one of the antigens resulted in the production of antibodies which strongly cross-react with other related branched polypeptides. If sensitized lymphocytes are cultured *in vitro* with the same antigens, a specific proliferative response is seen only in those cells sensitized to the antigen in question. For other polypeptide antigens, the distinction between the specificity of the T cell-controlled proliferative response and the degree of antibody cross-reactivity is not as sharp. When the random polymer poly(Glu^{52}Lys^{33}Tyr^{15}) is tested under the same experimental conditions, the degree of antibody and *in vitro* cell proliferative cross-reactivity to structurally similar antigen corresponds quite closely (55). The antibody produced to poly(Glu^{52}Lys^{33}Tyr^{15} may be slightly more efficient in recognizing small chemical differences between amino acids, but the molar ratios of amino acids and their charge, type, and isomeric configuration is capable of influencing the cross-reaction in both systems.

B. Mixed Lymphocyte Reaction

A number of cell-to-cell interactions also are controlled by genes present within the *B* region of the MHC. One of the most commonly employed measurement of this cellular interaction is the mixed lymphocyte reaction (MLR). This reaction may be an *in vitro* biological response of the lymphocyte to foreign antigens and represents histocompatibility differences *in vivo*. When immunocompetent lymphocytes from genetically different (allogeneic) individuals are incubated *in vitro*, they recognize cell surface antigenic differences and respond by undergoing transformation to lymphoblasts followed by cell division (3,5). Lymphocytes from unrelated individuals consistently stimulated each other, while lymphocytes from pairs of identical twins failed to respond (6). It was soon recognized that ability to respond is controlled by a gene(s) linked to the major histocompatibility system [for review, see Sørenson (73)]. In the rat, incompatibility at the major histocompatibility locus (*RT1*) is associated with positive MLR responses (17, 24,79).

In the rat, as in other species, the responding cell in the MLR is a small lymphocyte which requires the presence of an intact thymus (T cells) to mature. Cells derived from rats thymectomized at birth have a greatly diminished ability to respond to foreign lymphocytes in culture (79). Similarly, the separation of T and B cells into distinct populations *in vitro* has shown that T cells are efficient in producing, but ineffective at stimulating, a proliferative MLR response. The B cell population, in contrast, is ineffective at initiating a response but very effective as the stimulating cell in the culture system (53). In practical terms, the use of the MLR as a test for histocompatibility requires the inactivation of one cell population, either by irradiation or by treatment with mitomycin C. Once the system has been converted to a one-way response, the degree of histocompatibility in each direction can be determined. In humans and in mice the MLR measures one component of allogeneic tissue graft rejection, but the correlation is less than complete. It is possible that the MLR is a proliferative recognition phase and that the tissue destruction phase is mediated by the activation by T cells of a second subpopulation of cells with cytotoxic activity against the serologically defined histocompatibility antigens.

Experiments using rats as models for transplantation studies have been hampered by a lack of genetically characterized recombinations within the MHC. The result has been that various functions attributed to the MHC of the rat represent a summation of the effects of all the genes in the complex. It has not been possible until recently, for example, to determine the relative effect on allograft survival of such loci as those which control the expression of *A* region antigens versus the genes which control the MLR. The derivation of increasing numbers of strains with documented genetic recombinations within the MHC should result in a large volume of interesting data on allograft survival in the near future.

IV. ALLOGENEIC REACTIONS

The alloantigens coded for by genes in the MHC of the rat play a major role in the rejection of foreign tissue grafts. The success of these grafts is also influenced, however, by genes controlling the expression antigenic systems located in linkage groups that segregate independently from the complex. These loci usually are associated with weak histocompatibility effects, resulting in slow or chronic graft rejections. The number and strength of the loci that determine tissue incompatibilities differ with the strains tested and sensitivity of the grafting technique used. The LEW and BN strains, for example, differ for at least 14 histocompatibility loci (12). Since many weak loci can only be detected using small grafts (70) and many may be coded for by groups of linked genes, the true number may be substantially larger than previously reported.

Štark and his colleagues (45) have recently described two recombinant animals derived from breeding of LEW rats congenic for the *RT1*^{a}, *RT1*^{u}, and *RT1*^{n} haplotypes and that differ for a histocompatibility locus linked to the MHC. These animals display immunogenetic characteristics that indicate they are identical for their *A* and *B* region genes but reject skin grafts. This locus is linked to the *RT1* complex, does not elicit hemagglutinating antibodies, and is thought by the authors to represent a third (*C*) region of the MHC (see Fig. 3).

The *Ag-E* locus is a histocompatibility locus responsible for mediating acute skin rejection, but it has been of limited interest, however, because of its apparent strain restriction (54). This locus controls a histocompatibility difference between members of an inbred strain: strain A2 was derived from a Wistar rat stock and despite inbreeding by brother–sister mating for 72 generations approximately 50% of the skin grafts exchanged between strain members were rejected. Mating experiments conducted between members of this strain demonstrated the presence of a histocompatibility locus associated with a strong selective pressure favoring the implantation and survival of heterozygous animals. Little is known about the distribution of this system in other strains of rats or whether this locus is associated with alloantigenic differences.

Two additional histocompatibility loci, *H-3* and *H-4*, have been described by Štark and his co-workers (46). These two loci were discovered during the derivation of congenic rat lines and have not been studied extensively. The *H-4* locus codes for histocompatibility effects that have been demonstrated to be linked to the *albino* locus. This locus does not display any serologic activity, and its relationship to the serologically detectable *Ag-F* and *Pta* loci (also linked to the *albino* locus) remains to be established.

In the rat, as in the mouse, the male Y chromosome contains a histocompatibility (*H-Y*) locus. Skin grafts exchanged between male and female BN rats are associated with slow rejection of the male skin grafts by the females (12). The rejection is dependent upon the size of the graft (smaller grafts are more sensitive to rejection), and preimmunization of the females with male lymphocytes increases the capacity of the females to react (70).

V. IMMUNOGENETICS OF WILD POPULATIONS OF RATS

One of the most fascinating aspects of studies involving the major histocompatibility complex (*H-2*) of the mouse has been the demonstration that the genes coding for histocompatibility antigens are very numerous. Natural populations of wild mice [see Klein (44) for review] have many *H-2* alleles that are unique, suggesting that the number of alleles at these loci may number in the thousands. This type of genetic complexity (polymorphism) is unparalleled in other genetic systems and has been interpreted by many to be important in the function of the MHC. In man the same histocompatibility loci are extensively polymorphic, but the number of alleles at each locus currently numbers 30 or less (4) rather than thousands. This finding suggests that the degree of polymorphism seen in the mouse may not be a necessary prerequisite for appropriate MHC function.

Recent studies of natural populations of wild rats have demonstrated that the genetic complexity of the MHC in this species is distinctly different from that of the mouse, despite their close phylogenetic relationship. Rats from widely separated geographical locations have been examined for several phenotypical traits controlled by genes within the MHC, including their RT1 histocompatibility antigens, mixed lymphocyte response, Ia antigens, and immune responses to poly(Glu^{52}Lys^{33}Tyr^{15}). In each case the degree of genetic polymorphism for genes controlling these responses is greatly restricted when compared to the mouse.

The RT1 histocompatibility antigens of rats trapped in several locations within the United States and Europe were investigated by using monospecific antisera against ten major *RT1* haplotypes (20,68). These sera reacted in a specific fashion with the animals tested, and in the large majority of the rats it was possible to detect at least one, and many cases two, of the "private" antigens coded for by the various *RT1* haplotypes. In approximately 120 animals examined, 8 of 10 inbred *RT1* haplotypes were detected (20,67). The genetic and serological characteristics of these antigens were very similar, if not identical, to those found in inbred strains. Although animals that represent new *RT1* haplotypes were identified, they represented only a small portion of the total number of haplotypes examined. In contrast, antisera raised against inbred mice rarely detect the same histocompatibility antigens in wild mouse populations.

In addition to the restriction in the number of alleles that code for *A* region histocompatibility antigens, the immunogenetic characteristics of the mixed lymphocyte responses and immune response to synthetic antigens are very similar to the common laboratory inbred strains. A high percentage of the animals display one-way MLR identity reactions with one or two inbred *RT1* haplotypes, suggesting that substantial numbers of the alleles at the MLR locus in natural populations of rats are the same as those seen in the inbred animals (19). Immunization of these rats with the synthetic polypeptide poly(Gly^{52}Lys^{33}Tyr^{15}) is associated with distinct positive and negative antibody responses in individual animals (69). The genes that control both the MLR and immune responses are linked to the MHC. The difficulty in demonstrating differences between the MLR and Ir alleles in wild and inbred animals suggests that the restricted polymorphism of the rat histocompatibility alleles may extend to other loci within the complex.

VI. CONCLUSIONS

The rat has been widely used for a variety of studies in immunology, oncology, transplantation, genetics, and reproductive biology. The better definition of the immunogenetics of this species will enhance its usefulness greatly, and recent developments indicate that this genetic information is rapidly becoming available. The establishment of stabilized colonies of inbred rats, the development of congenic lines, the identification and isolation of recombinants, and the clearer definition of the loci controlling a variety of cell surface antigens are all progressing rapidly. The evidence to date indicates that the structure of the major histocompatibility complex in the rat may be more like that of the human than the mouse. The restricted polymorphism of this species in both wild and inbred populations offers an unique opportunity to study the MHC under conditions that may lead to a better understanding of how this important complex of genes functions.

ACKNOWLEDGEMENT

The research in the authors' laboratory was supported by grants from the National Institutes of Health (CA 18659, CA 25038, and HD 08662). Dr. D. V. Cramer is a recipient of the Research Career Development Award (AI00314).

REFERENCES

1. Armerding, D., and Rajewsky, K. (1969). The genetic control of immunological responsiveness to lactic dehydrogenase (LDH) isoenzymes. In "Protides of the Biological Fluids" (H. Peters, ed.), p. 185–188. Pergamon, Oxford.
2. Armerding, D., Katz, D. H., and Benacerraf, B. (1974). Immune response genes in inbred rats. II. Segregation studies of the GT and GA genes and their linkage to the major histocompatibility locus. Immunogenetics 1, 340–351.
3. Bach, F. H., and Hirschhorn, K. (1964). Lymphocyte interaction: A potential histocompatibility test in vitro. Science 143, 813–814.
4. Bach, F. H., and VanRood, J. J. (1976). The major histocompatibility complex-genetics and biology. N. Engl. J. Med. 295, 927–936.
5. Bain, B., Vas, M. R., and Lowenstein, L. (1963). A reaction between leukocytes in mixed peripheral blood cultures. Fed. Proc., Fed. Am. Soc. Exp. Biol. 22, 428.
6. Bain, B., and Lowenstein, L. (1964). Genetic studies on the mixed leukocyte reaction. Science 145, 1315–1316.
7. Barker, C. F., and Billingham, R. (1971). Histocompatibility requirements of heart and skin grafts in rats. Transplant. Proc. 3, 172–175.
8. Benacerraf, B., and McDevitt, H. O. (1972). Histocompatibility-linked immune response genes. Science 175, 273–279.
9. Benacerraf, B., and Katz, D. H. (1975). The histocompatibility-linked immune response genes. Adv. Cancer Res. 21, 121–175.
10. Bildsøe, P. (1972). Organ transplantation in the rat. The importance of the Ag-B (or H-1) locus. Acta Pathol. Microbiol. Scand. Sect. B 80, 221–230.
11. Bildsøe, P., Sørensen, S. F., Pettirossi, O., and Simonsen, M. (1970). Heart and kidney transplantation from segregating hybrid to parental rats. Transplant. Rev. 3, 36–45.
12. Billingham, R. E., Hodge, B. A., and Silvers, W. K. (1962). An estimate of the number of histocompatibility loci in the rat. Proc. Natl. Acad. Sci. U.S.A. 48, 138–147.
13. Bogden, A. E., and Aptekman, P. M. (1960). The R-1 factor. A histocompatibility antigen in the rat. Cancer Res. 20, 1372–1382.
14. Bogden, A. E., and Aptekman, P. M. (1962). Histocompatibility antigens and hemagglutinogens in the rat. Ann. N.Y. Acad. Sci. 97, 43–56.
15. Burhoe, S. O. (1947). Blood groups in the rat (Rattus norvegicus) and their inheritance. Proc. Natl. Acad. Sci. U.S.A. 33, 102.
16. Butcher, G. W., and Howard, J. C. (1977). A recombination in the major histocompatibility complex of the rat. Nature (London) 266, 362–364.
17. Cramer, D. V., Shonnard, J. W., and Gill, T. J., III. (1974). Genetic studies in inbred rats, II. Relationship between the major histocompatibility complex and mixed lymphocyte reactivity. J. Immunogenet. 1, 421–428.
18. Cramer, D. V., Shonnard, J. W., Davis, B. K., and Gill, T. J., III. (1977). The in vitro lymphocyte proliferative response to poly(Glu^{52}Lys^{33}Tyr^{15}) in inbred rats. Cell. Immunol. 28, 167–173.
19. Cramer, D. V., Shonnard, J. W., Davis, B. K., and Gill, T. J., III (1977). Polymorphism of the mixed lymphocyte response of wild Norway rats. Transplant. Proc. 9, 559–562.
20. Cramer, D. V., Davis, B. K., Shonnard, J. W., Štark, O., and Gill, T. J., III (1978). Phenotypes of the major histocompatibility complex in wild rats of different geographic origins. J. Immunol. 120, 179–187.
21. Davis, B. K., Shonnard, J. W., and Gill, T. J., III (1976). Genetic control of the immune response to poly(Glu^{52}Lys^{33}Tyr^{15}) in neonatally thymectomized high and low responder rats. Am. J. Pathol. 84, 55–67.
22. DeWitt, C. W., and McCullough, M. (1975). Ag-F: Serological and genetic identification of a new locus in the rat governing lymphocyte membrane antigens. Transplantation 19, 310–317.
23. Dorf, M. E., and Mauer, P. H., Merryman, C. F., and Benacerraf, B. (1976). Inclusion group systems and cis-trans effects in responses controlled by the two complementing Ir-GLφ genes J. Exp. Med. 143, 889–896.
24. Elves, M. W. (1967). The mixed leukocyte reaction in inbred strains of rats. Vox Sang. 13, 7–11.
24a. Fabre, J. W., and Morris, P. J. (1974). The definition of a lymphocyte-specific alloantigen system in the rat (Ly-1). Tissue Antigens 4, 238–246.
25. Feldman, J. D., and Lee, S. (1967). Renal homotransplantation in rats. I. Allogeneic recipients. J. Exp. Med. 126, 783–794.
26. Festing, M., and Staats, J. (1973). Standardized nomenclature for inbred strains of rats. Fourth listing. Transplantation 16, 221–245.
27. Frenzl, B., Křen, V., and Štark, O. (1960). Attempt to determine blood groups in rats. Folia Biol. (Praha) 6, 121–126.
28. Gasser, D. L., Newlin, C. M., Palm, J., and Gonatas, N. K. (1973). Genetic control of susceptibility to experimental allergic encephalomyelitis in rats. Science 181, 872–873.
29. Gasser, D. L., Silvers, W. K., Reynolds, H. M., Jr., Black, G., and Palm, J. (1973). Serum esterase genetics in rats: Two new alleles at ES-2, a new esterase regulated by hormonal factors, and linkage of these loci to the Ag-C blood group locus. Biochem. Genet. 10, 207–217.
30. Gill, T. J., III, Kunz, H. W., Stechschulte, D. J., and Austen, K. F. (1970). Genetic and cellular factors in the immune response. I. Genetic control of the antibody response to poly(Glu^{52}Lys^{33}Tyr^{15}) in the inbred rats strains ACI and F344. J. Immunol. 105, 14–28.
31. Gill, T. J., III, Cramer, D. V., and Kunz, H. W. (1978). The major histocompatibility complex—Comparison in the mouse, man and the rat. Am. J. Pathol. 90, 735–778.
32. Günther, E., and Rüde, E. (1975). Genetic complementation of histocompatibility-linked Ir genes in the rat. J. Immunol. 115, 1387–1393.

33. Günther, E., and Rüde, E. (1976). Cross-stimulation of antigens under histocompatibility-linked *Ir* gene control. *Immunogenetics* **3**, 261–269.

34. Günther, E., and Štark, O. (1977). The major histocompatibility system of the rat (*Ag-B* or *H-1* system). In ''The Organization of the Major Histocompatibility System in Animals'' (D. Götze, ed.), pp. 207–253. Springer-Verlag, Berlin.

35. Günther, E., Rüde, E., and Štark, O. (1972). Antibody response in rats to the synthetic polypeptide (T, G)-A—L genetically linked to the major histocompatibility system. *Eur. J. Immunol.* **2**, 151–155.

36. Günther, E., Rüde, E., Meyer-Delius, M., and Štark, O. (1973). Immune-response genes linked to the major histocompatibility system of the rat. *Transplant. Proc.* **5**, 1467–1469.

37. Günther, E., Mozes, E., Rüde, E., and Sela, M. (1976). Genetic control of immune responsiveness to poly(I Pro)–poly(I Lys)-derived polypeptides by histocompatibility-linked immune response genes in the rat. *J. Immunol.* **117**, 2047–2052.

38. Guttmann, R. D., Lindquist, R. R., Parker, R. M., Carpenter, C. B., and Merrill, J. P. (1967). Renal transplantation in the inbred rat. I. Morphologic, immunologic and functional alterations during acute rejection. *Transplantation* **5**, 668–681.

39. Hansen, C. T. (1974). ''Catalogue of NIH Rodents,'' DHEW Publ. No. (NIH) 74-606. US Govt. Printing Office, Washington, D.C.

40. Hausman, S., and Palm, J. (1973). Variable expression of Ag-B and non-Ag-B histocompatibility antigens in cultured rat cells of different histological origin. *Transplantation* **16,** 313–324.

41. Heslop, B. F. (1968). Histocompatibility antigens in the rat: The AS2 strain in relationship to the AS, BS and HS strains. *Aust. J. Exp. Biol. Med. Sci.* **46,** 479–491.

42. Howard, J. C., and Scott, D. W. (1974). The identification of sera distinguishing marrow-derived and thymus-derived lymphocytes in the rat thoracic duct. *Immunology* **27**, 903–922.

43. Katagiri, M., Natori, T., Tanigaki, N., Kreiter, V. P., and Pressman, D. (1975). Papain-solubilized Ag-B antigens. I. Isolation and characterization of two components composing Ag-B antigens. *Transplantation* **19**, 230–239.

44. Klein, J. (1975). ''Biology of the Mouse Histocompatibility-2 Complex.'' Springer-Verlag, Berlin and New York.

45. Kohoutová, M., Günther, E., Vojčík, L., and Štark, O. (1978). Further two recombinants in the major histocompatibility complex of the rat. *Folia Biol. (Prague)* **24,** 406–407.

46. Křen, V., Štark, O., Bilá, V., Frenzl, B., Krenová, D., and Křsiakova, M. (1973). Rat alloantigenic systems defined through congenic strain production. *Transplant. Proc.* **5**, 1463–1466.

47. Kunz, H. W., and Gill, T. J., III (1974). Genetic studies in inbred rats. I. Two new histocompatibility alleles. *J. Immunogenet.* **1**, 413–420.

48. Kunz, H. W., and Gill, T. J., III (1978). Red blood cell alloantigenic systems in the rat. *J. Immunogenet.* **5**, 365–382.

49. Kunz, H. W., Gill, T. J., III, and Borland, B. (1974). The genetic linkage of the immune response to poly(Glu^{52}Lys^{33}Tyr^{15}) to the major histocompatibility locus in inbred rats. *J. Immunogenet.* **1**, 277–287.

50. Kunz, H. W., Štark, O., and Gill, T. J., III (1977). Comparison of haplotypes of the major histocompatibility complex in the rat. II. Serological analysis of the haplogypes *H-1^a* (*Ag-B^4*), *H-1^d* (*Ag-B9*) and *H-1^f* (*Ag-B10*). *J. Immunogenet.* **4**, 411–422.

51. Lonai, P., and McDevitt, H. O. (1974). Genetic control of the immune response. *In vitro* stimulation of lymphocytes by (T,G)-A—L, (H,G)-A—L, and (Phe,G)-A—L. *J. Exp. Med.* **140**, 977–994.

51a. Lubaroff, D. M. (1977). Antigenic markers on rat lymphocytes. I. Characterization of A.R.T., an alloantigenic marker on rat thymus and thymus-derived cells. *Cell. Immunol.* **29**, 147–150.

52. Luderer, A. A., Mauer, P. H., and Woodland, R. T. (1976). Genetic control of the immune response in rats to the known sequential polypeptide (Tyr-Glu-Ala-Gly)$_n$. I. Antibody response. *J. Immunol.* **117**, 1079–1084.

53. Meo, T., David, C. S., and Schreffler, D. C. (1976). *H-2* associated MLR determinants: Immunogenetics of the loci and their products. *In* ''The Role of the Products of the Histocompatibility Gene Complex in Immune Responses'' (D. H. Katz and B. Benacerraf, eds.), pp. 167–178. Academic Press, New York.

54. Michie, D., and Anderson, N. F. (1966). A strong selective effect associated with a histocompatibility gene in the rat. *Ann. N.Y. Acad. Sci.* **129**, 88–93.

55. Minford, J. K., Cramer, D. V., and Gill, T. J., III (1976). Specificity of the *in vitro* immune response to synthetic polypeptide antigens in inbred rats. *J. Immunol.* **117**, 2264–2266.

56. Owen, R. D. (1962). Earlier studies of blood groups in the rat. *Ann. N.Y. Acad. Sci.* **97**, 37–42.

57. Palm, J. (1962). Current status of blood groups in rats. *Ann. N.Y. Acad. Sci.* **97**, 57–58.

58. Palm, J. (1964). Serological detection of histocompatibility antigens in two strains of rats. *Transplantation* **2**, 603–612.

59. Palm, J. (1971). Classification of inbred rat strains for Ag-B histocompatibility antigens. *Transplant. Proc.* **3**, 169–171.

60. Palm, J. (1971). Immunogenetic analysis of Ag-B histocompatibility antigens in rats. *Transplantation* **11**, 175–183.

61. Palm, J., and Black, G. (1971). Interrelationships of inbred rats strains with respect to Ag-B and non-Ag-B antigens. *Transplantation* **11**, 184–189.

62. Poloskey, P. E., Kunz, H. W., Gill, T. J., III, Shonnard, J. W., Hanson, C. T., and Dixon, B. D. (1975). Genetic studies in inbred rats. IV. Xenoantisera against rat erythrocyte (Ag-C) antigens. *J. Immunogenet.* **2**, 179–187.

63. Radka, S. F., Cramer, D. V., and Gill, T. J., III (1977). Ia antigens in inbred rats and their relationship to the MLR phenotype. *J. Immunol.* **119**, 2037–2044.

64. Rüde, E., and Günther, E. (1974). Genetic control of the immune response to synthetic polypeptides in rats and mice. *Prog. Immunol., Int. Congr. Immunol., 2nd, 1974* pp. 223–233.

65. Ruscetti, S. K., Gill, T. J., III, and Kunz, H. W. (1975). The genetic control of the antibody response in inbred rats. *Mol. Cell. Biochem.* **7**, 145–156.

66. Shinohara, N., Cullen, S. E., and Sachs, D. H. (1977). *Ag-B*-linked analogue of Ia antigens in the rat. *J. Immunol.* **118**, 2083–2087.

67. Shonnard, J. W., Cramer, D. V., Poloskey, P. W., Davis, B. K., and Gill, T. J. III (1976). Genetic studies in inbred rats. VI. Linkage relationships of mixed lymphocyte reactivity, serologically defined antigens (Ag-B, Ag-C) and the immune response to poly(Glu^{52}Lys^{33}Tyr^{15}). *J. Immunogenet.* **3**, 61–70.

68. Shonnard, J. W., Cramer, D. V., Poloskey, P. E., Kunz, H. W., and Gill, T. J., III (1976). Polymorphism of the major histocompatibility locus in the wild Norway rat. *Immunogenetics* **3**, 193–200.

69. Shonnard, J. W., Davis, B. K., Cramer, D. V., Radka, S. F., and Gill, T. J., III (1979). The association of immune responsiveness, mixed lymphocyte responses and Ia antigens in natural population of Norway rats *J. Immunol.* **123**, 778–783.

70. Silvers, W. K., Murphy, G., and Poole, T. W. (1977). Studies on the H-Y antigen in rats. *Immunogenetics* **4**, 85–100.

71. Simonian, S. J., Gill, T. J., III, and Gershoff, S. N. (1968). Studies on synthetic polypeptide antigens. XX. Genetic control of the antibody response in the rat to structurally different synthetic polypeptide antigens. *J. Immunol.* **101**, 730–742.

72. Snell, G. D., Dausset, J., and Nathenson, S. (1976). ''Histocompatibility.'' Academic Press, New York.

73. Sørenson, S. F. (1972). The mixed lymphocyte culture interaction. Techniques and immunogenetics. *Acta Pathol. Microbiol. Scand., Sect.*

B **280**, Suppl., 10–82.

74. Štark, O., Křen, V., and Günther, E. (1971). RtH-1 antigens in 39 rat strains and six congenic lines. *Transplant. Proc.* **3**, 165–168.

75. Štark, O., Günther, E., Kohoutová, M., and Vojčik, L. (1977). Genetic recombination in the major histocompatibility complex (*H-1, Ag-B*) of the rat. *Immunogenetics* **5**, 183–187.

76. Tada, N., Itakura, K., and Aizawa, M. (1974). Genetic control of the antibody response in inbred rats. *J. Immunogenet.* **1**, 265–275.

77. White, E., and Hildemann, W. H. (1968). Allografts in genetically defined rats: Differences in survival between kidney and skin. *Science* **162**, 1293–1295.

78. Williams, R. M., and Moore, M. J. (1973). Linkage of susceptibility to experimental allergic encephalomyelitis to the major histocompatibility locus in the rat. *J. Exp. Med.* **138**, 775–783.

79. Wilson, D. B. (1967). Quantitative studies on the mixed lymphocyte interaction in rats. I. Conditions and parameters of response. *J. Exp. Med.* **126**, 625–654.

80. Womack, J. E. (1973). Biochemical genetics of rat esterases: Polymorphism, tissue expression and linkage of four loci. *Biochem. Genet.* **9**, 13–24.

81. Würzburg, V. (1971). Correlation between the immune response to an enzyme and histocompatibility type in rats. *Eur. J. Immunol.* **1**, 496–497.

82. Würzburg, V., Schutt-Gerowitt, H., and Rajewsky, K. (1973). Characterization of an immune response gene in rats. *Eur. J. Immunol.* **3**, 762–766.

Chapter 11

Experimental Parasitology

Dickson D. Despommier

I. INTRODUCTION

The laboratory rat, *Rattus norvegicus,* has been used by a great many investigators for a wide variety of purposes in the field of experimental parasitology. Unfortunately, these studies are scattered throughout the world literature making it difficult for those interested in learning about rat–parasite interactions to gain an entrance into this fascinating field of biology.

The main purposes of this chapter, therefore, are to (a) acquaint the uninitiated reader with some of the protozoan and helminthic parasites which have been studied experimentally in the rat; (b) present, in abstract form, some of the experiments conducted with these parasites; and (c) discuss some important human parasitic infections with regards to the rat model by comparing and contrasting the infections in both hosts.

This chapter has excluded arthropods from discussion. However, the literature dealing with experimental entomology is extensive, and, in addition to the entomological journals, many useful citations can be obtained by consulting the "Index Catalogue of Medical and Veterinary Zoology" published by the United States Department of Agriculture.

II. SURVEY OF PARASITES WHICH HAVE BEEN USED EXPERIMENTALLY IN *Rattus norvegicus*

Table I in this section, has been constructed in such a way as

Table I

Survey of Parasites Which Have Been Used Experimentally in *Rattus norvegicus*

Parasite	Host characteristics[a]	Parasite stage and route of infection	Outcome of infection	Reference
Protozoa				
Babesia equi	Sprague-Dawley rat, splenectomized (only available information)	Parasitized erythrocytes (from a horse); 10 ml blood/kg body wt.	No infection resulted from exposure to the organism.	51
Babesia hylomysci	"Laboratory" rat (only available information)	Infected red cells; route not stated.	Rats became infected.	11
Eimeria contorta	Wistar and Sprague-Dawley, SPF	Sporulated oocyst, oral.	All stages of the parasite developed within the small intestinal epithelial cells between days 4 and 10 after infection; some rats died as the result of infection.	63
Eimeria falciformis	Wistar rat, SPF 30–150 gm	Sporulated oocyst, oral.	Schizogony I and II observed, but no oocysts produced; infection lasted 5 days.	62
Eimeria miyairii	"Albino" rat, 41–56 days old	Sporulated oocyst, 1500/rat/day for 5 days; oral.	Rats became infected and passed oocysts in feces on days 4 and 8 after infection.	14
Eimeria nieschulzi	F-344, inbred, SPF; female; 36–43 days old	Sporulated oocyst; oral with 3000–7000 oocysts/rat.	All rats were infected and shed oocysts in feces on days 7–11 after infection.	44
Eimeria nieschulzi	"Albino" rat (only available information)	Sporulated oocyst (6–7×10^6 rat); oral.	Rats were infected; mice fed sporulated oocysts did not become infected.	97
Eimeria separata	F-344, inbred, SPF; female; 36–43 days old	Sporulated oocysts; oral with 3000–7000 oocysts/rat.	All rats were infected and shed oocysts in feces on days 3–7 after infection.	44
Eimeria separata	"Laboratory" rat; SPF, male	Sporulated oocysts(4000/rat), oral.	Oocysts in feces at day 4–7 after infection; oocysts were infective for mice.	100
Eimeria vermiformis	"Laboratory" rat, young, weanling male; not treated	Sporulated oocyst oral.	Trophozoites only; no organisms found after 2 days.	167
Eimeria vermiformis	"Laboratory" rat, young, weanling male; dexamethasone treated	Sporulated oocyst oral.	Trophozoites and multinucleate schizonts present; second generation schizonts on day 6.	167
Entamoeba histolytica	Sherman strain, 40–43 gm	K9H strain trophozoite, intracecal.	No trophozoites recovered.	65
Entamoeba histolytica	Wistar strain, 18–29 gm	K9H strain trophozoite, intracecal.	86% of rats infected; 3/10 dead.	65
Entamoeba muris	"Albino" rat, 80–560 gm	Cyst, oral.	Trophozoites recovered from large intestine and cecum; 5–14 days after infection.	92
Giardia muris	"White" rat, adult and young (25–30 gm)	Trophozoite; laparotomy, small intestine.	15% of adult rats became infected; 25–100% of young rats became infected; 100% of young rats pretreated with atabrine became infected.	23
Hammondia hammondi	"Laboratory" rat (only available information)	Oocyst, oral.	Cysts were seen in skeletal muscle.	50
Leishmania tropica major	"White" rat; 14–30 days old	Leishman forms (from 10–17 day old culture from infected gerbils); subcutaneous.	Rats were infected for 280 days with cultural strain 3–4. No clinical signs or symptoms observed. No rats died, al-	109

(*continued*)

TABLE I (*Continued*)

Parasite	Host characteristics	Parasite stage and route of infection	Outcome of infection	Reference
			though organisms were seen on histological section at all times.	
(*Encephalitozoon*) (*Nosema*) *cuniculi*	State Vitamin Laboratory strain, male and female adult, tumor-resistant and non-tumor-resistant	Infected Yoshida rat ascites sarcoma cells, intraperitoneal.	All tumor-resistant rats survived the parasite infection; 80% of non-tumor-resistant rats died when given infected tumor cells; 99% of non-tumor-resistant rats died when given noninfected ascites tumor cells.	121
Nuttallia (=*Babesia*) *microti*	King's strain Albino rat (derived from Chester Beattie strain) male and female, 4–5 weeks old	Trophozoites in red blood cells, inoculated intravenously.	Trophozoites in red blood cells, peak parasitemia $\cong$ 16 days after inoculum. 20% of RBC infected.	111
Nuttallia (=*Babesia*) *rodhaini*	King's strain Albino rat (derived from Chester Beattie strain) female and male 4–5 weeks old	Trophozoites in red blood cells, inoculated intravenously.	Trophozoites in red blood cells, peak parasitemia $\cong$ 7 days after inoculation. 90% of RBC infected.	111
Plasmodium berghei	F-344, germ-free and conventional; young (90 gm) and adult (225 gm); splenectomized and normal	Infected red cells; 10^7 organisms/rat intraperitoneal.	Peak parasitemia (30 parasites per 1000 red cells) occurred between days 10 and 15 after infection. All rats survived the infection.	98
Plasmodium berghei killicki nov. ssp.	Wistar rat; 50–70 gm	Sporozoite; intraperitoneal, in a suspension of mosquito salivary gland material.	Preerythrocytic schizonts in liver matured within 50 hr after inoculation. No erythrocytic stages were looked for.	83
Plasmodium vinckei	Hebrew University strain, and hooded rat; young (21 days old); mature male (330 gm)	Parasitized mouse erythrocytes, young rats given 25–27. Mature rats given 140.	Parasitemia resulted in all rats. No mature rats died. Parasitemias were low and peaked at day 9 after infection. Three out of 24 young rats died; parasitemias were high and peaked on day 6.8 after infection.	193
Sarcocystis sp.	*Rattus norvegicus*, laboratory-bred; 22–28 days old	Sporocysts and oocysts (from feces of carpet python, *morelia spilotes variegata*); oral.	Cystlike bodies in muscle tissue of one rat 11 days after infection; two rats died on day 11, but no parasites found.	135
Trichomonas hominis	"Albino" rat; 2–3 months old; splenectomized	Trophozoite; oral; rectal, or intracecal.	Oral route produced no infection. Rectal route produced infections which last 1–2 months. Intracecal route produced some lethal infections after 5 days.	2
Toxoplasma gondii	Sprague-Dawley rat; female 150–250 gm	Trophozoite; intraperitoneally.	Rats became infected.	126
Trichomonas vaginalis	Wistar rat, female, 200 gm, untreated, or ovarectomized, given estrogen; then infected with *Candida albicans* intravaginally	Trophozoite (from axenic culture), intravaginally; (4–5 × 10^6/rat).	29% of untreated rats became infected; 92% of treated rats became infected.	101
Trypanosoma brucei brucei	"Laboratory" rat (only available information)	Blood forms (no pretreatment, or pretreated for 5 hr *in vitro*	No pretreatment; rats became infected with 48 hr after	127

(*continued*)

TABLE I (*Continued*)

Parasite	Host characteristics	Parasite stage and route of infection	Outcome of infection	Reference
		in normal human serum); intraperitoneal.	inoculation. Pretreated; no rats became infected.	
Trypanosoma congolense	"Laboratory" rat (only available information)	Bloodstream form (obtained from infected oxen); subcutaneous.	Rats which became infected died between 12 to 28 days after infection. About 50% of all rats inoculated with a variety of strains of parasite became infected.	19
Trypanosoma cruzi	"Albino" rat, 10–15 days old	Bloodstream form, from infected rat blood (only available information).	Most rats became infected and died on days 20–22 after infection.	32
Trypanosoma equiperdum	Wistar rat; male; 100–150 gm	Blood forms (2×10^7/rat); intravenous.	All rats died of the infection during the exponential growth phase of the parasite.	49
Trypanosoma evansi	"Albino" rat; male and female; 100–300 gm; black and white-hooded rat, weanling	Blood forms (4×10^5/rat); intravenous.	Blood forms in peripheral blood; peak parasitemia occurred on day 8; (20 organism/1000 red cells).	8
Trypanosoma brucei gambiense	Holtzman "albino" rat, female; 300–350 gm	Blood form (3×10^7/rat); intraperitoneal.	All rats died within 48 hr due to overwhelming infection.	154
Trypanosoma helogalei	"Laboratory" rat (only available information)	Metacyclic tryps (from culture of infected mongoose blood; peritoneal cavity or intravenous.	All rats became infected; blood forms demonstrated in peripheral blood.	61
Trypanosoma lewisi	"Holtzman" rat, male	Bloodstream form (10^6/rat); intraperitoneal.	All rats were infected; peak parasitemia occurred on days 10–12 after infection; all rats survived.	67
Trypanosoma brucei rhodesiense	"Laboratory" rat (only available information)	Blood forms (no pretreatment, or treated for 5 hr in normal human serum); intraperitoneal.	No pretreatment; rats became infected within 36 hr after inoculation. Pretreatment: rats became infected within 36–60 hr after inoculation.	127
Trypanosoma vivax viennei	"Laboratory" rat, pretreated with intraperitoneal injection of sheep serum	Blood form, intravenous.	Blood forms present in blood for 2–3 days only.	144
Nematoda				
Amplicaecum robertsi	"Laboratory" rat (only available information)	Infective egg, oral.	Third-stage larva, recovered from liver.	149
Ancyclostoma braziliense	"Albino" rat, 150–200 gm, female	Infective filariform larva, percutaneous.	All rats were infected, larvae were recovered up to day 28 in the skin. Evidence for lung migration was seen.	110
Ancylostoma caninum	"Laboratory" rat (only available information)	Infective filariform larva, oral.	Half of the larvae penetrated intestinal wall, migrated to liver, lungs, trachae, then muscle. A few larvae reached the small intestine, but none developed to adults. Larvae in muscle lived for up to 377 days.	81
Ancylostoma duodenale	"Laboratory" rat (only available information)	Infective filariform larva, oral.	Results the same as for *Ancylostoma caninum*.	81

(*continued*)

TABLE I (*Continued*)

Parasite	Host characteristics	Parasite stage and route of infection	Outcome of infection	Reference
Ancylostoma tubaeforme	"Albino" rat, 150–200 gm, female	Infective filariform larva, percutaneous.	Few larvae recovered from the skin up to day 1 after inoculation. Evidence for lung migration was seen.	110
Angiostrongylus cantonensis	"White" rat, laboratory-bred, 3–6 months old	Third-stage larva, oral.	Adult worms recovered from pulmonary arteries and heart. First-stage larvae found in feces and lung 4–6 months after infection.	17
Angiostrongylus mackerrasae	"White" rat, laboratory-bred, 3–6 months old	Third-stage larva, oral.	Adult worms recovered from pulmonary arteries and heart 4–6 months after infection; first-stage larvae found in feces and lungs.	17
Anisakis simplex (?) (from mesenteries of herring)	Glaxo strain, 170–200 gm	Larva (stage not known), oral.	Larvae penetrated stomach, lodged in adjacent mesenteries. Larvae molted on day 3, died on day 7.	189
Ascaris suum	"Albino" rats, Holtzman strain	Embryonated egg, oral.	Third-stage larvae only, found in lung.	22
Aspicularis tetraptera	Hooded Wistar rat, male and female	Infective egg, oral.	Larvae recovered on day 9 from colon; none present at day 12 after inoculum.	15
Breinlia booliati	Charles River-inbred; hooded inbred; outbred "laboratory" rat	Infective larva, from mosquito (*Aedes togdi* or *Armigeres subalbatus*) inoculation by incision, intrathoracic, intraperitoneal, subcutaneous, or ocular installation.	Charles River, inbred: all routes resulted in infection except ocular instillation, microfilariae and adults recovered. Hooded, inbred: only 2 rats became microfilaremic, one each by subcutaneous and ocular instillation. Outbred "Lab" rat: all routes produced microfilaremic infections. All results at 20 weeks after infection.	68
Breinlia sergenti	"Laboratory" rat (only available information)	Third-stage larva, injected subcutaneously.	No microfilariae present in blood up to 6 months after infection. Adult worms not looked for.	190
Brugia malayi (periodic)	"White" rat, bred at author's institute	Infective larva (obtained from *Anopheles barbirostris* or *Masonia uniformis*), subcutaneous.	No adults, no microfilariae.	82
Brugia malayi (semiperiodic)	"White" rat bred at author's institute	Infective larva (obtained from *Mansonia dives*), subcutaneous.	Microfilariae in one rat in heart blood 6 months after infection; fragments of adult worm from one rat.	82
Brugia pahangi	"Wistar" rat, randomly bred, less than 5 weeks, more than 5 weeks old, male and female; selected for susceptibility to infection	Infective larva, derived via dissection of mosquitoes (*Aedes aegypti*), injected into groin area via syringe.	Adult worms present in 36% of young male rats; 10% of old male rats; 5% of all ages of female rats; 71% of F_5 selected rats, microfilariae in 28% of F_5 selected rats.	160
Capillaria hepatica	"Laboratory" rats, 6 weeks old	Embryonated egg, oral.	Adult worms and eggs in liver tissue 42 days after infection.	173

(*continued*)

TABLE I (*Continued*)

Parasite	Host characteristics	Parasite stage and route of infection	Outcome of infection	Reference
Dirofilaria repens	"Albino" rat, 2 months old, female	Infective larva, subcutaneously. Adult worms, subcutaneously and intraperitoneally.	No worms recovered from rats receiving larvae. Microfilariae detected in blood of rats receiving adults on day 4 after ip and sc. and lasted to day 90. One live adult worm was recovered from each rat group 3 months after transplant.	70
Dipetalonema witei	"Laboratory" rat, 25–30 gm, and 70–80 gm	Infective larva, obtained from tick (*Ornithodorus tartakovsky* and *O. moubata*), subcutaneous. Immature worms, obtained from infected gerbils, subcutaneous.	Infective larva did not infect. Immature worms did not infect. Mature worms shed microfilariae in blood of recipients until day 49 after transplantation.	141
Dirofilaria roemeri	"Laboratory" rat, cortisone-treated	Adult worm, transplanted into peritoneal cavity.	Microfilaria, in bloodstream and in peritoneal cavity, persisted for 35 days. Adult worms lived for 62 days.	148
Eichinocephalus sinensis	*Rattus rattus* (?) (only available information)	Third-stage larva (from oyster), oral.	No rats were infected when examined 12–17 hr after administration of larvae.	80
Gnathostoma spinigerum	*Rattus norvegicus* var. *albinus*, adult	Third-stage larva, placed on skin.	Third-stage larvae recovered from muscle tissue 2 days after worms penetrated skin.	35
Gnathostoma spinigerum	*Rattus norvegicus* var. *albinus* (only available information)	Third-stage larva, oral.	Third-stage larvae recovered from muscle tissue of rat.	34
Gnathostoma spinigerum	*Rattus norvegicus* var. *albinus* (only available information)	Advanced third-stage larva, oral.	No adult worms found. Advanced third-stage larvae recovered from liver and muscle at 10 days after infection. Serial passage resulted in only one surviving larva at 160 days after first passage.	33
Hepatojarakus malayae	"Laboratory" rats (only available information)	Infective (third-stage?) larva, oral and percutaneous.	Larvae in lungs within 24 hr. Mature adults found in liver within 16 days after infection	13
Hexametra quadricornis	"Albino" rat (only available information)	Infective egg, oral.	Third-stage larva recovered from subcutaneous tissue and peritoneum at days 92 and 104 after infection.	122
Littosomoides carinii	"Albino" rat, 20 days old	Third-stage larva; natural route via bite of *Ornithonysus bacoti* (mite).	Fertile adults, microfilaria in blood after tenth week of infection.	12
Necator americanus	"Laboratory" rat (only available information)	Infective filariform larva, percutaneous.	Many larvae were recovered at 48 hr from lungs and subcutaneous tissues; few in muscle; none in GI tract. Larvae from the lungs of rats were infective for 2 out of 4 human volunteers.	116
Nematospiroides dubius	"Wistar" rat; male and female; gonadectomized, spayed, and normal.	Filariform larva, infected via stomach tube.	Adult worms were obtained from all groups of rats.	39
Neoaplectana glaseri	Sprague-Dawley rat, female, 100 gm	Third-stage larva, injected into the peritoneal cavity.	No mature adults seen; only third-stage larvae, mostly exsheathed.	72

(*continued*)

TABLE I (*Continued*)

Parasite	Host characteristics	Parasite stage and route of infection	Outcome of infection	Reference
Nippostrongylus brasiliensis	"Wistar" rat, male, albino, 100–150 gm	Filariform larva, injected subcutaneously. Adult worms introduced via laparotomy, intraduodenal.	Mature, fertile adults recovered from small intestine, in all groups of rats.	112
Oesophagostomum dentatum (?)	"Wistar" rat, 60–100 gm	Filariform larva, stomach tube.	Encapsulated infective larvae (0.02–0.64% of oral dose) were recovered from intestinal wall 7 days after oral infection.	73
Ophidascaris moreliae	"Laboratory" rat (only available information)	Infective egg (containing second-stage larva), oral.	Third-stage (?) larvae, found throughout the body, 78 weeks after administration of eggs.	150
Phocanema decipiens	Sprague-Dawley, 200 gm, female	Boring tooth larva, implanted in intraperitoneal cavity, or oral.	Larva were recovered from the peritoneum alive and had molted at 11 days; larvae died at day 18; Oral: on day 6, larvae recovered from stomach, intestine, caecum, and body cavity.	J. Bier, personal communication
Polydelphis anoura	"Laboratory" rat (only available information)	Infective egg, oral.	Third-stage (?) larva, isolated from liver, lungs, and intestinal wall 39–42 days after inoculation.	151
Setaria cervi	"White" rat (only available information)	Adult parasite (from slaughtered cattle) intraperitoneal.	Living adult worms recovered from peritoneum up to 45 days after transplant. Microfilaremia peaked on weeks 6 and 7 after transplant of adults.	5
Strongyloides ratti	Hebrew University "Sabra" strain, male, 60–70 gm	Filariform larva, percutaneous.	Adult female worms, recovered from small intestine 100 hr after infection.	191
Strongyloides venezuelensis	Hebrew University "Sabra" strain; male, 60–70 gm	Filariform larva, percutaneous.	Adult female worms, recovered from small intestine 110 hr after infection.	181
Syphacia oblevata	PVG/C inbred rat male and female, pretreated with piperizine	Embryonated egg, noninfected rats were exposed to naturally infected rats.	Eggs found in feces after exposure to naturally infected rats.	119
Terranova sp. (from muscle of codfish)	Glaxo strain, 170–200 gm	Larva (stage not known), oral.	Larvae penetrated to muscularis mucosa only, molted once, within 4 days; larvae were not recovered after day 7.	189
Toxocara canis	"Laboratory" rats (only available information)	Infective egg, oral.	Third-stage larva recovered.	40
Toxocara cati	"Laboratory" rat, 30 days old	Embryonated egg, oral.	On days 20 and 40 after infection, liver and lungs of infected rats showed pathological changes consistent with the migratory route of the larval stage of the parasite.	105
Toxocara mackerrasae	*Rattus norvegicus*, 12 days to 10 weeks old	Infective egg, oral. Adult worm, laporotomy, placed into lumen of stomach.	Fertile, mature adults recovered at day 111 after oral infection with eggs. Adults passed eggs 5–6 days after transplantation into stomach.	175

(*continued*)

TABLE I (*Continued*)

Parasite	Host characteristics	Parasite stage and route of infection	Outcome of infection	Reference
Trichinella spiralis	Black and white-hooded Lister, male, 40–60 gm; high and low protein diets	Muscle larva, oral.	Adults in small intestine, days 3 and 10 after infection; muscle larvae in muscle 21 and 30 days after infection. Mortality highest in low protein diet.	137
Trichosomoides crassicauda	Fisher 344, inbred, male and female, 100–150 gm	Egg, urine of infected rats permitted to drop onto noninfected animal; infection induced during preening.	All rats became infected with fertile adults within 6 weeks after exposure to contaminated rat urine.	26
Cestoda				
Diphyllobothrium sebego	"Albino" rat (only available information)	Single plerocercoid larva, oral.	No adult worm recovered.	103
Echinococcus granulosus canadensis	"White" rat (only available information	Embryonated egg, oral.	No rats were infected when examined 15 weeks after administration of eggs.	179
Echinococcus granulosus (Lebanon)	"White" rat (only available information)	Embryonated egg, oral.	No rats were infected when examined 15 or 36 weeks after oral administration of eggs.	179
Echinococcus multilocularis sibiricensis	"White" rat (only available information)	Embryonated egg, oral.	No rats were infected when examined 16 or 36 weeks after oral administration of eggs.	179
Hydatigera taeniaeformis (*cysticercus faserolaris*)	"Laboratory" rat, 45–61.5 gm	Embryonated egg, oral.	A mean of 191 cysticerci recovered from liver and lungs (1) on day 24 of infection.	133
Hymenolepis diminuta	Sprague-Dawley rats, male, young	Cysticercoid, oral.	Mature adults recovered.	125
Hymenolepis microstoma	Sprague-Dawley rat, weanling, 53 gm; Wistar rat, weanling and adult; hooded Lister rat, weanling	Cysticercus, oral. Adults, laparotomy, inserted into intestine.	Destrobilated adult worms recovered from all strains of weanling and adult rats 9 days after administration of cysticerci. Immunosuppression had no effect on the outcome of the infection. Adult worms took in transplant recipients, but did not grow.	56
Hymenolepis nana	University of Freiburg strain; male, 25–30 gm, newly weaned	Viable egg, oral, 200/rat.	Mature adults containing gravid proglottids were obtained at 20 days of infection.	77
Spirometra australian (?)	"Laboratory" rat, male, weanling, treated with propylthiouracil	Scolex (clipped), subcutaneous.	Same stage of infection remained under skin.	106
Spirometra malay	"Laboratory" rat, male, weanling, treated with propylthiouracil	Scolex (clipped), subcutaneous.	Same stage of infection remained under skin.	106
Spirometra mansonoides	"Holtzman" strain male rat rendered hypothyroid	Scolex (clipped), injected under dorsal skin.	Same stage of infection remained under skin.	107
Spirometra mansonoides	"Laboratory" rat, male weanling, 180 gm, hypophysectomized	Scolex (clipped); subcutaneous.	Same stage of infection remained under skin.	106
Spirometra taiwan (?)	"Laboratory" rat, male, weanling, treated with propylthiouracil	Scolex (clipped), subcutaneous.	Same stage remained under skin.	106

(*continued*)

TABLE I (*Continued*)

Parasite	Host characteristics	Parasite stage and route of infection	Outcome of infection	Reference
Taenia crassiceps	Wistar rat, male, pretreated with metacestode homogenate	30 metacestodes/rat; intraperitoneal.	Two rats each harbored 50,000 metacestodes 6 months after inoculum. These metacestodes infected nontreated rats.	27
Taenia taeniaeformis	"Laboratory" rat, 6 weeks old, 100 gm	Mature egg, oral.	Cysticercus, in liver at day 9 after infection; 0–10% of cysticerci viable at 9 days.	66
Trematoda				
Alaria marcianae	"Laboratory" rat (only available information)	Mesocercaria, oral or injected ip.	7–35 days after infection mesocercariae found in muscles, fatty tissues and glands of the throat and jaw region.	74
Clonorchis sinensis	"Albino" rat, 250–350 gm	Metacercaria (50–100/rat); stomach tube.	Adult worms (2–34) recovered from bile ducts and main hepatic ducts 30 days after infection.	161
Echinochasmus milvi	"Albino" rat, recently weaned sibling	Metacercaria, oral.	Matured to adults in small intestine. Uterus of worm contained eggs.	168
Echinochasmus schwartzi	"Laboratory" rat (only available information)	Metacercaria, oral.	Adult worms recovered 35 days after infection.	91
Echinostoma malayanum	"White" rat (only available information)	Metacercaria, oral.	Mature adults recovered at 14–17 days after infection.	88
Echinostoma murinum	"White" rat (only available information)	Metacercaria, oral.	Mature adult worms recovered from duodenum 7 months after infection.	89
Echinostoma paraensei	"Albino" rats (only available information)	Metacercaria, oral.	Adult worms containing eggs recovered.	90
Echonoparyplium flexum	"Albino" rat (only available information)	Metacercaria, oral.	Adult worm recovered (only available information).	108
Fasciola gigantica	"Laboratory" rat from East African Veterinary Research Institute	Metacercaria, oral.	Only one of fifty rats was infected; three nonfertile adults were recovered from abdominal cavity.	96
Fasciola hepatica	Wistar rat, male, 65–75 gm; specific pathogen-free	Metacercaria, oral.	Mature worms recovered from rats receiving low doses of metacercariae. Immature adult worms, some rats died by week 6 in those groups receiving high doses of metacercariae (80–160).	164
Haplorchis pumilio	"Laboratory" rat (only available information)	Metacercaria, oral.	Mature adult worm recovered.	120
Haplorchis taichui	"Laboratory" rat (only available information)	Metacercaria, oral.	Mature adult worm recovered.	120
Isthmiophora spiculator	"Laboratory" rat (only available information)	Metacercaria, oral.	Mature adult worms recovered 16–46 days after infection from the duodenum. Adult worm smaller in rat than in man.	41
Notocotylus stagnicolae	"Albino" rat (only available information)	Metacercaria, oral.	Adult worms obtained on days 11–14.	1
Orientobilharzia dattai	"White" rat (only available information)	Cercaria, no route stated.	Adult worms recovered from the liver 62 days after infec-	45

(*continued*)

TABLE I (*Continued*)

Parasite	Host characteristics	Parasite stage and route of infection	Outcome of infection	Reference
			tion. One rat had both sexes of worms, another only had males. 0.55–0.60% of all cercariae developed into adults.	
Ornithobilharzia turkestanicum	"Albino" rat (only available information)	Cercaria, route not stated.	2–3 months after infection, only 1 rat harbored 2 male parasites in the liver.	7
Paramphistomum explanatum	"Albino" rat (only available information)	Metacercaria, oral.	No infection occurred.	145
Paragonimus kellicotti	Holtzman, male, 120–200 gm	Metacercaria, oral.	Mature adult worms found in pleural cavity or in lung cysts 7 weeks after infection.	174
Paragonimus miuazakii	King strain, female, 100–200 gm	Metacercaria, oral.	Mature adults were recovered from the lungs.	162
Paragonimus ohirai	"Laboratory" rat (only available information)	Metacercaria, oral.	Mature adult worms were recovered from pleural cavity.	186
Paragonimus sadoensis	Wistar rat, male, 200 gm	Metacercaria, oral.	Mature, fertile adults were found in the pleural cavity 69–96 days after infection.	188
Paragonimus westermani	"Laboratory" rat, 150 gm	Metacercaria, oral.	Immature adult worms were recovered from the abdominal wall musculature, skeletal muscles, and pleural cavity 70–180 days after infection.	187
Quinqueserialis quinqueserialis	*Rattus norvegicus*, adult	Metacercaria, oral.	Mature adults recovered from intestine 32 days after oral infection.	79
Schistosoma intercalatum	*Rattus norvegicus* (only available information)	Cercaria, natural penetration of skin.	Immature adults recovered from venules at 65 days after infection. 6% recovery of worms.	184
Schistosoma japonicum	*Rattus norvegicus albus*, 220 gm	Cercaria, percutaneous.	Few mature worms, many immature worms recovered from liver and portal venules 5 weeks after infection; few worms remained after 15 weeks of infection.	69
Schistomsoma mansoni	Sprague-Dawley rat, mean weight 140 gm	Cercaria, percutaneous.	Nonfertile adults recovered; 13% of worms from portal–mesenteric system at week 8; none recovered at week 16.	176
Stellantchasmus aspinosus	"Laboratory" rat (only available information)	Metacercaria, oral.	Mature adult worms recovered.	120
Zygocotyle lunata	Sprague-Dawley rat, young	Metacercaria, oral.	Adults were produced within 54–237 days after administration of metacercaria.	9
Pentastomida				
Moniliformis dubius	CFHB strain, male, 150±10 gm	Cystacanth, oral.	Fertile adults began egg laying on day 38, and ended on day 144.	30
Acanthocephala				
Porocephalus crotali	"Albino" rat (only available information)	Egg, oral.	Development complete; infective for definitive host; the cotton mouth snake.	47

a Many of the rat strain or stock designations as given by the original authors are not in accordance with accepted current nomenclature. See Volume I, Chapters 1 and 3 for further information.

to allow the reader, in most instances, to gain some insight as to host susceptability and parasite development in *Rattus norvegius*. Experiments using other species of *Rattus* were excluded. Usually, a modern reference (1960 to present) is listed for each infectious organism, thus facilitating entry into the literature from which other studies may be derived. Some attention has been paid to experimental design and the outcome of the infection as it relates to a particular treatment or manipulation of the host. For this reason several listings for a single parasite are sometimes provided.

Two things are apparent from the data listed in this table. One is that the variety of parasites which are capable of infecting the rat is extensive, and the other is that a wide range of rat stocks have been used in experimental parasitology.

However, it is also evident from these data that some parasites are unable to complete their life cycles in the rat, while others may be able to complete only a portion of it in this host.

Unfortunately, many investigations have been carried out using ''albino'' or ''laboratory'' rats, instead of with well-defined stocks and strains of this host. It is now well established that most parasitic infections not only vary in their life cycles depending upon the species of host, but also vary with regards to the stock or strain of rat host. Further, such biological factors as the sex, age, weight, immunological status of the rat, the day–night cycle, crowding, food, and water all contribute to the outcome of any particular infection (48,78,117, 131,132,165).

However, in general, using the rat to its best advantage has been the hallmark of countless successful investigations into the biology of parasitism. In particular, *Rattus norvegicus* has been employed in the laboratory to (a) maintain various parasites without allowing further development, (b) study the interactions of a given parasite(s) with the mammalian host, (c) grow large quantities of parasites for use outside the host (i.e., the rat as a living test tube), and (d) attempt the completion of the life cycle of a given parasite, especially those ordinarily found in other host species.

Table I is repleat with many examples of each of the above listed categories of host uses.

III. *Rattus norvegicus* AS A MODEL FOR HUMAN PARASITIC DISEASES

The development of laboratory animal host–parasite models which are directly applicable to human disease equivalents has been a major goal of experimental parasitologists. No other laboratory animal, save the mouse, has been used more in attempts to bring this concept into reality than the laboratory rat.

Rattus norvegicus has been used as the definitive or intermediate host for a wide variety of parasitic protozoans and helminths which infect and cause disease in the human host.

In this section five major human parasitic diseases, namely, trichinellosis (*Trichinella spiralis*), schistosomiasis (*Schistosoma mansoni*, *S. haematobium*, and *S. japonicum*), hookworm disease (*Necator americanus* and *Ancylostoma duodenale*), malaria (*Plasmodium vivax*, *P. ovale*, *P. falciparum,* and *P. malariae*), and African trypanosomiasis (*Trypanosoma b. gambiense* and *T. b. rhodesiense*) will be discussed in relation to the rat model equivalents.

The above human diseases were chosen because they still cause significant health problems, and the rat has been used extensively in laboratory and field-based research on these infectious agents.

As will be seen, the rat has proved to be a variable host regarding these parasitic diseases, with some of them completely mimicking the human situation, and others stubbornly refusing to carry out a patent infection in this host. However, it will be pointed out that the rats' propensity for rejecting or showing extreme tolerances for the above mentioned parasites has been often used to good advantage by those investigators who have had the insight to exploit these life cycle anomalies.

One notable group of human parasitic diseases that has been omitted from this section is the filarial nematodes (*Wuchereria bancrofti*, *Brugia malayi,* and *Onchocerca volvulus*), the main causative agents of filariasis. While filariasis in its various forms is considered to be a major human health threat in many parts of the tropical and subtropical world, no satisfactory *Rattus norvegicus* model exists for this group of parasites (138).

A. Trichinellosis

1. The Human Infection

Trichinellosis is caused by the nematode *Trichinella spiralis* and is transmitted primarily from pig to human by the ingestion of infected undercooked or raw pork. This parasite was first described by Paget and Owen in 1835 by microscopic examination of diaphragmatic tissue taken from a patient at autopsy (118). The life cycle of *T. spiralis* was one of the first nematode life cycles to be described, with most of the experimental aspects of the infection being carried out in a wide variety of hosts by Virchow, Leidy, Herbst, Leuckart, and Cerfontaine before the turn of the century. Cerfontaine, in 1895 (25), was probably the first investigator to use rats in these early studies. Much later, Ducas (43) used rats for the maintenance of the infection and established that there was an acquired resistance which developed in rats upon reinfection. This finding contributed much to the then new field of immunoparasitology.

Since the above studies were carried out in different species of mammalian hosts it was established that *T. spiralis* was an exceptional parasite, since it lacked host specificity. *Trichinella spiralis* is an important human infection and on rare occasion results in death. Five to seven percent of the population in the United States, alone, harbor this parasite (192). It is, therefore, fortunate that much of the clinical and laboratory findings of human trichinellosis are mimicked by the infection in the rat.

In the human host, the infection can be conveniently divided clinically into three phases corresponding to the degree of development of the life cycle of the nematode. In phase I, the ingested muscle larvae are released in the stomach from their Nurse cells (124) and immediately penetrate intracellularly into the columnar epithelium in the small intestine, and remain there throughout the rest of their life cycle (53). The worms molt four times within 36 hr after ingestion, differentiating into adults (4). During this early morphogenesis, the infected individual may experience abdominal discomfort, and in moderate to heavy infections, diarrhea, as well.

The nematodes mate, and within 5 days the adult female begins depositing live newborn larvae (phase II), which then penetrate into the draining lymphatics (64). The larvae migrate passively through the thoracic duct into the bloodstream, and ultimately penetrate into skeletal muscle. For a recent discussion of this phase of the infection see Despommier (36). Twenty days after penetration into muscle, the larva matures (37) and is now infective for another host (171).

Calcification of the resulting Nurse cell–larva complex (phase III) usually occurs after 6 months to 2 years of the muscle infection.

The adults produce many larvae during their 3–5 week lifetime. While the precise number of larvae produced per female cannot be determined in humans, it is estimated that between 1000 and 1500 larvae are produced by each worm (21). The length of the adult worms' life span in the intestine is determined largely by the hosts' immune response (84,93), as is the damage which the host incurs during the migratory stage of the infection (58).

The clinical laboratory findings can be dramatic. A circulating eosinophilia of 89% has been recorded (147), but typically a 45–60% eosinophilia is observed. Antibodies may be detected in moderate infections during the later stages of phase II of the infection (i.e., 2–5 weeks after ingestion of larvae).

2. The Infection in the Rat

The quantitative aspects of the infection in the rat, as might be expected, vary depending upon the rat stock or strain used. Qualitatively, the infection proceeds well in most varieties of commercially available rats. Furthermore, the life cycle is surprisingly uniform with regards to the timing of the stages of the infection.

Fewer newborn larvae are produced per female worm in the rat than in human, averaging about 200–400 (94,95). However, since this aspect of the biology of the infection is also under the immunological control of the host (38), fecundity varies accordingly. For example, the WF strain of rat is able to limit adult worm fecundity, whereas the F_1 hybrid of the DA × LEW rat cannot affect this aspect of the worms' biology (D. D. Despommier, D. D. McGregor, E. D. Crum, and P. B. Carter, unpublished observations). Apparently, the ability to recognize protective antigens which are related to larval production is a function of the immunogenetic capabilities of the strain of rat in question; this concept is one which, in all likelihood, applies to much of the biology of any parasitic infection under investigation.

In addition to the immunological aspects of the infection (24, 84), the rat has served as a good model for the biochemical, pathological, histochemical, and hematological aspects of this disease (58).

As previously alluded to, an inherent criticism of the rat model is that some of the quantitative parameters, for example, the number of worms which establish as adults in the small intestine and the number of larvae produced per female worm during the first to third week of the infection, cannot be directly compared to the human situation. However, it can be stated, that apart from this above objection, the rat is an excellent model for the study of human trichinellosis.

B.   Schistosomiasis

1.   The Human Infection

Schistosomiasis occurs as three separate but related infections caused by *Schistosoma mansoni*, *S. haematobium*, and *S. japonicum*. The first two species are primarily found in Africa, South America, the Caribbean, and the Philippines, while *S. japonicum* is restricted to southeast Asia, China and Japan.

Bilharz, in 1852, discovered the adult of *S. haematobium* during an autopsy on a victim of the disease in Egypt (18). Even earlier, Fujii (52) described disease caused by *S. japonicum*, while Sambon (136) correctly distinguished between *S. haematobium* and another schistosome, which he named *Schistosomum mansoni*, while working in various parts of tropical Africa.

Nearly 200 million people, world-wide, harbor the adults of one or more of these three species of trematodes in their venules, and the incidence rate continues to rise in most endemic areas.

Briefly, the life cycle of all three blood flukes can be summarized as follows. The adults live as pairs of worms (*in copula*) in the venules surrounding a variety of organs, depending somewhat upon the species of worm in question. Eggs are constantly shed, with approximately half of them penetrating the tissues and leaving the host via either the feces or urine. The other 50% of the eggs are swept back into the liver and other organs, where they induce granuloma that eventually result in the destruction of the surrounding tissue via fibrotic replacement. The eggs reaching fresh water hatch, liberating the ciliated miracidium, which then penetrates an appropriate snail intermediate host. After much development and asexual reproduction, the next larval stage (i.e., the cercaria) emerges from the snail, still in fresh water, and actively swims about, seeking out its human (definitive) host.

After penetrating the unbroken skin, usually through a hair follicle, the schistosomula (immature adult) develops for several days in the dermis before migrating via the bloodstream into the lung, then the liver. After mating, the adults pair off and migrate out of the liver, against the venous flow, and into the venules surrounding the small and large intestine and the venus plexus of the bladder. Eggs are usually shed within 4–6 weeks after infection.

A remarkable feature of the adult worm is the age to which it survives in the human host, with some living for as long as 10–15 years.

2. The Infection in the Rat

A wide variety of experimental hosts have been used in attempts to develop laboratory models of the human disease (177). Unfortunately, the laboratory rat has proved to be of limited value in this respect (157). Most successful has been the use of the hamster for *S. haematobium* (104); the mouse for *S. mansoni* (157); and the guinea pig, rabbit, and mouse for *S. japonicum* (71). In all of these animal species, damage in the liver, bladder, and vascular system mimics the human disease, since each infection results in egg production.

In contrast, infection with all three species of schistosomes is aborted in the rat after the eighth to sixteenth week, prior to egg laying, thereby excluding the possibility of any pathology which might result from host exposure to ova. Despite this obvious defect in the model, the rat is still a useful host, and many studies involving cercarial penetration (158), and schistosomula development (28) have been carried out. However, the high death rate of schistosomulae in rat skin (28) restricts the use of this host for this aspect of the infection. Indeed, it is because of its high degree of innate resistance that the rat has been used in studies aimed at elucidating the mechanism(s) of immunity to this group of parasites (146).

Since the human schistosome species do not mature in the rat, the pathophysiology of the patent infection cannot be studied in this host. However, experiments dealing with the development of liver pathology caused by preadult and adult worms have been published (102).

The rat has been largely ignored as a model for studying egg-induced granuloma responses (178), and reflects the rela-

tive ease to which these reactions can be investigated in a permissive host, such as the mouse. In conclusion, the rat has been used only to a limited extent in the study of the pathology of schistosomiasis, but represents an excellent system for investigations into the mechanism of innate and acquired resistance to these important human diseases.

C. Hookworm Disease

1. The Human Infection

Human hookworm infection is primarily caused by two species of nematode parasites in the superfamily Ancylostomatoidea: *Necator americanus* and *Ancylostoma duodenale*. Dubini first described *A. duodenale* in 1843 (42), while Stiles identified *Necator americanus* in 1902 (155).

These two parasites are primarily distributed throughout the tropical and subtropical world, where they cause much disease in areas where marginal nutritional situations exist, especially in children. Stoll (159) estimated that over 457 million people harbored either *A. duodenale* or *N. americanus;* the prevalance rate for each has surely increased since then.

Both of these parasites live in the small intestine as adults where they feed on villus tissue (75). In addition, blood is sucked through the worms' digestive tract during the feeding process, and the patient, if heavily infected, suffers doubly from tissue damage and iron-deficiency anemia (128). Adult females pass fertilized, non-embryonated ova directly into the lumen where the eggs begin embryonation as the fecal mass forms in the large intestine. Upon being deposited in a warm, moist environment on the ground, the ova develop further until the rhabditiform larva hatches. After several molts, the filariform larva develops, and it is this stage (i.e., third larval stage) that is infective. The larvae migrate to the highest object in their immediate environment, and, after coming into direct contact with the epidermis, enter the unbroken skin. From the skin, larvae migrate via the circulation first to the heart, then to the capillaries of the lungs, where they break out into the alveolar spaces.

The larvae then traverse up the bronchi and arrive at the epiglottis, where they are swallowed, and establish themselves in the small intestine. After two more molts, the mature worms mate and egg production begins. However, some infective larvae may remain dormant in the gut or musculature, thereby serving as a potential reserve for the infection (139). Adult hookworms have survived in the small intestine for as long as 8 years, but usually live no more than 1–2 years.

As already mentioned, tissue damage and blood loss are the major pathological consequences of infection. The worm's habit of injecting anticoagulant (16), as well as its behavior of frequently relocating to new feeding sites compounds the damage, since abandoned points of attachment often continue to bleed for three or four days until normal repair mechanisms correct the worm-induced lesion.

Using epidemiological evidence, it has been tentatively concluded that there is acquired resistance to hookworm infection in endemic areas, because studies measuring fecal egg output on people living in these environs have established that the number of adult worms residing at any one time in any given individual remains fairly constant, despite frequent exposures to infective larvae in the soil (113).

2. The Infection in the Rat

While many attempts have been made to establish patent infections with *N. americanus* and *A. duodenale* in laboratory rats, they have been uniformly unsuccessful (81). The dog and cat are useful hosts for studying infections of *A. duodenale*, whereas *N. americanus* is best studied in the young dog and golden hamster (142).

When filariform larvae of *A. caninum* are introduced into the rat either percutaneously or subcutaneously, most larvae migrate into the skeletal muscle tissue, where they remain dormant but viable (76, 99), while some larvae find their way into the gut of the rat, but are quickly eliminated before developing into adults. The propensity of hookworm larvae to invade the musculature of the rat probably mimicks this aspect of the infection with *A. caninum* (86), and may even be similar to the situation in human hookworm infection.

The nematode, *Nippostrongylus brasiliensis* (superfamily Heligmosomatoidea) has been extensively studied in the rat as an alternative hookworm-like infection, and offers some advantages over the other animal models mentioned above because of the worm's comparatively brief life cycle, which is linked to a strong set of immunological responses on the part of the rat.

The life cycle of *N. brasiliensis* is similar to that of the human hookworm. The main differences between *N. brasiliensis* and the human equivalents are that adult *N. brasiliensis* live for only 14–16 days in the small intestine of rat, and the infection is not pathologically similar to the human hookworm infections, as *N. brasiliensis* is not an avid villus tissue feeding nematode; its diet primarily consists of blood and other tissue fluids.

Because of the rapid immune-mediated expulsion of this parasite from the gut of the rat, a great deal of attention has centered on determining the host and parasite factors involved in this relationship, and much useful data have been accumulated regarding the mechanism(s) of immune protection against this helminthic infection (114,115,165).

In conclusion, there is at present no laboratory rat model employing either *Necator americanus* or *Ancylostoma duodenale,* but studies with *Nippostrongylus brasiliensis* have yielded much information regarding immunobiological aspects of host interactions with gut-dwelling nematodes, and may

eventually be useful in explaining related problems in the human hookworm infections.

D. Malaria

1. The Human Infection

There are four major species of malarial parasites, namely, *Plasmodium vivax* (85), *P. malariae* (59), *P. falciparum*, (180), and *P. ovale* (152).

There is a rich and engrossing literature surrounding the discovery of the various species of malaria that is comprehensively reviewed by Garnham (54). Malaria is the most prevalent life-threatening infectious disease in the world, with an estimated 100 million new cases each year (183).

The basic life cycle of the malarial parasite is as follows. Two kinds of infection with *Plasmodium* are found in the intermediate (human) host; the exoerythrocytic (liver) and erythrocytic (red cell) cycle. The definitive host (mosquito) harbors the sexual stages of the infection and is also the vector.

The mosquito injects the sporozoites into the human host while taking a blood meal. The parasite is then carried via the bloodstream to the liver. There, the sporozoites penetrate parenchymal cells and undergo asexual division (exoerythrocytic cycle). Apparently, there is much variation in the duration of this phase of the infection both in people and other primates, depending upon the species of malaria in question. However, the same result is achieved, namely, the production of merozoites which are now capable of entering the bloodstream to begin their infection in red cells (erythrocytic cycle). The erythrocytic cycle results in more merozoites, the number produced per red cell, again, varying from malarial species to species. The erythrocyte-derived merozoites are capable of invading more red cells, but are unable to infect liver parencymal cells. Alternatively, the invaded red cell may end up harboring a sexual stage instead of merozoites, and it is this form which is now infective for the mosquito.

Development in the stomach of the mosquito, after the female insect has taken a blood meal from an infected individual, involves the formation of a zygote from the fusion of a spermlike microgamete (male) with a macrogamete (female). The oocysts produced lodge in the stomach wall and become filled with many sporozoites, which rupture into the hemocoel and eventually penetrate into the insect's salivary glands. The sporozoites are injected into the next host during the taking of a blood meal.

The major clinical features of malaria are fever, chills, splenomegaly, and anemia.

2. The Infection in the Rat

It should be stated from the outset that none of the human *Plasmodium* species are infective for the rat. The human malarias are infective for certain primates, with the most commonly studied ones being *P. falciparum* in the *Aotes* monkey, *P. malariae* in the chimpanzee, and *P. vivax* in the marmoset. However, there are two rat-adapted malarias (originally isolated from wild African rodent species) which have been used extensively as models for mammalian malaria, namely *Plasmodium berghei* (172) and *P. vinckei* (129).

The life cycles of *P. berghei* and *P. vinckei* follow the generalized life cycle described for the human *Plasmodium* species. However, the exoerythrocytic and erythrocytic cycles of *P. berghei* and *P. vinckei* differ from human infection with regards to their temporal relationships. Both *P. berghei* and *P. vinckei* are usually fatal infections in the young rat (130,143), but are not usually fatal in the adult host (194).

Splenectomy* of adult rats renders them completely susceptible to both species of rodent malaria, and most will die after this treatment (54). The erythrocytic stages of *P. berghei* tend to become sequestered in the bone marrow (3), thereby mimicking the deep vascular schizogany characteristic of *P. falciparum* and *P. knowlesi*.

The infections have been adapted to sporozoite transmission in the laboratory via laboratory-reared species of *Anopheles* mosquitos, with *A. stephensi* being a commonly used vector (169).

Because of the relative ease in inducing sporozoite-transmitted malaria, the rat serves as a good model for the exoerythrocytic cycle in the liver. Infections usually become patent in the rat within 2 days after allowing infected mosquitos to feed (10,185).

The rat, as well as the mouse, has been used extensively for studies on the pathophysiology, immunobiology, biochemistry, and cell biology of malaria infection.

One important disadvantage of the rat-adapted malarias is their propensity toward nonsynchronous infections, thereby forcing investigators to use other animal models, such as the duck with *P. lophurae* (29) and the rhesus monkey with *P. knowlesi* (156), when large batches of one specific stage of the erythrocytic cycle are required.

In conclusion, the rat and its adapted malarias have been very useful in the study of mammalian *Plasmodium* infections, while infections with human malaria have been limited to animal models employing primates.

E. African Trypanosomiasis

1. The Human Infection

African trypanosomiasis (sleeping sickness) occurs throughout most of tropical Africa and is caused by two species of

*Rats used in this procedure and in other experimental studies of blood parasites must be evaluated for presence of latent *Hemobartorella muris* (see Volume I, Chapter 12).

flagellated protozoans: *Trypanosoma brucei gambiense* (46) and *T. b. rhodesiense* (153). In addition, a similar disease in livestock is caused by *Trypanosoma brucei brucei* (123). All three infections are spread by various species of the dipteran vector, *Glossina* (tsetse fly).

Trypanosomiasis can be a fatal disease, occuring in epidemic fashion while existing in relatively low levels in most African countries. *Trypanosoma brucei gambiense* infects more people each year than *T. b. rhodesiense* (6,140).

The situation is equally serious with regards to livestock, and many countries suffer great economic loss each year due to trypanosome infection (57). Indeed, Williamson (182) points out that the geographic distribution of the tsetse fly corresponds, exactly, to those regions in Africa where meat consumption is lowest.

The infection is initiated through the bite of an infected tsetse fly. A chancre forms as the result of a local infection by the organisms in the subcutaneous tissues, and after several days the trypanosomes become systemic. They find their way to the draining lymph nodes, where they begin the acute phase of the infection. The organisms can readily be found in the bloodstream at this time, but within several months, their numbers become reduced below detection limits.

Eventually, the trypanosomes enter the cerebrospinal fluid initiating the chronic phase, where they continue to multiply in low numbers. It is this phase of the infection which eventually proves fatal, although the precise mechanism of death is not yet clear.

A characteristic of African trypanosome infections is the phenomenon referred to as antigenic variation (163,170). The organisms produce membrane-associated antigens, and with each new antigen type the population of trypanosomes increases, only to be reduced again by the induction of antibodies against that particular variant. These fluctuations in trypanosome populations occur in a predictable fashion, at least in cattle (60), and presumably also in human (170). Eventually, however, the parasites overwhelm the host and death ensues.

2. The Infection in the Rat

Rats can be easily infected with *T. b. gambiense, T. b. rhodesiense,* and *T. b. brucei* and are invariably killed by them, the time of death depending upon the virulence of the isolate.

In addition, the rat can also serve as the host for other species of trypanosomes (see Table I). Since the rat is particularly susceptible to the African trypanosomes, it is not well suited for investigations into the pathological correlates associated with the human or cattle diseases, the latter infections often remaining chronic for months or years. The infection in the rat with these three African trypanosome species is also different from the human situation in that the organisms remain in the bloodstream throughout the infection, and no chancre-like lesion is formed in the subcutaneous tissues. A possible exception to this might be primary isolates of *T. b. gambiense,* which can lead to chronic infections in the rat (134).

It can be concluded that in most instances the disease in the rat does not mimic either the human or cattle disease.

The rabbit appears to be a better host than the rat in which to induce chronic infections with African trypanosomes, since CNS involvement occurs somewhat more predictably (166).

While the rat is not a useful animal model for African trypanosomiasis, the high degree of susceptibility of the rat to these organisms has already proved useful in the diagnosis of the disease in cattle (55). Furthermore, large quantities of organisms can be harvested from infected rats in a relatively short time period, thereby facilitating studies on the biochemistry (20) and immuniology (eg. antigenic variation) of the parasite (31).

Of interest is the fact that the laboratory rat can be infected and survive an infection with *Trypanosoma lewisi* (87). *Trypanosoma lewisi* occurs in nature as a transient infection of wild rats of various species and is transmitted from rat to rat via the flea.

While *T. lewisi,* unlike the African trypanosomes, is considered nonpathogenic, much useful knowledge regarding trypanosome infections, in general, has been accumulated over the years with this host–parasite system.

ACKNOWLEDGMENTS

I would like to thank Drs. A. Agar, J. Bier, G.J. Jackson, J. McGlothlin, M. Müller, R. Nussenzweig, G. Schad, and E.J.L. Soulsby, and Ms. L. Aron, Ms. E. Mosenfelder, and Ms. T. Terilli for assisting me in many aspects of the preparation of this manuscript. I especially thank Drs. P. D'Alesandro, J. Vanderberg, S. Weisbroth, R. Williams, G.J. Jackson, and W.C. Campbell for their critical evaluation of various portions of this manuscript, and for suggesting many of the references or sources for references which proved invaluable to me throughout the writing of this text. This chapter contains references up to and including the year 1977.

REFERENCES

1. Acholonu, A.D. (1964). Life history of two Notocotylidae (Trematoda). *J. Parasitol.* **50,** Sect. 2, 28–29.
2. Al-Dabach, M.A., and Shafiq, M.A. (1970). Pathogenicity of *Trichomonas hominis* to splenectomized rats. *Trans. R. Soc. Trop. Med. Hyg.* **64,** 826–828.
3. Alger, N. (1963). Distribution of schizonts of *Plasmodium berghei* in rats, mice and hamsters. *J. Protozool.* **10,** 7–10.
4. Ali Kahn, Z. (1966). The post-embryonic development of *Trichinella spiralis* with special reference to ecdysis. *J. Parasitol.* **52,** 248–259.
5. Ansari, J.A. (1964). Studies on *Setaria cervi* (Nematoda: Filarioidea). Part II. Its peritoneal transplant and periodicity of the microfilariae in white rats. *Z. Parasitenkd.* **24,** 105–111.

6. Apted, F.I.C. (1970). The epidemiology of Rhodesian sleeping sickness. *In* "The African Trypanosomiases" (H.W. Mulligan, ed.), Wiley pp. 645–660. (Interscience), New York.
7. Arfaa, F., Sabaghian, H., and Ale-Dawood, H. (1965). Studies on *Ornithobilharzia turkestanicum* (Skrjabin, 1913), Price, 1919 in Iran. *Ann. Parasitol. Hum. Comp.* **40,** 45–50.
8. Assoku, R.K. (1975). Immunological studies of the mechanism of anaemia in experimental *Trypanosoma evansi* infection in rats. *Int. J. Parasitol.* **2,** 137–145.
9. Bacha, W.J., Jr. (1966). Viable egg production in *Zygocotyle lunata* following mono-metacercarial infections. *J. Parasitol.* **52,** 1216–1217.
10. Bafort, J.M. (1971). The biology of rodent malarial with particular reference to *Plasmodium vinckei vinckei. Ann. Soc. Belge Med. Trop.* **51,** 1–204.
11. Bafort, J.M., and Molyneux, D.H. (1970). Studies on the biology of a new rodent *Babesia* from the Congo. *J. Protozool.* **17,** Suppl., 27.
12. Bagai, R.C., and Subrahmanyam, D. (1968). Studies on the host-parasite relation in albino rats infected with *Littosomoides carinii. Am. J. Trop. Med.* **17,** 833–834.
13. Balasingam, E. (1965). Studies on *Hepatojarakus malayae* Yeh 1955 (Trichostrongylidae: Nematoda). *Med. J. Malaya* **20,** 68–69.
14. Becker, E.R., and Hall, P.R. (1933). Cross-immunity and correlation of oocyst production during immunization between *Eimeria miyairii* and *Eimeria separata* in the rat. *Am. J. Hyg.* **18,** 220–223.
15. Behnke, J.M. (1974). The distribution of larval *Aspicularis tetraptera* (Schulz) during a primary infection in *Mus musculus, Rattus norvegicus* and *Apodemus sylvaticus. Parasitology* **69,** 391–402.
16. Belding, P. (1965). "Textbook of Parasitology," p. 435. Meredith Publ. Co., New York.
17. Bhaibulaya, M. (1974). Experimental hybridization of *Angiostrungylus mackerrasae,* Bhaibulaya, 1968 and *Angiostrongylus cantonensis* (Chen, 1935). *Int. J. Parasitol.* **4,** 567–573.
18. Bilharz, T. (1852). A study on human helminthography, with brief observations by Bilharz in Cario, along with remarks by Siebald in Breslan. *Z. Wiss. Zool.* **4,** No. 1, 53–72.
19. Binns, H.R. (1938). Observations on the behaviour in laboratory animals of *Trypanosoma congolense* Borden, 1904. *Ann. Trop. Med. Hyg.* **32,** 425–430.
20. Bowman, I.B.R. (1974). Intermediary metabolism of pathogenic flagellates. *Trypanosomiasis Leishmaniasis Spec. Ref. Chagas' Dis., CIBA Found. Symp., 1973* No. 20 (new serv.), pp. 255–271.
21. Brown, H.W. (1969). "Basic Clinical Parasitology," 3rd ed., p. 108. Appleton, New York.
22. Campbell, W.C., and Timinski, S.F. (1965). Immunization of rats against *Ascaris suum* by means of nonpulmonary larval infections. *J. Parasitol.* **51,** 712–716.
23. Castellino, S., and de Carneri, I. (1963). Frequenza delle giardiasi in vari roditoria de laboratorio e attecchiomento delle giardie dell'hamster nel ratto. *Riv. Parassitol.* **24,** 231–235.
24. Catty, D. (1976). Immunity and acquired resistance to trichinosis. *In* "Immunology of Parasitic Infections" (S. Cohen and E.H. Sadun, eds.), pp. 359–379. Blackwell, Oxford.
25. Cerfontaine, P. (1895). Contribution à l'étude de la trichinose. *Arch. Beiol.* **13,** 125–145.
26. Chapman, W.H. (1969). Infection with *Trichosomoides crassicauda* as a factor in the induction of bladder tumors in rats fed 2-acetylaminofluorene. *Invest. Urol.* **7,** 154–159.
27. Chernin, J. (1974). Proceedings: Further studies on the effect of *Taenia crassiceps* on rats and mice. *Parasitology* **69,** 27.
28. Clegg, J.A., and Smithers, S.R. (1968). II. Penetration of mammalian skin by *Schistosoma mansoni. Parasitology* **58,** 111–128.
29. Coggeshall, L.T. (1938). *Plasmodium lophurae,* a new species of

malaria parasite pathogenic for the domestic fowl. *Am. J. Hyg.* **27,** 615-618.

30. Crompton, D.W., Arnold, S., and Barnard, D. (1972). The patent period and production of eggs of *Moniliformis dubius* (Acanthocephala) in the small intestine of male rats. *Int. J. Parasitol.* **2,** 319-326.

31. Cross, G.A.M. (1975). Identification, purification, and properties of clone-specific glycoprotein antigens constituting the surface coat of *Trypanosoma brucei. Parasitology* **71,** 393-417.

32. Culbertson, J.T., and Kolodny, M.H. (1938). Acquired immunity in rats against *Trypanosoma cruzi. J. Parasitol.* **24,** 83-90.

33. Daengsvang, S. (1968). Further observations on the experimental transmission of *Gnathostoma spinigerum. Ann. Trop. Med. Parasitol.* **67,** 88-94.

34. Daengsvang, S., Thienprasitthi, P., and Chomcherngpat, P. (1966). Further investigations on natural and experimental hosts of larvae of *Gnathostoma spinigerum* in Thailand. *Am. J. Trop. Med. Hyg.* **15,** 727-729.

35. Daengsvang, S., Sermswatsri, B., Youngyi, P., and Guname, D. (1970). Penetration of the skin by *Gnathostoma spinigerum* larvae. *Ann. Trop. Med. Parasitol.* **64,** 399-402.

36. Despommier, D.D. (1976). Musculature. *In* "Ecological Aspects of Parasitology" (C. Kennedy, ed.), pp. 269-285. North-Holland Publ., Amsterdam.

37. Despommier, D.D., Aron, L., and Turgeon, L. (1975). *Trichinella spiralis:* Growth of the Intracellular (muscle) Larva. *Exp. Parasitol.* **37,** 108-116.

38. Despommier, D.D., Campbell, W.C., and Blair, L.S. (1977). The *in vivo* and *in vitro* analysis of immunity to *Trichinella spiralis* in mice and rats. Parasitology, **74,** 109-119.

39. Dobson, C. (1966). Certain aspects of the host-parasite relationship of *Nematospiroides dubius* (Baylis). *Parasitology* **56,** 407-416.

40. Dobson, C., Campbell, R.W., and Webb, A.I. (1967). Anaphylactic and reagin-like antibody in rats infected with *Toxocara canis* larvae. *J. Parasitol.* **53,** 209.

41. Donges, J. (1964). Eine enheimische, fakultativ humanpathogene Echinostomatidenart (Trematoda) und der Infestationsverlauf beim Menschen. (Abstract of report before 2. Tog Deutsch. Gesellsch. Parasitol., München, Mar. 18-30. *Z. Parasitenkd.* **25,** 3.

42. Dubini, A. (1843). Nuovo verme intestinale umano (*Anchylostoma duodenale*), constitutente un sesto genere dei nematoidei proprii dell ecomo. *Ann. Univ. Med. Milano* **106,** 5-13.

43. Ducas, R. (1921). L'immunité dans la Trichinose. Ph.D. Thèse, Jouve et Cie, Paris.

44. Duszynski, D.W. (1972). Host and parasite interactions during single and concurrent infections with *Emeria nieschulzi* and *E. separata* in the rat. *J. Protozool.* **19,** 82-88.

45. Dutt, S.C., and Srivastava, H.D. (1962). Studies on the morphology and life history of the mammalian blood fluke *Orientobilharzia dattai* (Dutt and Srivastava, 1952). Dutt and Srivastava 1955. III. Definitive host specificity. *Indian J. Vet. Sci. Anim. Husb.* **32,** 260-268.

46. Dutton, J.E. (1902). Preliminary note upon a trypanosome occurring in the blood of man. *Thomas Yates Lab. Rep.* **4,** 455.

47. Esslinger, J.H. (1962). Development of *Porocephalus crotali* (Humbalt, 1808) (Pentastomidae) in experimental intermediate hosts. *J. Parasitol.* **48,** 452-456.

48. Fallis, A.M., ed. (1971). "Ecology and Physiology of Parasites." Univ. of Toronto Press, Toronto.

49. Foris, G. (1970). Growth curves of *Trypanosoma equiperdum* in rats treated with endotoxin of *Serratia marcescens. J. Infect. Dis.* **121,** 331-334.

50. Frenkel, J.K., and Dobey, J.P. (1975). *Hammondia hammondi:* A new coccidium of cats producing cysts in muscle of other mammals. *Science* **189,** 222-224.

51. Frerichs, W.M., Johnson, A.J., and Holbrook, A.A. (1969). Equine piroplasmosis—Attempts to infect laboratory animals with *Babesia equi. Am. J. Vet. Res.* **30,** 1333-1336.

52. Fujii, Y. (1847). Katayama Disease, Katayamaki, Chugai Iji Shimpo. No. 691, pp. 55-56.

53. Gardiner, C.H. (1976). Habitat and reproductive behavior of *Trichinella spiralis. J. Parasitol.* **62,** 865-870.

54. Garnham, P.C.C. (1966). "Malaria Parasites and Other Haemosparidia." Blackwell, Oxford.

55. Godfrey,D.G., and Killick-Kendrick, R. (1961). Bovine trypanosomiasis in Nigeria. I. The inoculation of blood into rats as a method of survey in the Donga Valley, Benue Province. *Ann. Trop. Med. Parasitol.* **55,** 287.

56. Goodall, R.I. (1972). The growth of *Hymenolepis microstoma* in the laboratory rat. *Parasitology* **65,** 137-142.

57. Goodwin, L.G. (1974). The African scene: Mechanisms of pathogenesis in trypanosomiasis. *Trypanosomiasis Leishmaniasis Spec. Ref. Chagas'Dis., Ciba Found. Symp.,* 20 (new ser.), pp. 107-124.

58. Gould, S.E. (1970). Clinical manifestations. *In* "Trichinosis in Man and Animals" (S.E. Gould, ed.), pp. 269-306. Thomas, Springfield, Illinois.

59. Grassi, B., and Feletti, R. (1892). Contribuzione allo studio dei parasiti malarici. *Atti Accad. Gioenia Sci. Nat. Catania, Mem.* [4] **5,** 1-80.

60. Gray, A.R. (1965). Antigenic variation in a strain of *Trypanosoma brucei* transmitted by *Glossina morsitans* and *G. palpalis. J. Gen. Microbiol.* **41,** 195-214.

61. Grewal, M.S. (1960). On a new trypanosome, *Trypanosoma helogalei* Grewal, 1956, from the blood of African mongoose, *Helogale undulata rufala* Peters (Peter's pigmy mongoose). *Indian J. Med. Res.* **48,** 418-432.

62. Haberkorn, A. (1970). Zur Empfanglichkeit micht spezifischer Wirte für Schizogonie-Stadien verschiedener Eimeria-Arten. *Z. Parasitenkd.* **35,** 156-161.

63. Haberkorn, A. (1971). Zur Wirtsspezifitat von *Eimeria contorta* n. sp. (Sporozoa; Eimeridae). *Z. Parasitenkd.* **37,** 303-314.

64. Harley, J.P., and Gallicchico, V. (1971). *Trichinella spiralis* migration of larvae in the rat. *Exp. Parasitol.* **30,** 11-12.

65. Healy, G.R., and Gleason, N.N. (1966). Studies on the pathogenicity of various strains of *Entamoeba histolytica* after prolonged cultivation, with observation of strain differences in the rats employed. *Am. J. Trop. Med.* **15,** 294-299.

66. Heath, D.D., and Pavloff, P. (1975). The fate of *Taenia taeniaformis* oncospheres in normal and passively protected rats. *Int. J. Parasitol.* **1,** 83-88.

67. Herbert, I.V., and Becker, E.R. (1961). Effect of cortisone and X-irradiation on the course of *Trypanosoma lewisi* infection in the rat. *J. Parasitol.* **47,** 304-308.

68. Ho, B.C., Singh, M., and Yap, E.H. (1976). Studies on the malayan forest rat filaria, *Breinlia booliati* (Filaroidea: Onchocercidae): Transmission to laboratory rats. *Int. J. Parasitol.* **6,** 113-116.

69. Ho, Yi-Ysün. (1963). On the host specificity of *Schistosoma japonicum. Chin. Med. J.* **82,** 405-414.

70. Ismail, M.M., Wijayaratnam, Y., and Amarasinghe, D.K.C. (1971). The eosinophil response in rats to extracts, larvae and adult worms of *Dirofilaria repens. Ceylon J. Med. Sci.* **20,** 15-27.

71. Ito, J. (1953). Studies on host-specificity of *Schistosoma japonicum.* 2. A comparison of the development of the worms in different laboratory mammalian hosts. *Nisshin Igaku* **40,** 518-522.

72. Jackson, G.J., and Bradbury, P.C. (1970). Cuticular fine structure and molting of *Neoaplectana glaseri* (Nematoda), after prolonged contact with rat peritoneal exudate. *J. Parasitol.* **56,** 108-115.

73. Jacobs, E.E., Dunn, A.M., and Walker, J. (1971). Mechanisms for the

dispersal of parasitic nematode larvae. 2. Rats as potential paratenic hosts for *Oesophagostomum* (Strongyloidea). *J. Helminthol.* **45,** 139–144.

74. Johnson, A.D. (1968). Life history of *Alaria marcianae* (La Rue, 1917) Walton, 1949 (Trematoda: Diplostomatidae). *J. Parasitol.* **54,** 324–332.

75. Kalkofen, U. (1970). Attachment and feeding behavior of *Ancylostoma caninum*. *Z. Parasitenkd.* **33,** 339–354.

76. Kamioka, S. (1938). On the development of *Ancylostoma caninum* in the body of non-specific host. *Keio Igaku* **18,** 55–70.

77. Katiyar, J.C., Tangri, A.N., Ghatak, S., and Sen, A.B. (1973). Serum protein patterns of rats during infection with *Hymenolepis* nana. *Indian J. Exp. Biol.* **11,** 188–190.

78. Kennedy, C.R. (1976). "Ecological Aspects of Parasitology." North-Holland Publ., Amsterdam.

79. Kinsella, J.M. (1971). Growth, development, and intraspecific variation of *Quinqueserialis equinquiserialis* (Trematode: Notocotylidae) in rodent host. *J. Parasitol.* **57,** 62–70.

80. Ko, R.C., Morton, B., and Wong, P.S. (1975). Prevalence and histopathology of *Echinocephalus sinensis* (Nematoda: Gnathostomatidae) in natural and experimental hosts. *Can. J. Zool.* **53,** 550–559.

81. Komiya, Y., and Yasuroaka, K. (1966). The biology of hookworms. *In* "Progress of Medical Parasitology in Japan" (K. Morishita, Y. Komiya, and H. Matsubayashi, eds.). Sankyo Bijyutsu Insatsu Co., Japan.

82. Laing, A.B.G., Edeson, J.F.B., and Wharton, R.H. (1961). Studies on Filariasis in Malaya: Further experiments on the transmission of *Brugia malayi* and *Wuchereria bancrofti*. *Ann. Trop. Med. Parasitol.* **55,** 86–92.

83. Landau, I., Michel, J.C., and Adam, J.P. (1969). Cycle biologique au laboratoire de *Plasmodium breghei killicki* n. sp. *Ann. Parasitol.* **43,** 545–549.

84. Larsh, J.E., Jr. (1967). The present understanding of the mechanism of immunity to Trichinella spiralis. *Am. J. Trop. Med. Hyg.* **16,** 123–132.

85. Laveran, A. (1891). "Paludiam" (transl. by J.W. Martin. London, New Sydenham Society, 1893).

86. Lee, K.T., Little, M.D., and Beaver, P.C. (1975). Intracellular (muscle fiber) habitat of *Ancylostoma caninum* in some mammalian hosts. *J. Parasitol.* **64,** 589–598.

87. Lewis, T.R. (1879). Flagellated organisms in the blood of healthy rats. *Q. J. Microsc. Sci.* **19,** 109.

88. Lie Kian Joe. (1963). The life history of *Echinostoma malayanum* Leiper, 1911. *Trop. Geogr. Med.* **15,** 17–24.

89. Lie Kian Joe. (1967). Studies on Echinostomatidae (Trematoda) in Malaya. XV. The life history of *Echinostoma murinum* (Tubanzui, 1931). *Proc. Helminthol. Soc. Wash.* **34,** 139–143.

90. Lie Kian Joe, and Basch, P.F. (1967). The life history of *Echinostoma paraensei* sp. n. (Trematoda: Echinostomatidae). *J. Parasitol.* **53,** 1192–1199.

91. Lillis, W.G., and Nigrelli, R.F. (1965). *Fundulus heteroclitus* (Killfish), second intermediate host for *Echinochasmus schwartzi* (Price, 1931) (Trematoda: Echinostomadae). *J. Parasitol.* **51,** Sect. 2, Suppl., 23 (abstr.).

92. Lin, T.M. (1971). Colonization and encystation of *Entamoeba muris* in the rat and the mouse. *J. Parasitol.* **57,** 375–382.

93. Love, R.J., Ogilvie, B.M., and McLaren, D.J. (1976). The immune mechanism which expels the intestinal stage of *Trichinella spiralis* from rats. *Immunology* **30,** 7–15.

94. McCoy, O.R. (1931). Immunity of rats to reinfection with *Trichinella spiralis*. *Am. J. Hyg.* **14,** 484–494.

95. McCoy, O.R. (1932). Size of infection as an influence on the persistence of adult trichinae in rats. *Science* **75,** 364–365.

96. Mango, A.M., Mango, C.K., and Esamal, D. (1972). A preliminary note on the susceptibility, prepatency and recovery of *Fasciola gigantica* in small laboratory animals. *J. Helminthol.* **46,** 381–386.

97. Marquardt, W.C. (1966). Attempted transmission of the rat coccidium *Eimeria nieschulzi* to mice. *J. Parasitol.* **52,** 691–694.

98. Martin, L.K., Einheber, A., Porro, R.F., Sadun, E.H., and Bauer, H. (1966). *Plasmodium berghei* infections in gnotobiotic mice and rats—Parasitologic, immunologic and histopathologic observations. *Mil. Med.* **131,** Suppl., 870–889.

99. Matsusaki, G. (1951). Studies on the life history of the hookworm. VII. On the development of *Ancylostoma caninum* in the abnormal host. *Yokohama Med. Bull.* **2,** 154–160.

100. Mayberry, L.F., and Marquardt, W.C. (1973). Transmission of *Eimeria separata* from the normal host, *Rattus*, to the mouse, *Mus musculus*. *J. Parasitol.* **59,** 198–199.

101. Meingassner, J.G., Georgopoulis, A.G., and Patoschka, M. (1975). Intravaginale infektionen der Ratte mit *Trichomonas vaginalis* und *Candida albicans*. *Tropenmed. Parasitol.* **26,** 395–398.

102. Meleney, H.E., Sandground, J.H., Moore, D.V., Most, H., and Carney, B.H. (1953). The histopathology of experimental Schistosomiasis. II. Bisexual infections with *S. mansoni*, *S. japonicum*, and *S. haematobium*. *Am. J. Trop. Med. Hyg.* **2,** 883–913.

103. Meyer, M.C., and Vik, R. (1963). The life cycle of *Diphyllobothrium sebago* (Ward, 1910). *J. Parasitol.* **49,** 962–968.

104. Moore, D.V., and Meleney, H.E. (1954). Comparative susceptibility of common laboratory animals to experimental infection with *Schistosoma haematobium*. *J. Parasitol.* **40,** 392–397.

105. Mossalam, I., Hosmey, Z., and Atallah, O.A. (1971). Larva migrans of *Toxocara cati* in visceral organs of experimental animals. *Acta Vet. Acad. Sci. Hung.* **21,** 405–412.

106. Mueller, J.F. (1968). Growth stimulating effect of experimental sparganosis in thyroidectomized rats, and comparative activity of different species of *Spirometra*. *J. Parasitol.* **54,** 795–801.

107. Mueller, J.F., and Reed, P. (1968). Growth stimulation induced by infection with *Spirometra mansonoides* sparga in propylthiouracil-treated rats. *J. Parasitol.* **54,** 51–54.

108. Nath, D. (1973). A note on the echinostomatid metacercarial fauna encountered in Indian fresh water snails. *Indian Vet. J.* **50,** 292–293.

109. Ni, G.V. (1966). Susceptability of some laboratory and wild mammals to infection with Uzbek strains of *Leishmania tropica major* isolated from gerbils (*Rhombormys opimus* Licht). *Med. Parazitol. Parazit. Bolezni* **35,** 270–274.

110. Norris, D.E. (1971). The migratory behavior of the infective stage larvae of *Ancylostoma braziliense* and *Ancylostoma tubaeforme* in rodent paratenic hosts. *J. Parasitol.* **57,** 998–1009.

111. Nowell, F. (1969). The blood picture resulting from Nuttalia- (=*Babesia*) *rodhaini* and *Nuttalia* (=*Babesia*) *microti* infections in rats and mice. *Parasitology* **59,** 991–1004.

112. Ogilvie, B.M. (1965). Use of cortisone derivatives to inhibit resistance to *Nippostrongylus brasiliensis* and to study the fate of parasites in resistant hosts. *Parasitology* **55,** 723–730.

113. Ogilvie, B.M. (1976). Immunity to nematode parasites of man with special reference to Ascaris, Hookworms and Filariae. *In* "Immunology of Parasitic Infections" (S. Cohen and E.H. Sadun, eds.), pp. 380–407. Blackwell, Oxford.

114. Ogilvie, B.M., and Jones, V.E. (1973). Immunity in the relationship between helminths and hosts. *Prog. Allergy* **17,** 93–144.

115. Ogilvie, B.M., and Love, R.J. (1974). Co-operation between antibodies and cells in immunity to a nematode parasite. *Transplant. Rev.* **19,** 147–168.

116. Ogino, A. (1963). Studies on infection made of hookworm oral infections in human host with larvae of *Necator americanus* isolated from

lungs of infected rats. *Jpn. J. Parasitol.* **12,** 56 (Engl. summ.).

117. Olsen, O.W. (1974). "Animal Parasites. Their Life Cycles and Ecology." Univ. Park Press, Baltimore, Maryland.

118. Owen, R. (1835). Description of a microscopic entozoan infecting the muscles of the human body. *Trans. Zool. Soc. London* **1,** 315–324.

119. Pearson, D.J., and Taylor, G. (1975). The influence of the nematode *Syphacia oblevata* on adjuvant arthritis in the rat. *Immunology* **29,** 391–396.

120. Pearson, J.C. (1964). A revision of the subfamily Haplorchinae Loose, 1899. (Trematoda: Heterphyidae). I. The *Haplorchis* Group. *Parasitology* **54,** 601–676.

121. Petri, M. (1966). The occurrence of *Nosema cuniculi* (*Encephalitozoon cuniculi*) in the cells of transplantable, malignant ascites tumours and its effect upon tumour and host. *Acta Pathol. Microbiol. Scand.* **66,** 13–30.

122. Petter, A.J. (1969). Observations sur la systématique et le aple de l' ascaride *Hexametra quadricornis* (Wedl, 1862). *Ann. Parasitol.* **43,** 655–691.

123. Plimmer, H.G., and Bradford, J.R. (1899). A preliminary note on the morphology and distribution of the organism found in the tsetse fly disease. *Proc. R. Soc. London, Ser. B.* **65,** 275.

124. Purkerson, M., and Despommier, D. (1975). Adaptive changes in muscle fibers infected with *Trichinella spiralis*. *Am. J. Pathol.* **78,** 477–496.

125. Read, C.P., and Kilejian, A.Z. (1969). Circadian migratory behavior of a cestode symbiote in the rat host. *J. Parasitol.* **55,** 574–578.

126. Remington, F.S., and Hackman, R. (1965). Changes in serum proteins of rats infected with *Toxoplasma gondii*. *J. Parasitol.* **51,** 865–870.

127. Rickman, L.R., and Robson, J. (1970). The testing of proven *Trypanosoma brucei* and *T. Rhodesiense* strains by the blood incubation infectivity test. *Bull. W. H. O.* **42,** 911–916.

128. Roche, M., and Layrisse, M. (1966). The nature and causes of "Hookworm Anemia." *Am. J. Trop. Med. Hyg.* **15,** 1032–1093.

129. Rodhain, J. (1952). *Plasmodium vinckei* n. sp. Un deuxième *Plasmodium* parasite de rongeurs sauvages au Katanga. *Ann. Soc. Belge Med. Trop.* **32,** 275–280.

130. Rodhain, J. (1954). Absence d' immunité croisée entre *Plasmodium berghei* et *Plasmodium vinckei* dans les infections chez les jeunes rats. *C. R. Seances Soc. Biol. Ses. Fil.* **148,** 1519–1520.

131. Rogers, W.P. (1962). "The Nature of Parasitism." Academic Press, New York.

132. Rogers, W.P., and Sommerville, R.I. (1968). The infectious process, and its relation to the development of early parasitic stages of nematodes. *Adv. Parasitol.* **4,** 107–186.

133. Rohde, R. (1966). Lung infections of rat with *Cysticercus fasciolaris*. *Med. J. Malaya* **20,** 355.

134. Roubaud, E., Stephanpoulo, G.J., and Duvolon, S. (1944). Etude chez le rat blanc d'une souche neurotrope de *Trypanosoma gambiense*. *Bull. Soc. Pathol. Exot.* **37,** 292–296.

135. Rzepczyk, C.M. (1974). Evidence of a rat-snake life cycle for sarcocystis. *Int. J. Parasitol.* **4,** 447–448.

136. Sambon, L.W. (1907). Remarks on *Schistosomum mansoni*. *J. Trop. Med. Hyg.* **10,** 303–304.

137. Saowakontha, S. (1974). The relationship between protein-calorie malnutrition and trichinosis. I. Studies on the number of intestinal worms and muscular larvae in rats fed low and high protein diets. *Southeast Asian J. Trop. Med. Public Health* **5,** 586–592.

138. Schacher, J. (1973). Laboratory models in filariasis: A review of filarial life cycle patterns. *Southeast Asian J. Trop. Med. Public Health* **4,** 336–349.

139. Schad, G.A., Chowdhury, A.B., Dean, A.B., Kochar, C.G., Nowalinski, V.K., Thomas, T.A., Jr., and Tonascia, J.A. (1973). Arrested

development in human hookworm infections: An adaptation to a seasonally unfavorable external environment. *Science* **180,** 502–504.

140. Scott, D. (1970). The epidemiology of Gambian sleeping sickness. *In* "The African Trypanosomiases" (H.W. Mulligan, ed.), pp. 614–644. Wiley (Interscience), New York.

141. Sen, A.B., and Bhattacharya, B.K. (1964). Studies on *Dipetalonema witei* infection in albino rats. *Indian J. Helminthol.* **16,** 142–150.

142. Sen, H.G. (1972). *Necator americanus:* Behavior in hamsters. *Exp. Parasitol.* **32,** 26–32.

143. Sergent, E., and Poncet, A. (1955). Etude experimental du paludisme des rongeurs a *Plasmodium berghei*. III. Resistance innee. *Arch. Inst. Pasteur Alger.* **33,** 287–306.

144. Shaw, J.J., and Laison, R. (1972). *Trypanosoma vivax* in Brazil. *Ann. Trop. Med. Parasitol.* **66,** 25–32.

145. Singh, M.D., and Pande, B.P. (1972). An experimental infection of guinea pigs, rabbits, and lambs with amphistome metacercaria: A histopathologic study. *Indian J. Anim. Sci.* **42,** 290–297.

146. Smithers, S.R. (1976). Immunity to Trematode infections with special reference to Schistosomiasis and Fascioliasis. *In* "Immunology of Parasitic Infections" (S. Cohen and E.H. Sadun, eds.), pp. 297–328. Blackwell, Oxford.

147. Spink, W.W., and Augustine, D.L. (1935). Trichinosis in Boston. *N. Engl. J. Med.* **213,** 527–531.

148. Spratt, D.M. (1972). Transplantation of adult *Dirofilaria roemeri* to grey kangaroos and laboratory rats. *J. Helminthol.* **48,** 81–89.

149. Sprent, J.F., and McKeown, A. (1967). A study on adaptation tolerance—Growth of Ascaridoid larvae in indigenous and nonindigenous hosts. *Parasitology* **57,** 549–554.

150. Sprent, J.F.A. (1970). Studies on ascaridoid nematodes in pythons: The life history and development of *Ophidascaris moreliae* in Australian pythons. *Parasitology* **60,** 97–122.

151. Sprent, J.F.A. (1970). Studies on ascaridoid hematodes in pythons: The life history and development of *Polydelphis anoura* in Australian pythons. *Parasitology* **60,** 375–397.

152. Stephens, J.W.W. (1922). A new malaria parasite of man. *Ann. Trop. Med. Parasitol.* **16,** 383–388.

153. Stephens, J.W.W., and Fantham, H.B. (1910). On the peculiar morphology of a trypanosome from a case of sleeping sickness and the possibility of its being a new species (*Trypanosoma rhodesiense*). *Proc. R. Soc. London, Ser. B* **83,** 28.

154. Stibbs, H.H., and Seed, J.R. (1975). Short-term metabolism of (14-C) tryptophan in rats infected with *Trypanosoma brucei gambiense*. *J. Infect. Dis.* **131,** 459–462.

155. Stiles, C.W. (1902). A new species of hookworm (*Uncinaria americana*) parasitic in man. *Am. Med.* **3,** 777–778.

156. Stinton, J.A., and Mulligan, H.W. (1932–1933). A critical review of the literature relating to the identification of the malarial parasites recorded from monkeys of the families Cercopithecidae and Colobidae. *Rec. Malar. Surv. India* **3,** 357–380 and 381–444.

157. Stirewalt, M.A., Kuntz, R.E., and Evans, A.S. (1951). The relative susceptabilities of the commonly used laboratory mammals to infection by *Schistosoma mansoni*. *Am. J. Trop. Med.* **31,** 57–82.

158. Stirewalt, M.A., and Hackey, J.R. (1956). Penetration of host skin by cercariae of *Schistosoma mansoni*. I. Observed entry into skin of mouse, hamster, rat, monkey, and man. *J. Parasitol.* **42,** 65–80.

159. Stoll, N.R. (1947). This wormy world. *J. Parasitol.* **33,** 1–18.

160. Sucharit, S., and Macdonald, W.W. (1973). *Brugia pahangi* in small laboratory animals—Attempts to increase susceptibility of white rats to *Brugia pahangi* by host selection. *Southeast Asian J. Trop. Med. Public Health* **4,** 71–77.

161. Sun, T., Chou, S.T., and Gibson, J.B. (1968). Route of entry of

Clonorchis sinensis to the mammalian liver. *Exp. Parasitol.* **22,** 346–351.

162. Tada, I. (1967). Physiological and serological studies of *Paragonimus miyazakii* infection in rats. *J. Parasitol.* **53,** 292–297.

163. Terry, R.J. (1976). Immunity to African trypanosomiasis. *In* "Immunology of Parasitic Infections" (S. Cohen and E.H. Sadun, eds.), pp. 204–221. Blackwell, Oxford.

164. Thorpe, E. (1965). Liver damage and the host–parasite relationship in experimental fascioliasis in the albino rat. *Res. Vet. Sci.* **6,** 498–509.

165. Thorson, R.E. (1970). Direct-infection nematodes. *In* "Immunity to Animal Parasites" (G.J. Jackson, R. Herman, and I. Singer, eds.), Vol. II, pp. 913–962. Meredith Corp., New York.

166. Tobie, E.J., and Highman, B. (1956). Influence of the amino nucleoside of puromycin on the course and pathology of trypanosome infections in rabbits and mice. *Am. J. Trop. Med. Hyg.* **5,** 504–515.

167. Todd, K.S., Lepp, D.L., and Trayser, C.V. (1971). Development of the asexual cycle of *Eimeria vermiformis*, Chobotar, and Hammond, 1971, from the mouse, *Mus musculus*, in dexamethazone-treated rats, Rattus norvegicus. *J. Parasitol.* **57,** 1137–1138.

168. Uzmann, J.R., and Hayduk, S.H. (1964). Larval *Echinochasmus* (Trematoda: Echinostomatidae) in rainbow trout, *Salmo gairdneri. J. Parasitol.* **50,** 586.

169. Vanderberg, J.P., Nussenzweig, R.S., and Most, H. (1968). Further studies on the Plasmodium *berghei-Anopheles stephensi*-rodent system of mammalian malaria. *J. Parasitol.* **54,** 1009–1016.

170. Vickerman, K. (1974). Antigenic variation in African trypanosomes. *Parasites Immunized Host: Mech. Surv. Ciba Found. Symp., 1974* No. 25 (new ser.), pp. 53–80.

171. Villella, J.B. (1970). Life cycle and morphology. *In* "Trichinosis in Man and Animals" (S.E. Gould, ed.), pp. 19–60. Thomas, Springfield, Illinois.

172. Vincke, I.H., and Lips, M. (1948). Un nouveau *Plasmodium* d'un rongeur sauvage du Congo, *Plasmodium berghei* n. sp. *Ann. Soc. Belge Med. Trop.* **28,** 97–104.

173. Waddell, A.H. (1969). Methyridine in the treatment of experimental *Capillaria hepatica* infection in the rat. *Ann. Trop. Med. Parasitol.* **63,** 63–65.

174. Waitz, A.J., McClay, P., and Thompson, P.E. (1964). Effects of Bithionol against *Paragoimus kellicotti* in rats. *J. Parasitol.* **13,** 584–588.

175. Warren, E.G. (1970). Observations on the life-cycle of *Toxocara mackerrasae. Syst. Pract. Med.* **1,** 17. *Parasitology* **60,** 239–253.

176. Warren, K.S., and Peters, P.A. (1967). Comparison of penetration and maturation of *Schistosoma mansoni* in the hamster, mouse, guinea pig, rabbit, and rat. *Am. J. Trop. Med.* **16,** 718–722.

177. Warren, K.S. (1973). "Schistosomiasis: The Evolution of a Medical Literature." MIT Press, Cambridge, Massachusetts.

178. Warren, K.S. (1976). Immunopathology due to cell-mediated (type IV) reactions. *In* "Immunology of Parasitic Infections" (S. Cohen and E.H. Sadun, eds.), pp. 448–463. Blackwell, Oxford.

179. Webster, G.A., and Cameron, T.W.M. (1961). Observations on experimental infections with *Echinococcus* in rodents. *Can. J. Zool.* **39,** 877–891.

180. Welch, W.H. (1897). Malaria: Definitions, synonyms, history, and parasitology. *In* Loomis and Thompson's *Syst. Pract. Med.* **1,** 17.

181. Wertheim, G. (1970). Growth and development of *Strongyloides venezuelensis* Brumpt 1934 in the albino rat. *Parasitology* **61,** 381–388.

182. Williamson, J. (1970). Review of chemotherapeutic and chemoprophylactic agents. *In* "The African Trypanosomiases" (H.W. Mulligan, ed.), pp. 126–221. Wiley (Interscience), New York.

183. World-Health Organization (1975). Six-monthly information on the world malaria situation, January–December 1973. *Week. Epidemiol. Rec.* No. 6, pp. 53–60; No. 7, pp. 76–86.

184. Wright, C.A., Southgate, V.R., and Knowles, R.J. (1972). What is *Schistosoma intercalatum* Fisher, 1934. *Trans. R. Soc. Trop. Med. Hyg.* **66,** 28–64.

185. Yoeli, M., and Most, H. (1965). Studies on sporozoite-induced infections of rodent malaria. I. The pre-erythrocytic tissue stage of *Plasmodium berghei. Am. J. Trop. Med. Hyg.* **14,** 700–714.

186. Yokogawa, M., Yoshimura, H., Sano, M., Okura, T., Tzuzi, M., and Takano, S. (1959). Studies on host-parasite relationship: Experimental infection in rats, mice, and guinea pigs with metacercariae of *Paragonimus ohirai* Miyazaki, 1939. *J. Parasitol.* **45,** Suppl., 20.

187. Yokogawa, M., Yoshimura, H., Sano, M., Okura, T., and Tsuzi, M. (1962). The route of migration of the larva of *Paragonimus westermani* in the final hosts. *J. Parasitol.* **48,** 525–531.

188. Yoshimura, K. (1972). Exposure of rats to a single metacercaria of *Paragonimus sadoensis. Jpn. J. Vet. Sci.* **34,** 269–274.

189. Young, P.C., and Lowe, D. (1969). Larval nematodes from fish of the subfamily Anisakinae and gastrointestinal lesions in mammals. *J. Comp. Pathol.* **79,** 301–313.

190. Zaman, V. (1972). Inability to transmit *Breinlia sergenti* to other laboratory animals. *Southeast Asian J. Trop. Med. Public Health* 3: 143.

191. Zamirdan, M., P.A.G. Wilson. (1974). *Strongyloides ratti:* Relative importance of maternal sources of infection. Parasitol. 69: 445–453.

192. Zimmerman, W.J. (1974). The current status of Trichinellosis in the United States. *In* "Trichinellosis" (C. Kim, ed.), pp. 603–610. Intext Ed. Pub., New York.

193. Zuckerman, A. (1958). Blood loss and replacement in Plasmodial infections. II. *Plasmodium vinckei* in untreated weanling and mature rats. *J. Infec. Dis.* 103: 205–224.

194. Zuckerman, A. (1970). Malaria of lower mammals. *In* "Immunity to Animal Parasites" (G. Jackson, R. Herman, and I. Singer, eds.) Vol. II, pp. 794–820 Meredith Corp., New York.

Chapter 12

Wild Rats in Research

Christine S.F. Williams

I. INTRODUCTION

Wild rats maintained in captivity have served as useful subjects for biomedical research, particularly in the areas of behavior, physiology, pathology, and epizootiology. They also have been studied for development of methods to control wild rodent populations which inflict economic losses by eating and contaminating foodstuffs, damaging man-made structures, and carrying zoonotic diseases. Comprehensive bibliographies and manuals on rodent pest biology and control have been published (36,66,118).

The choice of wild rats as laboratory subjects appears to be influenced by the personal interest of some investigators in these species or the unique role of certain wild rodents as vectors of zoonotic diseases. Some disease syndromes of interest as models for research, such as diabetes mellitus of sand

Table I

Classification of the Order Rodentia

Family	Subfamily	Tribe or group	Genus and species
Muridae	Murinae	Old World rats and mice	*Rattus norvegicus*
			R. rattus
			R. exulans
			Thamnomys surdaster
			Mastomys natalensis
Cricetidae	Cricetinae	New World rats	*Sigmodon* spp.
			Neotoma spp.
			Zygodontomys spp.
			Oryzomys spp.
			Tylomys nudicaudus
	Cricetinae	True hamsters	*Mystromys albicaudatus*
	Gerbillinae	Gerbils	*Psammomys obesus*
Heteromyidae	Dipodomyinae		*Dipodomys* spp.
Octodontidae			*Octodon degus*

rats and white-tailed rats, were discovered fortuitously when wild rats were kept in captivity for other reasons.

All of the rats mentioned in this chapter belong to the order Rodentia, which contains 35 families and exceeds all other orders in variety and actual numbers. For a listing of the respective families and subfamilies, see Table I, which is based on the classification according to Walker (110). Other authorities have not considered the Cricetidae to merit the status of a separate family, and have included its subfamilies in the Muridae family.

In the vernacular, there is no real distinction between rats and mice except for size. Small animals around 30 gm or less tend to be called mice, those of 100 gm or more tend to be called rats, and those in the intermediate range may be called rats or mice or both. For example, *Mastomys* is called both the "multimammate mouse" and the "multimammate rat." *Thamnomys,* while usually called a "tree rat," has been called a "forest mouse." Also, in common usage, some animals which are zoologically hamsters or gerbils, are called rats or mice, e.g., *Mystromys* is included in the tribe of true hamsters, yet it is called a "white-tailed rat," and *Psammomys* is in the subfamily Gerbillinae (gerbils), yet it is called a "sand rat."

II. LABORATORY MANAGEMENT OF CAPTIVE WILD RATS

While some management systems found to be successful for wild rats may seem unduly complicated, it should be remembered that they are the result of trial and error development under local conditions. It is relatively simple to keep wild rats alive, but more difficult to obtain satisfactory reproduction under laboratory conditions. Many of the complex homemade diets mentioned are empirical; it is not known whether they are nutritionally balanced. Neither is it known whether standard laboratory rat diet would be an improvement in all cases. Certainly every effort should be made to get a colony of wild rats onto a standard balanced diet unless there are specific contraindications.

If the original stock comes from a foreign country, two federal agencies should be contacted regarding import regulations. The Public Health Service (Center for Disease Control, Atlanta, Georgia) places restrictions on the importation of etiologic agents and vectors. Consequently, wild-caught rodents which might serve as natural hosts for diseases are prohibited from entry unless importation is authorized by permit. This only applies to wild-caught individuals since animals bred in captivity are not subject to these regulations. The Department of the Interior (U.S. Fish and Wildlife Service, Division of Law Enforcement, Washington, D.C.) has regulations which govern the importation of certain harmful or endangered rodents. Before rodents are trapped within the United States, inquiries should be made of the conservation agency in the state concerned to see if a trapping permit is necessary. Information on trapping methods and equipment can be found in a text by Giles (39). Once the animals arrive in the laboratory, consideration should be given to the fact that they may be carrying disease organisms which could be infectious to both personnel and laboratory animals, and appropriate precautions should be taken.

Systems of management which are designed specifically for wild rats must be adopted. Redfern and Rowe (85) have given details on the laboratory management of wild rats and house mice. Wild rats usually move quickly and are more active than laboratory rats. Heavy leather gloves afford some protection to the handler but make precise manipulations difficult. Butcher's boning gloves also have been used (18). A variety of restraint devices has been used, including wire sleeves (33) and tube devices which can also be used as chambers for the induction of inhalant anesthetics (16). Changing the cages of rats which jump actively can be difficult, and the task is easier if both the clean and dirty cages are placed in a high-walled changing container. Fall (34) used red light on both Norway and black rats and found that it markedly reduced or even eliminated the rats' nervousness and reactivity. Wild rats should be neither grouped together for bulk shipment nor paired or grouped indiscriminately after unpacking (23). Additional suggestions for management of wild rats in the laboratory will be found under the description of individual species.

III. WILD RATS USED IN RESEARCH

A. Norway Rat (*Rattus norvegicus*)

Norway rats, also known as brown rats (Fig. 1), belong to the family Muridae. They originated in the Far East and have been spread all over the world by commerce. They are large, weighing from 200 to 500 gm and are the progenitors of the laboratory rat.

King (59) kept wild Norway rats in cages for over 25 generations in a reenactment of domestication and found that the rats became more placid, attained greater body weight, had reduced levels of sterility and mortality, and reared their young more successfully with the progression of time. These changes were tentatively ascribed to genetic mutation. However, Calhoun (13) attributed such changes to selection for great tolerance to disturbance. He kept wild rats in a ¼ acre enclosure for 27 months to assess the social interactions and factors involved in the limitation of populations even when adequate food and harborage are supplied. The pen was enclosed with a

Fig. 1. Brown Norway rat (*Rattus norvegicus*). (Courtesy of Mr. Robert H. Benson, Veterinary Resources Branch, National Institutes of Health, Bethesda, Maryland.)

4-ft high mesh fence extending 2-ft underground with an inward directed 2-ft wide buried wire shelf. Electric fences on the inside prevented rats climbing out, on the top of the fence they prevented cats climbing in, and on the outside prevented dogs from coming too close. Traps around the perimeter and on top of the fence posts deterred other predators. The rats were fed commercial laboratory rodent diet with occasional garbage. Periodically, the rats were trapped. Calhoun found that it was easiest to get an old rat out of a trap by tipping it downward into a bag. Conversely, young rats ran out of the traps better if the traps were tilted up and the bag held above the trap level. Socially dominant Norway rats are neophobic and Calhoun found that there was a tendency for socially low-ranking rats to show reduced avoidance of traps. Other examples of low-rank attributes were slow growth rate, low adult weight, and tendencies to harbor in unfavorable sites and to form nonreproducing aggregates. Barnett (4) extensively reviewed rat behavior and such attributes as poison shyness and dietary self-selection which operate to the advantage of rats.

Boice and Boice (8) solved the problem of neophobia by trapping rats in a garbage dump which was bulldozed daily. Of the trapped rats, they selected those that did not have any scars or bite wounds because they found that unscarred rats would breed and rear offspring as efficiently as normal laboratory rats . They examined and sexed the rats after trapping by tipping them into heavy transparent plastic bags. If the offspring of rats were handled often enough, starting before weaning, it was possible to eventually handle an adult without gloves.

Thompson (107) kept individual wild rats in cages 12 × 12 × 7 inches and groups of rats in 36 × 24 × 12 inch cages. He gave straw for refuge to newly trapped rats and found that most acclimated within 1 week of trapping. However, some rats, particularly weanlings, would not adapt and starved to death. Connor (19) kept rats in 36-ft diameter metal corn cribs which had concrete footings 8 ft down into the ground. Straw was provided for cover, and the rats dug down 6 ft in the loose earth. He felt that putting a concrete floor in the structures would have created problems with drainage. Bowerman (11) kept captive rats in standard laboratory cages on standard food but found that the breeding efficiency was poor. He had also used larger cages (36 × 20 × 12 inches) with nest boxes for breeding, and kept large groups in pens with steel walls and 25 to 450 ft² of floor space. Jackson (54) kept rats in wire mesh cages or free living areas and fed them on commercial laboratory rat diet. He felt that breeding efficiency was improved by minimizing the intensity of social contact between rats. Polycarbonate or polypropylene plastic boxes were not used, since rats could easily chew their way out, once they got a start on a crack or damaged surface. Marsh (65) gave details of a multicompartment box for use in outdoor behavioral studies.

The majority of research involving Norway rats has been directed toward population control of these economically important pests. Calhoun (13) studied the social interactions limiting population size. The problem of resistance to anticoagulant rodenticides has been studied extensively (55, 63).

B. Black Rat (*Rattus rattus*)

The black or roof rat, which also belongs to the family Muridae, is found along Southern and Pacific coasts and waterways of the United States. It is stated to be more nervous, excitable, and faster, but less aggressive than the Norway rat. The black rat occurs in tropical–subtropical countries especially along the coasts and waterways because of its association with ships. Adult body weight is 115 to 350 gm.

In England, some black rats were captured at the docks and kept in captivity for several generations (58). Two foot square wire mesh cages with a wooden nest box inside were satisfactory for permanent pair breeding. Because of wartime restrictions invoked at the time, the rats were fed only on whole wheat grains, boiled potatoes, and table scraps, and it was necessary to keep food in front of them at all times to prevent chewing on the mesh. Kelway, who caught these rats by the base of the tail, was bitten many times, but maintained that the rats became gentler with succeeding generations. Grasping the body alarmed them too much so they were examined in small mesh traps. Powell (80) trapped black rats using sunflower seed and rolled oats as baits and found that they adapted well to laboratory life and thrived on commercial rat food. Metal mesh cages 8 × 16¼ × 9½ inches were adequate for breeding purposes. The provision of nesting material was not necessary. Sacher (90) found that laboratory rat diet was satisfactory as a maintenance diet. He was unable to improve the rather low weaning rates by feeding high protein dog food. Medina and associates (68) kept *Rattus rattus mindanensis*. They also found the weaning rate was low. Only 62% of offspring born were weaned, and average litter size was only 5. The rate was not improved by switching from a fruit and vegetable diet to one of high protein chicken diet. It was difficult to pair the rats, since the female usually killed strange males. They devised a system of putting one female with several males and then isolating the female and one of the males several days later. Black rats will not hybridize with either the wild or the laboratory Norway rat (15). They have been used in studies of behavior (80), anticoagulant rodenticides (40), and life shortening by chronic gamma irradiation (91).

C. Polynesian Rat (*Rattus exulans*)

The Polynesian rat similarly belongs to the family Muridae, occurs on nearly every Pacific island that lies within 30° of the Equator, and has also extended its range south to New Zea-

land. It causes great economic loss in cultivated sugar cane, rice, and coconut plantations. Williams (114) reviewed the ecology of the Polynesian rat and its relationship to *R. rattus* and *R. norvegicus*. Storer (102) edited the results of a Pacific island rat ecology study. The body weight of mature animals ranged from 70 to 150 gm. They have been kept successfully in the laboratory. Wirtz (116) kept breeding animals in hardware cloth cages, with nest boxes for pregnant females. The rats were fed on laboratory rat diet and greens and reproduced successfully. Egoscue (29) kept monogamous pairs, which were paired at 1 to 2 months and left together for the rest of their lives. Animals were housed in steel boxes with wood chip bedding and fed on commercial food plus a grain mixture. They were handled with rubber tip forceps by the base of the tail.

The main reason for keeping these rats in laboratories has been to develop control techniques. For example, Garrison and Johns (38) described the effect of an antifertility drug. Polynesian rats are vectors for the typhus-carrying mite (1) and the plague flea (32).

D. Tree Rat (*Thamnomys surdaster*)

This small rat belongs to the family Muridae, and the four species of this genera may be found from Ghana to Western Uganda. The body weight of mature animals is 55 to 65 gm. It has been raised successfully for many generations in laboratory cages. It is nocturnal in habit. Straw bedding with glass jar refuges has been employed successfully. It is reputed to be easy to handle. As long as the pairs are mated by about 6 weeks of age, they are compatible. The young are born in litters of 2 to 4 and hang onto the mother's nipples for about 2 weeks. This rat is the natural host for *Plasmodium berghei*, the most commonly used parasite for malaria research (123). It also has been used in the study of circadian rhythm (47).

E. Multimammate Rat [*Mastomys (Praomys) natalensis*]

This 40- to 80-gm rodent is called the multimammate mouse, by the South Africans who first developed it as a laboratory animal (21). It used to be included in the genus *Rattus*, but now *Mastomys* is considered the more appropriate genus. Other workers refer to it in the vernacular as *Praomys* or the multimammate rat. It belongs to the family Muridae and is found in Africa, particularly the sub-Sahara regions (Fig. 2).

There are 8 to 12 pairs of mammae distributed evenly along the abdomen from the pectoral to the inguinal region, and the female has a well-developed prostate gland. Standard rodent boxes with wood chip bedding and commercial laboratory rat food are satisfactory for keeping these animals. Some

Fig. 2. Multimammate rat (*Mastomys natalensis*). (Courtesy of Dr. John D. Strandberg, Johns Hopkins School of Medicine, Baltimore, Maryland.)

Fig. 3. Cotton rat (*Sigmodon* sp.). (Courtesy of Mr. Robert H. Benson, Veterinary Resources Branch, National Institutes of Health, Bethesda, Maryland.)

laboratories have fed carrots as a supplement. These animals reproduce well if kept in monogamous pairs and have litters of six to eight. They will mate at the postparturient estrus and produce 0.12 babies/female/day (22).

An attempt by Snell at inbreeding failed (96). Life span is around 2 to 3 years. Veenstra (109) indicated that these animals are not aggressive to their own or other species in the wild, but they will bite without provocation in the laboratory. They can be tamed, but this requires considerable and continued handling.

An early report (74) indicated an unusually high incidence of gastric adenocarcinomas in mature laboratory-raised animals; however, it has since been shown that these tumors are actually carcinoids (97). The frequency varies in different laboratories.

Snell (98) reviewed the spontaneous diseases occurring in these animals, which include degenerative joint disease (99), renal disease, thymomas, hematopoietic neoplasms, and polymyositis. This animal has also been used in the study of spontaneous epithelial skin tumors (89), pancreatic tumors (51), and prostatic tumors (50).

In its natural habitat, it ranks as a major agricultural pest. It is a reservoir of plague in Southern Africa and is used for routine plague testing there. In Western Africa, it is the only known natural host of the arenavirus which causes Lassa fever in man (71).

Since the discovery of the susceptibility of this animal to experimental schistosomiasis (64), it has been used extensively in parasitological studies, e.g., *Brugia* filariasis (78), *Litosomoides carinii* (rodent filariasis) (60), toxoplasmosis (113), and schistosomiasis (95).

F. Cotton Rat (*Sigmodon* spp.)

Cotton rats (Fig. 3) belong to the family Cricetidae, subfamily Cricetinae, and the New World rat tribe. They occur in a region extending across central America and into southern North America. The one most available for study in the United States is *Sigmodon hispidus*, which occurs in the southern states. Wild cotton rats weigh 90–110 gm, but they may reach 150–200 gm in captivity. A production colony of *Sigmodon hispidus* cotton rats from Florida was established by Meyer and Marsh (69) in support of a poliovirus project. They kept them in wire mesh cages with enclosed nest boxes and fed them on mixtures of one-quarter sunflower seeds and three-quarters rolled oats, and also one-half proprietary mouse food and one-half fox food. Fruit and vegetables were given as supplements. Permanent pairs were established at 6 to 7 weeks of age. The estrous cycle was found to be 4 to 8 days, gestation was 26 to 28 days, and the average litter size was 5. Mortality resulted if the rats temporarily ran out of food.

Baker (3) kept six species of cotton rats in wire mesh cages or standard plastic rat boxes. Tin cans for refuges were not given (cotton rats defecate in them), and the rats made nests out of commercial nesting products (Nestlets). *Sigmodon hispidus, S. mascotensis,* and *S. fulviventer,* all bred satisfactorily; *S. alleni,* and *S. leucotus* bred fairly well, but *S. ochrognathus* was not productive. Heavy gloves were used to handle the rats around the thorax, and cages were changed in a high-walled changing box because cotton rats jump actively, especially *S. hispidus saturatus* (23). If the young are allowed to remain with the adults until they reach puberty, there may be cannibalism. Crowded conditions and unsupervised pairing also will result in mortality. Individual cannibals exist among cotton rats, and these should be culled from breeding colonies.

The cotton rat has been used in studies on filariasis (82), echinococcosis (72,92,100), Venezuelan equine encephalitis (52), murine typhus (121), poliomyelitis (77), dental caries (35), calcinosis (9), and behavioral comparisons (80).

G. Wood Rat (*Neotoma* spp.)

Wood rats belong to the family Cricetidae, subfamily Cricetinae, and the New World rat tribe. The mature body weight of a wood rat is 200 to 430 gm. depending on the species. They also are called pack or trade rats, derived from their habit of carrying objects around, dropping them, and trading for others. This rat is found only in North America. According to Worth (119), *N. floridana* is easy to handle, and gloves are rarely necessary, but Dewsbury (26) used metal mesh gloves. Poole (79) found that they were lethargic and fearless during the daytime, but that at night they were very active, alert, and fast moving. Linsdale and Tevis (62) give detailed descriptions on how to handle wild-caught wood rats directly from the traps. Laboratory-maintained rats can be transferred from cage to cage via the refuge can. Plastic rat boxes or 20 × 22 × 14 inch wire mesh cages have been used successfully, and the rats may be maintained on standard laboratory rat diets. Fighting is infrequent, even in overcrowded or multigeneration cages, and males do not interfere with the young which are dragged around attached to the mother's teats for 2 to 3 weeks. The incisor teeth of nurslings are splayed out for the first 2 weeks of life, and the mother's nipple fits in the gap (45). These rats breed well in captivity, but since they only produce 2–3 young per litter, they are not as prolific as standard laboratory rats. They can be kept on reverse light cycles if behavioral studies must be conducted during normal working hours (27).

Wood rats have been used in the study of murine typhus (119), food preferences (46), karyological divergence (67), copulatory behavior (27), lens composition (83), renal physiology (93), and natural infection with *Brucella neotomae* (101). Detailed anatomical descriptions were made by Howell (53).

H. Cane Rat (*Zygodontomys* spp.)

Animals of the genus *Zygodontomys* may be called cane rats or cane mice and are found in Central America and Brazil. They belong to the family Cricetidae, subfamily Cricetinae, and New World rat tribe. Bates and Weir (5) kept *Z. microtinus* successfully on a homemade cereal–vegetable paste, and found the rats to be very healthy. Breeding was satisfactory, with sexual maturity occurring at 3 to 4 months. They had a gestation period of 28 days. Conception occurred at postparturient estrus, but Bates felt this led to the birth of unthrifty litters, so females were separated from the males as soon as pregnancy was obvious.

Worth (120) kept *Z. brevicauda* in cages 12 × 24 × 16 inches high with glass walls and mesh tops. Wood shavings were used for bedding and glass jars for refuges. Commercial fox diet supplemented with green vegetables was fed. Fighting was negligible, and the animals were not aggressive to handlers. Some difficulty in handling was experienced because of their habit of suddenly leaping out of the box. Females were

left with the male for 5 days and then isolated. Births occurred year round (outdoors in the tropics); average litter size was 3.7; and weaning time was 21 days. They live for 2 to 3 years, and may become obese if inactive.

Baker (3) kept cane rats in plastic rat boxes with refuge cans. Fed on laboratory rat food, the rats bred satisfactorily. Dawson (23) found the rats easy to handle, but recognized that the tail skin slips readily. Many rats in his colony had shortened stubby tails because the rats had chewed off the tail vertebrae following a tail-slip.

Cane rats have been used in the study of yellow fever (5), Venezuelan equine encephalomyelitis (28), Cocal virus (57), and Nariva virus (108).

I. Rice Rat (*Oryzomys* spp.)

The rice rat belongs to the family Cricetidae, New World rat tribe in the subfamily Cricetinae, and about 100 species have been described in all of the subgenera (110). Most of them are found in the Neotropics but the Atlantic and Gulf coast states of the United States also are inhabited. The weight may range from 25 to 150 gm. They can be kept in standard plastic rat boxes or mesh cages, should be provided with a shreddable nesting material, and fed on standard rat food (26). Worth (120) kept them in glass-walled boxes with refuge jars and fed them on fox diet and vegetables. Female rice rats often are belligerent toward the male, and two rats never share a nest or refuge jar. They are fairly "jumpy" animals. Litters average 3.5, and weaning can be done at 21 days. Worth (120) felt that they were troublesome, but not impossible, to rear. Rice rats have been used in the study of arboviruses (57), copulatory behavior (25), dental development (75), and murine typhus (119).

J. Climbing Rat (*Tylomys nudicaudus*)

The climbing rat belongs to the New World rat tribe in the family Cricetidae and inhabits junglelike forests of Central America and Ecuador. The body weight of a mature rat varies from 200 to 300 gm. These rats adapt easily to life in a laboratory. Standard-sized rat boxes are adequate for a breeding pair, with sawdust or woodchip bedding and cans for refuges. Laboratory rodent diet with supplements of fruit and vegetables is satisfactory (49,105). Bananas and apples are the favorite fruits. The climbing rat is very tame and can be handled easily by the tail; however, it resents being loosely held around the thorax and will bite. The young are born in litters of 1 to 4, with a gestation period of 39 days. They hang onto the mothers nipples for 2 to 3 weeks, and go wherever the mother goes since she does not build a nest.

These animals have been used in the study of spermatozoa (48), chromosomes (76), leishmaniasis (106), and arthropod-borne viruses (105).

K. African White-Tailed Rat (*Mystromys albicaudatus*)

This animal (Fig. 4) represents the only species of *Mystromys*, in the tribe of hamsters, which belong to the subfamily Cricetinae and family Cricetidae. The dry sandy region of Southern Africa is the habitat. The body weight of a mature animal varies from 75 to 185 gm. Breeding pairs should be mated when young, since the adults may fight. Standard plastic rodent boxes with processed corn cob bedding are satisfactory. Commercial rodent food can be offered *ad libitum*. Litter size averages 3, gestation is 38 days, and the young hang onto the nipples until 2 to 3 weeks of age. These rats should not be handled by the tail because it is fragile. Instead they should be picked up around the thorax, and the use of gloves is recommended (44).

The white-tailed rat has been used extensively in the study of spontaneous diabetes mellitus (87,104). Hyperglycemia is a consistent finding. Polyuria, polydipsia, glycosuria, and ketonuria are found less commonly, and obesity is not a characteristic. It also has been used as an experimental host for *Leishmania donovani* (70), the study of dental lesions (73), and spontaneous tumor development (84).

L. Sand Rat (*Psammomys obesus*)

The sand rat (Fig. 5) belonging to the family Cricetidae, subfamily Gerbillinae, weighs about 150 to 180 gm, and inhabits desert areas of the Middle East. In early attempts at laboratory rearing, efforts were made to duplicate the natural habitat and diet. Prange and associates (81) kept breeding pairs in 3 × 6 ft boxes with artificial tunnels, and nonbreeding animals in wire cages with activity wheels. In general, they felt that the animals lived better in solid-bottomed cages. The diet was 50 gm of vegetables (beet and spinach leaves) and 5 gm of laboratory rat food per animal per day. A 1.5% NaCl solution was offered as drinking water. Periodic vitamin supplements were felt to have beneficial results.

Strasser (103) kept his breeding animals in large boxes with a floor area of 2.25 m². The floor was interlaced with artificial tunnels. He fed them twice a day with mixed fruit, vegetables sprinkled with sea salt, chopped insects, and periodic vitamin supplements. Frenkel *et al.* (37) established a successful breeding program with rats housed in standard plastic boxes.

Fig. 4. African white-tailed rat (*Mystromys albicaudatus*). (a) Mother with 3-week-old pups. (b) Three-month-old female. (Courtesy of Dr. George Migaki, Armed Forces Institute of Pathology, Washington, D.C.)

Fig. 5. Sand rat (*Psammomys obesus*). (Courtesy of Drs. Knut Schmidt-Nielsen and Joseph Wagner, Duke University, Durham, North Carolina.)

Proestrous females, their condition detected by the vaginal smear technique, were placed with a male for 30 min, three times a day until mating took place. Sand rats are solitary creatures, and if the female is not receptive, serious fighting may occur between a breeding pair. Single sex pairs were housed in 35 × 35 cm boxes, and larger ones were used for females with litters. The diet was alfalfa pellets with 3% sea salt solution for drinking and weekly supplements of salt-bush, ground meat, and vitamins. In addition, one pellet of regular rat food was allowed per animal per day. The average litter size is 3.6, and gestation is 24 days.

The sand rat has been used in the study of diabetes mellitus (41,94). Animals kept in captivity with abundant food tend to develop mild to moderate obesity, and marked hyperglycemia with eventual ketoacidosis may also occur. Sand rats appear to do well on a low food–high exercise regimen and do not adapt well to an affluent environment. Renold (86) reviewed the pathology of sand rat diabetes. Lebovitz et al. (61) found that the sand rat pancreas secretes remarkably more insulin than most other species. Rouffignac et al. (88) have used the sand rat in the study of renal physiology.

M. Kangaroo Rat (*Dipodomys* spp.)

Kangaroo rats belong to the family Heteromyidae and live in arid or semiarid areas of the New World. The weight ranges from 30 to 140 gm depending on the species. They are nocturnal, have fur-lined external cheek pouches, and store food. Their fur gets matted and unkempt if they do not have access to dry earth or sand. The tail and hindlegs are extremely long, and they have been called "bipedal ricocheters." The nape of the neck is the safest place to hold them, but the hindlegs should be held as well to prevent kicking. If they are held by the tail, the skin is likely to slip off. The diet under laboratory conditions is usually a mixture of grains and seeds with lettuce as a water source. Eisenberg (30) advises against giving free water, since most animals ignore it, and a few may become compulsive drinkers.

There is a high degree of intraspecific aggressiveness, and the animals seem unable to tolerate each other, but they are not aggressive toward humans. They can be maintained in large cages, but it is extremely difficult to get them to breed. Bailey (2) managed to get one litter of *D. spectabilis* (banner-tailed kangaroo rat) in captivity. Culbertson (20) put six adult males of *D. nitroides* (Fresno kangaroo rat) in a 23 × 35 inch box; they fought for several hours and then settled down. He found that young ones did not fight so much when grouped, but he was unable to get any to breed in captivity.

Chew (17) tried to pair *D. merriami* (Merriam's kangaroo rat) in a 10 × 16 × 11 inch sand-floored cage. He switched from the traditional diet of pearl barley and sunflower seeds

with lettuce to a more complex diet. The kangaroo rats gained weight on this diet, and the biggest female produced a litter. Butterworth (12) succeeded in producing one litter of *D. deserti* by putting two animals in a sand-floored room 6 × 8 m. They fought for several hours, and after 5 months produced a litter, followed by six more litters. *Dipodomys deserti* is a very solitary animal and fights more than other kangaroo rats.

Jollie (56) tried to detect estrus using vaginal smears and then paired *D. ordii* (Ord's kangaroo rat) at estrus when the female was least aggressive toward the male. One litter was produced. The vaginal smear at estrus was then correlated with vulval enlargement, and any female showing vulval enlargement was presumed to be in heat and paired with a male in new territory. In this way, six females produced ten litters (24).

Eisenberg and Isaac (31) tried three methods of pairing. One method was putting two animals in a 5 × 7 × 2 ft cage until they bred. Another was placing two animals in a 6 ft² enclosure for 30 to 60 min. The third method involved maintaining the animals either side of a wire partition and opening the partition for a short while each day.

It can be seen that breeding kangaroo rats in captivity is not easy. The gestation period is 29 to 33 days, and 1 to 3 offspring are born in each litter.

Kangaroo rats show an unusual degree of economy in all aspects of water metabolism. They rarely drink even if water is available. Some species are able to survive on the metabolic water contained within air-dried food. They avoid the use of water as an evaporation–heat loss mechanism by staying out of the heat, living in long burrows, and being nocturnal.

Kangaroo rats have been used in the study of renal physiology and water conservation (93), whole-body irradiation (42), and psychotropic drug effects (43).

N. Trumpet-Tailed Rat (*Octodon degus*)

This rodent (Fig. 6) is commonly called "Degu"; it belongs to the family Octodontidae. It inhabits the coastal areas and foot hills of the Andes in South America. The body weight of a mature animal is 200 to 300 gm. It is gentle enough to be held without gloves, if it has been handled from birth. If handled ineptly by the tail, the skin may slip off and the handler is likely to be bitten. Males fight very little if housed together. Commercial diet with vegetable supplements is suitable, but the animals should not be fed *ad libitum,* because obesity is likely (112).

The gestation period is 90 days with 5 to 6 pups per litter (112) and two litters per year. The females have vaginal closure membranes which are open only during estrus and parturition. These animals have been used in investigating developmental patterns (122), reproductive and play behavior (115), the role of the thymus in immune reactions (10), genetic variation of enzymes (14), hystricomorph urine composition (6),

Fig. 6. Trumpet-tailed rat (*Octodon degus*). (a) Mother and weanling pup. (Courtesy Barbara J. Weir, Wellcome Institute of Comparative Physiology, London, England.) (b) Adult female. (Courtesy of Jerry Gordon, Division of Research Resources, National Institutes of Health, Bethesda, Maryland.)

congenital cataracts (117), and effect of chemotoxic agents on ocular tissue (111).

ACKNOWLEDGMENTS

The author wishes to thank the many investigators, particularly Dr. R. H. Baker of the Michigan State University Museum, who have been most helpful in supplying information regarding colonies, management practices, and personal experiences with wild rats in laboratories. Without their expert and willing assistance, it would have been impossible for her to write this chapter.

REFERENCES

1. Audy, J.R., and Harrison, J.R. (1951). A review of investigations on mite typhus in Burma and Malaya, 1945-1950. *Trans. R. Soc. Trop. Med. Hyg.* **44**, 371-404.

2. Bailey, V. (1931). Mammals of New Mexico. *North Am. Fauna* **53**, 247-259.

3. Baker, R.H. (1976). Personal communication.

4. Barnett, S.A. (1967). Rats. *Sci. Am.* **216**, 78-85.

5. Bates, M., and Weir, J.M. (1944). The adaptation of a cane rat (*Zygodontomys*) to the laboratory and its susceptibility to the virus of yellow fever. *Am. J. Trop. Med.* **24**, 35-37.

6. Bellamy, D., and Weir, B.J. (1972). Urine composition of some hystricomorph rodents confined to metabolism cages. *Comp. Biochem. Physiol. A,* **42**, 759-771.

7. Boice, R. (1971). Laboratorizing the wild rat (*Rattus norvegicus*). *Behav. Res. Methods Instrum.* **3**, 177-182.

8. Boice, R., and Boice C. (1968). Trapping Norway rats in a landfill. *J. Sci. Lab., Denison Univ.* **49**, 1-4.

9. Bone, J.F., Roffler, S.A., and Weswig, P.H. (1965). Incidence and histopathology of calcinosis in cotton rats. *Lab. Anim. Care* **15**, 81-93.

10. Boraker, D. (1975). Ontogenic studies of antigen-binding cells in the dual thymus glands of the South American rodent, *Octodon degus. Am. Zool.* **15**, 181-188.

11. Bowerman, A.M. (1976). Personal communication.

12. Butterworth, B.B. (1961). The breeding of *Dipodomys deserti* in the laboratory. *J. Mammal.* **42**, 413-414.

13. Calhoun, J.B. (1962). Ecology and sociology of wild Norway rat. *U.S., Public Health Serv. Pub.* **1008.**

14. Carter, N.D., Hill, M.R., and Weir, B.J. (1972). Genetic variation of phosphoglucose isomerase in some hystricomorph rodents. *Biochem. Genet.* **6**, 147-156.

15. Castle, W.E. (1947). The domestication of the rat. *Proc. Natl. Acad. Sci. U.S.A.* **33**, 109-117.

16. Caudill, C.J., and Gaddis, S.E. (1973). A safe and efficient handling device for wild rodents. *Lab. Anim. Sci.* **23**, 685-686.

17. Chew, R.M. (1958). Reproduction by *Dipodomys merriami* in captivity. *J. Mammal.* **39**, 597-598.

18. Cisar, C.F. (1973). The use of a butcher's boning glove for handling small laboratory animals. *Lab. Anim.* **7**, 139-140.

19. Connor, N.D. (1976). Personal communication.

20. Culbertson, A.E. (1946). Observations on the natural history of the Fresno kangaroo rat. *J. Mammal.* **27**, 189-203.

21. Davis, D.H.S., and Oettle, A.G. (1958). The multimammate mouse *Rattus (Mastomys) natalensis* Smith: A laboratory-adapted African wild rodent. *Proc. Zool. Soc. London* **131**, 293-299.

22. Davis, D.H.S. (1963). Wild rodents as laboratory animals and their contribution to medical research in South Africa. *S. Afr. J. Med. Sci.* **28**, 53-69.

23. Dawson, G.A. (1976). Personal communication.

24. Day, B.N., Egoscue, H.J., and Woodbury, A.M. (1956). Ord kangaroo rat in captivity. *Science* **124**, 485-486.

25. Dewsbury, D.A. (1970). Copulatory behaviour of rice rats (*Oryzomys palustris*). *Anim. Behav.* **18**, 266-275.

26. Dewsbury, D.A. (1974). The use of muroid rodents in the psychology laboratory. *Behav. Res. Methods Instrum.* **6**, 301-308.

27. Dewsbury, D.A. (1974). Copulatory behavior of *Neotoma floridana. J. Mammal.* **55**, 864-866.

28. Downs, W.G., Spence, L., and Aitken, T.H.G. (1962). Studies on the virus of Venezuelan equine encephalomyelitis in Trinidad, W. I. III. Reisolation of virus. *Am. J. Trop. Med. Hyg.* **11**, 841-843.

29. Egoscue, H.J. (1970). A laboratory colony of the polynesian rat, *Rattus exulans. J. Mammal.* **51**, 261-266.

30. Eisenberg, J.F. (1976). The heteromyid rodents. *In* "The UFAW Handbook on the Care and Management of Laboratory Animals" (Universities Federation for Animal Welfare, ed.), 5th ed., pp. 293-297. Churchill-Livingstone, London.

31. Eisenberg, J.F., and Isaac, D.E. (1963). The reproduction of heteromyid rodents in captivity. *J. Mammal.* **44,** 61–67.

32. Elbel, R.E., and Thaineua, M. (1957). A flea and rodent control program for plague prevention in Thailand. *Am. J. Trop. Med. Hyg.* **6,** 280–293.

33. Emlen, J.T. (1944). Device for holding live wild rats. *J. Wildl. Manage.* **8,** 264–265.

34. Fall, M.W. (1974). The use of red light for handling wild rats. *Lab. Anim. Sci.* **24,** 686–687.

35. Feller, R.P., Edmonds, E.J., Shannon, I.L., and Madsen, K.O. (1974). Significant effect of environmental lighting on caries incidence in the cotton rat. *Proc. Soc. Exp. Biol. Med.* **145,** 1065–1068.

36. Food and Agriculture Organization, World Health Organization (1973). Bibliography on Rodent Pest Biology and Control 1950–1959 and 1960–1969. FAO/WHO, Rome.

37. Frenkel, G., Shaham, Y., and Kraicer, P.F. (1972). Establishment of conditions for colony-breeding of the sand rat *Psammomys obesus. Lab. Anim. Sci.* **22,** 40–47.

38. Garrison, M.V., and Johns, B.E. (1975). Anti-fertility effects of SC-20775, 17-α-ethynyl-11-β-methylestradiol-3-cyclopentyl ether in Norway and Polynesian rats. *J. Wildl. Manage.* **39,** 26–29.

39. Giles, R.H., Jr., ed. (1972). "Wildlife Management Techniques," 3rd ed. Wildlife Society, Washington D.C.

40. Girish, G.K., Singh, K., Srivastava, P.K., and Krishnamurthy, K. (1972). Studies on rodents and their control. Part VIII. Susceptibility of *Rattus rattus* to different anticoagulants. *Bull. Grain Technol.* **10,** 113–115.

41. Hackel, D.B., Schmidt-Nielsen, K., Haines, H.B., and Mikat, E. (1965). Diabetes mellitus in the sand rat (*Psammomys obesus*). Pathologic studies. *Lab. Invest.* **14,** 200–207.

42. Haley, T.J., Lindberg, R.G., Flesher, A.M., Raymond, K., McKibben, W., and Hayden, P. (1960). Response of the kangaroo rat (*Dipodomys merriami* Mearns) to single whole-body X-irradiation. *Radiat. Res.* **12,** 103–111.

43. Haley, T.J., and Mavis, L. (1964). Effect of psychotropic drugs on the kangaroo rat. *Neuro-Psychopharmacol. Proc. Meet. Coll. Int. Neuro-Psychopharmacol. 3rd, 1962* pp. 55–58.

44. Hall, A., III, Persing, R.L., White, D.C., and Ricketts, R.T., Jr. (1967). *Mystromys albicaudatus* (the African white-tailed rat) as a laboratory species. *Lab. Anim. Care* **17,** 180–188.

45. Hamilton, W.J., Jr. (1953). Reproduction and young of the Florida wood rat *Neotoma f. floridana* (Ord). *J. Mammal.* **34,** 180–189.

46. Harriman, A.E. (1974). Self-selection of diet by Southern Plains wood rats (*Neotoma micropus*). *J. Genet. Psychol.* **90,** 53–61.

47. Hawking, F., Lobban, M.C., Gammage, K., and Worms, M.J. (1971). Circadian rhythms activity, temperature, urine and microfilariae in dog, cat, hen, *Thamnomys* and *Gerbillus. J. Interdiscip. Cycle Res.* **2,** 455–473.

48. Helm, J.D., III, and Bowers, J.R. (1973). Spermatozoa of *Tylomys* and *Ototylomys. J. Mammal.* **54,** 769–772.

49. Helm, J.D., III, and Dalby, P.L. (1975). Reproductive biology and postnatal development of the neotropical climbing rat, *Tylomys. Lab. Anim. Sci.* **25,** 741–747.

50. Holland, J.M. (1970). Prostatic hyperplasia and neoplasia in female *Praomys (Mastomys) natalensis. J. Natl. Cancer Inst.* **45,** 1229–1236.

51. Hosoda, S., Suzuki, H., and Suzuki, M. (1976). Spontaneous tumors and atypical proliferation of pancreatic acinar cells in *Mastomys natalensis. J. Natl. Cancer Inst.* **57,** 1341–1346.

52. Howard, A.T. (1974). Experimental infection and intracage transmission of Venezuelan equine encephalitis virus (subtype 1B) among cotton rats, *Sigmodon hispidus* (Say and Ord). *Am. J. Trop. Med. Hyg.* **23,** 1178–1184.

53. Howell, A.B. (1926). "Anatomy of the Wood Rat." Williams & Wilkins, Baltimore, Maryland.

54. Jackson, W.B. (1976). Personal communication.

55. Jackson, W.B., and Kaukeinen, D. (1972). The problem of anticoagulant rodenticide resistance in the United States. *Proc. Vertebr. Pest Control Conf., 5th, 1972* pp. 142–148.

56. Jollie, W.P. (1956). Rearing the pallid Ord kangaroo rat in the laboratory. *In* "Symposium on Ecology of Disease Transmission in Native Animals." pp. 54–56. Army Chem. Corps, Dugway, Utah.

57. Jonkers, A.H., Spence, L., Downs, W.G., and Worth, C.B. (1964). Laboratory studies with wild rodents and viruses native to Trinidad. II. Studies with the Trinidad Caraparu-like agent TRVL 34053-1. *Am. J. Trop. Med. Hyg.* **13,** 728–733.

58. Kelway, P. (1947). The black rat. *In* "The UFAW Handbook on the Care and Management of Laboratory Animals" (A.N. Worden, ed.), pp. 140–149. Williams & Wilkins, Baltimore, Maryland.

59. King, H.D. (1939). Life processes in gray Norway rats during fourteen years in captivity. *Am. Anat. Mem.* **17,** 1–77.

60. Laemmler, G., and Gruener, D. (1975). Chemotherapeutic studies on *Litosomoides carinii* infection of *Mastomys natalensis. Tropenmed. Parasitol.* **26,** 359–369.

61. Lebovitz, H.E., White, S., Mikat, E., and Hackel, D.B. (1974). Control of insulin secretion in the Egyptian sand rat (*Psammomys obesus*). *Diabetologia* **10,** Suppl., 679–684.

62. Linsdale, J.M., and Tevis, L.P., Jr. (1951). "The Dusky-Footed Wood Rat." Univ. of California Press, Berkley and Los Angeles.

63. Lund, M. (1967). Resistance of rodents to rodenticides. *World Rev. Pest Control* **6,** 131–138.

64. Lurie, H.I., and de Meillon, B. (1956). Experimental bilharziasis in laboratory animals. II. A comparison of the pathogenicity of *S. bovis,* South African and Egyptian *S. mansoni* and *S. hematobium. S. Afr. Med. J.* **30,** 79–82.

65. Marsh, R.E. (1968). A colony nest box for rearing wild rats. *Lab. Anim. Care* **18,** 639–641.

66. Marsh, R.E., and Howard, W.E. (1974). "Rodent Control Manual," insert in *Pest Control* **42.**

67. Mascarello, J.T., Warner, J.W., and Baker, R.J. (1974). A chromosome banding analysis of the mechanisms involved in the karyological divergence of *Neotoma phenax* (Merriam) and *Neotoma micropus* (Baird). *J. Mammal.* **55,** 831–834.

68. Medina, F.I.S., Madriaga, C.L., de la Cruz, B., Paguirigan, E., Padrelanan, E., Eslit, S., and Gregorio, J. (1973). A study on the adaptation and breeding of field rats (*Rattus rattus* sp.) in the laboratory. *Philipp. Agric.* **56,** 274–279.

69. Meyer, D.B., and Marsh, M. (1943). Development and management of a cotton rat colony. *Am. J. Public Health* **33,** 697–700.

70. Mikhail, J.W., and Mansour, W.S. (1973). *Mystromys albicaudatus,* the African white-tailed rat, as an experimental host for *Leishmania donovani. J. Parasitol.* **59,** 1085–1087.

71. Monath, T.P. (1975). Lassa fever. Review of epidemiology and epizootiology. *Bull. W.H.O.* **52,** 577–592.

72. Norman, L., and Kagan, I.G. (1961). The maintenance of *Echinococcus multilocularis* in gerbils (*Meriones unguiculatus*) by intraperitoneal injection. *J. Parasitol.* **47,** 870–874.

73. Ockerse, T. (1956). Experimental periodontal lesions in the white-tailed rat in South Africa. *J. Dent. Res.* **35,** 9–15.

74. Oettle, A.G. (1957). Spontaneous carcinoma of the glandular stomach in *Rattus (Mastomys) natalensis,* an African rodent. *Br. J. Cancer* **11,** 415–433.

75. Park, A.W. (1974). Biology of the rice rat (*Oryzomys palustris natator*) in a laboratory environment. *Acta Anat.* **87,** 433–446.

76. Pathak, S., Hsu, T.C., Shirley, L., and Helm, J.D., III (1973).

Chromosome homology in the climbing rats, genus *Tylomys* (Rodentia: Cricetidae). *Chromosoma* **42,** 215–228.

77. Perkins, F.T. (1957). The antibody response of monkeys, guinea pigs and cotton rats to British poliomyelitis vaccine. *Br. J. Exp. Pathol.* **38,** 542–547.

78. Petranyi, G., Mieth, H., and Leitner, I. (1975). *Mastomys natalensis* as an experimental host for *Brugia malayi* subperiodic. *Southeast Asian J. Trop. Med. Public Health* **6,** 328–337.

79. Poole, E.L. (1936). Notes on the young of the Allegheny wood rat. *J. Mammal.* **17,** 22–26.

80. Powell, R.E. (1973). Laboratory study of wild rats. *Bull. Psychon. Soc.* **1,** 119–120.

81. Prange, H.D., Schmidt-Nielsen, K., and Hackel, D.B. (1968). Care and breeding of the fat sand rat (*Psammomys obesus* Cretzschmar). *Lab. Anim. Care* **18,** 170–181.

82. Pringle, G., and King, D.F. (1968). Some developments in techniques for the study of the rodent filarial parasite. I. A preliminary comparison of the host efficiency of the multimammate rat, *Praomys (Mastomys) natalensis,* with that of the cotton rat, *Sigmodon hispidus. Ann. Trop. Med. Parasitol.* **62,** 462–468.

83. Ramos, F., and Smith, A.C. (1975). Protein differences in the lens nucleus from the desert wood rat (*Neotoma lepida*). *Psychol. Rep.* **37,** 219–222.

84. Rantanen, N.W., and Highman, B. (1970). Spontaneous tumors in a colony of *Mystromys albicaudatus* (African white-tailed rat). *Lab. Anim. Care* **20,** 114–119.

85. Redfern, R., and Rowe, F.P. (1976). Wild rats and house mice. *In* "The UFAW Handbook on the Care and Management of Laboratory Animals" (Universities Federation for Animal Welfare, ed.), 5th ed., pp. 218–228. Churchill-Livingstone, London.

86. Renold, A.E. (1968). Spontaneous diabetes and/or obesity in laboratory rodents. *Adv. Metab. Disord.* **3,** 49–84.

87. Riley, T., Stuhlman, R.A., Van Peenen, H.J., Esterley, J.A., and Townsend, J.F. (1975). Glomerular lesions of diabetes mellitus in *Mystromys albicaudatus. Arch. Pathol.* **99,** 167–169.

88. Rouffignac, C. de, Morel, F., Moss, N., and Roinel, N. (1973). Micropuncture study of water and electrolyte movements along the loop of Henle in *Psammomys* with special reference to magnesium, calcium and phosphorus. *Pfluegers Arch.* **344,** 309–326.

89. Rudolph, R., and Thiel, W. (1976). Pathological anatomy and histology of spontaneous epithelial skin tumors in *Mastomys natalensis. Zentralbl. Veterinaermed., Reihe A* **23,** 429–441.

90. Sacher, G.A. (1976). Personal communication.

91. Sacher, G.A., and Staffeldt, E. (1971). Species differences in sensitivity of myomorph and sciuromorph rodents to life shortening by chronic gamma irradiation. *Proc. Natl. Symp. Radioecol. 3rd, 1971,* pp. 1042–1047.

92. Sakamoto, T., Yamashita, J., Ohbayashi, M., and Orihara, M. (1965). Studies on echinococcosis. XVI. Effects of drugs upon scolices and daughter cysts of *Echinococcus multilocularis in vitro. Jpn. J. Vet. Res.* **13,** 127–136.

93. Schmidt-Nielsen, K. (1964). "Desert Animals." Oxford Univ. Press, London and New York.

94. Schmidt-Nielsen, K., Haines, H.B., and Hackel, D.B. (1964). Diabetes mellitus in the sand rat induced by standard laboratory diets. *Science* **143,** 689–690.

95. Schuster, J., Laemmler, G., Rudolph, R., and Zahner, H. (1973). Pathophysiological and toxicological aspects of *Schistosoma mansoni* infection in *Praomys (Mastomys) natalensis* under treatment with hycanthone. *Z. Tropenmed. Parasitol.* **24,** 487–499.

96. Snell, K.C., and Stewart, H.L. (1967). Neoplastic and non-neoplastic renal disease in *Praomys (Mastomys) natalensis. J. Natl. Cancer Inst.* **39,** 95–117.

97. Snell, K.C., and Stewart, H.L. (1969). Malignant argyrophilic gastric carcinoids of *Praomys (Mastomys) natalensis. Science* **163,** 470.

98. Snell, K.C., and Stewart, H.L. (1975). Spontaneous diseases in a closed colony of *Praomys natalensis. Bull. W.H.O.* **52,** 645–650.

99. Sokoloff, L., Snell, K.C., and Stewart, H.L. (1967). Degenerative joint disease in *Praomys (Mastomys) natalensis. Ann. Rheum. Dis.* **26,** 146–154.

100. Sousa, O.E., and Thatcher, V.E. (1969). Observations on the life cycle of *Echinococcus oligarthrus* (Diesing, 1863) in the Republic of Panama. *Ann. Trop. Med. Parasitol.* **63,** 165–175.

101. Stoenner, H.G., and Lackman, D.B. (1957). A new species of *Brucella* isolated from the desert wood rat, *Neotoma lepida* Thomas. *Am. J. Vet. Res.* **18,** 947–951.

102. Storer, T. I., ed. (1962). Pacific Island rat ecology. *Bull. Bishop Mus., Honolulu* **225,** 1–274.

103. Strasser, H. (1968). A breeding program for spontaneously diabetic experimental animals: *Psammomys obesus* (sand rat) and *Acomys cahirinus* (spiny mouse). *Lab. Anim. Care* **18,** 328–338.

104. Stuhlman, R.A., Srivastava, P.K., Schmidt, G., Vordbedk, M.L., and Townsend, J.F. (1974). Characterization of diabetes mellitus in South African hamsters (*Mystromys albicaudatus*). *Diabetologia* **10,** Suppl., 685–690.

105. Tesh, R.B., and Cameron, R.V. (1970). Laboratory rearing of the climbing rat, *Tylomys nudicaudatus. Lab. Anim. Care* **20,** 93–96.

106. Thatcher, V.E., Eisenmann, C., and Hertig, M. (1965). Experimental inoculation of panamanian mammals with *Leishmania braziliensis. J. Parasitol.* **51,** 842–844.

107. Thompson, H.V. (1947). Wild brown rats. *In* "The UFAW Handbook on the Care and Management of Laboratory Animals" (A.N. Worden, ed.), p. 138. Williams & Wilkins, Baltimore, Maryland.

108. Tikasingh, E.S., Jonkers, A.H., Spence, L., and Aitken, T.G.H. (1966). Nariva virus, a hitherto undescribed agent isolated from the Trinidadian rat, *Zygodontomys b. brevicauda* (J.A. Allen and Chapman). *Am. J. Trop. Med. Hyg.* **15,** 235–238.

109. Veenstra, A.J.F. (1958). Behaviour of the multimammate mouse, *Rattus (Mastomys) natalensis* (A. Smith). *Anim. Behav.* **6,** 195–206.

110. Walker, E.P. (1975). "Mammals of the World," 3rd ed., Vol. II. Johns Hopkins, Baltimore, Maryland.

111. Weinsieder, A., Briggs, R., Reddan, J., Rothstein, H., Wilson, D., and Harding, C.V. (1975). Induction of mitosis in ocular tissue by chemotoxic agents. *Exp. Eye Res.* **20,** 33–44.

112. Weir, B.J. (1970). The management and breeding of some more hystricomorph rodents. *Lab. Anim.* **4,** 83–97.

113. Werner, H., and Egger, I. (1974). Experimental study of the activity of the sulfamethoxypyrazine with pyrimethamine on the cyst development phase of *Toxoplasma gondii* in *Mastomys natalensis. Zentralbl. Bakteriol., Parasitenkd., Infektionskr. Hyg., Abt. I: Orig., Reihe A* **226,** 554–560.

114. Williams, J.M. (1973). The ecology of *Rattus exulans* reviewed. *Pac. Sci.* **27,** 120–127.

115. Wilson, S.C., and Kleiman, D.G. (1974). Eliciting play: A comparative study. *Am. Zool.* **14,** 341–370.

116. Wirtz, W.O. (1973). Growth and development of *Rattus exulans. J. Mammal.* **54,** 189–202.

117. Worgul, B.V., and Rothstein, H. (1975). Congenital cataracts associated with disorganized meridional rows in a new laboratory animal: The degu. (*Octodon degus*). *Biomed. Express* **23,** 1–4.

118. World Health Organization. "Information Circular on Vector and Reservoir Control and Ecology." World Health Organ., Geneva.

119. Worth, C.B. (1950). Observations on the behavior and breeding of captive rice rats and wood rats. *J. Mammal.* **31,** 421–426.

120. Worth, C.B. (1967). Reproduction, development and behavior of captive *Oryzomys laticeps* and *Zygodontomys brevicauda* in Trinidad. *Lab. Anim. Care* **17,** 355–361.

121. Worth, C.B., and Rickard, E.R. (1951). Evaluation of the efficiency of the common cotton rat ectoparasites in the transmission of murine typhus. *Am. J. Trop. Med.* **31,** 295–298.

122. Wright, J.W. (1975). The degu broadens its role as a laboratory animal. *Lab. Anim.* **4** (2), 17–18 and 44.

123. Yoeli, M., Alger, N., and Most, H. (1963). Tree rat, *Thamnomys surdaster,* in laboratory research. *Science* **142,** 1585–1586.

Appendix 1

Selected Normative Data

Henry J. Baker, J. Russell Lindsey, and Steven H. Weisbroth

Adult weight	
Male	300–400 gm
Female	250–300 gm
Life Span	
Usual	2.5–3 years
Maximum reported	4 yrs. 8 mo.
Surface area	0.03–0.06 cm²
Chromosome number (diploid)	42
Water consumption	80–110 ml/kg/day
Food consumption	100 gm/kg/day
Body temperature	99.5°F, 37.5°C
Puberty	
Male	50 ± 10 days
Female	50 ± 10 days
Breeding season	None
Gestation	21–23 days
Litter size	8–14 pups
Birth weight	5–6 gm
Eyes open	10–12 days
Weaning	21 days
Heart rate	330–480 beats/min
Blood pressure	
Systolic	88–184 mm Hg
Diastolic	58–145 mm Hg
Cardiac output	50 (10–80) ml/min
Blood volume	
Plasma	40.4 (36.3–45.3) ml/kg
Whole blood	64.1 (57.5–69.9) ml/kg
Respiration frequency	85.5 (66–114)/min
Tidal volume	0.86 (0.60–1.25) ml

(continued)

(Continued)

Minute volume	0.073 (0.05–0.101) ml
Stroke volume	1.3–2.0 ml/beat
Plasma	
pH	7.4 ± 0.06
CO_2	22.5 ± 4.5 mM/liter
CO_2 pressure	40 ± 5.4 mm Hg
Leukocyte counts	
Total	14 (5.0–25.0) $\times 10^3/\mu$l
Neutrophils	22 (9–34)%
Lymphocytes	73 (65–84)%
Monocytes	2.3 (0–5)%
Eosinophils	2.2 (0–6)%
Basophils	0.5 (0–1.5)%
Platlets	1240 (1100–1380) $\times 10^3/\mu$l
Packed cell volume	46%
Red blood cells	7.2–9.6 $\times 10^6$/mm^3
Hemoglobin	15.6 gm/dl
Maximum volume of single bleeding	5 ml/kg
Urine	
pH	7.3–8.5
Specific gravity	1.04–1.07

Appendix 2

Drugs and Dosages

Sam M. Kruckenberg

Table I

Drug	Classification	Dosage	Comment	Reference[a]
Acetylsalicylic acid	Analgesic	450 mg/kg po	Analgesic dose	22
α-Prodine	Analgesic	1.5–6 mg/kg sc	Analgesic activity dosages	33
Aminopyrineone	Analgesic	25–100 mg/kg sc	Analgesic activity dosages	33
Codeine	Analgesic	6.25–25 mg/kg sc	Analgesic activity dosages	33
Codeine	Analgesic	20 mg/kg sc	Hypnotic, analgesic, and antitussive drug	34
d-Propoxyphene	Analgesic	25 mg/kg ip	Analgesic dosage	34
Levorphan	Analgesic	0.125–0.5 mg/kg sc	Analgesic activity dosages	33
Meperidine	Analgesic	6.25–25 mg/kg sc	Analgesic activity dosages	33
Morphine	Analgesic	1.5–6 mg/kg sc	Analgesic activity dosages	33
Nalorphine	Analgesic	1.0–2.1 mg/kg sc	Analgesic—in writhing tests	34
Nalorphine	Analgesic	5 mg/kg im, sc	Analgesic ED_{50}	34
Phenazocine	Analgesic	0.25–0.5 mg/kg sc	Analgesic dosage	34
Phenylbutazone	Analgesic	25–100 mg/kg sc	Analgesic activity dosages	33
Sodium salicylate	Analgesic	25–100 mg/kg sc	Analgesic activity dosages	33
α-Chloralose	Anesthetic	55 mg/kg ip	Anesthetic dose	22
Anesthetic drugs	Anesthetic	Various drugs and dosages	14 Drugs, dosages, routes and references, p. 19	1
Chloral hydrate[b]	Anesthetic	300 mg/kg ip	Anesthetic dose	22
Hexobarbital	Anesthetic	100 mg/kg ip	Anesthetic dose	15
Inactin	Anesthetic	160 mg/kg ip	Anesthetic dose for 150–400 gm rat	8
Innovar-Vet	Anesthetic	0.2–0.3 ml/kg im	Anesthetic dose	17
Innovar-Vet	Anesthetic	0.13 ml/kg im	Analgesia, sedation and tranquilization	23
Ketamine + atropine	Anesthetic	44 mg/kg ketamine im	0.04 mg/kg atropine given with ketamine	44
Ketamine hydrochloride	Anesthetic	44 mg/kg im	Surgical anesthesia	46
Methitural	Anesthetic	120 mg/kg ip	Anesthetic dosage	34
Pentobarbital-chlorpromazine	Anesthetic	25 mg/kg Chl. 20 mg/kg Pen.	Chlopromazine 30 min before pentobarbital	5

(continued)

THE LABORATORY RAT, VOL. II

Table I (*Continued*)

Drug	Classification	Dosage	Comment	Reference[a]
Pentobarbital	Anesthetic	30–40 mg/kg ip	Adult rat dosage	15
Pentobarbital	Anesthetic	10–30 mg/kg ip	Dosage for rats under 50 gm	5
Carbon dioxide	Anesthetic gaseous	Gaseous anesthetic	Quick knockdown for short-term procedures	15
Halothane	Anesthetic gaseous	1% with oxygen	No rats died when oxygen was given with gas	18
Methoxyflurane (Metofane)	Anesthetic gaseous	Inhalation	Multiple rat anesthetic apparatus	28
Atropine	Anesthetic (pre-)	0.04 mg/kg im	Used with ketamine to prevent salivation	46
Atropine	Anesthetic (pre-)	None given	Rat has a high atropinase level	25
Atropine	Anesthetic (pre-)	1–5 mg/kg sc	Preanesthetic	34
Atropine sulfate	Anesthetic (pre-)	2.5 mg/rat sc	Give 30 min before gas anesthesia	15
Chlordiazepoxide	Anesthetic (pre-)	10 mg/kg po	Preanesthetic dosage	10
Mikedimide	Anesthetic–antagonist	1 mg/mg of pentobarbital	Barbiturate antagonist–respiratory stimulant	15
Meperidine	Anesthetic-narcotic	2–16 mg/kg ip	Analgesic dosage	34
Morphine	Anesthetic-narcotic	1.25–20 mg/kg/hr iv	Addictive dose	19
Morphine sulfate	Anesthetic-narcotic	20 mg/kg	Narcotic	14
Opium	Anesthetic-narcotic	0.0025 mg/kg	Narcotic analgesic expectorant dosage	34
Levallorphan	Anesthetic–narcotic antagonist	6 mg/kg sc	Narcotic antagonist	34
Levorphanol	Anesthetic–narcotic antagonist	3–5 mg/kg sc	Narcotic antagonist	34
Methadone	Anesthetic–narcotic antagonist	1 mg/kg ip	Dosage to stop shakes in morphine addiction	34
Phenobarbital	Anesthetic–sedative	15 mg/kg ip	Sedative dosage	34
Aminopyrine	Antiinflammatory	20 mg/kg ip	Antipyretic dosage	39
Aminopyrine	Antiinflammatory	20–40 mg/kg ip	Antiinflammatory	39
Benzydamine	Antiinflammatory	15–45 mg/kg sc	Analgesic and antiinflammatory	39
Chlorpromazine	Antiinflammatory	0.5–4 mg/kg iv	Antiedema dosages	31
Cortisone	Antiinflammatory	25 mg/kg sc	Prevent inflammation caused by brewers yeast	36
Indomethacin	Antiinflammatory	0.1–0.5 mg/rat	Inhibition of cotton pellet granuloma	39
Indomethacin	Antiinflammatory	4 mg/kg po	Analgesic and antiinflammatory drug	34
Methotrimeprazine	Antiinflammatory	0.1–1 mg/kg iv	Antiedema dosages	31
Methylpromazine	Antiinflammatory	0.5–4 mg/kg iv	Antiedema dosages	31
Oxyphenbutazone	Antiinflammatory	100 mg/kg po	Protective action in nicotinic acid deficiency	10
Phenylbutazone	Antiinflammatory	7.5–15 mg/kg sc	Antiinflammatory dosage	39
Phenylbutazone	Antiinflammatory	30 mg/kg po	Antiinflammatory dosage	39
Phenylbutazone	Antiinflammatory	100 mg/kg po daily	Inhibited primary arthritic lesions	29
Promethazine	Antiinflammatory	2–8 mg/kg iv	Antiedema dosages	31
Trimeprazine	Antiinflammatory	0.25–1 mg/kg iv	Antiedema dosages	31
6-Mercaptopurine	Antiinflammatory	2 mg/kg	Antiinflammatory in experimental arthritis	34
Chloramphenicol palmitate	Antimicrobial-antibiotic	20–50 mg/kg po	Treat once or twice a day	35
Chloramphenicol succinate	Antimicrobial-antibiotic	6.6 mg/kg im	Treat once or twice a day	35
Chlortetracycline	Antimicrobial-antibiotic	0.1% of water	Effective against *Corynebacterium kutscheri*	12
Chlortetracycline	Antimicrobial-antibiotic	6–10 mg/kg im	Treat once or twice a day	35
Chlortetracycline	Antimicrobial-antibiotic	0.025% of diet	Did not eliminate rhinitis, ear infection, etc.	13
Gentamicin	Antimicrobial-antibiotic	4.4 mg/kg im	Treat one or two times a day	35
Oxytetracycline	Antimicrobial-antibiotic	6–10 mg/kg IM	Treat one or two times a day	35
Penicillin g-procaine	Antimicrobial-antibiotic	20,000 units/kg po	Treat once daily	35
Tetracycline	Antimicrobial-antibiotic	0.1–.5% of diet	14-Day treatment for labrynthitis	34
Tetracycline	Antimicrobial-antibiotic	15–20 mg/kg po	Treat two or three times a day	35
Tylosin	Antimicrobial-antibiotic	2–4 mg/kg im	Treat once or twice a day	35
Amphotericin B	Antimicrobial-fungal	6.25 mg/kg sc daily	Fungal prevention in cortisoned rats	12

(continued)

Table I (*Continued*)

Drug	Classification	Dosage	Comment	Reference[a]
Griseofulvin	Antimicrobial–fungal	0.03% of feed	Fungal prevention in cortisoned rats	12
Pimaricin	Antimicrobial–fungal	100–200 ppm of feed	Eliminated yeast flora	26
Nitrofurantoin	Antimicrobial–furacin	12.5 mg/kg B.I.D. for 3–7 day	Treatment of *Proteus* or *E. coli* bladder infection	34
Sulfadiazine	Antimicrobial–sulfa	220 mg/kg then 110 mg/kg	Initial dose twice maintainance dose	34
Sulfadimethoxine	Antimicrobial–sulfa	20–50 mg/kg po	Treat once daily	35
Sulfamerazine	Antimicrobial–sulfa	0.025% of diet	Eliminated rhinitis, middle ear infection, etc.	13
Sulfamerazine	Antimicrobial–sulfa	50–80 mg/kg po	May be used in drinking water also	35
Sulfaquinoxaline	Antimicrobial–sulfa	0.025–0.1% of water	Therapeutic dose	35
Sulfaquinoxaline	Antimicrobial–sulfa	0.05% of diet	Therapeutic dose	35
Triple sulfas	Antimicrobial–sulfa	0.2% of water	Treatment for pneumonia	6
Bithional	Antiparasitic	35 mg/kg	Anthelmintic dosage for *Fasciola*	34
Carbarsone	Antiparasitic	0.05% of feed	Slight suppression of *Pneumocystis*	12
Carbon tetrachloride	Antiparasitic	800–1600 mg/kg po ip	Anthelmintic	34
Chloroquine	Antiparasitic	0.012% of feed	Ineffective against *Pneumocystis*	12
Dichlorvos (DDVP)	Antiparasitic	0.05–0.1% of diet	Used for 3–5 days not effective *T. crassicauda*	47
Emetine	Antiparasitic	5 mg/kg	Treatment for amoebiasis	34
Hexachloroethane	Antiparasitic	0.6–1.2 mg/kg po	Anthelmintic dosage	34
Hexachlorophene	Antiparasitic	10 mg/kg	Anthelmintic dose	34
Hydroxystilbamidine	Antiparasitic	0.17–1 mg/rat daily	Kills pneumocystis	12
Lead arsenate	Antiparasitic	0.5% of diet	Treatment of *Hymenolepis nana*	34
Malathion	Antiparasitic	0.125% dip	For control of external parasites	34
Methoxychlor	Antiparasitic	Topical application	Up to 2% dust for external parasites	34
Methyridine	Antiparasitic	100 mg/kg ip	Anthelmintic dose	34
Methyridine	Antiparasitic	200 mg/kg	Single dose effective against *Trichosomides*	32
Methyridine	Antiparasitic	65–125 mg/kg/day/4 days po	0.15% of drinking water for 4 days	34
Metronidazole	Antiparasitic	10–40 mg/day/rat	Slightly effective in killing *Pneumocystis*	12
Metronidazole	Antiparasitic	125 mg/kg ip	Trichomonacide	34
Niclosamide	Antiparasitic	100 mg/kg/day for 21 days	Effective against *Hymenolepis nana* and *H. diminuta*	16
Niclosamide	Antiparasitic	50–100 mg/kg po	Treatment for *Hymenolepis diminuta* infection	34
Niclosamide	Antiparasitic	30–90 mg/kg po in feed	Treat 3 days/off 3 days/treat 3 days in feed	35
Niridazole	Antiparasitic	100 mg/kg sc	Ineffective against *T. crassicauda*	47
Niridazole	Antiparasitic	0.05% of diet	Ineffective against *T. crassicauda*	47
Nitrofurantoin	Antiparasitic	0.2% of feed	Eradication of *Trichosomoides crassicauda*	4
Pentamidine isethionate	Antiparasitic	20 mg/kg sc daily	Causes tissue necrosis at injection site	12
Phenothiazine	Antiparasitic	1000 mg/kg po	Treatment for *Heterakis spumosa* infection	41
Piperazine	Antiparasitic	0.2% of water	Ineffective against *T. crassicauda*	47
Piperazine	Antiparasitic	200 mg/100 ml water	Anthelmintic dose for pinworm infections	34
Piperazine citrate	Antiparasitic	500–1000 mg/kg po-water	Treat week-off week-treat week	35
Pyrantel	Antiparasitic	50 mg/kg po	Ineffective against *T. crassicauda*	47
Pyrantel	Antiparasitic	6 mg/kg	Anthelmintic	34
Resorantel	Antiparasitic	20 mg/kg po	Treatment for *Hymenolepis diminuta* infection	34
SKF 29044	Antiparasitic	0.5% of diet-14 days	Partially effective against *T. crassicauda*	47
Tetramisole	Antiparasitic	40 mg/kg po	Partially effective against *T. crassicauda*	47
Tetramisole	Antiparasitic	12–25 mg/kg in water	Partially effective against *T. crassicauda*	47
Thiabendazole	Antiparasitic	100 mg/kg po	Ineffective against *T. crassicauda*	47
Thiabendazole	Antiparasitic	200 mg/kg for 5 days po	Anthelmintic	34
Thiabendazole	Antiparasitic	50 mg/kg ip	Partially effective against *T. crassicauda*	47

(*continued*)

Table I (*Continued*)

Drug	Classification	Dosage	Comment	Reference[a]
Pyrimethazine plus sulfadiazine	Antiparasitic–protozoal	0.03% of feed	Pneumocystosis treatment with sulfadiazine	12
Sulfadiazine plus pyrimethamine	Antiparasitic–protozoal	0.1% of feed	Pneumocystosis treatment with pyrimethazine	12
Metyrapone	Body function test	66 mg/kg ip	Pituitary and hypothalmic function test	34
Inulin	Body function test—kidney	90 mg/kg iv	Urinary excretion test for kidney function	8
Amaranth (azorubin s)	Body function test—liver	200 mg/kg iv	Plasma disappearance test for liver function	20
Bilirubin	Body function test—liver	25 mg/kg iv	Plasma disappearance test for liver function	20
Chlorothiazide	Body function test—liver	40 mg/kg iv	Plasma disappearance test for liver function	20
Dibromophthalein (DBSP)	Body function test—liver	96 mg/kg iv	Plasma disappearance test for liver function	20
Indocyanine green (ICG)	Body function test—liver	1–50 mg/kg iv	Plasma disappearance test for liver function	20
Ouabain octohydrate	Body function test—liver	5 mg/kg iv	Plasma disappearance test for liver function	20
PAEB	Body function test—liver	10 mg/kg iv	Procaine amide ethyl bromide liver function	20
Phenol red	Body function test—liver	30 mg/kg iv	Plasma disappearance test for liver function	20
Probenecid	Body function test—liver	75 mg/kg iv	Plasma disappearance test for liver function	20
Succinylsulfathiazole	Body function test—liver	40 mg/kg iv	Plasma disappearance test for liver function	20
Sulfobromophthalein BSP	Body function test—liver	120 mg/kg iv	Plasma disappearance test for liver function	20
Taurocholic acid-sodium	Body function test—liver	100 mg/kg iv	Plasma disappearance test for liver function	20
Iodine	Body function—thyroid	Iodine deficient diet	Triiodthyronine (T_3) and protein-bound iodine	45
Bretylium	CNS	5 mg/kg iv	Norepinephrine release inhibitor	1
Nervous system drugs	CNS and *peripheral* drugs	Various drugs and dosages	Review off peripheral nervous system drugs	3
Imipramine	CNS antidepressant	40 mg/kg po	Protective action in nicotinic acid deficiency	10
Imipramine	CNS antidepressant	20–40 mg/kg po	Depressant dosage	34
Imipramine	CNS antidepressant	Up to 10 mg/kg ip	Antidepressant	34
Dichloroisoproterenol	CNS beta blocking agent	0.1 mg/kg iv	Beta blocking agent	1
Nervous system drugs	CNS drugs	Various drugs and dosages	Review of CNS drugs	2
Mephenesin	CNS muscle relaxant	100 mg/kg po	Muscle relaxant	10
d-Tubocurarine tubarine	CNS nerve transmission	0.04–0.06 mg/kg iv	Neuromuscular junction affector	40
Decamethonium	CNS nerve transmission	1.25 mg/kg iv	Neuromuscular junction affector	40
Dimethine or Metubine	CNS nerve transmission	0.009 mg/kg iv	Neuromuscular junction affector	40
Laudolissin	CNS nerve transmission	1.05 mg/kg iv	Neuromuscular junction affector	40
Atropine	CNS parasympatholytic	10 mg/kg po 3 mg/kg sc	Anticholinergic parasympathathetic blockade	1
Acetylcholine	CNS parasympathomimetic	0.01 mg/kg iv	Direction acetylcholine sensitive sites	1
Dextroamphetamine	CNS stimulant	5 mg/kg po	CNS stimulant	10
Dextroamphetamine	CNS stimulant	1–2 mg/kg sc	Sympathomimetic and CNS stimulant	34
Pentylenetetrazol	CNS stimulant	50 mg/kg ip	Convulsive dose	34
Dihydrotestosterone	Hormone	125 μg/rat/day	Replacement therapy following castration	9
Epinephrine	Hormone		Sympathomimetic or adrenergic drug	34
Estradiol	Hormone	1–5 μg sc im	Estrogenic hormone	34
Estradiol benzoate	Hormone	.5–167 μg/rat/day	Replacement therapy following castration	9
Estradiol benzoate	Hormone	0.1–1 μg sc	Ovulation dosage for immature rats	48

(*continued*)

Table I (*Continued*)

Drug	Classification	Dosage	Comment	Reference[a]
Estrone	Hormone	20 μg im 200 units	Estrogenic hormone	34
Insulin	Hormone	Not over 3 units sc	Antidiabetogenic	34
Norethindrone	Hormone	15 μg oral	Progesteroid oral contraceptive	34
Oxytocin	Hormone	1 unit im or sc	Therapeutic dose	34
Parathyroid hormone	Hormone	30 units/kg ip		34
Pregnant mare's serum	Hormone	20 iu sc	Causes ovulation in immature rats	27
Progesterone	Hormone	3 mg sc im + estradiol	Causes eclampsia in pregnant rat	34
Prostaglandins	Hormone	0.5–1 μg iv	Three types of prostaglandins used	14
Testosterone	Hormone	125 μg/rat/day	Replacement therapy following castration	9
Testosterone	Hormone	0.5–1 mg/kg im sc	Androgenic hormone	34
Thyroxine	Hormone	30 μg/kg im sc		34
Adrenalectomized rat	Hormone replacement Rx	Various drugs and dosages	Hormone maintenance and replacement dosage, p. 278	1
Castrated rat	Hormone replacement Rx	Various drugs and dosages	Hormone maintenance and replacement dosage, p. 279	1
Hypophysectomized rat	Hormone replacement Rx	Various drugs and dosages	Hormone maintenance and replacement dosage, p. 277	1
Ovariectomized rat	Hormone replacement Rx	Various drugs and dosages	Hormone maintenance and replacement dosage, p. 279	1
Pancreatectomized rat	Hormone replacement Rx	Various drugs and dosages	Hormone maintenance and replacement dosage, p. 277	1
Parathyroidectomized rat	Hormone replacement Rx	Various drugs and dosages	Hormone maintenance and replacement dosage, p. 280	1
Thyroidectomized rat	Hormone replacement Rx	Various drugs and dosages	Hormone maintenance and replacement dosage, p. 280	1
Thyroparathyroidectomized	Hormone replacement Rx	Various drugs and dosages	Hormone maintenance and replacement dosage, p. 280	1
Dihydrotachysterol	Hormone synthetic	0.25–2 mg/rat sc po	Parathyroid-like hormone	34
Medroxyprogesterone	Hormone synthetic	300 μg sc	ED_{50} in delaying parturition	34
Acetylsalicylic acid	Miscellaneous	300 mg/kg	Analgesic and antipyretic	34
α-methyl-*p*-tyrosine	Miscellaneous	200 mg/kg po, 50 mg/kg ip	Catecholamine synthesis inhibitor	34
Apomorphine	Miscellaneous	1–1.5 mg/kg iv	Use to cause experimental agressiveness	34
Benadryl	Miscellaneous	100 mg/kg po	Toxic to rats at this dosage	10
Betaine hydrochloride	Miscellaneous	180–400 mg/kg po	Lipotropic agent and hydrochloric acid releaser	34
Carbon tetrachloride	Miscellaneous	800–1600/week po ip	Give weekly for liver necrosis	34
Chlorcyclizine HCl	Miscellaneous	25 mg/kg ip	Microsomal enzyme inducer	21
Cyclophosphamide	Miscellaneous	50–100 mg/kg ip	Antineoplastic drug	34
Diphenylhydantoin	Miscellaneous	20 mg/kg ip	Anticonvulsant	34
Evan's blue dye	Miscellaneous	2 mg/rat iv	Dosage to study vascular changes	31
Folinic acid	Miscellaneous	0.38–0.54 mg/kg daily	To treat pyrimethazine-sulfadiazine toxicity	12
γ-Chlordan	Miscellaneous	50 mg/kg ip	Microsomal enzyme inducer	21
Glycidol	Miscellaneous	15 mg/kg	Antifertility agent	34
Hydralazine	Miscellaneous	2 mg/kg ip	Hypotensive agent	34
Hydroxyurea	Miscellaneous	250 mg/kg ip	Antineoplastic drug	34
Iodoacetamide	Miscellaneous	50–100 mg/100 ml water	Ulcerogenic agent	34
Iproniazid	Miscellaneous	25–100 mg/kg ip	Monoamine oxidase inhibitor	34
Isoproterenol	Miscellaneous	Various dosages and routes	Adrenergic or sympathomimetic drug	34
Isosorbide	Miscellaneous	1–2 gm/kg	Osmotic diuretic	34
Lysergide-LSD	Miscellaneous	5–300 μg po ip	Psychopharmaceutical	34
Methotrexate	Miscellaneous	0.65–1 mg/kg po or ip	Antineoplastic drug	34
Methoxamine	Miscellaneous	2–3 mg/rat	Sympathomimetic or adrenergic drug	34
Metopirone (SU-4885)	Miscellaneous	10 mg 2$\times$/day for 42 days	Causes hypertension	7
Minerals	Miscellaneous	Various dosages and routes	Chapter on the nutrient requirements of rats	30

(*continued*)

Table I (*Continued*)

Drug	Classification	Dosage	Comment	Reference[a]
Nicotinamide	Miscellaneous	10 mg/kg po	Protective action in nicotinic acid deficiency	10
Norethandrolone	Miscellaneous	0.0002–0.004% of feed	Anabolic agent	34
Nutrients	Miscellaneous	Various dosages and routes	Chapter on the nutrient requirements of rats	30
Oubain	Miscellaneous	0.05–0.09 mg/kg iv	Dosage to increase cardiac output	34
Pentazocine	Miscellaneous	2 mg/kg sc	Analgesic	34
Phenacetin	Miscellaneous	200 mg/kg	Antipyretic action lasts at least 4 hrs	34
Phenobarbital	Miscellaneous	75 mg/kg ip for 4 days	Microsomal enzyme inducer	20
Phentolamine	Miscellaneous	25 mg/kg ip	Alpha adrenergic blocking agent	34
Phenylbutazone sodium	Miscellaneous	125 mg/kg ip	Microsomal enzyme inducer	21
Physostigmine	Miscellaneous	0.02 mg/kg iv	Parasympathomimetic and anti-cholinesterase drug	34
Physostigmine	Miscellaneous	0.1–1 mg/kg im	Parasympathomimetic and anti-cholinesterase drug	34
Pilocarpine	Miscellaneous	0.5 mg/kg iv	Parasympathomimetic and cholinergic drug	34
Proadifen	Miscellaneous	25–50 mg/kg ip	Drug potentiator—liver microsome inhibitor	34
Procarbazine	Miscellaneous	30–250 mg/kg ip	Antineoplastic drug	34
Procyclidine	Miscellaneous	15 mg/kg	Anticholinergic	34
Promethazine	Miscellaneous	12.5 mg/kg ip, 1–5 mg/kg sc	Antihistamine	34
Prostaglandins	Miscellaneous	5–10 μg/kg	Experimental parenteral dosage	34
Ribaminol	Miscellaneous	25–50 mg/kg po	Learning and memory enhancer—experimental	34
Spironolactone	Miscellaneous	5–10 mg/rat B.I.D. sc or po	Diuretic—Aldosterone antagonist	34
Streptokinase	Miscellaneous	90,000 units po	Antiinflammatory	34
Teratogenic agents	Miscellaneous	Varous dosages and routes	649 drugs and agents tested—teratology	37
Thio-TEPA	Miscellaneous	1.8–8.4 mg/kg iv, ip	Antineoplastic drug	34
Tolbutamide	Miscellaneous	250–2000 mg/day	Hypoglycemic agent	34
Tranylcypromine	Miscellaneous	2.5–10 mg/kg ip po	Monoamine oxidase inhibitor	34
Tremorine	Miscellaneous	5–20 mg/kg po iv ip im sc	Parasympathomimetic or cholinergic	34
Triamterene	Miscellaneous	30 mg/kg po	Diuretic—naturetic	34
Tripelennamine	Miscellaneous	6 mg/kg iv	Antihistamine	34
Tyramine	Miscellaneous	0.5 mg/kg iv	Vasopressor drug	34
Uridine	Miscellaneous	10 mg/kg iv	Nucleoside	34
Vinblastine	Miscellaneous	0.4–0.65 mg/kg iv	Antineoplastic	34
Vitamins	Miscellaneous	Various dosages and routes	Chapter on the nutrient requirements of rats	30
2-Thiouracil	Miscellaneous	0.01% of ration	Thyroid inhibitor	34
6-Mercaptopurine	Miscellaneous	20–100/rat/day po or ip	Antineoplastic drug	34
Methylcholanthrene	Miscellaneous carcinogen	100–600 mg/kg	Treat weekly for 50–60 days for carcinogen	34
Zedalin-streptozotocin	Miscellaneous carcinogen	15:85 mixture of 50 mg/kg iv	To produce experimental renal tumors	34
Carrageenin	Miscellaneous irritant	0.05 ml of a 1% solution id	Edema and erythema following hindpaw injection	39
Compound 48/80	Miscellaneous irritant	0.1 ml id of 10 μg/ml	Hindpaw edema producing dosage	31
Croton oil and air	Miscellaneous irritant	25 ml air + oil sc	Granuloma pouch method for inflammation	24
Dextran	Miscellaneous irritant	0.1 ml id of 60 μg/ml	Hindpaw edema producing dosage	31
Egg white	Miscellaneous irritant	0.1 ml id of 0.5% solution	Hindpaw edema producing dosage	31
Formalin	Miscellaneous irritant	1 mg/kg of a 1% solution ip	Peritonitis and pleuritis following ip injection	39
Formalin	Miscellaneous irritant	0.1 ml of a 3% solution id	Edema and erythema following hindpaw injection	39
Histamine	Miscellaneous irritant	0.1 ml id of 1 mg/ml solution	Hindpaw edema producing dosage	31

(*continued*)

Table I (*Continued*)

Drug	Classification	Dosage	Comment	Reference[a]
Mustard	Miscellaneous irritant	0.1 ml of 2.5% solution	Edema and erythema following hindpaw injection	43
Phenylquinone	Miscellaneous irritant	0.25 ml of 0.02% solution	Peritonitis and pleuritis following ip injection	38
Silver nitrate	Miscellaneous irritant	0.2 ml of a 1% solution	Silver nitrate injected under Achilles tendon	39
5-Hydroxytryptamine	Miscellaneous irritant	0.1 ml id of 0.5 μg/ml	Hindpaw edema producing dosage	31
Atropine	Miscellaneous poison Rx	17 mg/kg + oximes 30 mg/kg	Treatment for organophosphate poisoning	34
Pralidoximes	Miscellaneous poison Rx	30 mg/kg atropine 17 mg/kg	Cholinesterase reactivator in organophosphate	34
Aminopterin	Miscellaneous—cytostatic	0.4–0.8 mg/kg po 3×/week	Dose prevented body weight gain	12
Azathioprine	Miscellaneous—cytostatic	0.02% of feed	Dose prevented body weight gain	12
Chlorambucil	Miscellaneous—cytostatic	8 mg/kg po 3×/week	Dose prevented body weight gain	12
Cyclophosphamide	Miscellaneous—cytostatic	20–40 mg/kg po 3×/week	Dose prevented body weight gain	12
Nitrogen mustard	Miscellaneous—cytostatic	0.2–0.4 mg/kg ip 3×/week	Dose prevented body weight gain	12
Vinblastine	Miscellaneous—cytostatic	0.25–0.50 mg/kg ip 3×/week	Dose prevented body weight gain	12
6-Mercaptopurine	Miscellaneous—cytostatic	75 mg/kg po 3×/week	Dose prevented body weight gain	12
Aldosterone	Steroid	20–40 μg/rat sc	Given B.I.D. to produce hypertension	11
Corticosterone	Steroid	40 mg/kg po 10 mg/kg im	Protective action in nicotinic acid deficiency	10
Cortisone	Steroid	0.25–1.25 mg/day sc, im, po	Antiinflammatory and glucocorticoid	34
Cortisone acetate	Steroid	135 mg/kg sc 2–3×/week	Dosage to "cortsone-condition" rats	12
Deoxycorticosterone	Steroid	40 mg/kg sc	Protective action in nicotinic acid deficiency	10
Dexamethazone	Steroid	0.05 mg/kg sc	Protective action in nicotinic acid deficiency	10
Hydrocortisone	Steroid	5–20 mg/kg po, im, or sc	Antiinflammatory and glucocorticoid	34
Prednisolone	Steroid	2–10 mg/kg im, sc, po	Immunosuppressant dosage	34
Triamcinolone	Steroid	0.01 mg/kg sc	Protective action in nicotinic acid deficiency	10
Chlordiazepoxide	Tranquilizer	3–10 mg/kg ip	CNS depressant and tranquilizer	34
Chlorpromazine	Tranquilizer	20 mg/kg po	Protective action in nicotinic acid deficiency	10
Guanethidine	Tranquilizer	5 mg/kg iv	Norepinephrine depleting agent	1
Haloperidol	Tranquilizer	0.35 mg/kg sc	Tranquilizer	34
Meprobamate	Tranquilizer	100 mg/kg po	Tranquilizer	10
Meprobamate	Tranquilizer	15–150 mg/kg ip im	Tranquilizer	34
Methotrimeprazine	Tranquilizer	10 mg/kg im	Tranquilizer	34
Oxazepam	Tranquilizer	20 mg/kg	Tranquilizer	34
Perphenazine	Tranquilizer	1 mg/kg po	ED_{50} for tranquilization	34
Phenylcyclidine HCl	Tranquilizer	10 mg/kg ip	Excited with ataxia	42
Phenylcyclidine HCl	Tranquilizer	50 mg/kg ip	Cataleptoid with tremors	42
Phenylcyclidine HCl	Tranquilizer	2 mg/kg ip	Slightly ataxic	42
Promazine	Tranquilizer	0.5–1 mg/kg im	Tranquilizer	34
Reserpine	Tranquilizer	5 mg/kg sc	Norepinephrine depleting agent	1
Reserpine	Tranquilizer	1 mg/kg po	Stimulates corticosteroid output	10
Reserpine	Tranquilizer	10 mg/kg po	Toxic dose to rat	10
Reserpine	Tranquilizer	0.4 mg/kg iv	Norepinephrine depleting agent	1
Reserpine	Tranquilizer	5 mg/kg po	Stimulates corticosteroid output	10
Thalidomide	Tranquilizer	150 mg/kg po	Given during pregnancy but few malformations	10

[a] Key to references:
1. Barnes, C. D., and Eltherington, L. G. (1973). "Drug Dosage in Laboratory Animals," Rev. Ed. Univ. of California Press, Berkeley, California.
2. Bowen, J. M. (1976). Drugs acting on the central nervous system. *In* "Handbook of Laboratory Animal Science" (E. C. Melby, Jr. and N. H. Altman, eds.), Vol. 3, pp. 65-95. CRC Press, Cleveland, Ohio.
3. Bowen, J. M., and Butrram, J. M. (1976). Drugs affecting the peripheral nervous system. *In* "Handbook of Laboratory Animal Science" (E. C. Melby, Jr. and N. H. Altman, eds.), Vol. 3, pp. 96-118. CRC Press, Cleveland, Ohio.

Table I (*Continued*)

4. Chapman, W. H. (1964). The incidence of a nematode, *Trichosomoides crassicauda* in the bladder of laboratory rats: Treatment with nitrofurantoin and preliminary report of the influence on the urinary calculi and experimental bladder tumor. *Invest. Urol.* **2**, 52–57.

5. Charles River Digest (1966). "Anesthesia in Small Laboratory Animals," Vol. 5, No. 3. Charles River Breed. Lab., Inc., Wilmington, Massachusetts.

6. Charles, R. T., and Rees, O. (1958). Use of sulphonamides in the treatment of pleuro-pneumonia-like organisms in rats. *Nature (London)* **181**, 1213.

7. Colby, D., Skelton, F. R., and Brownie, A. C. (1970). Metopirone-induced hypertension in the rat. *Endocrinology* **86**, 620–628.

8. Danielson, R. A., and Schmidt-Nielsen, B. (1972). Recirculation of urea analogs from renal collecting ducts of high- and low-protein-fed rats. *Am. J. Physiol.* **223**, 130–139.

9. Feder, H. H., Naftolin, F., and Ryan, K. J. (1974). Male and female sexual responses in male rats given estradiol benzoate and 5α-ancrostan-17β-1,3-one propionate. *Endocrinology* **94**, 136–141.

10. Fratta, I. D. (1969). Nicotinamide deficiency and thalidomide: Potential teratogenic disturbances in Long-Evans rats. *Lab. Anim. Care* **19**, 727–732.

11. Fregly, M. J., Kim, K. J. and Hood, C. I. (1969). Development of hypertension in rats treated with aldosterone acetate. *Toxicol. Appl. Pharmacol.* **15**, 229–243.

12. Frenkel, J. K., Good, J. T., and Schultz, J. A. (1966). Latent *Pneumocystis* infection of rats, relapse, and chemotherapy. *Lab Invest.* **15**, 1559–1576.

13. Habermann, R. T., Williams, F. P., McPherson, C. W., and Every, R. R. (1963). The effect of orally administered sulfamerazine and chlortetracycline on chronic respiratory disease in rats. *Lab. Anim. Care* **13**, 28–40.

14. Hedge, G. A., and Hanson, S. D. (1972). The effects of prostaglandins on ACTH secretion. *Endocrinology* **91**, 925–933.

15. Hoar, R. M. (1965). Anesthetic technics of the rat and guinea pig. *In* "Experimental Animal Anesthesiology" (D. C. Sawyer, ed.), pp. 325–344. USAF School of Aerospace Medicine, Brooks Air Force Base, Texas.

16. Hughes, H. C., Jr., Barthel, C. H., and Lang, C. M. (1973). Niclosamide as a treatment for *Hymenolepis nana* and *Hymenolepsis diminuta* in rats. *Lab. Anim. Care* **23**, 72–73.

17. Jones, J. B., and Simmons, M. L. (1968). Innovar-Vet as an intramuscular anesthetic for rats. *Lab. Anim. Care* **18**, 642–643.

18. Kaczmarczyk, G., and Reinhardt, H. W. (1975). Arterial blood gas tensions and acid–base status of Wistar rats during thiopental and halothane anesthesia. *Lab. Anim. Sci.* **25**, 184–190.

19. Khazan, N., and Colasanti, B. (1972). Protracted rebound in rapid eye movement sleep time and electroencephalogram voltage output in morphine dependent rats upon withdrawal. *J. Pharmacol. Exp. Ther.* **183**, 23–30.

20. Klassen, C. D. (1970). (A). Effects of phenobarbital on the plasma disappearance and biliary excretion of drugs in rats. *J. Pharmacol. Exp. Ther.* **175**, 289–300.

21. Klaassen, C. D. (1970). (B). Plasma disappearance and biliary excretion of sulfobromophthalein and phenol-3,6-dibromphthalein disulfonate after microsomal enzyme induction. *Biochem. Pharmacol.* **19**, 1241–1249.

22. Latt, R. H. (1976). Drug dosages for laboratory animals. *In* "Handbook of Laboratory Animal Science" (E. C. Melby, Jr. and N. H. Altman, eds.), Vol. 3, pp. 561–568. CRC Press, Cleveland, Ohio.

23. Lewis, G. E., and Jennings, P. B. (1972). Effective sedation of laboratory animals using Innovar-Vet. *Lab. Anim. Sci.* **22**, 430–432.

24. Llaurado, J. G. (1961). The effects of some 21-methyl-substituted corticoids on inflammation, liver glycogen and electrolyte-regulating activity in the rat. *Acta Endocrinol. (Copenhagen)* **38**, 137–150.

25. Lumb, W. V. (1965). Pre- and postanesthetic agents. *In* "Experimental Animal Anesthesiology" (D. C. Sawyer, ed.), pp. 48–66. USAF School of Aerospace Medicine, Brooks Air Force Base, Texas.

26. Manten, A., and Hoogerheide, J. C. (1958). The influence of a new antifungal antibiotic, pimaricin, on the yeast flora of the gastrointestinal tract of rats and mice during tetracycline administration. *Antibiot. Chemother. (Washington, D.C.)* **8**, 381–386.

27. McCormack, C. E., and Meyer, R. K. (1968). Evidence for the release of ovulating hormone in PMS-treated immature rats. *Proc. Soc. Exp. Biol. Med.* **128**, 18–23.

28. Molello, J. A., and Hawkins, K. (1968). Methoxyflurane anesthesia of laboratory rats. *Lab. Anim. Care* **18**, 581–583.

29. Newbould, B. B. (1965). Suppression of adjuvant-induced arthritis in rats with 2-butoxycarbonylmethylene-4-oxothiazolidine. *Br. J. Pharmacol. Chemother.* **24**, 632–640.

30. "Nutrient Requirements of Laboratory Animals" (1972). 2nd rev. ed., No. 10. N.A.S., Washington, D.C.

31. Parratt, J. R., and West, G. B. (1958). Inhibition by various substances on oedema formation in the hind-paw of the rat induced by 5-hydroxytryptamine, histamine, dextran, eggwhite and compound 48/80. *Br. J. Pharmacol. Chemother.* **13**, 65–70.

32. Peardon, D. L., Tufts, J. M., and Eschroeder, H. C. (1966). Experimental treatment of laboratory rats naturally infected with *Trichosomoides crassicauda*. *Invest. Urol.* **4**, 215–219.

33. Randall, L. O., and Selitto, J. J. (1957). A method for measurement of analgesic activity of inflamed tissue. *Arch. Int. Pharmacodyn. Ther.* **111**, 409–419.

34. Rossoff, I. F. (1974). "Handbook of Veterinary Drugs." Springer-Verlag, Berlin and New York.

35. Schuchman, S. M. (1974). Individual care and treatment of mice, rats, guinea pigs, hamsters and gerbils. *Curr. Vet. Ther.* **5**, 610–614.

36. Selitto, J. J., and Lowell, O. R. (1954). Screening of antiinflammatory agents in rats. *Fed. Proc., Fed. Am. Soc. Exp. Biol.* **13**, 403–404.

37. Shepard, T. H. (1973). "Catalog of Teratogenic Agents." Johns-Hopkins Univ. Press, Baltimore, Maryland.

38. Siegmund, E., Cadmus, R., and Go, L. (1957). A method for evaluating both non-narcotic and narcotic analgesics. *Proc. Soc. Exp. Biol. Med.* **95**, 729–731.

39. Silvestrini, B., Garan, A., Pozzatti, C., and Cioli, V. (1965). Pharmacological research on benzydamine—A new analgesic-anti-inflammatory drug. *Arzneim.-Forsch.* **16**, 59–63.

40. Spector, S. (1956). "Handbook of Biological Data." Saunders, Philadelphia, Pennsylvania.

41. Steward, J. S. (1955). Anthelmentic studies. *Parasitology* **45**, 231–241.

42. Stoliker, H. E. (1965). The physiologic and pharmacologic effects of sernylan: A review. *In* "Experimental Animal Anesthesiology" (D. C. Sawyer, ed.), pp. 148–184. USAF School of Aerospace Medicine, Brooks Air Force Base, Texas.

Table I (*Continued*)

43. Stucki, J. C., and Thompson, C. R. (1962). Anti-inflammatory activity of amidines of substituted triphenylethylenes. *Toxicol. Appl. Pharmacol.* **4**, 362–384.

44. Stunkard, J. A., and Miller, J. C. (1974). An outline guide to general anesthesia in exotic species. *Vet. Med. Small Anim. Clin.* **69**, 1181–1186.

45. Volpert, E. M., and Werner, S. C. (1972). Serum triiodothyronine concentration in the iodine-deficient rat. *Am. J. Anat.* **135**, 187–190.

46. Weisbroth, S. H., and Fudens, J. H. (1972). Use of ketamine hydrochloride as an anesthetic in laboratory rabbits, rats, mice, and guinea pigs. *Lab. Anim. Sci.* **22**, 904–906.

47. Weisbroth, S. H., and Scher, S. (1971). *Trichosomoides crassicauda* infection of a commercial rat breeding colony. 2. Drug screening for anthelmentic activity and field trials with methyridine. *Lab. Anim. Sci.* **21**, 213–219.

48. Ying, S., and Greep, R. O. (1971). Effect of a single low dose of estrogen on ovulation, pregnancy, and lactation in immature rats. *Fertil. Steril.* **22**, 165–169.

[b] Important complications have been reported from use of this drug in rats (Fleischman, R. W., McCracken, D., and Forbes, W. (1977). *Lab. An. Sci.* **27**, 238–243).

Subject Index

A

Acute toxicity studies, 114
Adrenalectomy, 26
African white-tailed rat, 251
Aged rats
   cost of production, 157
   sources, 157
Agricultural chemicals, evaluation of toxicity, 116
Alcian blue stain, 89
Alizarin red stain, 89
Allele, definition of, 214
Alloantigens
   cell surface, 214–217
   definition of, 214
Allograft, definition of, 214
Allophenic rats, 91
Anal cups, 12
Analgesics, *see* Appendix 2
Anemia, left ventricular hypertrophy caused by, 173
Anesthesia, 22–24
   ether, 23
   halothane, 23
   inhalation, 22–23
   Innovar-vet, 23
   Ketamine, 23
   methoxyflurane, 23
   pentobarbital, 23
Anesthetics, *see* Appendix 2
Antibiotics, *see* Appendix 2
Antibody, 219
Antigen-free diets, 49
Antiteratogen, 87
Aortic banding, *see* Banding of aorta
Arboviruses, 250
Arteriosclerosis, intensified by breeding, 171–172
Arthritis, *Salmonella*-associated, 204
Artificial respiration, 23–24
Aspicularis tetraptera, 229
Associating rats, 48
Atherosclerosis, 174
Autoimmune complex nephritis, 218
Autoimmune disease, 204
   encephalitis, 204
   glomerulonephritis, 204
   myasthenia gravis, 204
   thyroiditis, 204
Autoimmune encephalitis, 204
Autopsy, 28
Axenic, definition, 44

B

Banding of aorta, left ventricular hypertrophy caused by, 173
Bile collection, 14
Bile duct cannulation, 14–15
Biting loads, 61
Black rat, 248
Blastocyst, 76
Blood collection, 5–10
   aorta, 10
   cardiac puncture, 6–7
   carotid vein, 10
   decapitation, 10
   dorsal metatarsal vein, 7–8
   femoropopliteal vein, 8
   jugular vein, 9–10
   neonates, 6
   retroorbital plexus, 7
   saphenous vein, 8
   tail, 8–9
   terminal, 10
   vacuum devices, 9
   vena cava, 10
   volume, 6
Blood pressure, 165–166
Blood pressure measurement
   indirect, 27
   intraarterial, direct, 27
   small vessels, 27
Blood volume measurement, 28
Body development, 82
Bollman cage, 3
Bone marrow sampling, 13
Breast cancer, *see* Rat tumor models
Breeding colony
   management, 85
   records, 86
Brucella neotomae, 250